Dietary Polyphenols in Human Diseases

Diets rich in plant polyphenols or dietary polyphenols are the subject of increasing scientific interest because of their diverse range of health benefits and medicinal uses. This book, *Dietary Polyphenols in Human Diseases: Advances and Challenges in Drug Discovery*, fills the gap that exists in the current knowledge by presenting the latest information in the area of polyphenol-based drug discovery research. The book focuses on the current understanding of the beneficial effects on human health of diets rich in polyphenols and/or polyphenolic compounds derived from dietary sources (plant-based foods) and their possible preventive role in the management of chronic human diseases. In addition, biochemical mechanisms involved in the antioxidative effect of dietary polyphenols along with their bioavailability, pharmacokinetic, and toxicological considerations are also discussed herein.

Key Features

- Presents health benefits and medicinal importance of dietary polyphenols having antioxidative effects and their possible preventive role against oxidative stress (OS)-induced chronic human diseases.
- Summarizes the latest understanding of the biochemical mechanism(s) involved in the antioxidative action of dietary polyphenols along with their bioavailability, pharmacokinetic, and toxicological considerations.
- Highlights novel approaches of drug discovery from dietary polyphenols through computational screening of bioactive phytochemical components.

With contributions by a global team of experts (academics, scientists, and researchers), this book is intended to be a useful resource for a wider audience, particularly those working in the area of drug discovery and development research, such as discovery scientists, pharmaceutical scientists (R&D), formulation scientists, medicinal chemists, pharmacologists, toxicologists, phytochemists, biochemists, clinicians, researchers, students, teachers, and many others.

Mithun Rudrapal, PhD, FIC, is Associate Professor and Head at the Department of Pharmaceutical Chemistry, Rasiklal M. Dhariwal Institute of Pharmaceutical Education and Research (affiliated with Savitribai Phule Pune University), Pune, Maharashtra, India. Dr. Rudrapal works in Medicinal Chemistry, CADD, Drug Repurposing, Phytochemistry, Herbal Drugs, and Nanophytotherapeutics.

Dietary Polyphenols in
Human Diseases

Advances and Challenges in Drug Discovery

Edited by
Mithun Rudrapal

CRC Press
Taylor & Francis Group
Boca Raton London New York

CRC Press is an imprint of the
Taylor & Francis Group, an **informa** business

Cover image credit: ©Shutterstock

First edition published 2023
by CRC Press
6000 Broken Sound Parkway NW, Suite 300, Boca Raton, FL 33487-2742

and by CRC Press
4 Park Square, Milton Park, Abingdon, Oxon, OX14 4RN

CRC Press is an imprint of Taylor & Francis Group, LLC

Library of Congress Cataloging-in-Publication Data
Names: Rudrapal, Mithun, editor.
Title: Dietary polyphenols in human diseases : advances and challenges in
drug discovery / edited by Mithun Rudrapal.
Description: First edition. | Boca Raton, FL : CRC Press, 2023. |
Includes bibliographical references and index.
Identifiers: LCCN 2022030535 (print) | LCCN 2022030536 (ebook) |
ISBN 9781032170381 (hbk) | ISBN 9781032170411 (pbk) |
ISBN 9781003251538 (ebk)
Subjects: LCSH: Polyphenols. | Polyphenols–Physiological effect. |
Polyphenols–Therapeutic use.
Classification: LCC QK898.P764 D54 2023 (print) |
LCC QK898.P764 (ebook) | DDC 572/.2–dc23/eng/20220916
LC record available at https://lccn.loc.gov/2022030535
LC ebook record available at https://lccn.loc.gov/2022030536

ISBN: 9781032170381 (hbk)
ISBN: 9781032170411 (pbk)
ISBN: 9781003251538 (ebk)

DOI: 10.1201/9781003251538

Typeset in Times New Roman
by Newgen Publishing UK

Contents

Preface

Polyphenols are naturally occurring compounds found largely in dietary sources such as fruits, vegetables, spices, cereals, and beverages. Fruits like grapes, apple, pears, cherries, and berries contain up to 200–300 mg of polyphenols per 100 grams fresh weight. Polyphenols, including flavonoids and other food phenolics, are the subject of increasing scientific interest because of their diverse range of health benefits and medicinal uses. Epidemiological studies and associated meta-analyses strongly suggest that long-term consumption of diets rich in plant polyphenols offer significant protection against development of cancer, cardiovascular diseases, diabetes, arthritis, and neurodegenerative diseases.

This book focuses on the current understanding of the beneficial effects on human health of diets rich in plant polyphenols and/or polyphenolic compounds derived from dietary sources (plant-based foods), and their possible preventive role in the management of chronic human diseases. Considerable attention has been given to dietary polyphenolic phytochemicals having antioxidative properties that can effectively scavenge elevated reactive oxygen species (ROS) levels and subsequently reduce cellular oxidative stress (OS) in disease/pathological conditions. Dietary polyphenols (for example, curcumin, resveratrol, catechins, quercetin and so forth) that possess antioxidant properties, and thereby exhibit preventive potential against cellular or DNA damages involved in various health implications/disease conditions, have been covered in this book. In addition, biochemical mechanisms involved in the antioxidative effect of dietary polyphenols along with their bioavailability, pharmacokinetic, and toxicological considerations have been delineated.

The proposed book is expected to be a valuable resource for research studies in the domain of drug discovery and drug delivery from phytopolyphenols. The exceptional design, well-defined structure and the rich content of the book is timely and need of the hour. It is, therefore, believed that through this book many stakeholders from pharmaceutical and biotechnology sectors/industries, including discovery scientists, formulation scientists, medicinal chemists, biotechnologists, pharmacologists, toxicologists, phytochemists, biochemists, clinicians, health-care professionals, researchers, students, and teachers will certainly be benefitted.

About the Editor

Mithun Rudrapal, PhD, FIC, FICS, CChem, is Associate Professor and head at the Department of Pharmaceutical Chemistry, Rasiklal M. Dhariwal Institute of Pharmaceutical Education and Research (affiliated with Savitribai Phule Pune University), Pune, Maharashtra, India. Dr. Rudrapal has been actively engaged in teaching and research in the field of Pharmaceutical and Allied Sciences for more than 12 years. He has over a hundred publications in peer-reviewed international journals to his credit and has filed a number of Indian/International patents. In addition, Dr. Rudrapal is the author or editor of a dozen published or forthcoming books. He works in the area of Medicinal Chemistry, CADD, Drug Repurposing, Phytochemistry, Herbal Drugs, and Nanophytotherapeutics. His research interests include discovery and development of drugs for infectious diseases, diabetes, and inflammatory disorders from synthetic as well as plant sources.

Contributors

Muhammad Zeeshan Ahmed
School of Biochemistry and Biotechnology,
University of the Punjab, Lahore, Pakistan;
and Department of Biochemistry, Bahauddin
Zakariya University, Multan 60800, Pakistan

Jyotirmoi Aich
School of Biotechnology and Bioinformatics,
DY Patil Deemed to be University, CBD
Belapur, Navi Mumbai, Maharashtra, India

Poonam Arora
Department of Pharmacognosy and
Phytochemistry, SGT College of Pharmacy,
SGT University, Gurugram, Haryana, India

Quratulain Babar
Department of Biochemistry, Government
College University, Faisalabad, Pakistan

Merin Babu
Synthite Industries, Synthite Valley,
Kadayiruppu, Kerala, India

Mustansir Bhori
School of Biotechnology and Bioinformatics,
DY Patil Deemed to be University, CBD
Belapur, Navi Mumbai, Maharashtra, India

Baby Kumaranthara Chacko
Synthite Industries, Synthite Valley,
Kadayiruppu, Kerala, India

Amrita Chatterjee
Department of Pharmaceutical Sciences and
Technology, Birla Institute of Technology,
Mesra, Ranchi, Jharkhand, India

Dipak Chetia
Department of Pharmaceutical Sciences,
Faculty of Science and Engineering,
Dibrugarh University, Dibrugarh,
Assam, India

Behrokh Daei-Hasani
Department of Biology, Payame-Noor
University, Tehran, Iran

Prashanta Kumar Deb
Department of Pharmaceutical Sciences and
Technology, Birla Institute of Technology,
Mesra, Ranchi, Jharkhand, India

Fahima Dilnawaz
Laboratory of Nanomedicine, Institute of Life
Sciences, Nalco Square, Bhubaneswar,
Odisha, India

Jayati Dwivedi
Bioorganic Research Laboratory, Department
of Chemistry, University of Allahabad,
Prayagraj, Uttar Pradesh, India

Yeganeh Farnamian
Student Research Center, Urmia University of
Medical Sciences, Urmia, Iran

Aline Priscilla Gomes da Silva
Department of Biosystems and Agricultural
Engineering, Michigan State University, USA

Viju Jacob
Synthite Industries, Synthite Valley,
Kadayiruppu, Kerala, India

Dolly Jain
Adina College of Pharmacy, Sagar, Madhya
Pradesh, India

Sapna Joshi
Institute of Pharmacy, Ram-Eesh Institute of
Vocational and Technical Education, Greater
Noida, Uttar Pradesh, India

Veda Joshi
School of Biotechnology and Bioinformatics,
DY Patil Deemed to be University, CBD
Belapur, Navi Mumbai, Maharashtra, India

Himadri Kalita
Department of Zoology, Life Sciences Division,
Assam Don Bosco University, Guwahati,
Assam, India

Payal Kesharwani
Institute of Pharmacy, Ram-Eesh Institute
    of Vocational and Technical Education,
    Greater Noida, Uttar Pradesh, India

Swapnil Devidas Khamborkar
Synthite Industries, Synthite Valley,
    Kadayiruppu, Kerala, India

Johra Khan
Department of Medical Laboratory Sciences,
    College of Applied Medical Sciences,
    Majmaah University, Al Majmaah,
    Saudi Arabia

Satish Kumar
Department of Pharmaceutical Sciences and
    Technology, Birla Institute of Technology,
    Mesra, Ranchi, Jharkhand, India

H. Lalthanzara
Department of Zoology, Pachhunga University
    College, Aizawl, Mizoram, India

Aastha Mahapatra
Department of Pharmaceutical Chemistry,
    School of Pharmaceutical Sciences, Siksha
    'O'Anusandhan (Deemed to be University),
    Bhubaneswar, Odisha, India

Siddhartha Maji
Institute of Pharmacy, Ram-Eesh Institute
    of Vocational and Technical Education,
    Greater Noida, Uttar Pradesh, India

Lalit Mohan Nainwal
Department of Pharmaceutical Chemistry,
    School of Pharmaceutical Education and
    Research, Jamia Hamdard, New Delhi, India

Neelesh Kumar Nema
Synthite Industries, Synthite Valley,
    Kadayiruppu, Kerala, India

Ipsa Padhy
Department of Pharmaceutical Chemistry,
    School of Pharmaceutical Sciences, Siksha
    'O'Anusandhan (Deemed to be University),
    Bhubaneswar, Odisha, India

Ghanshyam Parmar
Department of Pharmacy, Sumandeep
    Vidyapeeth, Vadodara, Gujarat, India

Vaishali Patel
Department of Pharmaceutics, Laxminarayndev
    College of Pharmacy, Bharuch,
    Gujarat, India

Arpita Paul
Department of Pharmaceutical Sciences,
    Faculty of Science and Engineering,
    Dibrugarh University, Dibrugarh,
    Assam, India

Shiv Kumar Prajapati
Institute of Pharmacy, Ram-Eesh Institute
    of Vocational and Technical Education,
    Greater Noida, Uttar Pradesh, India

Aneena Peter
Synthite Industries, Synthite Valley,
    Kadayiruppu, Kerala, India

Safa Rafique
Department of Biochemistry and
    Biotechnology, University of Gujarat,
    Gujarat, Pakistan; and School of
    Biochemistry and Biotechnology, University
    of the Punjab, Lahore, Pakistan

Nayana Rajan
Synthite Industries, Synthite Valley,
    Kadayiruppu, Kerala, India

Yousef Rasmi
Cellular and Molecular Research Center,
    Cellular and Molecular Medicine
    Institute, Urmia University of Medical
    Sciences, Urmia, Iran; and Department of
    Biochemistry, School of Medicine, Urmia
    University of Medical Sciences, Urmia, Iran

Sahar Rezaei
Department of Biochemistry, School of
    Medicine, Urmia University of Medical
    Sciences, Urmia, Iran

Mithun Rudrapal
Department of Pharmaceutical Chemistry,
Rasiklal M. Dhariwal Institute of
Pharmaceutical Education and Research,
Chinchwad, Pune, Maharashtra, India

Sachithra Sabu
Synthite Industries, Synthite Valley,
Kadayiruppu, Kerala, India

Rajdeep Saha
Department of Pharmaceutical Sciences and
Technology, Birla Institute of Technology,
Mesra, Ranchi, Jharkhand, India

Ngurzampuii Sailo
Department of Pharmaceutical Sciences,
Faculty of Science and Engineering,
Dibrugarh University, Dibrugarh,
Assam, India

Linson Cheruveettil Sajan
Synthite Industries, Synthite Valley,
Kadayiruppu, Kerala, India

Biswatrish Sarkar
Department of Pharmaceutical Sciences and
Technology, Birla Institute of Technology,
Mesra, Ranchi, Jharkhand, India

Suparna Roy Sarkar
Department of Pharmaceutical Sciences and
Technology, Birla Institute of Technology,
Mesra, Ranchi, Jharkhand, India

Smitha Sarojam
Synthite Industries, Synthite Valley,
Kadayiruppu, Kerala, India

Ashish Shah
Department of Pharmacy, Sumandeep
Vidyapeeth, Vadodara, Gujarat, India

Tripti Sharma
Department of Pharmaceutical Chemistry,
School of Pharmaceutical Sciences, Siksha
'O'Anusandhan (Deemed to be University),
Bhubaneswar, Odisha, India

Ramendra K. Singh
Bioorganic Research Laboratory, Department
of Chemistry, University of Allahabad,
Prayagraj, Uttar Pradesh, India

Vishal K. Singh
Bioorganic Research Laboratory, Department
of Chemistry, University of Allahabad,
Prayagraj, Uttar Pradesh, India

Swarangi Tambat
School of Biotechnology and Bioinformatics,
DY Patil Deemed to be University, CBD
Belapur, Navi Mumbai, Maharashtra, India

Kanchanlata Tungare
School of Biotechnology and Bioinformatics,
DY Patil Deemed to be University, CBD
Belapur, Navi Mumbai, Maharashtra, India

Aditya K. Yadav
Bioorganic Research Laboratory, Department
of Chemistry, University of Allahabad,
Prayagraj, Uttar Pradesh, India

Zonunmawii
Department of Pharmaceutical Sciences,
Faculty of Science and Engineering,
Dibrugarh University, Dibrugarh,
Assam, India

James H. Zothantluanga
Department of Pharmaceutical Sciences,
Faculty of Science and Engineering,
Dibrugarh University, Dibrugarh,
Assam, India

1 Dietary Polyphenols, Antioxidant Effects, and Human Diseases

Himadri Kalita, Prashanta Kumar Deb, Rajdeep Saha, Amrita Chatterjee, Suparna Roy Sarkar, Satish Kumar, and Biswatrish Sarkar*
*E-mail: kalitahimadri123@gmail.com; shaandeb2010@gmail.com; rajboy277773@gmail.com; chatterjee.amrita95@gmail.com; srromiroy@gmail.com; 8709489546satish@gmail.com; biswatrishsarkar@bitmesra.ac.in

CONTENTS

1.1 INTRODUCTION

The human body physiologically experiences continuous assaults from reactive oxygen species (ROS) and reactive nitrogen species (RNS) generated by the cellular respiration, exercise, environmental pollutant, and xenobiotics. These reactive species, predominantly ROS, play a significant role in the regulation of cell function and tissue viability. At the same time, free-radical mediated chain reactions impact nucleic acids and proteins adversely and in turn create oxidative stress. Accelerating ROS production is one of the tipping points of cellular homeostatic disruption and onset of oxidative stress. To counter such imbalances, the physiological system and vital organs

DOI: 10.1201/9781003251538-1

are heavily dependent on versatile enzymatic and non-enzymatic antioxidant defense systems in our body, like superoxide dismutase (SOD), catalase and glutathione peroxidase (GPx), as well as vitamins like E, C, and carotenoids. These cellular systems are capable of trapping and controlling the free-radical chain reactions to regulate the overall ROS levels and maintain physiological red-ox homeostasis. However, the production and function of the endogenous antioxidant macromolecules become limited in various pathological conditions and need reinforcement from external sources like antioxidant-rich dietary sources and antioxidant-enriched herbal supplements.

In the beginning of the twenty-first century, the National Academy of Sciences defined a dietary antioxidant as "a dietary substance in food that significantly decreases the adverse effects of free radicals on normal physiological functions in humans." Phytoantioxidants or dietary antioxidants are the dietary phytochemicals present in food that significantly decrease the adverse effects of ROS on normal physiological functions in the human body. Inhibition of ROS-induced oxidative damages by supplementation of phytoantioxidants, the so-called "antioxidant therapy," has become an attractive therapeutic strategy to reduce the risk of ROS-related diseases and has led to flourishing research in the past decade. Various dietary antioxidants such as vitamin C, E, carotenoids, phenolic, and flavonoids are well known for providing protection against oxidative stress-induced damage to different important organs (Pisoschi et al., 2015).

Phenolics, also known as polyphenols, are one of the largest and most extensively dispersed families of natural compounds, and their involvement in human health have piqued the interest of nutritionists, food scientists, and consumers (Tsao, 2010; Lê and Tappy, 2006). Although polyphenols are technically defined as compounds with phenolic structural properties, they are a broad collection of natural products that includes various sub-groups of phenolic compounds. Polyphenols are abundant in fruits, vegetables, whole grains, and other foods and in beverages such as tea, chocolate, and wine (Tsao, 2010). Flavonoids, the major groups under the polyphenols family, are ubiquitously present in every vegetable and fruit which are part of the human diet. Like carotenoids, a wide array of biological activities has been attributed to the antioxidant activity of flavonoids. Flavonoids are secondary metabolites found in high concentrations in plants, fruits, and seeds, and are responsible for colour, aroma, and flavor (Dias et al., 2021). Flavonoids have a variety of activities in plants, including controlling cell growth, attracting pollinators, and defending against biotic and abiotic stressors (Luna et al., 2020). Similarly, a range of therapeutic benefits of the polyphenols, including flavonoids, have been documented globally especially from the antioxidant potential point of view. However, many questions dealing with the correlation between chemical and biological activities, the bioavailability and "non-antioxidant" effects of polyphenolic antioxidants, are still under debate. Nevertheless, despite all the drawbacks in terms of bioavailability and other physicochemical point of view, the health benefits of phenolics and flavonoids are enormous and may direct us to newer arenas for development of multiple strategies to formulate effective therapy to manage various oxidative stress-associated complications.

1.2 DIETARY POLYPHENOLS

1.2.1 GENERAL CHEMISTRY, STRUCTURE AND SOURCES OF POLYPHENOLS

Polyphenols mostly contain repeating phenolic moieties of pyrogallol, phloroglucinol, pyrocatechol, and resorcinol, connected by esters or more stable C-C bonds. These compounds frequently have functional groups other than hydroxyl groups, such as ether ester bonds and carboxylic acids (Quideau et al., 2011).

Polyphenols are classified chemically into two groups: flavonoids and non-flavonoids, based on the number and arrangement of distinct phenolic subunits as well as the hydroxyl moiety's attachment to the phenolic skeleton (Durazzo et al., 2019). Flavonoids are considered as one of the largest groups of polyphenolic compounds, comprising more than ten molecules found in the plant

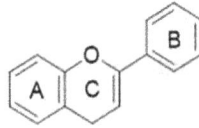

FIGURE 1.1   Benzo -α-pyrone ring (Flavonoid backbone).

kingdom. Flavonoids have 15 carbon atoms (C6-C3-C6) in their backbone, with two aromatic rings A and B joined by a C ring with three carbon atoms (Khan et al., 2021) (Figure 1.1).

Based on the degree of oxidation of the heterocyclic ring, the flavonoid class of chemicals is divided into several subclasses. Flavonoids that are connected at position 3 of the ring C are denoted as isoflavones (e.g., daidzein, genistein, glycitein), and those connected in position 4, known as neoflavonoids (Panche et al., 2016). The most common flavonoids found in dietary sources are flavonols, which contain a ketone group, present in position 3 of the C ring, and a hydroxyl group present in position 3 of the C ring (e.g., (+)-catechin, (-)- epicatechin, (-)-epigallocatechin gallate (Iwashina, 2013). Another important group, flavanones are the flavonoids that are structurally identical to flavonols but have the C ring saturated at the 2 and 3 positions. The neoflavonoids are formed when the 4-phenylchromen backbone has no hydroxyl group substitution at position 2 and instead has the hydroxyl group linked to position 3 of the C ring (Khan et al., 2021). Brightly coloured flavonoids with distinct blue, black, red, and pink colours that are modified by pH and methylation or acylation of the hydroxyl groups on the A and B rings are categorized as anthocyanins (Ayash et al., 2020).

On the other hand, the most common non-flavonoid polyphenols are the phenolic acids, which are split into hydroxybenzoic acids (C1-C6 backbone) and hydroxycinnamic acids (C3-C6 backbone) and structurally characterized by a carboxylic acid group attached to the phenolic ring (Tsao, 2010). Apart from that, other non-flavonoid polyphenols are stilbenes, coumarins, and lignans. All these polyphenols are abundantly distributed in the plant kingdom and forms part of our daily diet. The classification, structure and source of these polyphenols is summarized in Tables 1.1 and 1.2.

1.3   DIETARY POLYPHENOLS AS ANTIOXIDANTS

1.3.1   Antioxidant Mechanisms of Dietary Polyphenols

Polyphenols may exhibit antioxidant properties through several mechanisms: (a) direct neutralization of various types of free radicals or ROS like superoxide radical, OH·, peroxynitrite radical and so forth; (b) chelation of metallic ions; (c) increased production of endogenous antioxidant enzymes, such as GSH, SOD, and catalase via translocation of nuclear transcription factor; (d) inhibition of cellular ROS generating enzymes such as xanthine oxidase, myeloperoxidase (Dias et al., 2021; Nijveldt et al., 2001; Akhlaghi et al., 2009).

Neutralizing or scavenging various free radicals is the most common mechanism of antioxidant actions of polyphenolic compounds. The presence of hydroxyl group (-OH) on the phenolic ring system increases the polyphenol's propensity to donate hydrogen atoms to free radicals, which mediates the formation of stable intermediate and eventually quenches reactive free radicals (Fraga et al., 2010; Lü et al., 2010).The participation of л-electrons, bond dissociation enthalpy, and hydrogen atom transfer in a series of biological events can be rationalized thermodynamically as the mechanism behind radical scavenging properties of polyphenolic compounds (Meo et al., 2013; Leopoldini et al., 2011). However, the free radical quenching and antioxidant mechanisms of polyphenols may be influenced by the number and position of -OH, as well as by the coupling of other functional groups to the structural skeleton (Dias et al., 2021; Durazzo et al., 2019). Polyphenols,

TABLE 1.1

Classification of polyphenolic compounds and main sources with examples

Class of polyphenols	Sub-class	Examples	Sources from foods and dietary botanicals
Flavonoids	Flavone	Apigenin, baicalein, chrysin, luteolin	Citrus fruits, garlic, green leafy vegetables, fenugreek, herbal tea, onion, pepper
	Flavonol	Fisetin, myricetin, rutin, quercetin	Apples, broccoli, berries, chocolate, cocoa, kale, tea, onions, scallions
	Isoflavone	Biochanin A, daidzein, genistein, glycitein	Chickpea, dairy products, egg, pea nut, legumes, meat, seafood, soy products
	Flavanol	(+)-catechin,(-)- epicatechin, (-)-epigallocatechin gallate (EGCG), theaflavins	Apple, chocolate, cocoa, grapes, tea
	Flavanonol	Aromadendrin, engeletin, taxifolin	Milk thistle seeds, citrus fruits
	Flavanone	Eriodictyol, hesperidin, hesperetin, naringenin, naringin	Citrus fruits, grapefruit, pomegranate, tomatoes
	Chalcone	Butein, cardamomin, isoliquiritigenin, isosalipurposide, 4-hydroxyderricin, xanthoangelol	Licorice, shallots, and bean sprout tomatoes
	Dihydro-chalcones	Aspalathin, nothofagin, phlorizin	Apple, apple cider vinegar, rooibos tea
	Anthocya-nidins	Cyanidin, delphinidin, malvidin, peonidin, pelargonidin, petunidin	Black soybeans, cherries, berries, red wines, red grapes, oranges, purple/black rice, hibiscus sp., onions, red potatoes, purple cabbage
Non-flavonoids	Phenolic acids	Caffeic acid, *p*-coumaric acid, ferulic acid, gallic acid, protocatechuic acid, shikimic acid, sinapic acid	Apples, berries, cherries, citrus fruits, corn flours, coffee, green tea, kiwi, rice bran, mangoes, wheat, passion fruit
	Lignans	Ecoisolariciresinol, lariciresino, matairesinol, medioresinol, pinoresinol, sylibin, silymarin, sesamin, syringaresinol	Apricots Broccoli, flaxseed, cabbage, soybeans, milk thistle, and strawberries
	Stilbenes	Resveratrol	Grapes skin, peanuts, mulberries, and red wine
	Coumarins	Osthole, dicumarol	Bison grass, carrot, cinnamon, green tea

like various flavonoids, have been reported to reduce the amount of ROS produced during various events of mitochondrial respiration by blocking enzymes and chelating trace elements involved in the electron transport chain reaction (Sandoval-Acuña et al., 2014; Khan et al., 2021).

Chelation of metal ions is another important mechanism of polyphenolic antioxidants (Pandey et al., 2009). Often, metal chelating activity of polyphenols aids in avoiding Fe or Cu catalyzed free radical production in biological systems. Polyphenols construct stable coordination complexes with transitional metallic ions present in the biological system (Lakey-Beitia et al., 2021). This mechanism is also applicable for the counter of ROS generating enzymes like xanthine oxidase, myeloperoxidase, and so forth, by binding with the metallic co-factor resulting in the arrest of ROS generation via enzyme catalysis (Samoylenko et al., 2013).

Different polyphenols have been found to activate intracellular antioxidant signalling pathways to produce endogenous antioxidants such as GSH, SOD, and catalase (Clifford et al., 2021). During oxidative stress, nuclear factor erythroid 2 (Nrf2) acts as a key cellular machinery to produce endogenous antioxidants. In its normal physiological state, Nrf2 binds to KEAP1 protein in the

TABLE 1.2
Classification of polyphenols and examples of each class along with structures

Class of Polyphenols	Examples
Flavone	Apigenin · Baicalein · Luteolin
Flavonol	Quercetin · Myricetin · Fisetin
Isoflavone	Daidzein · Genistein · Glycitein
Flavanol	(+)-catechin · (-)-epicatechin · (-)-epigallocatechin gallate
Flavanonol	Aromadendrin · Taxifolin · Engeletin

(continued)

TABLE 1.2 (Continued)
Classification of polyphenols and examples of each class along with structures

Class of Polyphenols   Examples

Flavanone · Naringenin · Naringin

Chalcone · Phloretin · Phlorizin

Anthocyanidin · Delphinidin · Malvidin · Cyanidin-3-O-glucoside

Kelch domain of KEAP1 and degrades spontaneously in the cytoplasm (Clifford et al., 2021; Suraweera et al., 2020). Polyphenolic compounds have been shown to reduce the spontaneous degradation of Nrf2 protein by inhibition of Nrf2-KEAP1 protein-protein interactions in the cytoplasm. Polyphenols, such as various flavonoids, compete with Keap1 for the Nrf2 binding site, resulting in Nrf2 protein translocation into the nucleus and activation of downstream proteins HO1 and NQO1 (Khan et al., 2021; Scapagnini et al., 2011). The up-regulation of antioxidant genes like GSH, SOD, and catalase is directly influenced by the activation of these downstream proteins. For example, flavonoids such as quercetin, luteolin, baicalin, genistein, wogonin, and others have been found to protect various organ toxicities against oxidative stress via activation of the Nrf2-Keap1/ARE pathway (Suraweera et al., 2020). Table 1.3 presents brief summary of antioxidant mechanisms and pharmacological actions exerted by various classes of polyphenols.

1.3.2 METABOLISM AND BIOAVAILABILITY OF DIETARY POLYPHENOLS

Despite numerous health benefits of the polyphenols in various pathological conditions, due to poor bioavailability and metabolic stability concerns, many polyphenolic compounds fail to go forward

TABLE 1.3
Antioxidant mechanisms of dietary polyphenol

Polyphenol	Pharmacological effects	In vitro/in vivo models	Antioxidant Mechanism	Reference
Luteolin	Hepatoprotection	Lead acetate induced hepatotoxicity in Wistar albino rats	↓ROS production, ↑Nrf2, ↑HO-1, ↑ NQO-1, ↑SOD, ↑GPx, ↑CAT, ↑GR, ↓MDA	(Li et al., 2019; Al-Megrin et al., 2020; Liu et al., 2015)
	Cardio protection	• in vitro study in H9C2 cells with injury caused by high glucose level • Streptozotocin induced diabetic cardio myopathy model in C57BL/6 mice		
	Neuroprotection	• Marmarou's weight-drop model of brain injury in mice • in vitro scratch model in primary cultured neurons cell		
Apigenin	Cardio protection	• Primary neonatal rat cardiomyocytes • Doxorubicin induced cardio toxicity model in male Wistar rats	↑AMPK signalling pathway, ↑ERK/CREB/BDNF pathway, ↓BACE1 and β-CTF, ↑GSH, ↑SOD, ↓MDA, ↑GPx, ↑GR, ↑CAT, ↓cytochrome P450 2E1, ↑Nrf2-ARE pathway	(S. Li et al., 2018; Ting et al., 2013; Zhou et al., 2020)
	Neuroprotection	Anti-Alzheimer's model of APP/PS1 double transgenic mice model to assesses cognitive function		
	Hepatoprotection	Ethanol-induced rat hepatocytes and D-galactosamine/lipopolysaccharide induced rat BRL cells		
Vitexin	Neuroprotection	Aβ$_{25-35}$ induced Alzheimer's disease in Neuro-2a cell line	↑(MAPK)-Nrf2-ARE pathway along with ↓NF-κB, ↓ROS and RNS production, ↑GSH, ↑SIRT-1 and ↑PPAR-γ	(Malar et al., 2018; Ganesan et al., 2020; Shang et al., 2021)
	Anti-diabetic	Oxidative stress induced β-cell dysfunction in rat insulinoma INS-1 cell line		
	Skin protection	Hydrogen peroxide (H$_2$O$_2$) induced vitiligo in human epidermal melanocyte cell line PIG1		
Isoorientin	Hepatoprotection	In vitro study in Benzo[a]pyrene induced HL-7702 cell line	↓ROS production, Inhibition of ROS/NFκB/NLRP3/Caspase-1 signalling pathway, Oxidative stress inhibition via ↑JNK and ERK, ↑AMPK/AKT/Nrf2 signalling pathway, ↑SIRT1/SIRT6/Nrf-2 Pathway,	(Shang et al., 2021; A. et al., 2020)
	Neuroprotection	6-hydroxydopamine induced SH-SY5Y cell line to assess neurotoxicity		

(continued)

TABLE 1.3 (Continued)
Antioxidant mechanisms of dietary polyphenol

Polyphenol	Pharmacological effects	In vitro/in vivo models	Antioxidant Mechanism	Reference
	Nephrotoxicity	• Cisplatin-induced nephrotoxicity model in mouse renal tubular epithelial cells • Cisplatin-induced nephrotoxicity model in C57BL/6 WT mice		
Quercetin	Hepatoprotection	Oxidative stress induced in human normal liver L-02 cells by clivorine, acetaminophen, ethanol and carbon tetrachloride.	↓ROS production, ↑Nrf2-ARE pathway, ERK and JNK/activator protein-1 pathway activation, ↑pi3k, ↑Akt, ↑creb and ↑bdnf in the hippocampus.	(Pang et al., 2016; Zuo et al., 2016; Xia et al., 2015)
	Ocular damage protection	H_2O_2 induced oxidative damages in ARPE-19 cell line		
	Neuroprotective	High fat diet induced hippocampus dependent cognitive disorder in male Chinese Kunming (KM) mice		
Naringenin	Hepatoprotection	• 12-O-Tetradecanoylphorbol-13-acetate (TPA)-induced hepatotoxicity in human hepatoma cells and murine embryonic liver cells • Doxorubicin induced liver impairment in Wistar rats	↓ROS production, ↓ROS-induced lipid peroxidation, and ↑antioxidant enzymes level ↓ERK MAPKinase and PI3K/Akt leading to ↑ Nrf2-ARE pathway, ↑IGF-1R, activation of NF-κB	(Yen et al., 2015; Wali et al., 2020; Jin et al., 2019; Tseng et al., 2021; Tan et al., 2011)
	Neuroprotection	• H_2O_2 induced neurotoxicity in human neuroblastoma SH-SY5Y cells • methylglyoxal induced neurotoxicity in NSC34 motor neuron cell line		
	Airway protection	Human neutrophil elastase induced airway epithelial cell line		
Rutin	Hepatoprotection	CCl_4 induced hepatotoxicity in BALB/cN mice	↑Nrf2-ARE pathway, ↓iNOS, ↓ROS production, ↑ERK1, ↑CREB, and ↓MDA, ↑SOD, ↑CAT, ↑GSH-Px, ↓MAPK	(Singh et al., 2019; Domitrović et al., 2012)
	Neuroprotection	Assessment of spinal cord injury in Sprague Dawley rats using Allen's model		
	Nephroprotection	Vancomycin induced nephrotoxicity in male Wistar rat model		

for drug development. Pharmacokinetic profiles of many flavonoids have been found poor from the dietary matrixes (Kamiloglu et al., 2021). At the same time, metabolism of dietary polyphenols are an extremely overwhelming area of research. Nevertheless, bioavailability and metabolism of these compounds can be altered to be suitable for therapeutic uses by employing various pharmaceutical techniques (Thilakarathna et al., 2013). Therefore, researchers around the globe are showing very much interest towards the absorption, bio-distribution, metabolism, and excretion in a short ADME profile of polyphenols from various dietary materials.

Absorption of dietary polyphenols in the gut varies with the structure of the individual polyphenols. The most addressed structural parameters responsible for the variation of gut absorption considered are molecular weight, glycosylation, and esterification (Manach et al., 2004). Proanthocyanidins are flavonoid polymers with higher molecular weight (578 or higher), and varying degrees of polymerization contribute to its low bioavailability. Most flavonoids in food are glycosylated, except catechins and proanthocyanidins, and this glycosylation modulates absorption through the gut barrier (Scalbert et al., 2002). Often the aglycon and a few glycosides are absorbed in the gut, but a majority of the flavonoids are absorbed in the small intestine after undergoing deglycosylation (Khan et al., 2021; Dias et al., 2021). The conjugated flavonoids that are not broken down by colonic microflora in the intestine gets eliminated with the feces (Sandoval et al., 2020). However, the conjugation and the position of the sugar moiety in the flavonoid ring system plays a vital role in this deglycosylation process (Gonzales et al., 2015). At the same time gut microbiota also plays significant role in the conversation of parent polyphenolic compounds from dietary sources (Khan et al., 2021). This phenomenon is often bidirectional where polyphenols undergo conversion by the influence of enzymes secrets by gut microflora and on the other side some polyphenols stimulate gut microorganism to release vital enzymes (Corrêa et al., 2019).

The polyphenols are prone to metabolic conversion by action of sulfotransferases, catechol-O-methyltransferases (COMT), uridine-50-diphosphate glucuronosyltransferases (UGT), and glutathione transferase enzymes (Lee et al., 2014). Similarly, CYP450 enzymes influence significantly in the hepatic metabolism of the flavonoid type of polyphenols. Mostly CYP1A2 and CYP3A4 have been demonstrated and are the most common isoforms of CYP450 involved in the microsomal metabolism of flavonoids (Villegas et al., 2007). Most metabolites of polyphenol are eliminated from the circulation through urinary excretion. Metabolic conversion not only facilitates the excretion of flavonoids from the human body, but it also aids interactions at the molecular therapeutic target.

1.4   ROLE OF POLYPHENOLS IN HUMAN DISEASE MANAGEMENT

Plant-based foods are the main source of polyphenols. The impact of polyphenols on different biological processes, ranging from simple gallic acid to the most complex proanthocyanidins, has been indubitably demonstrated by numerous studies (Tresserra-Rimbau et al., 2018). Many epidemiological studies have found that eating a polyphenol-rich diet has a negative impact on the etiology of chronic human diseases. The capacity of polyphenolic groups to take an electron and create relatively stable phenoxyl radicals interrupts free radical chain events within biological components. It is also well established that polyphenol-rich foods and beverages are responsible for an increment in plasma antioxidant capacity (Pandey et al., 2009). Studies also proved that, as antioxidants, polyphenols can protect cell constituents against oxidative damage and, therefore, limit the risk of various progressive diseases associated with oxidative stress.

In this regard it is worth stating that experimental models *in vitro* and *in vivo* are critical for elucidating the mechanisms of action underlying the health benefits of polyphenols seen in human studies. However, due to significant inter-individual variability and other external influences, their findings are not always simply generalized to humans. Potential effect-modulating variables such as sex, age, smoking habits, BMI, and hormone levels, as well as the influence of other foods, nutrients, and even culinary styles, must all be found (Rimbau et al., 2018; Landberg et al., 2019). This section is thus devoted to the various protective properties of polyphenols against human diseases.

1.4.1 PROTECTIVE ROLE OF DIETARY POLYPHENOLS IN METABOLIC SYNDROME (METS)

In today's world, metabolic syndrome (MetS) is the most frequent clinical syndrome that manifests as a number of aberrant medical consequences, including central obesity, hypertension, dyslipidemia, and excessive blood sugar (Figure 1.2). MetS is a major contributor to the onset and progression of type 2 diabetes (T2D) and cardiovascular disease (CVDs). The most prominent pathophysiology of MetS is oxidative stress (OS), which is caused by the formation of free radicals. OS in the body triggers several diseases, such as diabetes (Sottero et al., 2015), CVDs like atherosclerosis (Wu et al., 2004), Alzheimer's disease (Fukui et al. 2007), neurodegenerative disease, and even cancer (Zuo et al., 2016). MetS is now best managed by eating a regular and normal diet rich in fruits and vegetables, limiting high carbohydrate and high-fat foods, exercising regularly, and limiting alcohol use (Grundy, 2016). As a result, a novel preventative strategy is required to reduce MetS morbidity and death. Intake of single/multiple polyphenolic compounds or particular phenolic extracts has been shown to have protective and preventative effects on MetS in cells and animal models in several studies. Tian et al. suggested that by inhibiting extracellular signal-regulated kinase (ERK) 1/2-peroxisome proliferator-activated receptor (PPAR) γ-adiponectin pathway, green tea (GT), a potent source of polyphenol, reduced fat formation in high fat-fed rats (Tian et al., 2013). They also reported that intake of green tea polyphenols reduced the risk of coronary heart disease in middle-aged and elderly Chinese people to some degree (Tian et al., 2016). In a dose-dependent manner, resveratrol (RSV, a kind of non-flavonoid) administration reduced the hyperpermeability and overexpression of cav-1 produced by high glucose. The elevated expressions of vascular endothelial growth factor (VEGF) and kinase insert domain receptor (KDR or VEGF receptor-2) induced by high hyperglycemia were proven to be downregulated by RSV (Tian et al., 2013). Faghihzadeh et al. discovered

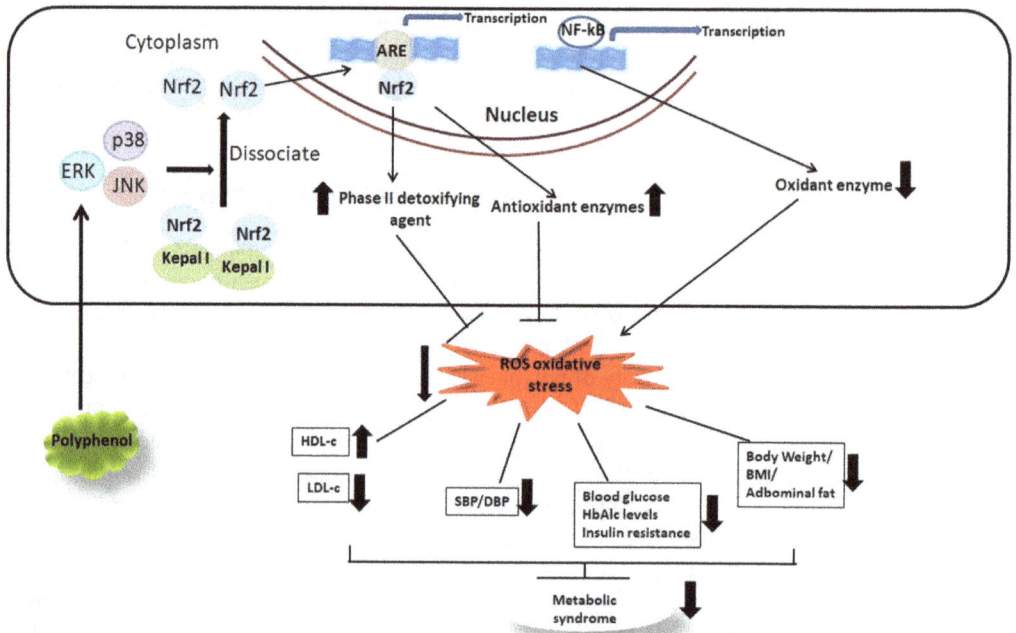

FIGURE 1.2 Schematic representation of the relationship between antioxidative property of polyphenol and MetS (Dembinska-Kiec et al. 2008).

that RSV supplementation for 12 weeks significantly reduced alanine aminotransferase (ALT) levels and stopped hepatic steatosis in nonalcoholic fatty liver disease (NAFLD) patients compared to the control group ($P < 0.05$) (Faghihzadeh et al., 2015). As a result, polyphenols, which are powerful antioxidants found only in fruits and vegetables, are being evaluated as a potential MetS treatment candidate (Amiot et al., 2016).

1.4.2 Protective role of dietary polyphenols in CVDs

Coronary heart disease (CHD), hypertensive heart disease (HHD), heart failure, and stroke are the top causes of death worldwide (Wang et al., 2016). CVDs could be caused by hypertension, dyslipidemia, atherosclerosis, oxidative stress, inflammation, and enteric dysbacteriosis (Benjamin et al., 2017). Several synthetic medicines have been used to treat CVDs, but they have side effects such gastrointestinal upset, hyperkalemia, and arrhythmias (Ziff et al., 2016). On the other hand, it was found that certain antioxidant natural compounds may be a safe and effective option for the prevention and treatment of CVDs (Miller et al., 2017).

Natural products are high in dietary fibres, polyphenols, vitamins, minerals, and other beneficial components, and they have a wide range of bioactivities, including antioxidant, anti-inflammatory, anticancer, antidiabetic, antiobesity, hepatoprotective, immunoregulatory, antibacterial, and cardiovascular protection (Shang et al., 2021). According to epidemiological research, people who eat more fruits, vegetables, teas, cereals, and nuts had a lower incidence of CVDs, and the antioxidants in these natural items were thought to be the key contributors (Lule et al., 2019). Experiments also revealed that several antioxidant natural products and their active components can prevent and treat CVDs via various mechanisms of action (Olagunju et al., 2018). Furthermore, clinical trials offered more credible human evidence for the prevention and treatment of CVDs using antioxidant natural compounds (Azimi et al., 2016).

1.4.3 Hepatoprotective effects of dietary polyphenols

Liver serves vital protective functions against potentially hazardous chemicals as it is able to convert lipophiles into more water-soluble metabolites that can be easily removed from the body via the urine (Torre et al., 2021). Liver disease, which includes fibrosis, cirrhosis and hepatocellular cancer, is a major health concern affecting people all over the world (Forbes et al., 2016). Polyphenols have shown potential in treating hepatic disorders because of their remarkable ability to reduce oxidative stress, lipid metabolism and inflammation via complex molecular mechanistic pathways (Domitrović et al., 2016).

Quercetin, a natural flavonoid, greatly protected the liver from CCl_4-induced hepatotoxicity. Its antioxidant and anti-inflammatory mechanisms were attributed to the suppression of Toll-like receptor 2 (TLR2) and Toll-like receptor 4 (TLR4) activations, as well as MAPK phosphorylation, which resulted in NF-κB inactivation and, as a result, a reduction in hepatic inflammatory cytokines (Ma et al., 2015). Quercetin protected non-alcoholic fatty liver disease by up-regulating Nrf2-mediated heme oxygenase-1 (HO-1) expression leading to hepatic macrophage polarization (Kim et al., 2016). A well-known flavonoid, baicalin, can treat acetaminophen-induced liver inflammation by inhibiting the ERK signalling pathway (Miller et al., 2017). By activating the Nrf2/ARE pathway and blocking pro-inflammatory cytokines like NF-κB and others, nobiletin (a methoxyflavone found mainly in the peels of oranges, lemon and other citrus fruits) was able to protect the liver from lipopolysaccharide (LPS)/D-galactosamine-induced damage (Torre et al., 2021). Resveratrol, commonly found in peanuts, grapes, wines, blueberries and cocoa, down regulated NF-κB and CYP2E1 expression in thioacetamide-induced hepatotoxicity (Torre et al., 2021). Another study suggesting that AMPK activation by proanthocyanidins were involved in liver protection against alcohol-induced hepatic lesions, where proanthocyanidins were found to suppress the inflammatory gene expressions like interleukin 1β (IL-1β), IL-6, and TNF-α (Pang et al., 2016). Hepatoprotection has been exhibited by kaempferol by inhibiting NF-κB in nucleus and cytoplasm, which further reduced TNF-α and IL-6 expressions (Sun et al., 2015; Torre et al., 2021). Polyphenols like apigenin, quercetin, curcumin, resveratrol, and epigallocatechin -3 gallate inhibits hepatic stellate cells (HSCs) activation (Li et al., 2018). Morin, quercetin and gallic acid have been shown to have anti-fibrotic properties through increasing apoptosis in activated HSCs, which is predominantly linked to NF-κB and TNFα signalling (Torre et al., 2021). Curcumin, the main flavonoid component of turmeric (*Curcuma longa*), demonstrated to reduce the severity of LPS-induced liver failure by inhibiting activation of the MAPK/P38/c-JNK cascade in rat liver. Furthermore, it suppressed the PI3K/AKT signalling pathway and inhibited CREB/caspase production, lowering blood cytokines, including IL-6, IL-1, and TNF-α (Inzaugarat et al., 2017). It also modulated the oxidative stress-related signalling pathway in LPS-treated mice, as evidenced by lower CYP2E1/Nrf2/ROS protein expression (Torre et al., 2021). Isoorientin stimulates phase II detoxifying enzymes like glutathione peroxidase (GSH-Px), glutathione (GSH), catalase (CAT), haeme oxygenase-1 (HO-1) and superoxide dismutase (SOD) which potentially inhibited HEPG-2 cell proliferation (Yuan et al., 2016). Flavones with 2-phenylchromen-4-one backbone like baicalein, vitexin, Isoorientin, chrysin, and luteolin were found to promote apoptosis by decreasing cell proliferation in many hepatocellular carcinoma cell lines. Anticancer mechanisms of these flavones were found to be JNK pathways, caspase-dependent, caspase-independent apoptotic signalling pathways and mTOR pathway (S. Li et al., 2018).

1.4.4 The Chemopreventive Role of Dietary Antioxidant Polyphenols

Polyphenols can have biological effects in vitro and in vivo by interacting with a variety of cellular targets implicated in carcinogenesis (Torre et al., 2021). The benefit of using dietary polyphenol's anti-oxidant property is to decrease the side effects and toxicity of radio- and chemotherapy in healthy cells, which outweigh the dangers of affecting the treatment's effectiveness in cancer cells (Russo et al., 2017). In the cells ROS and reactive nitrogen species (RNS) are created constantly by a range of endogenous and external mechanisms (Inzaugarat et al. 2017). Antioxidant defense regulates their levels, reducing the high reactivity of ROS/RNS and preserving redox balance (Torre et al., 2021). Numerous results from both in vitro and animal trials demonstrated that polyphenols can aid in the formation of relatively stable radicals (Abarikwu et al., 2017). As a result, polyphenols' anti-oxidant activities in normal cells or pre-cancerous cells with abnormal ROS/RNS levels are thought to be advantageous in cancer prevention (Torre et al., 2021). Cellular metabolites of polyphenols, particularly hydroquinones and quinones (electrophilic form) can oxidise Keap1 (kelch-like ECH-associated protein 1) on cysteine residues at low concentrations, allowing nuclear factor (erythroid-derived 2)-like 2 (Nrf2) to translocate into the nucleus. The translocation of Nrf2 activates anti-oxidant defense genes, allowing polyphenols to indirectly activate anti-oxidant responses at low concentrations and creating non-toxic levels of quinone-like intermediates (Russo et al.,2017; Forman et al., 2014). In in vitro and in vivo models, polyphenols such as epigallocatechin-3-gallate (EGCG) from green tea, quercetin, and curcumin from turmeric have been shown to reduce tumour growth and stop carcinogenesis (in vitro and in vivo). These effects can be linked to these antioxidants' non-scavenging functions, which include induction of apoptosis, growth arrest, DNA synthesis inhibition, and signal transduction pathway regulation (Alothman et al., 2009).

EGCG prevents the generation of metal-catalyzed free radicals via Fenton reactions. In cultured cells, EGCG has been found to block numerous key signalling pathways, as well as the activation of the redox-sensitive transcription factors NFκ-B and AP-1 (Sah et al., 2004; Khan et al., 2006; Na et al., 2006). EGCG suppresses EGFR, the first kinase in the EGF signalling cascade, and increasing ERK1/2 and AKT activity. This inhibition is linked to decreased phosphorylation levels, and activity of ERK1/2 and AKT downstream substrates, leading in G1 arrest and increased apoptosis (Sah et al.,

2004). Beyond cell proliferation and death, EGCG inhibits telomerase activity, which is essential for the onset and development of cervical lesions (Yokoyama et al., 2004; Noguchi et al., 2006). It also decreases anchorage-independent growth, cell proliferation, and transformation caused by IGF-IR overexpression. IGF-IR found to inhibits downstream signalling and expression of the cervical cancer cell phenotype (Villegas et al., 2007). Polyphenols also auto-oxidize and generate hydrogen peroxide (50 M EGCG produces 1 M H_2O_2 in Jurkat cells) in the culture medium at high concentrations contributing to cancer-cell cytotoxicity (Nakagawa et al., 2004).

The aryl hydrocarbon receptor (AhR), which regulates the expression of cytochrome P450 (CYP) enzymes (capable of activating procarcinogens) and a transcription factor that can be activated by polycyclic aromatic hydrocarbons (PAHs) implicated in lung cancer, has been discovered to be a potent antagonist (Denison et al., 2002; Denison et al., 2003; Murakami et al., 2008). Quercetin, which may be found in onions and apples, has been reported to have an inverse relationship with lung-cancer risk, particularly in squamous cell carcinoma (Russo et al., 2007). It is also established that lower dietary flavonoid consumption can raise the incidence of lung cancer in a population-based case control study (1061 cases and 1425 controls) (Christensen et al., 2012).

Curcumin (1,7-bis (4-hydroxy 3-methoxy phenyl)-1,6-heptadiene-3,5-dione) reduces all three phases of carcinogenesis: initiation, promotion, and progression, according to several researchers. Suppression of NFκ-Band following down regulation of multiple NFκ-B -related pro-inflammatory pathways are most likely the main targets for its effectiveness (Goel et al., 2010). Curcumin inhibits NFκ-B activation via IB breakdown and AP-1 binding, resulting in COX-2 down regulation and enhanced tumour cell death. It also inhibits a variety of critical components in cellular signal transduction pathways, such as c-Jun/AP-1 activation and protein kinase-catalyzed phosphoryl-ation processes (Wood et al., 2008). Curcumin also suppresses angiogenesis, the FGF-2-induced angiogenic response, matrix metalloprotease expression, and cyclooxygenase-2 expression in vivo (Plummer et al., 1999; Plummer et al., 2001; Mohan et al., 2000). This flavonoid can also inhibit HPV-18 transcription in HeLa cells, diminish AP-1 binding activity selectively, and reverse c-fos and fra-1 expression dynamics (Prusty et al., 2005). Among various molecular pathways by which curcumin reduces telomerase activity, the Ras and ERK signalling pathway, cyclin D1, COX-2, and iNOS activity is found in cervical cancer cells (Singh et al., 2009; Babbar et al., 2011). When combined with ionizing radiation (IR), curcumin, induced a considerable rise in ROS, which led to persistent ERK 1/2 activation, boosting the radiotherapeutic action on cervical cancer cells (Javvadi et al., 2008). It is also found that the thioredoxin reductase enzyme is involved in curcumin's radiosensitizing effect in cervical cancer cells (Javvadi et al., 2010). Curcumin delivered during paclitaxel therapy serves as a down-regulator of paclitaxel pro-survival pathways linked to NFκ-B and Akt activation in cervical cancer in in vitro and in vivo models of cervical cancer(Bava et al., 2011; Sreekanth et al., 2011). Table 1.4 summarizes few mechanisms of chemoprevention by few selective flavonoids in vitro.

1.4.5 PROTECTIVE ROLE OF DIETARY ANTIOXIDANT POLYPHENOLS IN NEURODEGENERATIVE DISORDERS

Neurodegenerative diseases, including Parkinson's and Alzheimer's disease (PD and AD), are often associated with dysfunction of the central nervous system. Atrophic changes in brain is hall-mark of AD, while loss of dopaminergic neurons from mid- brain essentially affects the muscular movements causing loss of balance and walking difficulties in AD. AD has links with the accu-mulation of improperly folded proteins (β- amyloid and tau proteins) around the neurons in the form of plaques. Over- accumulation and modification of α-synuclein in Parkinson's disease leads to the loss of neurons (Wei et al., 2013). Both diseases have been linked to high levels of expres-sion of pro-inflammatory markers, mitochondrial dysfunction, and oxidative stress, accompanied by malfunctioning of proteins (Bhullar et al., 2013). Pro-inflammatory cytokines and reactive species

TABLE 1.4

Chemopreventive mechanisms of Epigallocatechin-3-gallate, Quercetin and Curcumin

Polyphenol	Mode of experiment	Antioxidant mechanism	Reference
Epigallocatechin-3-gallate	In vitro	Activation of the redox-sensitive transcription factors NFκ-B and AP-1	(Sah et al., 2004; Khan et al., 2006; Na et al., 2006)
	In vitro	Suppresses EGFR increasing ERK1/2 and AKT activity; decreased phosphorylation levels, and activity of ERK1/2 and AKT downstream substrates, leading in G1 arrest and increased apoptosis	(Sah et al., 2004)
	In vitro	Inhibits telomerase activity	(Noguchi et al., 2006; Yokoyama et al., 2004)
	In vitro	COX-1; COX-2 iNOS; HSP7; TRX-R; ROS production	(Javvadi et al., 2008; Javvadi et al., 2010)
	In vitro	G1 phase arrest; WAF-1/ p21; KIP-1/p27; cyclin; CD	(Yokoyama et al., 2004; Noguchi et al., 2006)
Quercetin glucuronides	In vitro	Decrease cell viability (dose and time dependent) Arrest cell cycle at G2/M phase via caspase-3 cascade	(Yang et al., 2006)
	In vitro	antagonist polycyclic aromatic hydrocarbons	(Denison et al., 2002; Denison et al., 2003; Murakami et al., 2008)
Curcumin	In vitro	Inhibit cell proliferation Induce fork head box protein O1 (FOXO1) expression Inhibit cell growth Induce G1/S arrest via activating CDK inhibitor genes p21 and p27	(Guo et al., 2014; Saha et al., 2010)
	In vitro	COX-1; COX-2 iNOS; HSP7; TRX-R; ROS production	(Javvadi et al., 2008; 2010; Singh et al., 2009)

may accelerate neuron and glial cell death and enhanced blood-brain barrier permeability, resulting in brain injury (Sochocka et al., 2017). Similarly, microglia and astrocyte activation are increased and persistent during neuroinflammation.

Polyphenols particularly act against the neurodegeneration by mostly exerting its antioxidant and anti-inflammatory properties (Khan et al., 2021). Different classes of flavonoids and its consumption as diet has reduced the burden of neurodegeneration. Curcumin has been found to reduce lipid peroxidation and free radical generation in homocysteine-induced neurotoxicity in rat models, which prevented the mitochondrial complex I dysfunction, improved memory, cognitive and motor functions. EGCG was able to ameliorate the neurodegeneration in vitro in neurotoxicity models (H 19-7, HT22 cell lines) and in animal models like C57BL/6 mice with transient global cerebral ischemia, SOD1-G93A transgenic mice (Torre et al., 2021). The effects were mainly via boosting the endogenous antioxidant system, restoring mitochondrial functions, minimizing MDA, NF-κβ activation, caspase 3 cleavage, while activating HIF-1α pathway, along with the modulation of some common signalling pathways like PI3K/MAPK, Akt, ERK1/2, PKC regulating the cell survival and apoptosis (Bava et al., 2011). The effect of resveratrol has also been studied in various in vitro and rodent models for neurotoxicity, ischemic cerebral stroke, neurodegenerative conditions, where it prevented ROS accumulation, induces endogenous antioxidants, preserves mitochondrial functions, neurons, along with regulating pathways like, ERK, SIRT-1, PI3K/Akt, Nrf-2/ARE, and MAPK as well. Polyophenols also prevent tau amyloidiopathies by regulating α, β, γ secretase activity, inhibiting β-amyloid protein oligomerisation, reducing its concentration and deposition in the brain, exhibiting marked ameliorative effect on both β- amyloid protein induced neurotoxicity and neuronal inflammation, as reported by actions of myricitin, quercetin, kaempferol, apigenin, EGCG, and polyphenols from red wine (Ebrahimi et al., 2012). EGCG has been reported to activate Keap1/ARF/Nrf-2 pathway, while some other polyphenols act via promoting the expression of neurotrophic factors like BDNF, GDNF, NGF, along with Bcl-2, Bcl-xL via tyrosine receptor kinase pathway, and suppression of TNF- α and IL-6, all converging to their neuroprotective effect. The diminished expression of TLR4, along with IRF3, and MyD88 may just be another pathway for curcumin to be of benefit in PD (Meo et al., 2020). Various clinical trials concerning AD demonstrated the use of polyphenols like green tea, fortified fruit juice, cocoa and so forth, which reported better cognitive functions, increased antioxidant capacity of plasma, and better visual memory in treated patients as compared to placebo groups, enforcing the health benefits of polyphenols in neurodegenerative diseases (Colizzi, 2019). Other polyphenols with anti-inflammatory action on the central nervous system include curcumin and apigenin, which have been studied in vitro and in animal models (Venigalla et al., 2015). Flavonoids, like quercetin, genistein, luteolin, daidzein and rutin have shown neuroprotective properties in vitro and in vivo models through lowering neuroinflammation (Del Rio et al., 2013).

1.4.6 DIETARY PHYTOCHEMICALS IN GASTROINTESTINAL DISEASE AND MICROBIAL INFECTIONS

Dietary polyphenols constitute a significant source of antioxidants consumed by human beings. In vivo studies show that in inflammatory bowel disease (IBD), oxidative stress plays a vital role in its pathogenesis, as polyphenols possess not just only antioxidant properties but also antiviral, antibacterial, anti-inflammatory and anticarcinogenic effects, in addition to the capability to regulate specific signalling pathways like NF-κB activation. Green tea leaf extract rich in polyphenol ameliorated the colonic inflammation in a 2,4,6-dinitrobenzene sulphonic acid-induced animal model. By regulating the levels of inductive enzymes that defend against oxidative stress, decrease in levels of TNF-α, expression of intercellular adhesion molecule-1, and increase in levels of heme oxygenase-1 was observed in this study (Gerald et al., 2014). Another potential mechanism independent of antioxidant property for the improvement of IBD demonstrated by EGCG in a fetal intestinal epithelial cell line (IEC-6), by inhibiting activation of NF-κB and preventing its binding to a cytokine promoter sequence in the nucleus (Ahmad et al., 2000).

Polyphenols have a wide range of antimicrobial effects through various mechanisms that are typically similar to those of traditional antibiotics, and hence could be useful in the development of antimicrobial treatments. Quercetin diminishes cell motility, an essential factor of bacterial virulence, by increasing cell membrane through rising dissipation in membrane potential. Various in vitro studies of Gnemonol B and Gnetin E, a dietary polyphenol, found to have intense antibacterial activity against methicillin-resistant *Staphylococcus aureus* (MRSA) and vancomycin-resistance *Enterococci* (VRE). Another in vitro study of Hydroxytyrosol, a dietary polyphenol, also found an anti-mycoplasmal activity in *M. hominis, M. fermentans* and *M. pneumoniae* (Han et al., 2007).

When administered orally with a dose of 140mg/kg, quercetin protects against *Shigella* infection and *E.coli* gyrase inhibition by inhibiting the ATPase GyrB subunit (Marín et al., 2015). The study also shows a significant decrease in gastric ulcers by quercetin when ulcerated rats are treated with a dose of 25 and 50 mg/kg body weight (Bakr, 2020). *H.pylori* causes gastric lesions leading to chronic infection of gastric mucosa through increasing expression of pro-inflammatory cytokines, mainly IL-1b, IL-8, TNF-α and IL-6, which activate neutrophil that leads to leukocyte infiltration. Allylpyrocatechol found to reduce the gastric lesions by inhibiting particular inflammatory signalling pathway, that is, MAPK signalling that contribute to transcription and translation process of proteins responsible for the inflammatory process (Farzaei et al., 2015). Myeloperoxidase, a peroxidase enzyme found abundantly in neutrophils also a biomarker for oxidative damage caused in IBD via production of ROS that triggers GI mucosal damage that can be prevented by calycosin, a dietary polyphenol found in black beans, lentils and pista through downregulating pro-inflammatory cytokines, myeloperoxidase activity and also by inhibiting NF-kB pathway and JNK- signalling (Arya et al., 2020). Resveratrol, a dietary polyphenol found in nuts and red grapes, have a protective effect against gastritis caused by *H. pylori* through their antioxidant activity anti-inflammatory activity mainly by inhibition of IL-8, iNOS (Inducible nitric oxide synthase), blocking the activation of NF-kB and increase in Nrf2 (nuclear erythroid 2-related factor 2) in the infected mucosal layer (Bakr, 2020).

1.4.7 INTERACTION OF POLYPHENOLS WITH GUT MICROBIOTA, NEURODEGENERATION AND COGNITION

Neuroinflammation and oxidative stress are closely connected processes that are frequently implicated in neurodegenerative diseases. Dietary polyphenols are beneficial in reducing neuroinflammation and oxidative stress while improving memory and cognitive function, through indirect mechanisms, such as regulation of the microbiota-gut-brain axis (Kennedy, 2014; Sandhu et al., 2017). The role of gut microbiota and its association with various disease conditions are well known. Polyphenols are metabolized within the intestine and thus undergo extensive interaction with gut microbiota. Polyphenols can stimulate the growth of gut microorganisms recognized as prebiotic targets. In fact, a bidirectional interaction is observed, in which polyphenols modulate the gut microbiota and, conversely, microbiota can modulate the activity of the phenolic compounds and as a result causes different effects on host health.

Dietary polyphenols, such as plant lignans, influence the gut–brain axis, which includes gut bacterial metabolism converting polyphenolic chemicals into physiologically active and neuroprotector molecules (known as human lignans); some of them have a higher permeability of the blood–brain barrier (Meo et al., 2020). These gut bacterial metabolites protect against a variety of neurodegenerative disorders like AD, PD, traumatic brain damage and autism spectrum disorder (ASD) (Serra et al., 2019). The therapeutically important polyphenols enterolactone and enterodiol are generated as lignans' secondary gut bacterial metabolites. These chemicals are also acetylcholinesterase inhibitors, which mean they might be used to treat AD and other neurological disorders. Resveratrol is a phytoalexin derived from grapes, has been demonstrated to block key pro-inflammatory signalling pathways, including the JAK/STAT, AP-1 and NF-kB pathways. In a variety of in vitro and in vivo studies, resveratrol reduces the production of several pro-inflammatory cytokines and pro-oxidants like COX-2, IL-12, TNF-α, IL-6 and iNOS and thereby counteracts neuroinflammation (Renaud et al., 2014; Spencer et al., 2012). A clinical trial also backs up preclinical findings, confirming the neuroprotective potential of resveratrol in AD (Moussa et al., 2017).

A variety of phenolic phytochemicals help with memory and cognition by influencing aspects like neurogenesis in hippocampus and cerebral blood flow as well (Kennedy, 2014; Spencer, 2008). After treatment with a blueberry beverage high in flavonoids, an increase in cerebral blood flow was detected in certain parts of patients' brains, like parietal lobes, frontal lobe and occipital cortex in a clinical investigation on healthy young and elderly individuals. This had a favorable impact on their cognitive capacities (Rendeiro et al., 2015). Resveratrol has been demonstrated to be useful in reducing autistic behavior and the suppression of biochemical alterations in the brain of autistic patients was seen in several ASD in vivo models (Bakheet et al., 2017; Bhandari et al., 2017). Xylo-oligosaccharide (XOS) and quercetin have been successfully used as a therapeutic intervention against gut dysbiosis mediated cognitive decline (Roy et al., 2020). The XOS and quercetin diet substantially reduced antibiotic-associated gut dysbiosis, resulting in reduced oxidative stress, pro-inflammatory cytokines and chemokines, and protection of hippocampus neurons, according to the study (by revival of *Lactobacillus*, *Bifidobacterium*, *Firmicutes* and *Clostridium* sp. in gut) and improved fear conditioning, spatial and recognition memory in mice.

1.4.8 DIETARY POLYPHENOLS IN ARTHRITIS AND INFLAMMATIONS

Polyphenols are found in various dietary plants, including fruits, vegetables, legumes, herbs, cocoa, beverages, spices, and so forth, as secondary metabolites with eight thousand different structural variants, bearing aromatic ring(s) with the presence of one or more hydroxyl group. Their structure greatly varies from a simple phenolic acid to condensed tannins that are highly polymerized compounds (Yoon et al., 2005). Recent studies reveal that dietary polyphenols have an inhibitory effect on inflammation and also have efficacy for rheumatoid arthritis through their effective free radical scavenging property due to the presence of a para-hydroxyl group. Dietary polyphenols are having significant effect in amelioration of rheumatoid arthritis, mainly by the tendency to interfere with the inflammatory pathways by suppressing the various inflammatory signals, such as NF-κb, ERK1/2, IL-1, IL-6, TNF-α, RANKL, and so forth, in addition to its antioxidant abilities (Rahman et al., 2006).

There are various signalling pathways associated with rheumatoid arthritis-like Mitogen-activated protein kinase pathway (MAPK) and NF-κB signalling pathway that controls and regulates the production of pro-inflammatory cytokines (IL-1, IL-2, IL-6, IL-8, TNF-alpha, and chemokines); all these lead to bone destruction, cartilage destruction, matrix degradation, and angiogenesis. Studies have revealed that polyphenol-rich extract of *Pergularia domenia*, when given to in collagen-induced arthritic (CIA) rats, caused a reduction in bone destruction, and white blood cell (WBC) production was observed by interfering with NF-Kb signalling pathway. Also, when A-glucosyl hesperidin (found in citrus fruits) given in a dose of 3 mg/0.3 ml thrice a week for 31 days to CIA rats, a remarkable decrease in the production of TNF-α was observed through the inhibition of mitogen-activated protein kinase signalling mainly by ERK1/2, JNK, and p38/SARK signals (Behl et al., 2021). These in vivo and in vitro studies suggested a strong connection between dietary polyphenols and prevention of arthritis related to musculoskeletal inflammation mainly due to an underlying mechanism of inhibiting the various inflammatory signals produced by our body, such as the release of auto-antibodies production, macrophages activation, osteoclast formation, and pro-inflammatory cytokines such as IL-1, IL-6, and TNF-α: these all contribute to reduction in inflammatory signals (Shen et al., 2012). Decreased production of H_2O_2, TNF-α, IL-17, and IgG was observed in CIA-induced mice by grape seed procyanidin extract due to the inhibition of the RANKL pathway that leads to oesteoclast differentiation and cytokine (IL-1beta, IL-6, TNF-α) which, in turn, contribute to fibroblast-like synovial cell proliferation. Both can further lead to inflammation, pannus formation, oesteoclastic activity, and degradation of cartilage (Christman et al., 2020).

Great progress has been made on the studies towards effectiveness of dietary polyphenols to mitigate arthritis. However, there are still some gaps of pure understanding of the poor absorption of dietary polyphenols and also about extensive phase-II metabolism. For example, quercetin absorbed in the body gets converted into quercetin sulfate, quercetin glucuronide, and methylated quercetin.

1.5 FUTURE PERSPECTIVES AND CONCLUSION

Polyphenols' ability to defend against oxidative stress has been confirmed as a possible mechanism for their bioprotective effects in a variety of disorders. Protective properties are the most significant antioxidant activity, but these properties have various synergistic mechanisms in *in vitro* and *in vivo* conditions. Polyphenols, or polyphenolic diets, have gradually become an important part of the dietary prevention of numerous diseases, including MetS. Improving polyphenol bioavailability in order to increase its efficiency is a difficult task. Researchers have concluded that polyphenols have a low bioavailability/high bioactivity conundrum based on multiple animal models and clinical investigations (Rio et al., 2013). The daily dose of resveratrol in humans, for example, is substantially smaller than in mice (Baur et al., 2006); humans, on the other hand, have roughly 2–23 times higher plasma resveratrol levels than mice (Lagouge et al., 2006). These disparities could be explained by the fact that humans and mice have varied resveratrol metabolism rates, implying that resveratrol has different effects in various mammals (Timmers et al., 2011). Polyphenols have the ability to reduce MetS characteristics, according to current findings. However, more research is needed to determine the therapeutic potential of polyphenols in the progression of MetS. In addition, several undesirable properties of polyphenols have been disclosed, including genotoxicity (Barbonetti et al., 2016), thyroid damage (Jomaa et al., 2015), hormone disorder (Mumford et al., 2015), antinutritional activity (Bitzer et al., 2016), and interactions with other drugs (Boušová et al., 2015).

A polyphenol-rich diet provides significant protection against pathogenesis of various diseases. While these pharmacological effects can be explained based on epidemiological studies, the mechanism of action of all classes of polyphenols is not entirely deciphered. An improved information of some variables of polyphenol (e.g., bioavailability, including ADME, cytotoxicity and synergisms etc.) will facilitate the design of such studies. The effect of polyphenols in human health is still a little-understood but rich area of research. Thus, founded on our present scientific knowledge, polyphenols offer abundant hope as an important prevention against a plethora of human diseases.

REFERENCES

Abarikwu, S.O., Olufemi, P.D., Lawrence, C.J., Wekere, F.C., Ochulor, A.C., and Barikuma, A.M. 2017. Rutin, an antioxidant flavonoid, induces glutathione and glutathione peroxidase activities to protect against ethanol effects in cadmium-induced oxidative stress in the testis of adult rats. *Andrologia* 49: e12696.

Ahmad, A., Abuzinadah, M., Alkreathy, H., Kutbi, H., Shaik, N., Ahmad, V., Saleem, S., and Husain, A. 2020. A novel polyherbal formulation containing Thymoquinone attenuates carbon tetrachloride-induced hepatorenal injury in a rat model. *Asian Pac J Trop Biomed* 10: 147–155.

Ahmad, N., Gupta, S., and Mukhtar, H. 2000. Green tea polyphenol epigallocatechin-3-gallate differentially modulates nuclear factor κB in cancer cells versus normal cells. *Arch BiochemBiophys* 376: 338–346.

Akhlaghi, M., and Bandy, B. 2009. Mechanisms of flavonoid protection against myocardial ischemia–reperfusion injury. *J Mol Cell Cardiol* 46: 309–317.

Al-Megrin, W.A., Alkhuriji, A.F., Yousef, A.O.S., Metwally, D.M., Habotta, O.A., Kassab, R.B., Abdel Moneim, A.E. and El-Khadragy, M.F. 2020. Antagonistic efficacy of luteolin against lead acetate exposure – associated with hepatotoxicity is mediated via antioxidant, anti-inflammatory, and anti-apoptotic activities. *Antioxidants* 9: 1–18.

Alothman, M., Bhat, R., and Karim, A.A. 2009. Antioxidant capacity and phenolic content of selected tropical fruits from Malaysia, extracted with different solvents. *Food Chem* 115: 785–788.

Aminger, W., Hough, S., Roberts, S.A., Meier, V., Spina, A.D., Pajela, H., McLean, M. and Bianchini, J.A. 2021. Preservice secondary science teachers' implementation of an NGSS practice: Using mathematics and computational thinking. *J Sci Teach Educ* 32: 188–209.

Amiot, M.J., Riva, C., and Vinet, A. 2016. Effects of dietary polyphenols on metabolic syndrome features in humans: A systematic review. *Obes Rev* 17: 573–586.

Arya, V.S., Kanthlal, S.K., and Linda, G. 2020. The role of dietary polyphenols in inflammatory bowel disease: A possible clue on the molecular mechanisms involved in the prevention of immune and inflammatory reactions. *J Food Biochem* 44: e13369.

Azimi, P., Ghiasvand, R., Feizi, A., Hosseinzadeh, J., Bahreynian, M., Hariri, M., and Khosravi-Boroujeni, H. 2016. Effect of cinnamon, cardamom, saffron and ginger consumption on blood pressure and a marker of endothelial function in patients with type 2 diabetes mellitus: A randomized controlled clinical trial. *Blood Press* 25: 133–40.

Babbar, N., Oberoi, H.S., Uppal, D.S., and Patil, R.T. 2011. Total phenolic content and antioxidant capacity of extracts obtained from six important fruit residues. *Int Food Res J* 44: 391–396.

Bakheet, S.A., Alzahrani, M.Z., Ansari, M.A., Nadeem, A., Zoheir, K., Attia, S.M., Al-Ayadhi, L.Y., and Ahmad, S.F. 2017. Resveratrol ameliorates dysregulation of Th1, Th2, Th17, and T regulatory cell-related transcription factor signaling in a BTBR T+ tf/J mouse model of autism. *MolNeurobiol* 54: 5201–5212.

Barbonetti, A., Castellini, C., Di Giammarco, N., Santilli, G., Francavilla, S., and Francavilla, F. 2016. In vitro exposure of human spermatozoa to bisphenol A induces pro-oxidative/apoptotic mitochondrial dysfunction. *ReprodToxicol* 66: 61–67.

Baur, J.A., Pearson, K.J., Price, N.L., Jamieson, H.A., Lerin, C., Kalra, A., Prabhu, V.V., Allard, J.S., Lopez-Lluch, G., Lewis, K., and Pistell, P.J. 2006. Resveratrol improves health and survival of mice on a high-calorie diet. *Nature* 444: 337–342.

Bava, S.V., Sreekanth, C.N., Thulasidasan, A.K.T., Anto, N.P., Cheriyan, V.T., Puliyappadamba, V.T., Menon, S.G., Ravichandran, S.D., and Anto, R.J. 2011. Akt is upstream and MAPKs are downstream of NF-κB in paclitaxel-induced survival signaling events, which are down-regulated by curcumin contributing to their synergism. *Int J BiochemCell Biol* 43: 331–341.

Behl, T., Mehta, K., Sehgal, A., Singh, S., Sharma, N., Ahmadi, A., Arora, S., and Bungau, S. 2021. Exploring the role of polyphenols in rheumatoid arthritis. *Crit Rev Food SciNutr* 1–22.

Benjamin, E.J., Blaha, M.J., Chiuve, S.E., Cushman, M., Das, S.R., Deo, R., De Ferranti, S.D., Floyd, J., Fornage, M., Gillespie, C., and Isasi, C.R. 2017. Heart disease and stroke statistics – 2017 update: A report from the American Heart Association. *Circulation* 135: e1-e458

Bhandari, R., and Kuhad, A. 2017. Resveratrol suppresses neuroinflammation in the experimental paradigm of autism spectrum disorders. *NeurochemInt* 103: 8–23.

Bhullar, K.S., and Rupasinghe, H.P. 2013. Polyphenols: Multipotent therapeutic agents in neurodegenerative diseases. *Oxid Med Cell Longev* 2013: 1–18.

Bitzer, Z.T., Elias, R.J., Vijay-Kumar, M., and Lambert, J.D. 2016. (-)-Epigallocatechin-3-gallate decreases colonic inflammation and permeability in a mouse model of colitis, but decreases macronutrient digestion and exacerbates weight loss. *MolNutr Food Res* 60: 2267–2274.

Boušová, I., Skálová, L., Souček, P., and Matoušková, P. 2015. The modulation of carbonyl reductase 1 by polyphenols. *Drug Metab Rev* 47: 520–533.

Christensen, K.Y., Naidu, A., Parent, M.É., Pintos, J., Abrahamowicz, M., Siemiatycki, J., and Koushik, A. 2012. The risk of lung cancer related to dietary intake of flavonoids. *Nutr. Cancer* 64: 964–974.

Christman, L.M., and Gu, L. 2020. Efficacy and mechanisms of dietary polyphenols in mitigating rheumatoid arthritis. *J Funct Foods* 71: 104003.

Clifford, T., Acton, J.P., Cocksedge, S.P., Davies, K.A.B., and Bailey, S.J. 2021. The effect of dietary phytochemicals on nuclear factor erythroid 2-related factor 2 (Nrf2) activation: A systematic review of human intervention trials. *MolBiol Rep* 48: 1745–1761.

Colizzi, C., 2019. The protective effects of polyphenols on Alzheimer's disease: A systematic review. *Alzheimer's Dement Transl Res ClinInterv* 5: 184–196.

Corrêa, T.A.F., Rogero, M.M., Hassimotto, N.M.A., and Lajolo, F.M. 2019. The two-way polyphenols-microbiota interactions and their effects on obesity and related metabolic diseases. *Front Nutr* 6: 188.

de la Torre, A.M. 2021. Nutrition in clinical practice. *Med Clin* 157: 385–387.

DeFries, R. and Rosenzweig, C. 2010. Toward a whole-landscape approach for sustainable land use in the tropics. *PNAS* 107: 19627–19632.

Del Rio, D., Rodriguez-Mateos, A., Spencer, J.P., Tognolini, M., Borges, G. and Crozier, A. 2013. Dietary (poly) phenolics in human health: Structures, bioavailability, and evidence of protective effects against chronic diseases. *Antioxid Redox* Signal 18: 1818–1892.

Dembinska-Kiec, A., Mykkänen, O., Kiec-Wilk, B., and Mykkänen, H. 2008. Antioxidant phytochemicals against type 2 diabetes. *Br J Nutr* 99: ES109–17.

Denison, M.S., and Nagy, S.R. 2003. Activation of the aryl hydrocarbon receptor by structurally diverse exogenous and endogenous chemicals. *Annu Rev PharmacolToxicol* 43: 309–334.

Denison, M.S., Pandini, A., Nagy, S.R., Baldwin, E.P. and Bonati, L. 2002. Ligand binding and activation of the Ah receptor. *ChemBiol Interact.* 141: 3–24.

Di Meo, F., Lemaur, V., Cornil, J., Lazzaroni, R., Duroux, J.L., Olivier, Y. and Trouillas, P. 2013. Free radical scavenging by natural polyphenols: Atom versus electron transfer. *J PhyChemA* 117: 2082–2092.

Di Meo, F., Valentino, A., Petillo, O., Peluso, G., Filosa, S. and Crispi, S. 2020. Bioactive polyphenols and neuromodulation: Molecular mechanisms in neurodegeneration. *Int J MolSci* 21: 2564.

Dias, M.C., Pinto, D.C. and Silva, A. 2021. Plant flavonoids: Chemical characteristics and biological activity. *Molecules* 26: 5377.

Domitrović, R. and Potočnjak, I. 2016. A comprehensive overview of hepatoprotective natural compounds: Mechanism of action and clinical perspectives. *Arch Toxicol.* 90: 39–79.

Domitrović, R., Jakovac, H., Vasiljev Marchesi, V., Vladimir-Knežević, S., Cvijanović, O., Tadić, Ž.,Romić, Ž. and Rahelić, D. 2012. Differential hepatoprotective mechanisms of rutin and quercetin in CCl4-intoxicated BALB/cN mice. *ActaPharmacol Sin* 33: 1260–1270.

Dryden, G.W., Song, M. and McClain, C. 2006. Polyphenols and gastrointestinal diseases. *CurrOpinGastroenterol* 22: 165.

Durazzo, A., Lucarini, M., Souto, E.B., Cicala, C., Caiazzo, E., Izzo, A.A., Novellino, E. and Santini, A. 2019. Polyphenols: A concise overview on the chemistry, occurrence, and human health. *Phytother Res* 33: 2221–2243.

Ebrahimi, A. and Schluesener, H. 2012. Natural polyphenols against neurodegenerative disorders: Potentials and pitfalls. *Ageing Res Rev* 11: 329–345.

Elsoadaa, S.S., Bakr, E.S., Hijazi, H.H., and Baz, S. 2018. Quercetin and resveratrol effects on peptic ulcer in experimental rats. *Life Sci* 15 .

Faghihzadeh, F., Adibi, P. and Hekmatdoost, A. 2015. The effects of resveratrol supplementation on cardiovascular risk factors in patients with non-alcoholic fatty liver disease: A randomised, double-blind, placebo-controlled study. *Br J Nutr* 114: 796–803.

Farzaei, M.H., Abdollahi, M. and Rahimi, R. 2015. Role of dietary polyphenols in the management of peptic ulcer. *World J Gstroenterol* 21: 6499.

Fischer, S., Macare, C. and Cleare, A.J. 2017. Hypothalamic-pituitary-adrenal (HPA) axis functioning as predictor of antidepressant response – Meta-analysis. *NeurosciBiobehav Rev* 83: 200–211.

Foley, J.A., DeFries, R., Asner, G.P., Barford, C., Bonan, G., Carpenter, S.R., Chapin, F.S., Coe, M.T., Daily, G.C., Gibbs, H.K., and Helkowski, J.H. 2005. *Glob cons land Use Sci* 309: 570–574.

Forbes, S.J. and Newsome, P.N. 2016. Liver regeneration – mechanisms and models to clinical application. *Nat Rev GastroenterolHepatol* 13: 473–485.

Forman, H.J., Davies, K.J. and Ursini, F. 2014. How do nutritional antioxidants really work: Nucleophilic tone and para-hormesis versus free radical scavenging in vivo. *Free Rad Biol Med* 66: 24–35.

Fraga, C.G., Galleano, M., Verstraeten, S.V. and Oteiza, P.I. 2010.Basic biochemical mechanisms behind the health benefits of polyphenols. *Mol Asp Med* 31: 435–445.

Fukui, H., Diaz, F., Garcia, S. and Moraes, C.T. 2007. Cytochrome c oxidase deficiency in neurons decreases both oxidative stress and amyloid formation in a mouse model of Alzheimer's disease. *ProcNatlAcadSci* 104: 14163–14168.

Ganesan, K., Ramkumar, K.M. and Xu, B. 2020. Vitexin restores pancreatic β-cell function and insulin signaling through Nrf2 and NF-κBsignaling pathways. *EurPharmacol* 888: 173606.

Goel, A. and Aggarwal, B.B. 2010. Curcumin, the golden spice from Indian saffron, is a chemosensitizer and radiosensitizer for tumors and chemoprotector and radioprotector for normal organs. *Nutr Cancer* 62: 919–930.

Gonzales, G.B., Smagghe, G., Grootaert, C., Zotti, M., Raes, K. and Camp, J.V. 2015. Flavonoid interactions during digestion, absorption, distribution and metabolism: A sequential structure–activity/property relationship-based approach in the study of bioavailability and bioactivity. *Drug Metabol Rev* 47: 175–190.

Grundy, S.M. 2016. Metabolic syndrome update. *Trends CardiovascMed* 26: 364–373.

Guo, C., Zhang, C., Li, L., Wang, Z., Xiao, W. and Yang, Z. 2014. Hypoglycemic and hypolipidemic effects of oxymatrine in high-fat diet and streptozotocin-induced diabetic rats. *Phytomedicine* 21: 807–814.

Han, X., Shen, T. and Lou, H. 2007. Dietary polyphenols and their biological significance. *Int J MolSci* 8: 950–988.

Inzaugarat, M.E., De Matteo, E., Baz, P., Lucero, D., García, C.C., Gonzalez Ballerga, E., Daruich, J., Sorda, J.A., Wald, M.R. and Cherñavsky, A.C. 2017. New evidence for the therapeutic potential of curcumin to treat nonalcoholic fatty liver disease in humans. *PLoS One* 12: 0172900.

Iwashina, T. 2013. Flavonoid properties of five families newly incorporated into the order Caryophyllales. *BullNatlMus Nat Sci* 39: 25–51.

Javvadi, P., Hertan, L., Kosoff, R., Datta, T., Kolev, J., Mick, R., Tuttle, S.W. and Koumenis, C. 2010.Thioredoxin reductase-1 mediates curcumin-induced radiosensitization of squamous carcinoma cells. *Cancer Res* 70: 1941–1950.

Javvadi, P., Segan, A.T., Tuttle, S.W. and Koumenis, C. 2008. The chemopreventive agent curcumin is a potent radiosensitizer of human cervical tumor cells via increased reactive oxygen species production and overactivation of the mitogen-activated protein kinase pathway. *MolPharmacol* 73: 1491–1501.

Jin, Y. and Wang, H. 2019. Naringenin inhibit the hydrogen peroxide-induced SH-SY5Y cells injury through Nrf2/HO-1 pathway. *Neurotox* Res 33: 796–805.

Jomaa, B., de Haan, L.H., Peijnenburg, A.A., Bovee, T.F., Aarts, J.M. and Rietjens, I.M. 2015.Simple and rapid in vitro assay for detecting human thyroid peroxidase disruption. *ALTEX* 32: 191–200.

Kamiloglu, S., Tomas, M., Ozdal, T. and Capanoglu, E. 2021. Effect of food matrix on the content and bioavailability of flavonoids. *Trends Food SciTechnol* 117: 15–33.

Kennedy, D.O. 2014. Polyphenols and the human brain: Plant "secondary metabolite" ecologic roles and endogenous signaling functions drive benefits. *AdvNutr* 5: 515–533.

Khan, J., Deb, P.K., Priya, S., Medina, K.D., Devi, R., Walode, S.G. and Rudrapal, M. 2021. Dietary flavonoids: Cardioprotective potential with antioxidant effects and their pharmacokinetic, toxicological and therapeutic concerns. *Molecules* 26: 4021.

Khan, N., Afaq, F., Saleem, M., Ahmad, N. and Mukhtar, H. 2006. Targeting multiple signaling pathways by green tea polyphenol (−)-epigallocatechin-3-gallate. *Cancer Res* 66: 2500–2505.

Kim, C.S., Choi, H.S., Joe, Y., Chung, H.T. and Yu, R. 2016. Induction of heme oxygenase-1 with dietary quercetin reduces obesity-induced hepatic inflammation through macrophage phenotype switching. *Nutr Res Pract* 10: 623–628.

Lagouge, M., Argmann, C., Gerhart-Hines, Z., Meziane, H., Lerin, C., Daussin, F., Messadeq, N., Milne, J., Lambert, P., Elliott, P. and Geny, B. 2006. Resveratrol improves mitochondrial function and protects against metabolic disease by activating SIRT1 and PGC-1α. *Cell* 127: 1109–1122.

Lakey-Beitia, J., Burillo, A.M., La Penna, G., Hegde, M.L. and Rao, K.S. 2021. Polyphenols as potential metal chelation compounds against Alzheimer's disease. *J Alzheimer's Dis* 82: S335-S357.

Landberg, R., Manach, C., Kerckhof, F.M., Minihane, A.M., Saleh, R.N.M., De Roos, B., Tomas-Barberan, F., Morand, C. and Van de Wiele, T. 2019. Future prospects for dissecting inter-individual variability in the absorption, distribution and elimination of plant bioactives of relevance for cardiometabolic endpoints. *Eur J Nutr* 58: 21–36.

Lê, K.A. and Tappy, L. 2006. Metabolic effects of fructose. *Curr Opin Clin Nutr Metab Care* 9: 469–475.

Lee, B.H., Lee, C.C. and Wu, S.C. 2014. Ice plant (*Mesembryanthemumcrystallinum*) improves hyperglycaemia and memory impairments in a Wistar rat model of streptozotocin-induced diabetes. *J Sci Food Agri* 94: 2266–2273.

Leopoldini, M., Russo, N. and Toscano, M. 2011. The molecular basis of working mechanism of natural polyphenolic antioxidants. *Food Chem* 125: 288–306.

Li, L., Luo, W., Qian, Y., Zhu, W., Qian, J., Li, J., Jin, Y., Xu, X. and Liang, G. 2019. Luteolin protects against diabetic cardiomyopathy by inhibiting NF-κB-mediated inflammation and activating the Nrf2-mediated antioxidant responses. *Phytomedicine* 59: 152774.

Li, S., Tan, H.Y., Wang, N., Cheung, F., Hong, M. and Feng, Y. 2018.The potential and action mechanism of polyphenols in the treatment of liver diseases. *Oxid Med Cellu Long* 2018: 1–25.

Li Shen, C., Smith, B.J., Lo, D.F., Chyu, M.C., Dunn, D.M., Chen, C.H. and Kwun, I.S. 2012. Dietary polyphenols and mechanisms of osteoarthritis. *J NutrBiochem* 23: 1367–1377.

Liu, J., Willför, S. and Xu, C. 2015. A review of bioactive plant polysaccharides: Biological activities, functionalization, and biomedical applications. *BioacCarbo Diet Fibr* 5: 31–61.

Lü, J.M., Lin, P.H., Yao, Q. and Chen, C. 2010. Chemical and molecular mechanisms of antioxidants: Experimental approaches and model systems. *J Cell Mole Med* 14: 840–860.

Lule, S.A., Namara, B., Akurut, H., Lubyayi, L., Nampijja, M., Akello, F., Tumusiime, J., Lule, S.A., Namara, B., Akurut, H., Lubyayi, L., Nampijja, M., Akello, F., Tumusiime, J.,Aujo, J.C., Oduru, G., Mentzer et al. 2019. Blood pressure risk factors in early adolescents: Results from a Ugandan birth cohort. *J Hum Hyperten* 33: 679–692.

Ma, J.Q., Li, Z., Xie, W.R., Liu, C.M. and Liu, S.S. 2015.Quercetin protects mouse liver against CCl4-induced inflammation by the TLR2/4 and MAPK/NF-κB pathway. *Intern Iimmuno* 28: 531–539.

Malar, D.S., Suryanarayanan, V., Prasanth, M.I., Singh, S.K., Balamurugan, K. and Devi, K.P. 2018. Vitexin inhibits Aβ25-35 induced toxicity in Neuro-2a cells by augmenting Nrf-2/HO-1 dependent antioxidant pathway and regulating lipid homeostasis by the activation of LXR-α. *Toxi in Vitro* 50: 160–171.

Manach, C., Scalbert, A., Morand, C., Rémésy, C. and Jiménez, L. 2004. Polyphenols: Food sources and bioavailability. *Amer J ClinNutri* 79: 727–747.

Marín, L., Miguélez, E.M., Villar, C.J. and Lombó, F. 2015. Bioavailability of dietary polyphenols and gut microbiota metabolism: Antimicrobial properties. *BioMed Res Intl*: 1–18.

Meo, Francesco Di, Anna Valentino, Orsolina Petillo, Gianfranco Peluso, Stefania Filosa, and Stefania Crispi. 2020. Bioactive Polyphenols and Neuromodulation: Molecular Mechanisms in Neurodegeneration. *Int J of Mol Sci*.

Meyfroidt, P., Rudel, T.K. and Lambin, E.F. 2010.Forest transitions, trade, and the global displacement of land use. *Proceedings of the National Academy of Sciences* 107: 20917–20922.

Miller, V., Mente, A., Dehghan, M., Rangarajan, S., Zhang, X., Swaminathan, S., Dagenais, G., Gupta, R., Mohan, V., Lear, S. et al. 2017. Fruit, vegetable, and legume intake, and cardiovascular disease and deaths in 18 countries (PURE): A prospective cohort study. *Lancet* 390: 2037–2049.

Mohan, R., Sivak, J., Ashton, P., Russo, L.A., Pham, B.Q., Kasahara, N., Raizman, M.B. and Fini, M.E. 2000.Curcuminoids inhibit the angiogenic response stimulated by fibroblast growth factor-2, including expression of matrix metalloproteinase gelatinase B. *J Bio Chem* 275: 10405–10412.

Moussa, C., Hebron, M., Huang, X., Ahn, J., Rissman, R.A., Aisen, P.S. and Turner, R.S. 2017. Resveratrol regulates neuro-inflammation and induces adaptive immunity in Alzheimer's disease. *J Neuroinflammation* 14: 1–10.

Mumford, S.L., Kim, S., Chen, Z., Barr, D.B. and Louis, G.M.B. 2015. Urinary phytoestrogens are associated with subtle indicators of semen quality among male partners of couples desiring pregnancy. *The J Nutri* 145: 2535–2541.

Murakami, A., Ashida, H. and Terao, J. 2008. Multitargeted cancer prevention by quercetin. *CancLett* 269: 315–325.

Na, H.K. and Surh, Y.J. 2006. Intracellular signaling network as a prime chemopreventive target of (–)-epigallocatechingallate. *MolNutri Food Res* 50: 152–159.

Nakagawa, H., Hasumi, K., Woo, J.T., Nagai, K. and Wachi, M. 2004. Generation of hydrogen peroxide primarily contributes to the induction of Fe (II)-dependent apoptosis in Jurkat cells by (−)-epigallocatechingallate. *Carcinogenesis* 25: 1567–1574.

Nassour, R., Ayash, A., and Al-Tameemi, K. 2020. Anthocyanin pigments: Structure and biological importance. *J Chem Pharm Sci* 13: 45–57.

Nijveldt, R.J., Van Nood, E.L.S., Van Hoorn, D.E., Boelens, P.G., Van Norren, K. and Van Leeuwen, P.A. 2001. Flavonoids: A review of probable mechanisms of action and potential applications. *The Amer J ClincNutri* 74: 418–425.

Noguchi, M., Yokoyama, M., Watanabe, S., Uchiyama, M., Nakao, Y., Hara, K. and Iwasaka, T. 2006. Inhibitory effect of the tea polyphenol,(−)-epigallocatechingallate, on growth of cervical adenocarcinoma cell lines. *CancLett* 234: 135–142.

Olagunju, A.I., Omoba, O.S., Enujiugha, V.N., Alashi, A.M. and Aluko, R.E. 2018. Antioxidant properties, ACE/renin inhibitory activities of pigeon pea hydrolysates and effects on systolic blood pressure of spontaneously hypertensive rats. *Food SciNutri* 6: 1879–1889.

Panche, A.N., Diwan, A.D. and Chandra, S.R. 2016. Flavonoids: An overview. *J NutritSci* 5: 1–15.

Pandey, K.B., and Rizvi. S.I. 2009. Plant polyphenols as dietary antioxidants in human health and disease. *Oxid. Med. Cell Longev* 2: 270–278.

Pang, C., Zheng, Z., Shi, L., Sheng, Y., Wei, H., Wang, Z. and Ji, L. 2016. Caffeic acid prevents acetaminophen-induced liver injury by activating the Keap1-Nrf2 antioxidative defense system. *Free Rad Bio Med* 91: 236–246.

Pisoschi, A.M. and Pop, A. 2015. The role of antioxidants in the chemistry of oxidative stress: A review. *Euro J Med Chem* 97: 55–74.

Plummer, S.M., Hill, K.A., Festing, M.F., Steward, W.P., Gescher, A.J. and Sharma, R.A. 2001. Clinical development of leukocyte cyclooxygenase 2 activity as a systemic biomarker for cancer chemopreventive agents. *Cancer Epidemiol BiomarkersPrev* 10: 1295–1299.

Plummer, S.M., Holloway, K.A., Manson, M.M., Munks, R.J., Kaptein, A., Farrow, S. and Howells, L. 1999. Inhibition of cyclo-oxygenase 2 expression in colon cells by the chemopreventive agent curcumin involves inhibition of NF-κB activation via the NIK/IKK signalling complex. *Oncogene* 18: 6013–6020.

Prusty, B.K. and Das, B.C. 2005. Constitutive activation of transcription factor AP-1 in cervical cancer and suppression of human papillomavirus (HPV) transcription and AP-1 activity in HeLa cells by curcumin. *Int J Cancer Res* 113: 951–960.

Quideau, S., Deffieux, D., Douat-Casassus, C. and Pouységu, L. 2011. Plant polyphenols: Chemical properties, biological activities, and synthesis. *Angew Chem Int Ed* 50: 586–621.

Rahman, I., Biswas, S.K. and Kirkham, P.A. 2006. Regulation of inflammation and redox signaling by dietary polyphenols. *Biochem Pharmacol* 72: 1439–1452.

Renaud, J. and Martinoli, M.G. 2014. Resveratrol as a protective molecule for neuroinflammation: A review of mechanisms. *CurrPharm Biotechnol* 15: 318–329.

Rendeiro, C., Rhodes, J.S. and Spencer, J.P. 2015. The mechanisms of action of flavonoids in the brain: Direct versus indirect effects. *Neurochem Int* 89: 126–139.

Rodríguez De Luna, S.L., Ramírez-Garza, R.E. and Serna Saldívar, S.O. 2020. Environmentally friendly methods for flavonoid extraction from plant material: Impact of their operating conditions on yield and antioxidant properties. *Sci World J* 2020: 1–38.

Roy Sarkar, Suparna, Papiya Mitra Mazumder, and Sugato Banerjee. 2020. Probiotics Protect against Gut Dysbiosis Associated Decline in Learning and Memory. *J of Neur*.

Rupasinghe, H.P.V., Dellaire, G. and Xu, Z. 2020. Regulation of nrf2/ARE pathway by dietary flavonoids: A friend or foe for cancer management? *Antioxidants* (Basel) 9: 973

Russo, G.L. 2007. Ins and outs of dietary phytochemicals in cancer chemoprevention. *Biochem Pharmacol* 74: 533–544.

Russo, G.L., Vastolo, V., Ciccarelli, M., Albano, L., Macchia, P.E. and Ungaro, P. 2017. Dietary polyphenols and chromatin remodeling. *Crit Rev Food SciNutr* 57: 2589–2599.

Sah, J.F., Balasubramanian, S., Eckert, R.L. and Rorke, E.A. 2004. Epigallocatechin-3-gallate inhibits epidermal growth factor receptor signaling pathway: evidence for direct inhibition of ERK1/2 and AKT kinases. *J BiolChem* 279: 12755–12762.

Saha, A., Kuzuhara, T., Echigo, N., Fujii, A., Suganuma, M. and Fujiki, H. 2010. Apoptosis of human lung cancer cells by curcumin mediated through up-regulation of "growth arrest and DNA damage inducible genes 45 and 153." *Biol Pharm Bull* 33: 1291–1299.

Samoylenko, A., Hossain, J.Al., Mennerich, D., Kellokumpu, S., Hiltunen, J.K. and Kietzmann, T. 2013. Nutritional countermeasures targeting reactive oxygen species in cancer: From mechanisms to biomarkers and clinical evidence. *Antioxid Redox Signal* 19: 2157–2196.

Sandhu, K.V., Sherwin, E., Schellekens, H., Stanton, C., Dinan, T.G. and Cryan, J.F. 2017. Feeding the microbiota-gut-brain axis: Diet, microbiome, and neuropsychiatry. *Transl Res* 179: 223–244.

Sandoval, V., Sanz-Lamora, H., Arias, G., Marrero, P.F., Haro, D. and Relat, J. 2020. Metabolic impact of flavonoids consumption in obesity: From central to peripheral. *Nutrients* 12: 2393.

Sandoval-Acuña, C., Ferreira, J. and Speisky, H. 2014. Polyphenols and mitochondria: An update on their increasingly emerging ROS-scavenging independent actions. *Arch BiochemBiophys* 559: 75–90.

Sarkar, S.R., Mazumder, P.M. and Banerjee, S. 2020. Probiotics protect against gut dysbiosis associated decline in learning and memory. *J Neuroimmunol* 348: 577390.

Scapagnini, G., Vasto, S., Abraham, N.G., Caruso, C., Zella, D. and Fabio, G. 2011. Modulation of Nrf2/ARE pathway by food polyphenols: A nutritional neuroprotective strategy for cognitive and neurodegenerative disorders. *MolNeurobiol* 44: 192–201.

Shang, A., Gan, R.Y., Xu, X.Y., Mao, Q.Q., Zhang, P.Z., Li, H.B. 2021. Effects and Mechanisms of Edible and Medicinal Plants on Obesity: An Updated Review. *Critical Reviews in Food Science and Nutrition* 61: 2061–2077.

Singh, M. and Singh, N. 2009. Molecular mechanism of curcumin induced cytotoxicity in human cervical carcinoma cells. *Mol CellBiochem* 325: 107–119.

Singh, N., Baby, D., Rajguru, J.P., Patil, P.B., Thakkannavar, S.S. and Pujari, V.B. 2019. Inflammation and cancer. *Ann Afr Med* 18: 121.

Sochocka, M., Diniz, B.S. and Leszek, J. 2017. Inflammatory response in the CNS: Friend or foe?. *MolNeurobiol* 54: 8071–8089.

Sottero, B., Gargiulo, S., Russo, I., Barale, C., Poli, G. and Cavalot, F. 2015. Postprandial dysmetabolism and oxidative stress in type 2 diabetes: Pathogenetic mechanisms and therapeutic strategies. *MedRes Rev* 35: 968–1031.

Spencer, J.P.E. 2008. Flavonoids: Modulators of brain function?. *BrJNutr* 99: ES60–ES77.

Spencer, J.P.E., Vafeiadou, K., Williams, R.J. and Vauzour, D. 2012. Neuroinflammation: Modulation by flavonoids and mechanisms of action. *MolAspects Med* 33: 83–97.

Sreekanth, C.N., Bava, S.V., Sreekumar, E., Anto, R.J. 2011. Molecular evidences for the chemosensitizing efficacy of liposomal curcumin in paclitaxel chemotherapy in mouse models of cervical cancer. *Oncogene* 30: 3139–3152.

Sun, R.Q., Wang, H., Zeng, X.Y., Chan, S.M.H., Li, S.P., Jo, E., Leung, S.L., Molero, J.C., Ye, J.M. 2015. IRE1 impairs insulin signaling transduction of fructose-fed mice via JNK independent of excess lipid. *Biochim Biophys Acta – Mol Basis Dis* 1852: 156–165.

Suraweera, T.L., Rupasinghe, H.P.V., Dellaire, G., and Xu, Z. 2020. "Regulation of Nrf2/Are Pathway by Dietary Flavonoids: A Friend or Foe for Cancer Management?" *Antioxidants* 9: 973.

Tan, S., Xu, Q., Luo, Z., Liu, Z., Yang, H., Yang, L. 2011. Inquiry of water-soluble polysaccharide extraction conditions from grapefruit skin. *Engineering* 3: 1090–1094

Thilakarathna, S.H. and Rupasinghe, H.P.V. 2013. Flavonoid bioavailability and attempts for bioavailability enhancement. *Nutrients* 5: 3367–3387.

Tian, C., Huang, Q., Yang, L., Légaré, S., Angileri, F., Yang, H., Li, X., Min, X., Zhang, C., Xu, C. et al. 2016. Green tea consumption is associated with reduced incident CHD and improved CHD-related biomarkers in the Dongfeng-Tongji cohort. *Sci Rep* 6: 24353.

Tian, C., Ye, X., Zhang, R., Long, J., Ren, W., Ding, S., Liao, D., Jin, X., Wu, H., Xu, S. et al. 2013. Green tea polyphenols reduced fat deposits in high fat-fed rats via erk1/2-PPARγ-adiponectin pathway. *PloS one* 8: e53796.

Tian, C., Zhang, R., Ye, X., Zhang, C., Jin, X., Yamori, Y., Hao, L., Sun, X. and Ying, C. 2013. Resveratrol ameliorates high-glucose-induced hyperpermeability mediated by caveolae via VEGF/KDR pathway. *Genes Nutr* 8: 231–239.

Timmers, S., Konings, E., Bilet, L., Houtkooper, R.H., van de Weijer, T., Goossens, G.H., Hoeks, J., van der Krieken, S., Ryu, D., Kersten, S. et al. 2011. Calorie restriction-like effects of 30 days of resveratrol supplementation on energy metabolism and metabolic profile in obese humans. *Cell Met* 14: 1–22.

Tresserra-Rimbau, A., Lamuela-Raventos, R.M., Moreno, J.J. 2018. Polyphenols, food and pharma. Current knowledge and directions for future research. *BiochemPharmacol* 156: 186–195.

Trisurat, Y., Alkemade, R., Verburg, P.H. 2010. Projecting land-use change and its consequences for biodiversity in Northern Thailand. *Environ Manage* 45: 626–639.

Tsao, R. 2010. Chemistry and biochemistry of dietary polyphenols. *Nutrients* 2: 1231–1246.

Tseng, Y.T., Hsu, H.T., Lee, T.Y., Chang, W.H., Lo, Y.C. 2021.Naringenin, a dietary flavanone, enhances insulin-like growth factor 1 receptor-mediated antioxidant defense and attenuates methylglyoxal-induced neurite damage and apoptotic death. *NutrNeurosci* 24: 71–81.

Venigalla, M., Gyengesi, E., Münch, G. 2015. Curcumin and Apigenin – novel and promising therapeutics against chronic neuroinflammation in Alzheimer's disease. *Neural Regen Res* 10: 1181–1185.

Villegas, R., Liu, S., Gao, Y.T., Yang, G., Li, H., Zheng, W., Shu, X.O. 2007. Prospective study of dietary carbohydrates, glycemic index, glycemic load, and incidence of type 2 diabetes mellitus in middle-aged Chinese women. *ArchIntern Med* 167: 2310–2316.

Wali, A.F., Rashid, S., Rashid, S.M., Ansari, M.A., Khan, M.R., Haq, N., Alhareth, D.Y., Ahmad, A. and Rehman, M.U. 2020. Naringenin regulates doxorubicin-induced liver dysfunction: Impact on oxidative stress and inflammation. *Plants*(Basel) 9: 550.

Wang, H., Naghavi, M., Allen, C., Barber, R.M., Bhutta, Z.A., Carter, A., Casey, D.C., Charlson, F.J., Chen, A.Z., Coates, M.M. et al. 2016. Global, regional, and national life expectancy, all-cause mortality, and cause-specific mortality for 249 causes of death, 1980–2015: A systematic analysis for the Global Burden of Disease Study 2015. *Lancet* 388: 1459–1544.

Wei, Q., Ren, X., Jiang, Y., Jin, H., Liu, N. and Li, J. 2013. Advanced glycation end products accelerate rat vascular calcification through RAGE/oxidative stress. *BMC CardiovascDisord* 13: 1–10.

Wood, E.C., Tappan, G.G., Hadj, A. 2004.Understanding the drivers of agricultural land use change in south-central Senegal. *J Arid Environ* 59: 565–582.

Wu, L., Ashraf, M.H.N., Facci, M., Wang, R., Paterson, P.G., Ferrie, A. and Juurlink, B.H.J. 2004. Dietary approach to attenuate oxidative stress, hypertension, and inflammation in the cardiovascular system. *PNAS* 101: 7094–7099.

Xia, S.F., Xie, Z.X., Qiao, Y., Li, L.R., Cheng, X.R., Tang, X., Shi, Y.H. and Le, G.W. 2015. Differential effects of quercetin on hippocampus-dependent learning and memory in mice fed with different diets related with oxidative stress. *PhysiolBehav* 138: 325–331.

Yamamoto, M., Miyamoto, S., Moon, J.H., Murota, K., Hara, Y., Terao, J. 2006. Effect of dietary green tea catechin preparation on oxidative stress parameters in large intestinal mucosa of rats. *BiosciBiotechnolBiochem* 70: 286–289.

Yang, J.H., Hsia, T.C., Kuo, H.M., Lee Chao, P.D., Chou, C.C., Wei, Y.H. and Chung, J.G. 2006.Inhibition of lung cancer cell growth by quercetinglucuronides via G2/M arrest and induction of apoptosis. *Drug MetabDispos* 34: 296–304.

Yen, H.R., Liu, C.J., Yeh, C.C. 2015.Naringenin suppresses TPA-induced tumor invasion by suppressing multiple signal transduction pathways in human hepatocellular carcinoma cells. *ChemBiol Interact* 235: 1–9.

Yokoyama, M., Noguchi, M., Nakao, Y., Pater, A. and Iwasaka, T. 2004. The tea polyphenol,(−)-epigallocatechingallate effects on growth, apoptosis, and telomerase activity in cervical cell lines. *GynecolOncol* 92: 197–204.

Yoon, J.H. and Baek, S.J. 2005. Molecular targets of dietary polyphenols with anti-inflammatory properties. *Yonsei Med J* 46: 585–596.

Yuan, L., Wang, J., Wu, W., Liu, Q. and Liu, X. 2016. Effect of isoorientin on intracellular antioxidant defence mechanisms in hepatoma and liver cell lines. *BiomedPharmacother* 81: 356–362.

Ting, Z., Guang- hua, M., Min, Z., Fang, Li., Ye, Z., Wei, Z., Da-heng, Z., Liu-qing, Y., Xiang-yang, WU. 2013. Anti-diabetic effects of polysaccharides from ethanol-insoluble residue of *Schisandrachinensis* (Turcz) Baill on alloxan-induced diabetic mice. *Chem Res Chin Univ* 29: 99–102.

Zhou, R.J., Zhao, Y., Fan, K. and Xie, M.L. 2020. Protective effect of apigenin on d-galactosamine/LPS-induced hepatocellular injury by increment of Nrf-2 nucleus translocation. *NaunynSchmiedebergs Arch Pharmacol* 393: 929–936.

Ziff, O.J. and Kotecha, D. 2016. Digoxin: The good and the bad. *Trends Cardiovasc Med* 26: 585–595.

Zuo, T., Zhu, M. and Xu, W. 2016. Roles of oxidative stress in polycystic ovary syndrome and cancers. *Oxid Med Cell Longev* 2016: 1–14.

2 Biochemical, Molecular, Pharmacokinetic, and Toxicological Aspects of Dietary Polyphenols

Yousef Rasmi, Aline Priscilla Gomes da Silva, Sahar Rezaei, Safa Rafique, and Muhammad Zeeshan Ahmed
*E-mail: rasmiy@umsu.ac.ir; alinepgsilva@gmail.com; saharrezaei1475504@gmail.com; safa.sandhu@gmail.com; mzeeshanahmed121@gmail.com

CONTENTS

DOI: 10.1201/9781003251538-2

2.1 INTRODUCTION

Polyphenols are natural antioxidants with a different chemical structure and occur mainly in fruits, tea, extra virgin olive oil, vegetables, chocolate, cocoa products, and wine (Han et al., 2007). According to the complexity and diversity of wide polyphenol variation in nature, they can be classified generally as flavonoids and non-flavonoids, which are subdivided into many sub-classes based on their structural characteristics. These compounds are mainly derivatives and/or isomers of flavonols, flavones, flavanones and flavanonols, anthocyanins, anthocyanidins, phenolic acids, and polyphenol amides (da Silva, 2021, Singla et al., 2019).They possess different biological characteristics, for instance anti-inflammation, anti-oxidant, -apoptosis, -carcinogen, -atherosclerosis, -aging, cardio-vascular protection, improvement of endothelial function, and angiogenesis inhibition (Han et al., 2007). The biological activities of most of these compounds can be accounted for by their intrinsic antioxidant characteristics. Nutritional polyphenols may provide indirect protection in part via activating endogenous defense mechanisms and via balancing cell-signaling pathways, including NF-κB activation, biosynthesis of glutathione, Mitogen-Activated Protein Kinases (MAPKs) proteins (JNK, ERK, and P38) activation, and PI3-kinase/Akt pathway (Chen et al., 2000, Han et al., 2007, Molina et al., 2003).

2.2 CHEMISTRY AND BIOLOGICAL IMPORTANCE

2.2.1 CLASSIFICATION

Dietary polyphenols are among the broadest and most important natural product classes, accounting for over 8,000 compounds in, fruits, vegetables, beverages like tea and other products, such as olive oil and chocolate. These compounds are synthesized from secondary plant metabolites and are composed of one or more hydroxyl substituents and two phenyl rings (Ambriz-Pérez et al., 2016) (Figure 2.1). Due to the complexity and diversity of a wide variety of polyphenols in nature, their classification can be categorized purely as non-flavonoids and flavonoids and further subdivided into many sub-classes based on the following: (1) molecular structure includes the number of phenol units; (2) group replacement; and (3) linkage-type between phenol units (da Silva, 2021, Singla et al., 2019). In addition, besides chemical structure, some classifications also consider their origin and biological function (Tsao, 2010). More details concerning polyphenol classifications, including their chemical structure, are described below.

2.2.2 CHEMICAL STRUCTURE

2.2.2.1 Flavonoids

As indicated previously, polyphenols are synthesized from secondary plant metabolites and are abundantly distributed in plant tissues, where they commonly occur in glycoside forms. Although

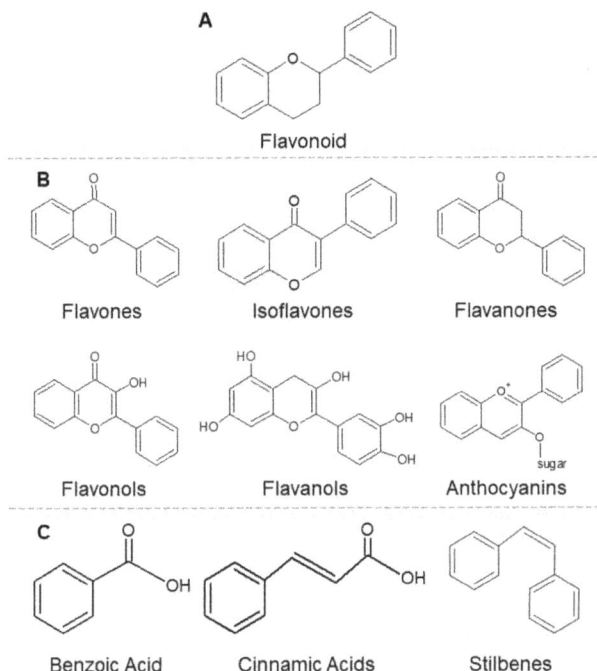

FIGURE 2.1 (A) Typical flavonoid structure; (B) General structures of the main polyphenol classes; (C) General structures of the main non-flavonoids.

basic flavonoid structures comprise aglycones (glycosides that are non-sugar fragments), flavonoids are found as aglycones and glycosides. All flavonoids share the main molecular skeleton structure C6-C3-C6 (diphenyl propanes), along a heterocyclic ring called ring C, generally a closed pyran, linking two phenolic rings (rings A and B), as displayed in Figure 2.1 (Kumar and Pandey, 2013). The diverse classes of flavonoids differ in hydroxylation pattern degree and C-ring behavior. In contrast, individual compounds within the class differ in A- and B-ring substitution behavior. These characteristics provide a diverse range of derivatives, known as flavones, flavonols, flavanones, flavanonols, anthocyanidins, anthocyanins, flavanols, proanthocyanidins, and isoflavones. This structural diversity also indicates different biological activities between dietary polyphenol classes and sub-classes (Yang et al., 2021a).

2.2.2.1.1   Flavones, flavonols, flavanones and flavanonols

Flavones and flavonols, along with flavanones and flavanonols, comprise the largest and most popular subclasses of polyphenols. Flavones are characterized by: (1) the link of the B ring to C2; (2) double bond between C2 and C3 (Jiang et al., 2016). In parallel, an oxygen atom is placed in the C4 position, and ring B (in isoflavones) is linked to the heterocyclic ring (C3 position) instead of C2, as in other classes. Meanwhile, flavonols contain the 3-hydroxy derivatives of flavones, which differ from flavones in having a -OH group in the C3 position. Luteolin and apigenin are the most well-known and important flavones (Verma and Pratap, 2010), while kaempferol and quercetin are the same for flavonols (in the aglycone form) (Figure 2.1). These compounds are the most common and predominant because they include methoxides, glycosides, and other acylated products on all three rings in their combinations (Singla et al., 2019, Tsao, 2010).

Flavanones and flavanonols are characterized as presenting a saturated three-carbon chain without a -OH group at the C3 position of the heterocyclic ring. Flavanonols consist of the 3-hydroxy derivatives of flavanones. Well-known flavanones include naringenin and hesperetin, while taxifolin is a well-known flavanonos. Both are found in aglycone forms, mostly in citrus fruits (Tomas-Barberan and Clifford, 2000) (Table 2.1).

2.2.2.1.2 Anthocyanins and anthocyanidins

Anthocyanins and anthocyanidins contain two double bonds in their heterocyclic rings, differing from other flavonoids. Anthocyanins comprise the glycosylated form of anthocyanidins (a no sugar attached combination to the side groups of flavylium ion) and are expressed by methoxylation and hydroxylation patterns on ring B (Figure 2.1). The change in the position and number of methoxy and/or -OH groups andthe origin and amount of linked sugar portions (glycosylation) and/or attached aliphatic or aromatic acids (acylation) are responsible for various anthocyanins (Singla et al., 2019, Sinopoli et al., 2019). The most observed in nature are cyanidin, delphinidin, pelargonidin, peonidin, and malvidin. These compounds have been studied and characterized because they confer the pigments (blue, red and purple) of most vegetables, fruits, leaves, and flower petals, correlated with bioactive properties (Krga and Milenkovic, 2019). Furthermore, they have been revealed as conferring beneficial health effects through the regulation and modulation of different molecular metabolic disease pathways, such as obesity, diabetes, and metabolic syndrome (Table 2.1) (Martin and Ramos, 2021).

2.2.2.1.3 Flavanols and proanthocyanidins

Among flavonoids, flavanols (flavan-3-ol) comprise one of the unique classes. In contrast to other flavonoids, flavanol molecules comprise (1) only aglycone forms; (2) display no double bond between C2 and C3, only a linkage by a linear three-carbon oxygenated heterocycle, hydroxylated at C3 between two benzene rings (Martin and Ramos, 2021), and (3) display hydroxylation at C3, producing two chiral behaviors (C2 and C3), possibility diastereoisomers (Figure 2.1b). Thus, these compounds display both a trans (catechin) and cis (epicatechin) configuration, allowing for several stereoisomers (e.g. (+)-catechin, (−)-catechin, (+)-epicatechin, and (−)-epicatechin) (Tsao, 2010). Flavanols are usually categorized as monomers (catechins and epicatechin). Oligomers and polymeric forms, called tannins (proanthocyanidins), are defined by monomer type and connection with monomers. Epicatechin and catechin are found in numerous fruits, especially in the skins of berries, tea, apple, and cocoa, among others.

Recently, proanthocyanidins (polymerized forms called *condensed tannins*) have attracted attention, as they display attributes involved in several potential health benefits. Proanthocyanidins are classified as prodelphinidins, propelargonidins, and procyanidins, depending on oxidation level and substitution pattern, with procyanidins comprising the most common polymerized molecule form in plants, formed from an oligomer and monomeric mixture (Yang et al., 2021a). Procyanidins can be further subclassified into three different types of procyanidins (A, B, and C), depending on interflavanic connections. Type A displays a C2–O–C7 or C2–O–C5 configuration binding to monomers; type B comprises C4–C6 or C4–C8, with type C1 being a trimer. Procyanidins are complex and diverse structures, and their formation depends on several factors, such as spatial configuration, methylation or esterification modifications, flavanol units, and their bonds. They may also be classified by polymerization degree (number of units), with a higher polymerization degree comprising compounds with several molecular weights and many oligomerization shapes. Seeds and grapes, as well as nuts, and beans, are well-known proanthocyanidin sources (Table 2.1).

2.2.2.2 Non-flavonoids

The main structure of non-flavonoids is composed of a single aromatic ring. This class consists of phenolic acids, lignans, and stilbenes, with the phenolic acids, mainly cinnamic and benzoic acid derivatives, the most studied compounds to date (Figure 2.1). These compounds can exist in nature

TABLE 2.1
Dietary polyphenol classification and sources

Main classes	Sub-classes	Dietary polyphenols	Sources	References
Flavonoid	Flavones	Luteolin and apigenin	Carrot, peppers, and celery	(Shankar et al., 2017)
	Flavonols	Quercetin and kaempferol	Berries, kale, and spinach	(Dabeek and Marra, 2019)
	Flavanones	Naringenin and hesperetin	Grapefruits and oranges	(Zuiter, 2014)
	Flavanonols	Taxifolin	Citrus	(Tsao, 2010)
	Anthocyanins and anthocyanidins	Cyanidin and delphinidin	Berries, plums, and apples	(Krga and Milenkovic, 2019)
	Flavanols	Catechin and epicatechin	Cocoa, tea, and apples	(Martin and Ramos, 2021)
	Proanthocyanidins	Procyanidin B1 and Procyanidin B2	Fruit seeds, nuts, and beans	(Yang et al., 2021a)
Non-flavonoids	Phenolic acids	Gallic and sinapic acids	Teas, and citrus	(El-Seedi et al., 2012)
	Stilbenes	Resveratrol	Red and white wines, peanuts	(Meng et al., 2021)
	Lignans	Secoisolariciresinol and matairesinol	Flaxseed and sesame seeds	(Durazzo et al., 2018)
	Polyphenol Amides	Capsaicinoids and avenanthramides	Chili peppers and oats	(Zeng et al., 2021)

in both free or bound forms (e.g., with other polyphenols, sugars, and acids, or as original structural components of plants) (El-Seedi et al., 2012).

Phenolic acids are categorized as hydroxybenzoic and hydroxycinnamic acids, and what determines the difference in the skeletons of these acids are the location and -OH groups numbers on the aromatic rings (Figure 2.1). For instance, Ferulic and caffeic acids are the most prevalent hydroxycinnamic acids, extensively present in food items (El-Seedi et al., 2012), characterized by the presence of a three-carbon side chain (C6–C3) in their aromatic ring. On the other hand, gallic, syringic acids, and protocatechuic are the major popular and common hydroxybenzoic acids present in both free and bound forms (linked with esters, or glycosides) (da Silva, 2021).

Unlike phenolic acids, which comprise many compounds and derivatives, stilbenes, another non-flavonoid class, are categorized as a comparatively modest class. Their molecular structure is distinguished by two phenyl moieties attached together via a two-carbon methylene group (Tsao, 2010). Similar other dietary polyphenols, these compounds occur in both *cis* and *trans* configurations, and also in free and glycosylated forms. Resveratrol, one example of a well-known stilbene, has been extensively studied in grapes and other berries due to its potential as a bioactive compound able to enhance gut microbiota health and as a preventive factor in metabolic disorders, for instance metabolic syndrome, diabetes, and obesity (Meng et al., 2021).

Lignans comprise a non-flavonoid found in vegetables, seeds, and oils (Durazzo et al., 2018). In contrast to other polyphenols, these compounds are found mainly in free form. This class is characterized by two carbon C6-C3 skeleton molecules (propylbenzene units) linked together between the β-position in C8 of the propane side chains, with substitutions in bonds C9 and C9` positions conferring various patterns, resulting in various structural forms (Durazzo et al., 2018).

2.2.2.3   Polyphenol amides

This class may exhibit an N-containing functional substituent. Capsaicinoids (chili peppers) are mostly responsible for the hotness of the chili peppers, and, alongside avenanthramides found in oats, comprise the major groups representing the compounds class (Xiang et al., 2021, Zeng et al.,

2021). These compounds have been described as participating in the oxidative defense system and exhibiting anti-inflammatory and antioxidant effects (Xiang et al., 2021).

2.2.3 MEDICINAL IMPORTANCE

Dietary polyphenols possess many beneficial medicinal uses and applications, resulting in extensive research concerning both *in vitro* and *in vivo* tests (Yang et al., 2021a). These reports have demonstrated the amelioration of some metabolic dysfunctions, for instance arthritis, diabetes, cancer, cardiovascular disease, and neurodegenerative disorders. Moreover, these compounds have been described as exhibiting anti-oxidant, -proliferative, -inflammatory, -tumor, -microbial activities. For instance, flavonoids affect several enzymes that regulate multiple cell-signaling pathways and can alter various cellular-signaling pathways frequently present in many chronic disorders (Singh et al., 2014). Therefore, dietary polyphenols may be an essential tool for developing new therapeutic agents.

Currently, growing global demand for natural products that provide health benefits is noted (Sridhar et al., 2021). However, the use of polyphenols is not limited only to health properties, as an increased number of applications in the beverage, food and pharmaceutical sectors has been observed in recent years, causing the polyphenol market to rise in estimated value to USD 873.7 million in 2018, with a 6.1 percent growth rate due to elevating demands and size of market (Ameer et al., 2017). One of the leading products in this market comprises grape seeds which, due to their antioxidant characteristics, also play a considerable role in beauty-market supplements. Similarly, increasing social awareness for natural products instead of synthetic drugs is believed to have improved the demand for polyphenols. Furthermore, the development of new extraction technologies focusing on human and animal applications, both cost-effective and environmentally friendly, has been estimated to boost a significant market expansion over the coming years.

2.3 POLYPHENOLS AND OS

2.3.1 BIOCHEMISTRY OF OS

OS is the disproportion between the antioxidants and oxidants in the biological system that disrupts the redox balance, control, and signaling and cause molecular damage. Structural disruption may cause system failure and diseases (Birben et al., 2012). The main idea depends on redox homeostasis, which has a set of principles – called redox code. In biological systems, redox code describes the spatial position, time, and functions of the disulfide, thiol, nicotinamide adenine dinucleotide (NAD), and thiol redox proteome. The redox code depends on the four main principles – their mechanisms and effects are depicted in Figure 2.2. Moreover, oxygen-dependent life depends on adaptation, development, and differentiation (Jones and Sies, 2015).

Oxygen is essential for the existence of life. Oxygen has two unpaired electrons in its valance shell; this peculiar property supports free radical formation. About 1-4 percent of the taken oxygen converts into FRs. The oxygen in mitochondria is completely reduced in water; however, due to some flaws, reactive intermediates are produced by the univalent oxygen reduction (Bardaweel et al., 2018). Further, the reductive environment in the cell supports the partial univalent reduction leads to oxygen toxicity. Various types of reactive oxygen species (ROS) (Table 2.2) are produced by homolytic fission, photolysis, radiolysis, and redox reactions that are involved in different intracellular processes, for instance cell signaling and homeostasis (Das and Roychoudhury, 2014). Primarily, FRs are formed by three significant steps, including initiation, propagation, and termination steps. The initiation step starts with high temperatures, superoxides, or UV radiation and produces a distinct number of FRs. In the next phase, an adequate number of FRs are produced and

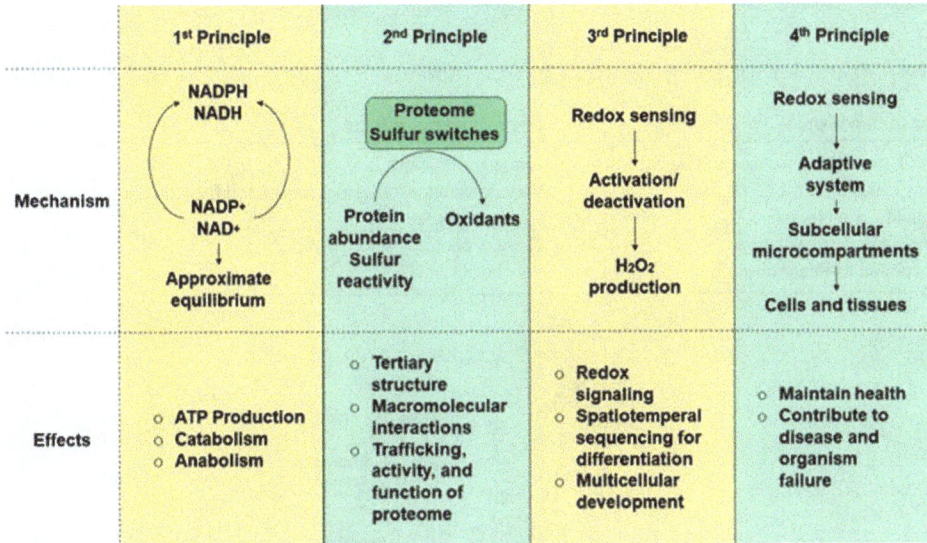

FIGURE 2.2   Mechanisms and effects of the redox code principles.

TABLE 2.2
Reactive oxygen species, their types, and examples

Reactive oxygen species	Oxygen-free radicals	$^{\bullet}OH$ (hydroxyl radical)
		$O_2^{\bullet-}$ (superoxide anion)
		$^1O_2$ (singlet oxygen)
		$HO_2^{\bullet}$ (hydroperoxyl radical)
	Iron-oxygen complexes	$Fe = O^{2+}$ (ferryl-oxygen complex)
		$Fe = O^{3+}$ (perferryl-oxygen complex)
	Reactive nitrogen species	$NO^{\bullet}$ (nitric oxide)
		$NO_2^{\bullet}$ (nitric dioxide)
	Organic free radicals	$R–O^{\bullet}$ (alkoxy radical)
		$R\text{-}OO^{\bullet}$ (peroxide radicals)
		$^{\bullet}QH$ (semiquinone radical)
		$^{\bullet}QH^-$ (semiquinone anionradical)
	Non-radical oxidants	H_2O_2 (hydrogen peroxide)
		$ONOOH$ (peroxynitrous acid)
		$HClO$ (hypochlorous acid)
		$HOSCN$ (hypothiocyanous acid)
	Other	HIO (hypoiodous acid)
		$HClO$ (hypochlorous acid)
		$HBrO$ (hypobromous acid)

are accelerated by the previously produced FRs. In the termination step, FRs combine to produce stable species (Lobo et al., 2010). In biological systems, cellular metabolism and environmental effects are the primary sources of FRs (Phaniendra et al., 2015). Some examples for the generation of FRs are listed in Table 2.3.

ROS are the by-products of biochemical reactions in cytochrome P450, mitochondria, peroxisomes, and other cellular compartments (Janku et al., 2019). Environmental factors activate the various mitochondrial molecules that start producing FRs and ATP (adenosine triphosphate).

TABLE 2.3
Different examples for the generation of free radicals

Cellular metabolism	Environmental effects
H_2O_2 or O_2 production for the oxidase enzymes	Ionizing radiations
Electrons leakage for the ETC	Reactions related to cytochrome P450 for drug metabolism
Prostaglandin's synthesis	Cigarette smoke
Lipid membrane peroxidation	Oxygen photolysis by UV-radiation
NO production from arginine	Alcohol

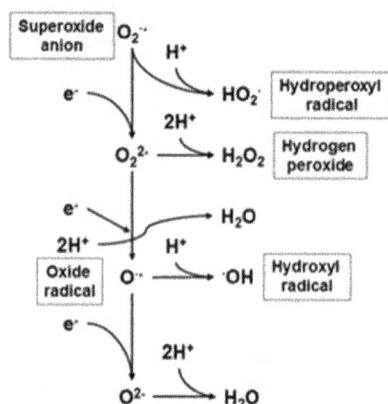

FIGURE 2.3   Production of oxygen radicals from superoxide anion.

When electrons pass through the mitochondrial complexes I, II, III, and IV of ETC (electron transport chain) to oxygen to form water, superoxide anion and ATP are produced. Complex I is mainly involved in the superoxide production of other complexes because it oxidizes the succinate, augmenting the cell's superoxide production. Superoxide anion produces further ROS, shown in Figure 2.3. Although the role of complex II in OS is minimal, but cannot be ignored. Complex II is succinate dehydrogenase that produces a superoxide anion by the oxidation of flavin semiquinone radical in hypoxic conditions. Similarly, ROS production is linked to redox-signaling pathways in complex III (Pruchniak et al., 2016).

In contrast to mitochondrial complexes, certain enzymes and organelles are also involved in ROS production. NADPH oxidase converts oxygen to a superoxide anion. Cholesterol desmolase converts cholesterol to pregnenolone; by-products of this conversion produce superoxides by reacting with oxygen. Microsomal electron transport chain and peroxisomes produce a series of ROS (Table 2.2). In mitochondria, nitric radicals and reactive nitrogen species are also produced in hypoxic conditions (Harwell, 2007, Phaniendra et al., 2015).

ROS constantly interact with the biomolecules, especially proteins, to disrupt the cell-signaling pathways and generate OS in cells and tissues. OS has adverse consequences in cells and tissues (Schieber and Chandel, 2014). For instance, glyceraldehyde-3-phosphate dehydrogenase inhibition limits ATP production; inactivation of calcium pumps elevates the cytosolic Ca2+ ion; DNA damage; depolarization increases the lipid bilayer permeability and others. When OS persists for a long time, apoptosis or necrosis occurs in the cell (Bayir and Kagan, 2008). In contrast to harmful effects, ROS produced from the respiratory burst and macrophages activates the phagocytes and bactericidal effect (Herb and Schramm, 2021, Slauch, 2011).

2.3.2 BIOCHEMICAL/MOLECULAR MARKERS AS TARGETS FOR POLYPHENOLS

Polyphenols as the most abundant antioxidants compounds in the human diet, interact with the FRs by transferring hydrogen and electrons to reduce the OS (Shabbir et al., 2021). Polyphenols diminish the hydroxyl radical production and chelation mechanism; promote the antioxidant enzymes; and scavenge ROS, RNS, and gene expression (Wu et al., 2021). Polyphenols activate the redox-responsible transcription factors that modulate coding antioxidants, anti-apoptotic Bcl-2 and survival neurotrophic factors. Moreover, apoptosis modulation by polyphenols promotes or prevents autophagic degradation, mitochondrial biogenesis, and dynamic (fission and fusion) of mitochondria (Naoi et al., 2019). The scavenging ability of polyphenols depends on the number and location of -OH groups linked with aromatic rings (Wojtunik-Kulesza et al., 2020). This effective FRs scavenging ability can interrupt the lipid autoxidation chain reactions (Zhang and Tsao, 2016). Other groups such as C=O, –COOH, –NR2, –PO3H2, –S-, and –SH in polyphenols also stimulate the metal ion chelation process (Russo et al., 2020). In the process of metal chelation, metal pro-oxidants and hydroperoxides are converted into stable compounds (Zhang and Tsao, 2016). Gay and co-workers (Gay et al., 2020) reported a positive effect of butein, isoliquiritigenin, and scopoletin against H_2O_2-induced cell death, and ROS. In another report (Zhao et al., 2019), the combined effect of callistephin and isoflurane suppressed ROS, and NO production, along with excessive caspase 3/7 activity and engulfment.

2.3.3 MOLECULAR MECHANISMS OF POLYPHENOLS IN OS

It has been discussed that the structures and functional groups in polyphenols reduce the OS. The most attentive action mechanism of polyphenols against OS is the FRs scavenging property in which hydrogen atoms and electrons are transferred. The mechanisms of different polyphenols and their effects are mentioned in Table 2.4.

Polyphenols are classified into two groups due to the action mechanisms: ROS that affect reducers and preventers Three principal mechanisms are involved in the primary (chain-breaking) antioxidants' ability to scavenge FRs, and each has a critical impact on determining the particular radical scavenging activity due to the surrounding statuses. These mechanisms are (1) electron transfer-proton transfer (SETPT), (2) hydrogen-atom transfer (HAT), and (3) sequential proton loss electron transfer (SPLET) (Fereidoon and Ying, 2010, Litwinienko and Ingold, 2007, Ross et al., 2009). In SETPT and SPLET, primary antioxidants donate hydrogen and receive an electron from two different lipid radicals to form a stable radical-free product. While, in HAT, OH bond after homolytic cleavage converted into antioxidant radical and harmless hydroperoxides that are less reactive than lipid radicals. The delocalization of unpaired electrons around the aromatic rings of polyphenols forms a stable resonating hybrid structure. These structures interrupt the propagation and initiation of new radical formation (Fereidoon and Ying, 2010, Litwinienko and Ingold, 2007).

Polyphenols employ several mechanisms to obstruct oxidation by inhibiting and suppressing the oxidation promoters such as pro-oxidative enzymes, singlet oxygen, metal ions, and other oxidants. Some polyphenols are metal chelators that reduce the metal ions concentration (Chobot et al., 2014). Polyphenols metal chelators stop the metals reacting with H_2O_2, generating FRs (Kurutas, 2016). Similarly, the combined effects of two or more antioxidants enhance the antioxidant effects and reduce the OS. For instance, a research reported a combined protective effect of polyphenols and carotenoids against photo-induced skin cells damage. Polyphenols and carotenoids release the cytokine IL-6 and inhibit the activity of UVB-induced NFκB in keratinocytes, thus reducing the UV-induced skin cells damage (Calniquer et al., 2021).

Earlier researchers have revealed the effects of the polyphenols on transcriptional factors and signaling proteins (Karunaweera et al., 2015). In which polyphenols target the cascade proteins that induce the inflammation and expression of genes and metabolites in cell death. During the OS, a nuclease transcriptional factor kappa beta (NF-κβ) expresses that binds to the DNA and activates

TABLE 2.4

The mechanisms of different polyphenols in antioxidant activity and their effects

Polyphenols	Functional groups	Mechanisms	Effects	References
Flavonoids	o-dihydroxyl Unsaturation carbonyl	Metal chelation, ↑ROS-removing enzymes, ↓ROS-producing enzymes	Antioxidant activity, signal cascades modulation, gene expression induction	(Ho et al., 2018, Van Giau et al., 2018, Vogt et al., 2018)
Curcumin	β-diketo, hydroxyl, and methylene	Scavenge hydroxyl radical, H_2O_2, and peroxynitrite, ↓Fe^{3+}, ↓lipid peroxidation	Anti-apoptotic, anticancer, anti-diabetic, anti-inflammatory, antioxidant	(Cory et al., 2018, Zheng et al., 2018)
Phenolic acids	Electron-donating 4-hydroxy, and 3-methoxy groups	Terminate radical chain reactions, scavenge superoxide, hydroxyl radicals, and peroxynitrite, prevent lipid peroxidation, ↑regulate protective genes, such as heat shock protein 70, heme oxygenase-1 (HO-1), extracellular signal-related kinase 1/2 (ERK1/2)	Preventive against cardiovascular diseases, diabetes, cancer, skin diseases, and neurodegenerative disorders	(Cory et al., 2018, Zhang et al., 2020)
Resveratrol	Hydroxyl groups	Scavenge hydroxyl radicals, activate Nrf2/ARE pathway, and induce SOD, and HO-1	Anti-aging, anti-inflammatory, antimicrobial, antioxidant, anti-tumor, cardio- and neuro-protective, activities	(Gianchecchi and Fierabracci, 2020, Salehi et al., 2018)
Astaxanthin	C=C, hydroxyl, and keto moieties	↓lipid peroxidation, ↑regulate the protective gene expression, such as Bcl-2, scavenge superoxide and H_2O_2	Antioxidant, anti-inflammatory, anti-tumor, and anti-apoptosis activities	(Hormozi et al., 2019, Kohandel et al., 2022)
Sesame lignans	Hydroxyl groups	Scavenge superoxide, ↓lipid peroxidation, ↓H_2O_2, ↑antioxidant enzymes, ↓oxidative enzymes	Anti-hyperlipidemic, anti-hypertensive, and anti-oxidative activities	(Mithul Aravind et al., 2021, Shabbir et al., 2021)

the pro-inflammatory cytokines such as IL-6, TNF-α, cyclooxygenase 2, and inducible NO synthase (Stockert and Hall, 2021). Studies have reported the complete blockage and reduction of NF-κβ activity by different polyphenols (Karunaweera et al., 2015, Ruiz et al., 2007, Ruiz and Haller, 2006). Although some polyphenols activate the transcriptional factor Nrf2, that increases the production of OS protecting genes and ROS-removing enzymes, thus reducing the inflammation and OS (Ahmed et al., 2017, Zhou et al., 2019). Primarily, polyphenols decrease OS by different mechanisms depending upon the particular situations, degree, and efficacy as an antioxidant. The efficiency of antioxidants is primarily influenced by their structural, reaction sites, medium acidity, pro-oxidants, synergists, and oxidizable substrate. Antioxidants' efficiency can be optimized by considering all such characteristics and their effects (Losada-Barreiro and Bravo-Díaz, 2017) (Table 2.4).

2.4   ANTI-INFLAMMATORY ACTIVITIES

Recently, the hyperlink between nutrition and inflammation has grown to be more and more apparent. Excessive macronutrient intake has an inflammatory response in humans, and dietary polyphenols have an incidence of inflammatory diseases (Di Lorenzo et al., 2013). Pathogenesis of various diseases involves inflammation. Macrophages constitute one of the main immune cells responsible for the control of inflammation. Inflammatory mediators such as prostaglandin E2 (PGE$_2$), nitric oxide (NO), IL-6, interleukin-1β (IL-1β), and tumor necrosis factor-α (TNF-α) (Boscá et al., 2005, Kaplanski et al., 2003). There are many different types of cytokines produced by cells of the immune system.. Cytokines influence the properties of many types of cells and serve as communication signals between the immune system cells (Ware, 2005). Increased and uncontrolled production of these inflammatory cytokines can lead to serious complications, such as septic shock, tissue damage, and microcirculatory dysfunction, which can lead to life-threatening outcomes (Ulevitch and Tobias, 1995).

An oxygenated heterocycle is produced when two aromatic rings are joined together by three carbon atoms, forming the flavonoids (Manach et al., 2004). Dietary polyphenols and polyphenol supplements derived from fruits contain high amounts of various polyphenols, so the mechanisms of intake and metabolite production are very complex due to individual variations of microbiota composition (Kay et al., 2017). People get polyphenols through their diet, much like the Mediterranean diet (Casas et al., 2014). Importantly, flavonoids are nutritionally the most abundant polyphenols with beneficial effects on metabolism, body weight, chronic diseases, and neuroendocrine immunity regulation (Cory et al., 2018, Laganà et al., 2019, Tresserra-Rimbau et al., 2018). Mechanisms include the following:

2.4.1 REGULATION OF NF-κB

Some flavonoids inhibit Iκκ activation early in the NFκB activation process, while others prevent NFκB binding to DNA late in the process (Ichikawa et al., 2004, Lin and Lin, 1997, Mackenzie et al., 2004).

2.4.2 MITOGEN-ACTIVATED PROTEIN KINASES (MAPKs) REGULATION

MAPKs regulate the activity of transcription factors related to gene transcription and inflammation. Of these, extracellular signaling kinases (ERK) 1 and 2, amino-terminal kinases cJun (JNK) 1/2/3, p38MAPK, and ERK5 can interact with NFκB, indicating the complexity of the MAPK pathway. Both quercetin and EGCG have been shown to reduce the production of TNFα and IL12 in immune and non-immune cells through their interaction with MAPK signaling systems (Cho et al., 2003, Wadsworth et al., 2001).

2.4.3 ARACHIDONIC ACID REGULATION

Polyphenols also exert anti-inflammatory effects by inhibiting the arachidonic acid (AA) pathway. Phospholipase A breaks down membrane phospholipids to release AA, COX and LOX convert AA into prostaglandins (PG) and thromboxane A2, respectively (Chandrasekharan et al., 2002).

2.5 ANTI-OXIDANT ACTIVITIES

The production of FRs in living cells is a normal process. Many diseases, however, are caused by excess FR formation from either endogenous or exogenous sources. OS and FRs are involved in many degenerative and chronic diseases, including atherosclerosis, aging, cancer, immunity suppression, diabetes, and neurodegenerative diseases (Young and Woodside, 2001). Different cellular antioxidant systems protect cells against FR-mediated OS, including enzymes that interact with ROS, such as peroxidases, superoxide dismutases, catalases; and other enzymes that generate reduced antioxidants; and low molecular mass antioxidants (glutathione, tocopherols, ascorbic acid) (Blokhina et al., 2003). According to epidemiological studies, fruits and vegetables help prevent diseases related to OS. Plant tissue contains various antioxidants, such as flavonoids, lignin, and tannins precursors, which trap ROS (Blokhina et al., 2003). The polyphenol structure is responsible for properties such as antioxidant activity, and tendency to degrade by oxidation (Rice-Evans et al., 1996) and absorption in both the UV and visible regions of the spectrum (Cheynier et al., 2006). The flavonoids antioxidant activity is highly dependent on the number and location of -OH groups in the molecule. The presence of catechin dehydroxylated rings, unsaturation of C rings, and tetra-oxo are involved to increase antioxidant capacity (Harborne and Williams, 2000, Heim et al., 2002, Rice-Evans et al., 1996).

2.6 ANTI-TUMOR ACTIVITY

Cancer is the second leading cause of death from non-communicable diseases. Although cancer deaths have decreased by approximately 31 percent over the past 30 years, as healthy lifestyles improve health, it continues to remain a main global challenge for world health systems (Sung et al., 2021). For systemic cancer treatment, chemotherapy drugs prevail. Most cause DNA damage, preventing cells from killing or dividing faster. Chemotherapeutic agents are given as a single dose or short-term treatment at the maximum acceptable dose, followed by a treatment time that must be maintained to permit normal cells to recover (Nurgali et al., 2018). The affordability, health advantages, economic importance and safety of polyphenols relative to synthetic drugs make them excellent candidates for exploring possible therapeutic effects for prevention or treatment of various types of cancer due to its ability to regulate multiple signaling pathways, including PI3K/Akt and MAPK, and important proteins involved in development of cancer, such as p53 and RAS, to induce a promising expectation for these compounds (Karimi et al., 2015).

Hydroxycinnamic acid is an aromatic carboxylic acid with an unsaturated (usually modified) side chain and is more abundant than hydroxybenzoic acid (El-Seedi et al., 2012). Hydroxycinnamic acid is considered to be a potent anticancer agent due to the presence of α,β-unsaturated. (De et al., 2011). Phenolic acid has a variety of health benefits. Chlorogenic acid has shown anticancer activity that induces differentiation by increasing KHSRP, p53, and p21, decreasing poorly differentiated cMyc and CD44 genes, and inhibiting oncogenic miRNA17 family members in cancer cell lines (Huang et al., 2019a). Also, other mechanisms are involved in epigenetic regulation. Several epigenetic targets have been introduced that play a substantial role in the GADD45 signaling pathway (Weng et al., 2018).

2.7 ANTIAGING ACTIVITIES

The general understanding of diet has evolved from the earlier concept of "eating to live" to promote health and inhibit disorders. Critical dietary intervention research and epidemiological investigations have shown that eating a diet rich in fruits and vegetables has significant effects on health (Willett, 1994). The proportion of the population aged 60 and over is estimated to increase from 12 percent in 2015 to 22 percent in 2050 (Barha et al., 2019). Thus, delaying or preventing the inception of age-related disorders caused by cell damage and decreased function could significantly improve life quality and longevity, and reduce the burden caused by the exciting health system (Barha et al., 2019). Chemicals involved in aging processes might activate similar signaling pathways, and a single chemical class may activate several pathways.

2.8 CARDIO-PROTECTING EFFECTS

The statistics of the Eurostat in 2016, the European Heart Network in 2017, and the World Health Organization (WHO), reveal an increase in deaths due to cardiovascular diseases. It should be noted that, in some cases, the worldwide mortality rate is higher than that of malignant neoplasm cancer (26.0%). It has been suggested that polyphenols, which possess vasodilatory properties, lead to the prevention and/or treatment of cardiovascular diseases (Otręba et al., 2020). Many important biological activities of polyphenols are known – for instance, anti-allergic, antioxidant, anti-inflammatory, antibacterial, antithrombotic, cardioprotective, and vasodilatory effects (Balea et al., 2018). The cardioprotective effects of polyphenols are due to their antioxidant, anticoagulant, and antiplatelet effects, fibrinolytic activity; activation of nitric oxide synthase, AMP-activated protein kinase, and sirtuin 1; inhibition of geotensin converting enzymes and phosphate diesterases; and improvement of endothelial cell functions (Chu, 2018). In addition, flavonoids improve insulin resistance, ventricular health, and plasma lipid markers. Flavonoids have an anti-inflammatory effect. They lower blood pressure, improve overall vascular health, block cholesterol oxidation, lower LDL levels, reduce atherosclerosis, and ultimately decrease the risk of cardiovascular disease(Cory et al., 2018, Grootaert et al., 2015). There are many mechanisms explaining polyphenols' anti-inflammatory activity (especially flavonoids), including antioxidants, radical scavengers, and modulating inflammation-related cellular activity, including alterations in endothelial function.

2.9 NEURO-PROTECTIVE EFFECTS

It has been revealed that biologically active plant compounds have a positive effect on the functioning of the central nervous system, such as metabolic regulation and the activity of certain neurotransmitters (Bhullar and Rupasinghe, 2013). Some neurodegenerative and emotional disorders, such as mood disorders, Parkinson's disease (PD), and Alzheimer's disease (AD), are caused by disorders in the metabolism of gamma-aminobutyric acid (GABA), serotonin (5HT), glutamate (Glu), and acetylcholine (ACh) (Bhullar and Rupasinghe, 2013, Panche et al., 2016). Studies show ACh is involved in the distribution of Aβ plaques and cholinergic signaling in the brain (Dhanasekaran et al., 2015).The phytochemical's sedative and anxiolytic effects or cognitive enhancement may be caused via direct binding to $GABA_A$ receptors (Ren et al., 2010). Polyphenols have antidepressant properties that inhibit monoamine oxidase resulting in increased levels of 5HT, dopamine (DA), or norepinephrine (Xu et al., 2010). It also has been shown that some polyphenols may be used in the neurodegenerative diseases by interacting with α7nAChR and increasing mAChR expression and surface density. According to in vitro studies, human neuroblastoma cells pretreated with curcumin polyphenols showed a significant increase in choline-induced Ca^{2+} transients. As a result, the α7 nicotinic acetylcholine receptor is activated (Delbono et al., 1997).

2.10 PHARMACOKINETICS PROFILE

The following metabolic processes help to understand the pharmacokinetics of polyphenols (Figure 2.4).

2.10.1 POLYPHENOL INTERACTION WITH SALIVA

Polyphenols exist abundantly in food, and the major route of delivery is via oral intake. They change when they come into contact with saliva for the first time. Saliva, dietary nutrients, gingival crevicular fluid, xenobiotics, indigenous microbial flora, and medications all interact in a variety of ways. Albumin and mucin, two salivary proteins, have been revealed to significantly boost polyphenol antioxidant activity (Soares et al., 2011).

An important contributing factor to the high redox potential of the oral cavity comes from salivary antioxidants working along with other antioxidants in food and drink (Ginsburg et al., 2012).

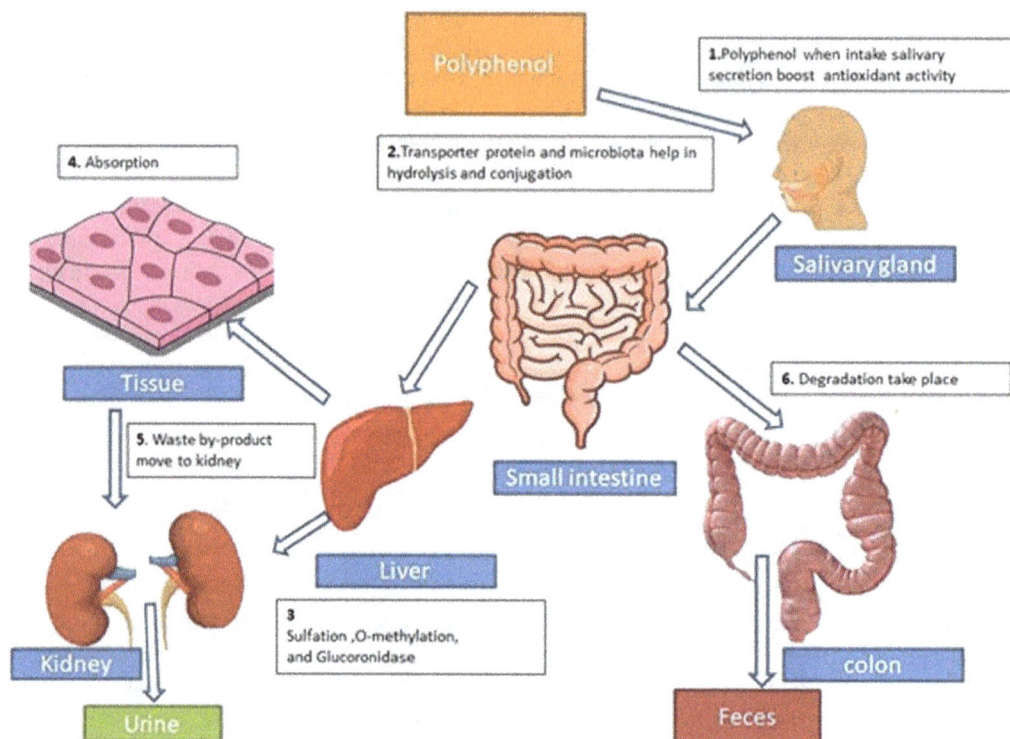

FIGURE 2.4   Pharmacokinetic profile of polyphenol.

It has been experimentally examined by Hoda et al. (2019) that apple incubation in saliva causes phloretin and quercetin hydrolysis into their respective aglycosides and sugar moieties. Thus, saliva, being the main site of contact for polyphenols might function as metabolizers and solubilizers of lipophilic substances, converting them into effective antioxidants (Ginsburg et al., 2012).

2.10.2   POLYPHENOL TRANSIT IN SMALL INTESTINE

The second step of polyphenol absorption occurs in the gastrointestinal tract where these are absorbed or changed by colonic bacteria in the stomach, small intestine, and colon. Polyphenols that have been metabolized or are unabsorbed are expelled out into the faeces. Different factors impact intestinal absorption, metabolism, and bioavailability, including transporter proteins and gut microbiota.

2.10.2.1   Protein transporters mediated polyphenol transit

Polyphenol transfer through the gut barrier might be owing to active transport and passive diffusion, with the former being the more common. Active transport is heavily controlled by numerous transporter proteins and polyphenol types. Passive diffusion is greatly influenced the physicochemical properties of polyphenols. In the intestine, the ATP binding cassette (ABC) superfamily, which includes solute carrier superfamilies (SLC), P-glycoprotein (P-gp), multiple drug resistance proteins (MDR), and breast-cancer resistance protein (BRCP) having monocarboxylate transporters and organic anion polypeptide has been identified as playing an effective role in polyphenols absorption

TABLE 2.5
List of intestinal transporter proteins involved in the transport of polyphenols

Transporter protein	Polyphenols	Reference
BRCP	Resveratrol	(Kaldas et al., 2003)
LPH	Daidzein-7-glucoside, Genistein-7-glucoside, Quercetin-3-glucoside, Quercetin-4 '-glucoside	(Morand et al., 2000)
MRP1	Epigallocatechin (EGCG)	(Hong et al., 2003)
OATP	Apigenin	(Falé et al., 2013)
MRP2	Chlorogenic acids	(Erk et al., 2014)
	(+)-Catechin	
	4″-O-methyl EGCG 4′, 4″-di-O-methyl EGCG	
OATP2B1	Resveratrol	(Kondo et al., 2017)
	Theaflavin	
	Resveratrol-3-O-4″ disulfate	
OATP1B3	Resveratrol-3-O-sulfate	(Riha et al., 2014)
	Esveratrol-3-O-4″ disulfate,	
P-gp	Chlorogenic acids	(Erk et al., 2014, Falé et al., 2013)
	Rosmarinic acid	
	Puerarin (A)	
SGLT1	Quercetin-4′-glucoside	(Day et al., 2003)

Abbreviations: Breast cancer resistance protein (BRCP), Lactase phlorizin hydrolase (LPH), Organic anion polypeptide transporters (OATP2B1); Sodium-dependent glucose transporter (SGLT1),

(Table 2.5). It has been shown that the inclusion of methyl groups in the polyphenols improves their intestinal absorption (Bohn, 2014).

Hoda et al. (2019) found that emodin, polyphenols, resveratrol, and apigenin are more quickly taken up by the body in the gastrointestinal tract than chrysophanol. The product of these polyphenols, that is, sulfates and glucuronides are subsequently moved towards the vascular side of the small intestine. P-gp, SGLT1, MRP2 as well as other transporters, are thought to influence the property of bioavailability of curtained polyphenol molecules (Teng and Chen, 2019).

2.10.3 GUT MICROBIOTA-MEDIATED POLYPHENOL TRANSIT

Complex polyphenols are transferred and metabolized by the bacteria in the intestines. Streptoccococci, Proteobacteria, Lactobacilli, Firmicutes, Enterococci, and bifidobacterium, are only a few of the 100 trillion bacteria that populate the gastrointestinal system. Their metabolic capacity is determined by several parameters, such as the host physiology, nutritional supply, and the quantity of polyphenol available (Ray and Mukherjee, 2021). Polyphenols may either be conjugated with sugars (rutin) and organic acids or nonconjugated oligomers (condensed tannins) to maximize absorption and bioavailability (hesperitin). The polyphenol is hydrolyzed, cleaved, or reduced by the gut bacteria through a metabolic process (Table 2.6). It has been shown that ellagic acid and ellagitannins can be converted into beneficial health metabolites (Bohn, 2014).

There is another process of polyphenol degradation in which C-C bond breakage or demethylation occurs due to the activation of anaerobic bacteria like Clostridium with oxygen (Yang et al., 2021b). Hoda et al. (2019) present an in-depth review of the metabolism chemistry of polyphenol, including hydrogenation, decarboxylation, A- and C-ring cleavages, and reduction (Table 2.6). Gut microbes also reduced the polyphenol via NADPH-dependent curcumin/dihydrocurcumin reductase (Kawabata et al., 2019).

TABLE 2.6
List of polyphenols metabolized by gut microbiota

Polyphenol class	Polyphenol	Metabolites	Reactions	Reference
Anthrones	Emodin, Rhein, Sennoside, Chrysophanol	64 metabolites	Hydroxylation Oxidation Demethylation	(Huang et al., 2019b)
Anthroquinones	Aloe emodin	64 metabolites	Demethylation Dehydroxylation, Dehydroxylation	(Huang et al., 2019b)
Biflavone	Chamaechromone	24 metabolites	Methylation Dehydroxylation hydration, Glucuronidation	(Lou et al., 2014)
Flavanoid glycoside	Orientin	Deoxygenated orientin	Deoxygenation	(Lou et al., 2015)
Iridoid glucoside	Catalpol	Bacteroides Enterococcus Human intestinal flora	Methylation Hydroxylation Acetylation hydrolysis	(Tao et al., 2016)
Isoflavone	Irisolidone Kakkalide	17 metabolites	Hydrolysis, hydroxylation dehydroxylation, demethoxylation, hydroxylation, decarbonylation,	(Zhang et al., 2014)
Saponin	Anemoside B4	Luteolin	oxygenation and deglycosylation	(Wan et al., 2017)

2.10.4 CONJUGATION AND NATURE OF METABOLITES

After intestinal absorption glucuronidation, methylation, and sulfation are the most common kinds of conjugation that polyphenols undergo. A methyl group may be transferred from S-adenosyl-L-methionine to polyphenols containing an o-diphenolic (catechol) moiety by the enzyme catalyzing the reaction. This is the first time that Bresciani et al. (2019) have shown that cyanidin is converted to peonidin in humans. Polyphenol methylation happens mostly at the 3' position, although a small amount of 4'-O-methylated molecule is also produced (Reygaert, 2018). There are specific enzymes that may transfer the sulfate portion from 3-phosphoadenosine-5'-phosphosulfate to an -OH group on diverse substrates, such as glutathione and glutamine (Polyphenols, steroids, bile acids, etc.). The precise locations of sulfation for the different polyphenols have not even been properly defined, however; sulfation happens mostly in the liver (Murota, 2020).

Glucuronic acid is transferred from UDP-glucuronic acid to polyphenols, steroids and bile acids, and hundreds of dietary components and xenobiotics via membrane-bound enzymes called UDP glucuronosyltransferases (UGT). He et al. (2017) give evidence showing that polyphenols initially undergo glucuronidation in enterocytes before any more cross-linking in the liver. This indicates that the liver is the final destination for these metabolically altered compounds after they have been perfused throughout the small intestine with glucuronidated compounds (He et al., 2017).

2.10.5 PLASMA TRANSPORT AND LIPID STRUCTURAL PARTITIONING

Polyphenol metabolites do not flow readily through the blood in solitary form; instead, they become attached to certain proteins or lipids. The comparative affinity of metabolites for plasma proteins is high as compared to lipids. Quercetin, a flavonoid, is shown to be linked to 99 percent of plasma proteins at rates of up to 15 mol/L, and only 0.5 percent with VLDL (Reis et al., 2021). Albumin is the primary protein involved in binding. Different polyphenols have different affinities for albumin, depending on their structural chemistry. In contrast to quercetin, kaempferol, and isorhamnetin, which have a different B-ring substitution than quercetin, exhibit an extra affinity for albumin (Grabska-Kobylecka et al., 2020). (Figueira et al., 2017)

2.10.6 Plasma concentrations

After the intake of polyphenols, plasma concentrations became high, depending on the kind of polyphenol and type of source for a dietary supplement. After ingesting 0.08-0.1g equivalent quercetin in the form of onions, apples, or plant-rich meals, they are on the order of 0.03–0.075 mol/dlL (Murphy et al., 2018). Anroedh et al. (2018) measured that 130–220mg dose of orange juice release the highest plasma concentration of hesperetin at 1.3–2.2 mol/L after 5–7 hours post-ingestion.

2.10.7 Tissue uptake

The bioavailability of different polyphenols differs according to the absorption site and duration. The concentrations of various polyphenols in the skin, bone, heart, spleen, kidney, pancreas, bladder, uterine, prostate, mammary gland, ovary, and testes tissues varied from 0.03 to 3μg aglycone equivalents/g tissue, contingent on the provided amount, type of tissue (Pressman et al., 2017).

2.10.8 Elimination

Polyphenol metabolites may be excreted through two routes: biliary or urine. Bile is considered as a preferred route that cause conjugation of metabolites, whereas urine is preferable route for smaller conjugates, like mono-sulfates. Various polyphenols have different effects on laboratory animals in terms of the ratio of urinary to biliary excretion (Allaoui et al., 2020).

2.11 TOXICITY OF THE POLYPHENOL

The majority of polyphenol reports focused on determining the preventive properties of polyphenols against illnesses or hazardous medications, and only a few researchers have looked into their potential toxicity. In labs, it is tested that pomegranate juice contains punicalagin, which has been shown to have no acute harmful effects on mice or rats when administered at 60 g/day of food in the form of proanthocyanidin extract of a pomegranate seed (Kerimi et al., 2017). However, high dosages of quercetin (2% or 4%) in the diet of rats caused chronic skin disease, (Shi et al., 2019). Blagosklonny (2021) research mentioned that quercetin (0.1 percent) dramatically shortened the expected life of mice, although no influence was detected on overall survival. At high dosages, certain polyphenols may cause genotoxicity or carcinogenesis There are some diseases and disorders that show the toxic cause of polyphenol when taken at a high concentration (Table 2.7).

TABLE 2.7
List of Polyphenols and their Toxic Effect on the Body

Polyphenols	Toxic effects on the body	Reference
Caffeic acid	Tumor of intestine and Kidney	(Montis et al., 2021).
Catecholestrogens	Kidney damage	(Nakagomi et al., 2018)
Quercetin	Carcinogenic, kidney problem	(Lu et al., 2018)
Vitexin	Goiter	
Glyceollin	Thyroid problems, increase antithyroid peroxidase level	(Domínguez-López et al., 2020)
Proanthocyanidins	Less protein consumption, and inhibition of enzyme	(Chew et al., 2019)
Caffeine	Low iron absorption	(Hart et al., 2017)

2.11.1 INTESTINAL AND KIDNEY DYSFUNCTIONS

The overdose of caffeic acid in the diet causes cancerous tumors of the intestines and kidneys (Montis et al., 2021). Extending these findings to typical dietary amounts suggests a significant danger. Catecholestrogens is another example related to development the estradiol-induced kidney tumorigenesis. It raises the amounts of 2-hydroxyestrodiol by 80 percent in the kidneys by inhibiting the O-methylation of catecholestrogens. Extending these findings to typical dietary amounts suggests a significant danger. The induction of kidney malignancies by estradiol is also thought to be mediated by catecholestrogens (Nakagomi et al., 2018).

Quercetin belongs to the flavonoid group and has been studied as an effective reducer of cancer cell proliferation in previous studies. However, its high dose has been reported to boost cellular proliferation at modest concentrations (1–5mol/L) of green tea catechins in animals(Lu et al., 2018). It is involved in the inhibition of O-methylation of catecholestrogens and raised kidney concentrations of 2- and 4-hydroxyestrodiol by 60-80 percent. That results in catecholestrogen redox cycling and estradiol-induced carcinogenesis.

2.11.2 THYROID DYSFUNCTION

In the same way as synthetic antioxidants, different flavonoids are used to boost thyroid hormone production (free radical iodination) and inhibit thyroid peroxidase. In vitro injection of vitexin, C-glycosylflavone, which is abundantly present in millet, raised thyroid weight while decreasing the level of the plasma thyroid hormone (Chin and Pang, 2020). However, Elnour et al. (1997) studied the role of millet in the West Sudan diet, judging that due to excess use of millet a lot of people in West Sudan suffered from endemic goiter.

Furthermore, supplementation with genistein is also considered effective in reducing the activity of thyroid peroxidase in rats. These effects of genistein on the thyroid are more noticeable in conditions of iodine insufficiency (Barrasa et al., 2018). This is especially concerning in the case of babies who have been fed soybean in large dosages. Soybean is considered an effective source of seven polyphenols, glyceollin afromosin, glyceofuran coumestrol, formononetin, phaseol, and isotrifoliol. It is improved by two clinical investigations on adults that found thyroid hormone levels did not fluctuate significantly by consuming isoflavone-rich soy protein for three to six months (Domínguez-López et al., 2020).

2.11.3 ANTI-NUTRITIONAL EFFECTS

2.11.3.1 Low iron absorption

A high amount of polyphenols in one's diet may also result in anti-nutritional problems. Tea drinking has been shown to reduce nonheme-iron absorption; nonetheless, high polyphenol consumption may increase the danger of iron depletion in populations of people with poor iron status. A crucial element in this context is the absence of vitamin C in essential sources of polyphenols, such as coffee, tea, and wine, which are often consumed with meals (Hart et al., 2017).

2.11.3.2 Low protein digestion absorption

Ellagitannins and proanthocyanidins (condensed tannins) have been recognized as antinutritional compounds, important in animal nutrition due to their capacity to inhibit several enzymes and bond with proteins. It is seen that when rats are given meal at a high dose (10 g/kg diet), but not at a lower one, had an adverse effect on their development and digestibility (Chew et al., 2019). Eating fava beans high in proanthocyanidins lowered net protein consumption in Egyptian boys; this was recovered by dehulling the cover of beans. Importantly, these particular effects are unlikely to occur in the typical Indian diet, which includes a substantial tannin diet (Van Buiten and Elias, 2021).

2.12 DRUG PHARMACOKINETICS AND BIOAVAILABILITY

Polyphenols may potentially affect the pharmacokinetics and bioavailability of certain drugs. Because of CYP3A4 inhibition, the bioavailability of some drugs, such as benzodiazepines and terfenadine, may be tripled when combined with grapefruit juice (which has a high concentration of naringenin). Moreover, cyclosporine has a limited therapeutic range; these effects, which may be attributed in part to psoralens and naringenin, are clinically significant (Ahmadi et al., 2019).

In vitro and animal research has indicated many of these effects; however, it has not been shown that these effects exist in a large population of humans. There is a very low probability of observational epidemiology related to the carcinogenic effect of polyphenols. Dietary intake is generally lower than the doses used in these reports, and food type can also affect the impact of polyphenols. However, the results of the experiment need to be taken as seriously as the positive benefits. As a result of documented endocrine- and carcinogenic-related effects of some polyphenols, it is considered unethical to use a high dose of polyphenol. The synergistic interaction of polyphenol with other nutritional supplements should be studied in this perspective for the treatment and prevention of medical problems (Santana-Gálvez et al., 2020). Furthermore, developing bioprocesses to produce next-generation functional drinks and foods is critical for reaching the market and providing people with the necessary polyphenol benefits. Recent research suggests that postharvest abiotic stressors applied to horticulture crops to raise the concentration of polyphenol compounds, followed by their subsequent processing into processed foods utilizing non-thermal processing methods, and that might be an efficient strategy for producing shelf-stable goods with a high concentration of antioxidant polyphenols (Jacobo-Velázquez et al., 2021).

2.13 CONCLUSION AND FUTURE PROSPECTIVE

Polyphenols are secondary plant metabolites with a wide range of health benefits. More than 8,000 polyphenols have been discovered in nature. Polyphenols have abundant physiological functions, including growth factors and chemical intermediates for the production of other molecules. Polyphenols with wide range of anti-inflammatory, anti-aging, and antitumor effects, are used in the prevention and treatment of many diseases, including cardiac and nervous system problems. In excess amounts, they cause some toxic effects, especially on thyroids, kidneys, and the small intestine. It is becoming more common to examine the bioactivities of phenolic compounds isolated from various plant species. There is a constant stream of fresh scientific evidence that polyphenol may prevent a variety of degenerative and chronic disorders. Identifying and analyzing the bioactivities of polyphenol's natural sources should be the focus of future researches. Since the bioactivity of many polyphenols has now been determined, subsequent research efforts might concentrate on producing unique food items and supplement formulations to commercialize the developed foundational knowledge. The development of effective nutritional mixtures in the form of foodstuffs, drinks, and nutritional supplements, which might be consumed not only to protect but also to cure chronic illness, is an exciting new field of study.

REFERENCES

Ahmadi, Z., Mohammadinejad, R. and Ashrafizadeh, M. 2019. Drug delivery systems for resveratrol, a nonflavonoid polyphenol: Emerging evidence in last decades. *J Drug Deliv Sci Technol.* 51: 591–604.

Ahmed, S.M.U., Luo, L., Namani, A., Wang, X.J. and Tang, X. 2017. Nrf2 signaling pathway: Pivotal roles in inflammation. *Biochim Biophys Acta Mol Basis Dis.* 1863: 585–597.

Allaoui, S., Naciri Bennani, M., Ziyat, H., Qabaqous, O., Tijani, N. and Ittobane, N. 2020. Kinetic study of the adsorption of polyphenols from olive mill wastewater onto natural clay: Ghassoul. *J. Chem.* 2020.

Ambriz-Pérez, D.L., Leyva-López, N., Gutierrez-Grijalva, E.P. and Heredia, J.B. 2016. Phenolic compounds: Natural alternative in inflammation treatment. A review.*Cogent food agric.* 2: 1131412.

Ameer, K., Shahbaz, H.M. and Kwon, J.H. 2017. Green extraction methods for polyphenols from plant matrices and their byproducts: A review. *Compr. Rev. Food Sci. Food Saf.* 16: 295–315.

Anroedh, S., Hilvo, M., Akkerhuis, K.M., Kauhanen, D., Koistinen, K., Oemrawsingh, R., Serruys, P., van Geuns, R.-J., Boersma, E. and Laaksonen, R. 2018. Plasma concentrations of molecular lipid species predict long-term clinical outcome in coronary artery disease patients. *J. Lipid Res.* 59: 1729–1737.

Balea, Ş.S., Pârvu, A.E., Pop, N., Marín, F.Z. and Pârvu, M. 2018. Polyphenolic compounds, antioxidant, and cardioprotective effects of pomace extracts from Fetească Neagră Cultivar. *Oxidative med cell longev.* 2018.

Bardaweel, S.K., Gul, M., Alzweiri, M., Ishaqat, A., Alsalamat, H.A. and Bashatwah, R.M. 2018. Reactive oxygen species: The dual role in physiological and pathological conditions of the human body. *Eurasian J Med* 50: 193–201.

Barha, C.K., Hsu, C.-L., Ten Brinke, L. and Liu-Ambrose, T. 2019. Biological sex: A potential moderator of physical activity efficacy on brain health. *Front Aging Neurosci.* 11: 329.

Barrasa, G.R.R., Cañete, N.G. and Boasi, L.E.V. 2018. Age of postmenopause women: Effect of soy isoflavone in lipoprotein and inflammation markers. *J Menopausal Med.* 24: 176–182.

Bayir, H. and Kagan, V.E. 2008. Bench-to-bedside review: Mitochondrial injury, oxidative stress and apoptosis – There is nothing more practical than a good theory. *Crit Care.* 12: 1–11.

Bhullar, K.S. and Rupasinghe, H. 2013. Polyphenols: multipotent therapeutic agents in neurodegenerative diseases. *Oxid med cell longev.* 2013: 891748.

Birben, E., Sahiner, U.M., Sackesen, C., Erzurum, S. and Kalayci, O. 2012. Oxidative stress and antioxidant defense. *World Allergy Organ J.* 5: 9–19.

Blagosklonny, M.V. 2021.The goal of geroscience is life extension. *Oncotarget* 12: 131.

Blokhina, O., Virolainen, E. and Fagerstedt, K.V. 2003. Antioxidants, oxidative damage and oxygen deprivation stress: A review. *Ann Bot.* 91: 179–194.

Bohn, T. 2014. Dietary factors affecting polyphenol bioavailability. *Nutr Rev.* 72: 429–452.

Boscá, L., Zeini, M., Través, P.G. and Hortelano, S. 2005. Nitric oxide and cell viability in inflammatory cells: A role for NO in macrophage function and fate. *Toxicology* 208: 249–258.

Bresciani, L., Angelino, D., Vivas, E.I., Kerby, R.L., García-Viguera, C., Del Rio, D., Rey, F.E. and Mena, P. 2019. Differential catabolism of an anthocyanin-rich elderberry extract by three gut microbiota bacterial species. *J Agric Food Chem.* 68: 1837–1843.

Calniquer, G., Khanin, M., Ovadia, H., Linnewiel-Hermoni, K., Stepensky, D., Trachtenberg, A., Sedlov, T., Braverman, O., Levy, J. and Sharoni, Y. 2021. Combined effects of carotenoids and polyphenols in balancing the response of skin cells to UV irradiation. *Molecules* 26: 1931.

Casas, R., Sacanella, E. and Estruch, R. 2014. The immune protective effect of the Mediterranean diet against chronic low-grade inflammatory diseases. *Endocr Metab Immune Disord Drug Targets.* 14: 245–254.

Chandrasekharan, N., Dai, H., Roos, K.L.T., Evanson, N.K., Tomsik, J., Elton, T.S. and Simmons, D.L. 2002. COX-3, a cyclooxygenase-1 variant inhibited by acetaminophen and other analgesic/antipyretic drugs: Cloning, structure, and expression. *Proc Natl Acad Sci.* 99: 13926–13931.

Chen, C., Yu, R., Owuor, E.D. and Kong, A.-N.T. 2000. Activation of antioxidant-response element (ARE), mitogen-activated protein kinases (MAPKs) and caspases by major green tea polyphenol components during cell survival and death. *Arch Pharm Res.* 23: 605–612.

Chew,B., Mathison, B., Kimble, L., McKay, D., Kaspar, K., Khoo, C., Chen, C.-Y.O. and Blumberg, J. 2019. Chronic consumption of a low calorie, high polyphenol cranberry beverage attenuates inflammation and improves glucoregulation and HDL cholesterol in healthy overweight humans: A randomized controlled trial. *Eur J Nutr.* 58: 1223–1235.

Cheynier, V., Duenas-Paton, M., Salas, E., Maury, C., Souquet, J.-M., Sarni-Manchado, P. and Fulcrand, H. 2006. Structure and properties of wine pigments and tannins. *Am J Enol Vitic.* 57: 298–305.

Chin, K.-Y. and Pang, K.-L. 2020. Skeletal effects of early-life exposure to soy isoflavones: A review of evidence from rodent models. *Front Pediatr.* 8.

Cho, S.-Y., Park, S.-J., Kwon, M.-J., Jeong, T.-S., Bok, S.-H., Choi, W.-Y., Jeong, W.-I., Ryu, S.-Y., Do, S.-H. and Lee, C.-S. 2003. Quercetin suppresses proinflammatory cytokines production through MAP kinases and NF-κB pathway in lipopolysaccharide-stimulated macrophage. *Mol Cell Biochem.* 243: 153–160.

Chobot, V., Hadacek, F. and Kubicova, L. 2014. Effects of selected dietary secondary metabolites on reactive oxygen species production caused by iron(II) autoxidation. *Molecules* 19: 20023–20033.

Chu, A. 2018. Cardioprotection by bioactive polyphenols: A strategic view. *Austin J Cardiovasc Dis Atheroscler.* 5: 1034.

Cory, H., Passarelli, S., Szeto, J., Tamez, M. and Mattei, J. 2018. The role of polyphenols in human health and food systems: A mini-review. *Front Nutr.* 5, 87.

da Silva, A.P.G. 2021. Fighting coronaviruses with natural polyphenols. *Biocatal Agric Biotechnol.* 37: 102179.

Dabeek, W.M. and Marra, M.V. 2019. Dietary quercetin and kaempferol: bioavailability and potential cardiovascular-related bioactivity in humans. *Nutrients* 11.

Das, K. and Roychoudhury, A. 2014. Reactive oxygen species (ROS) and response of antioxidants as ROS-scavengers during environmental stress in plants. *Front Environ Sci.* 2: 53.

Day, A.J., Gee, J.M., DuPont, M.S., Johnson, I.T. and Williamson, G. 2003. Absorption of quercetin-3-glucoside and quercetin-4'-glucoside in the rat small intestine: The role of lactase phlorizin hydrolase and the sodium-dependent glucose transporter. *Biochem Pharmacol.* 65: 1199–1206.

De, P., Baltas, M. and Bedos-Belval, F. 2011. Cinnamic acid derivatives as anticancer agents: A review: *Curr Med Chem* 18: 1672–1703.

Delbono, O., Gopalakrishnan, M., Renganathan, M., Monteggia, L.M., Messi, M. and Sullivan, J. 1997. Activation of the recombinant human α7 nicotinic acetylcholine receptor significantly raises intracellular free calcium. *Journal of Pharmacology and Experimental Therapeutics* 280: 428–438.

Dhanasekaran, S., Perumal, P. and Palayan, M. 2015. In-vitro screening for acetylcholinesterase enzyme inhibition potential and antioxidant activity of extracts of Ipomoea aquatica Forsk: Therapeutic lead for Alzheimer's disease. *J. Appl. Pharm. Sci.* 5: 12–6.

Di Lorenzo, C., Dell'Agli, M., Colombo, E., Sangiovanni, E. and Restani, P. 2013. Metabolic syndrome and inflammation: A critical review of in vitro and clinical approaches for benefit assessment of plant food supplements. *Evid Based Complementary Altern Med.* 2013.

Domínguez-López, I., Yago-Aragón, M., Salas-Huetos, A., Tresserra-Rimbau, A. and Hurtado-Barroso, S. 2020. Effects of dietary phytoestrogens on hormones throughout a human lifespan: A review. *Nutrients* 12: 2456.

Durazzo, A., Lucarini, M., Camilli, E., Marconi, S., Gabrielli, P., Lisciani, S., Gambelli, L., Aguzzi, A., Novellino, E., Santini, A., Turrini, A. and Marletta, L. 2018. Dietary lignans: Definition, description and research trends in databases development. *Molecules* 23 .

El-Seedi, H.R., El-Said, A.M., Khalifa, S.A., Goransson, U., Bohlin, L., Borg-Karlson, A.-K. and Verpoorte, R. 2012. Biosynthesis, natural sources, dietary intake, pharmacokinetic properties, and biological activities of hydroxycinnamic acids. *J Agric Food Chem.* 60: 10877–10895.

Elnour, A., L\u00e9idén, S.-A., Bourdoux, P., Eltom, M., Khalid, S. and Hambraeus, L. 1997. The goitrogenic effect of two Sudanese pearl millet cultivars in rats. *Nutr Res.* 17: 533–546.

Erk, T., Hauser, J., Williamson, G., Renouf, M., Steiling, H., Dionisi, F. and Richling, E. 2014. Structure–and dose–absorption relationships of coffee polyphenols. *Biofactors* 40: 103–12.

Falé, P.L., Ascensão, L. and Serralheiro, M.L. 2013.Effect of luteolin and apigenin on rosmarinic acid bioavailability in Caco-2 cell monolayers. *Food Func.* 4: 426–431.

Fereidoon, S. and Ying, Z. 2010. Lipid oxidation and improving the oxidative stability. *Chem Soc Rev.* 39: 4067–4079.

Figueira, I., Garcia, G., Pimpão, R. C., Terrasso, A., Costa, I., Almeida, A., Tavares, L., Pais, T., Pinto, P. and Ventura, M. 2017. Polyphenols journey through blood-brain barrier towards neuronal protection. *Scientific Reports* 7: 1–16.

Gay, N. H., Suwanjang, W., Ruankham, W., Songtawee, N., Wongchitrat, P., Prachayasittikul, V., Prachayasittikul, S. and Phopin, K. 2020. Butein, isoliquiritigenin, and scopoletin attenuate neurodegenerationviaantioxidant enzymes and SIRT1/ADAM10 signaling pathway. *RSC Adv.* 10: 16593–16606.

Gianchecchi, E. and Fierabracci, A. 2020. Insights on the effects of resveratrol and some of its derivatives in cancer and autoimmunity: A molecule with a dual activity. *Antioxidants.* 9: 91.

Ginsburg, I., Kohen, R. and Koren, E. 2012. Saliva: a 'solubilizer' of lipophilic antioxidant polyphenols. *Oral Dis.* 19: 321–322.

Grabska-Kobylecka, I., Kaczmarek-Bak, J., Figlus, M., Prymont-Przyminska, A., Zwolinska, A., Sarniak, A., Wlodarczyk, A., Glabinski, A. and Nowak, D. 2020. The presence of caffeic acid in cerebrospinal fluid: Evidence that dietary polyphenols can cross the blood-brain barrier in humans. *Nutrients.* 12: 1531.

Grootaert, C., Kamiloglu, S., Capanoglu, E. and Van Camp, J. 2015. Cell systems to investigate the impact of polyphenols on cardiovascular health. *Nutrients.* 7: 9229–9255.

Han, X., Shen, T. and Lou, H. 2007. Dietary polyphenols and their biological significance.*Int J Mol Sci.* 8: 950–988.

Harborne, J.B. and Williams, C.A.2000. Advances in flavonoid research since 1992. *Phytochemistry.* 55: 481–504.

Hart, J.J., Tako, E. and Glahn, R.P. 2017. Characterization of polyphenol effects on inhibition and promotion of iron uptake by Caco-2 cells. *J Agric Food Chem.* 65: 3285–3294.

Harwell, B. 2007. Biochemistry of oxidative stress. *Biochem Soc Trans.* 35: 1147–1150.

He, X., Song, Z.-J., Jiang, C.-P. and Zhang, C.-F. 2017. Absorption properties of luteolin and apigenin in genkwa flos using in situ single-pass intestinal perfusion system in the rat. *Am J Chinese Med.* 45: 1745–1759.

Heim, K.E., Tagliaferro, A.R. and Bobilya, D.J. 2002. Flavonoid antioxidants: chemistry, metabolism and structure-activity relationships. *J Nutr Biochem.* 13: 572–584.

Herb, M. and Schramm, M. 2021. Functions of ros in macrophages and antimicrobial immunity. *Antioxidants* 10: 1–39.

Ho, L., Ono, K., Tsuji, M., Mazzola, P., Singh, R. and Pasinetti, G.M. 2018. Protective roles of intestinal microbiota derived short chain fatty acids in Alzheimer's disease-type beta-amyloid neuropathological mechanisms. *Expert Rev Neurother* 18: 83–90.

Hoda, M., Hemaiswarya, S. and Doble, M. 2019. Pharmacokinetics and pharmacodynamics of polyphenols. *Role of Phenolic Phytochemicals in Diabetes Management (Book)*: Springer, 159–173.

Hong, J., Lambert, J.D., Lee, S.-H., Sinko, P.J. and Yang, C.S. 2003.Involvement of multidrug resistance-associated proteins in regulating cellular levels of (−)-epigallocatechin-3-gallate and its methyl metabolites. *Biochem Biophys Res Commun.* 310: 222–227.

Hormozi, M., Ghoreishi, S. and Baharvand, P. 2019. Astaxanthin induces apoptosis and increases activity of antioxidant enzymes in LS-180 cells. *Artificial Cells, Nanomed Biotechnol.* 47: 891–895.

Huang, S., Wang, L.-L., Xue, N.-N., Li, C., Guo, H.-H., Ren, T.-K., Zhan, Y., Li, W.-B., Zhang, J. and Chen, X.-G. 2019a. Chlorogenic acid effectively treats cancers through induction of cancer cell differentiation. *Theranostics* 9: 6745.

Huang, Z., Xu, Y., Wang, Q. and Gao, X. 2019b. Metabolism and mutual biotransformations of anthraquinones and anthrones in rhubarb by human intestinal flora using UPLC-Q-TOF/MS. *J Chromatogr B.* 1104: 59–66.

Ichikawa, D., Matsui, A., Imai, M., Sonoda, Y. and Kasahara, T. 2004. Effect of various catechins on the IL-12p40 production by murine peritoneal macrophages and a macrophage cell line, J774. 1. *Biol Pharm Bull.* 27: 1353–1358.

Jacobo-Velázquez, D.A., Santana-Gálvez, J. and Cisneros-Zevallos, L. 2021. Designing next-generation functional food and beverages: Combining nonthermal processing technologies and postharvest abiotic stresses. *Food Eng. Rev.* 13: 592–600.

Janku, M., Luhová, L. and Petrivalský, M. 2019. On the origin and fate of reactive oxygen species in plant cell compartments. *Antioxidants* 8: 105.

Jiang, N., Doseff, A.I. and Grotewold, E. 2016. Flavones: From biosynthesis to health benefits. *Plants* (Basel) 5.

Jones, D.P. and Sies, H. 2015. The redox code. *Antioxid Redox Signal.* 23: 734–746.

Kaldas, M.I., Walle, U.K. and Walle, T. 2003. Resveratrol transport and metabolism by human intestinal Caco-2 cells. *J Pharm pharmacol.* 55: 307–312.

Kaplanski, G., Marin, V., Montero-Julian, F., Mantovani, A. and Farnarier, C. 2003. IL-6: a regulator of the transition from neutrophil to monocyte recruitment during inflammation. *Trends Immunol.* 24: 25–9.

Karimi, A., Majlesi, M. and Rafieian-Kopaei, M. 2015. Herbal versus synthetic drugs; beliefs and facts. *J Nephropharmacol.* 4: 27.

Karunaweera, N., Raju, R., Gyengesi, E. and Munch, G. 2015. Plant polyphenols as inhibitors of nf-Kb induced cytokine production – A potential anti-inflammatory treatment for Alzheimer's disease? *Front Mol Neurosci.* 8.

Kawabata, K., Yoshioka, Y. and Terao, J. 2019. Role of intestinal microbiota in the bioavailability and physiological functions of dietary polyphenols. *Molecules* 24: 370.

Kay, C.D., Pereira-Caro, G., Ludwig, I.A., Clifford, M.N. and Crozier, A. 2017. Anthocyanins and flavanones are more bioavailable than previously perceived: A review of recent evidence. *Ann Rev Food Sci Technol.* 8: 155–180.

Kerimi, A., Nyambe-Silavwe, H., Gauer, J.S., Tomás-Barberán, F.A. and Williamson, G. 2017. Pomegranate juice, but not an extract, confers a lower glycemic response on a high–glycemic index food: Randomized, crossover, controlled trials in healthy subjects. *Am J Clin Nutr.* 106: 1384–1393.

Kohandel, Z., Farkhondeh, T., Aschner, M., Pourbagher-Shahri, A.M. and Samarghandian, S. 2022. Anti-inflammatory action of astaxanthin and its use in the treatment of various diseases. *Biomedicine and Pharmacotherapy* 145: 112179.

Kondo, A., Narumi, K., Ogura, J., Sasaki, A., Yabe, K., Kobayashi, T., Furugen, A., Kobayashi, M. and Iseki, K. 2017. Organic anion-transporting polypeptide (OATP) 2B1 contributes to the cellular uptake of theaflavin. *Drug MetabPharmacokinet.* 32: 145–150.

Krga, I. and Milenkovic, D. 2019. Anthocyanins: From sources and bioavailability to cardiovascular-health benefits and molecular mechanisms of action. *J Agric Food Chem.* 67: 1771–1783.

Kumar, S. and Pandey, A.K. 2013.Chemistry and biological activities of flavonoids: an overview. *Sci World J.:* 162750.

Kurutas, E.B. 2016. The importance of antioxidants which play the role in cellular response against oxidative/nitrosative stress: Current state. *Nutr J.* 15: 71.

Laganà, P., Anastasi, G., Marano, F., Piccione, S., Singla, R.K., Dubey, A.K., Delia, S., Coniglio, M.A., Facciolà, A. and Di Pietro, A. 2019.*Phenolic Substances in Foods: Health Effects as Anti-inflammatory and Antimicrobial Agents.* Oxford University Press.

Lin, Y.-L. and Lin, J.-K. 1997. (−)-Epigallocatechin-3-gallate blocks the induction of nitric oxide synthase by down-regulating lipopolysaccharide-induced activity of transcription factor nuclear factor-κB. *Mol Pharmacol.* 52: 465–472.

Litwinienko, G. and Ingold, K.U. 2007. Solvent effects on the rates and mechanisms of reaction of phenols with free radicals. *Acc Chem Res.* 40: 222–230.

Lobo, V., Patil, A., Phatak, A. and Chandra, N. 2010. Free radicals, antioxidants and functional foods: Impact on human health. *Pharmacogn Rev.* 4: 118–126.

Losada-Barreiro, S. and Bravo-Díaz, C. 2017. Free radicals and polyphenols: The redox chemistry of neurodegenerative diseases. *Eur J Med Chem.* 133: 379–402.

Lou, Y., Zheng, J., Hu, H., Lee, J. and Zeng, S. 2015. Application of ultra-performance liquid chromatography coupled with quadrupole time-of-flight mass spectrometry to identify curcumin metabolites produced by human intestinal bacteria. *J Chromatogr B.* 985: 38–47.

Lou, Y., Zheng, J., Wang, B., Zhang, X., Zhang, X. and Zeng, S. 2014. Metabolites characterization of chamaechromone in vivo and in vitro by using ultra-performance liquid chromatography/Xevo G2 quadrupole time-of-flight tandem mass spectrometry. *J ethnopharmacol.* 151: 242–252.

Lu, H., Wu, L., Liu, L., Ruan, Q., Zhang, X., Hong, W., Wu, S., Jin, G. and Bai, Y. 2018. Quercetin ameliorates kidney injury and fibrosis by modulating M1/M2 macrophage polarization. *Biochem Pharmacol.* 154: 203–212.

Mackenzie, G.G., Carrasquedo, F., Delfino, J.M., Keen, C.L., Fraga, C.G. and Oteiza, P.I. 2004. Epicatechin, catechin, and dimeric procyanidins inhibit PMA-induced NF-κB activation at multiple steps in Jurkat T cells. *FASEB J.* 18: 167–169.

Manach, C., Scalbert, A., Morand, C., Rémésy, C. and Jiménez, L. 2004. Polyphenols: food sources and bio-availability. *American J Clin Nutr.* 79: 727–747.

Martin, M.A. and Ramos, S. 2021. Impact of dietary flavanols on microbiota, immunity and inflammation in metabolic diseases.*Nutrients* 13.

Meng, T., Xiao, D., Muhammed, A., Deng, J., Chen, L. and He, J. 2021. Anti-inflammatory action and mechanisms of resveratrol. *Molecules* 26.

Mithul Aravind, S., Wichienchot, S., Tsao, R., Ramakrishnan, S. and Chakkaravarthi, S. 2021. Role of dietary polyphenols on gut microbiota, their metabolites and health benefits. *Food Res Int.* 142: 110189.

Molina, M.F., Sanchez-Reus, I., Iglesias, I. and Benedi, J.2003.Quercetin, a flavonoid antioxidant, prevents and protects against ethanol-induced oxidative stress in mouse liver. *Biol Pharm Bull.* 26: 1398–1402.

Montis, A., Souard, F., Delporte, C., Stoffelen, P., Stévigny, C. and Van Antwerpen, P. 2021. Coffee leaves: An upcoming novel food? *Planta Med.*

Morand, C., Manach, C., Crespy, V. and Remesy, C. 2000. Quercetin 3-O-β-glucoside is better absorbed than other quercetin forms and is not present in rat plasma. *Free Radic Res.* 33: 667–676.

Murota, K. 2020. Absorption pathway of dietary flavonoids: the potential roles of the lymphatic transport in the intestine. *Funct Foods Health Dis.* 10: 274–289.

Murphy, N., Achaintre, D., Zamora-Ros, R., Jenab, M., Boutron-Ruault, M.C., Carbonnel, F., Savoye, I., Kaaks, R., Kühn, T. and Boeing, H. 2018. A prospective evaluation of plasma polyphenol levels and colon cancer risk. *Int J Cancer.* 143: 1620–1631.

Nakagomi, M., Suzuki, E., Saito, Y. and Nagao, T. 2018. Endocrine disrupting chemicals, 4-nonylphenol, bisphenol A and butyl benzyl phthalate, impair metabolism of estradiol in male and female rats as assessed by levels of 15α-hydroxyestrogens and catechol estrogens in urine. *J Appl Toxicol.* 38: 688–695.

Naoi, M., Wu, Y., Shamoto-Nagai, M. and Maruyama, W. 2019. Mitochondria in neuroprotection by phytochemicals: bioactive polyphenols modulate mitochondrial apoptosis system, function and structure. *Int J Mol Sci.* 20: 2451–2451.

Nurgali, K., Jagoe, R.T. and Abalo, R. 2018. Adverse effects of cancer chemotherapy: Anything new to improve tolerance and reduce sequelae? *Front Pharmacol.* 9: 245.

Otręba, M., Kośmider, L., Stojko, J. and Rzepecka-Stojko, A. 2020. Cardioprotective activity of selected polyphenols based on epithelial and aortic cell lines. A review. *Molecules* 25: 5343.

Panche, A., Diwan, A. and Chandra, S. 2016. Flavonoids: An overview. *J Nutr Sci.* 5.

Phaniendra, A., Jestadi, D.B. and Periyasamy, L. 2015. Free radicals: Properties, sources, targets, and their implication in various diseases. *Indian Journal of Clin Biochem.* 30: 11–26.

Pressman, P., Clemens, R.A. and Hayes, A.W. 2017. Bioavailability of micronutrients obtained from supplements and food: A survey and case study of the polyphenols. *Toxicol Res Application* 1: 2397847317696366.

Pruchniak, M.P., Araźna, M. and Demkc, U. 2016. Biochemistry of oxidative stress. *Advances in Experimental Medicine and Biology*: *Adv Exp Med Biol.* 9:19.

Ray, S.K. and Mukherjee, S. 2021. Evolving interplay between dietary polyphenols and gut microbiota – An emerging importance in healthcare. *Front Nutr.* 8.

Reis, A., Perez-Gregorio, R., Mateus, N. and de Freitas, V. 2021. Interactions of dietary polyphenols with epithelial lipids: Advances from membrane and cell models in the study of polyphenol absorption, transport and delivery to the epithelium. *Critic Rev Food Sci Nutr.* 61: 3007–3030.

Ren, L., Wang, F., Xu, Z., Chan, W.M., Zhao, C. and Xue, H. 2010. GABAA receptor subtype selectivity underlying anxiolytic effect of 6-hydroxyflavone. *Biochem Pharmacol.* 79: 1337–1344.

Reygaert, W.C. 2018. Green tea catechins: Their use in treating and preventing infectious diseases. *BioMed Res Int.* 2018.

Rice-Evans, C.A., Miller, N.J. and Paganga, G. 1996. Structure-antioxidant activity relationships of flavonoids and phenolic acids. *Free Radic Biol Med.* 20: 933–956.

Riha, J., Brenner, S., Böhmdorfer, M., Giessrigl, B., Pignitter, M., Schueller, K., Thalhammer, T., Stieger, B., Somoza, V. and Szekeres, T. 2014. Resveratrol and its major sulfated conjugates are substrates of organic anion transporting polypeptides (OATPs): Impact on growth of ZR-75-1 breast cancer cells. *Mol Nutr Food Res.* 58: 1830–1842.

Ross, L., Barclay, C. and Vinqvist, M.R. 2009. Phenols as antioxidants. *PATAI'S Chemistry of Functional Groups*: John Wiley.

Ruiz, P.A., Braune, A., Hölzlwimmer, G., Quintanilla-Fend, L. and Haller, D. 2007. Quercetin inhibits TNF-induced NF-κB transcription factor recruitment to proinflammatory gene promoters in murine intestinal epithelial cells. *J Nutr.* 137: 1208–1215.

Ruiz, P.A. and Haller, D. 2006. Functional diversity of flavonoids in the inhibition of the proinflammatory NF-κB, IRF, and Akt signaling pathways in murine intestinal epithelial cells. *J Nutr.* 136: 664–671.

Russo, G.L., Spagnuolo, C., Russo, M., Tedesco, I., Moccia, S. and Cervellera, C. 2020. Mechanisms of aging and potential role of selected polyphenols in extending healthspan. *Biochem Pharmacol* 173: 113719–113719.

Salehi, B., Mishra, A.P., Nigam, M., Sener, B., Kilic, M., Sharifi-Rad, M., Fokou, P. V.T., Martins, N. and Sharifi-Rad, J. 2018. Resveratrol: A double-edged sword in health benefits. *Biomedicines* 6: 91.

Santana-Gálvez, J., Villela-Castrejón, J., Serna-Saldívar, S.O., Cisneros-Zevallos, L. and Jacobo-Velázquez, D.A. 2020. Synergistic combinations of curcumin, sulforaphane, and dihydrocaffeic acid against human colon cancer cells. *Int J Mol Sci.* 21: 3108.

Schieber, M. and Chandel, N.S. 2014. ROS function in redox signaling and oxidative stress. *Curr Biol.* 24: R453.

Shabbir, U., Tyagi, A., Elahi, F., Aloo, S.O. and Oh, D.H. 2021. The potential role of polyphenols in oxidative stress and inflammation induced by gut microbiota in Alzheimer's disease.*Antioxidants* 10: 1370.

Shankar, E., Goel, A., Gupta, K. and Gupta, S. 2017. Plant flavone apigenin: An emerging anticancer agent. *Curr Pharmacol Rep.* 3: 423–446.

Shi, Y., Xu, H., Gu, Z., Wang, C. and Du, Y. 2019. Sensitive detection of caffeic acid with trifurcate PtCu nanocrystals modified glassy carbon electrode. *Colloids Surf A: Physicochem Eng Asp.* 567: 27–31.

Singh, M., Kaur, M. and Silakari, O. 2014. Flavones: an important scaffold for medicinal chemistry. *Eur J Med Chem.* 84: 206–239.

Singla, R.K., Dubey, A.K., Garg, A., Sharma, R.K., Fiorino, M., Ameen, S.M., Haddad, M.A. and Al-Hiary, M. 2019. Natural polyphenols: Chemical classification, definition of classes, subcategories, and structures. *J AOAC Int.* 102: 1397–1400.

Sinopoli, A., Calogero, G. and Bartolotta, A. 2019. Computational aspects of anthocyanidins and anthocyanins: A review. *Food Chem.* 297: 124898.

Slauch, J.M. 2011.How does the oxidative burst of macrophages kill bacteria? Still an open question. *Mol Microbiol.* 80: 580–583.

Soares, S., Vitorino, R., Osorio, H., Fernandes, A., Venancio, A., Mateus, N., Amado, F. and de Freitas, V. 2011. Reactivity of human salivary proteins families toward food polyphenols. *J Agric Food Chem.* 59: 5535–5547.

Sridhar, A., Ponnuchamy, M., Kumar, P.S., Kapoor, A., Vo, D.N. and Prabhakar, S. 2021. Techniques and modeling of polyphenol extraction from food: A review. *Environ Chem Lett.* 1–35.

Stockert, A. and Hall, S. 2021. Therapeutic potential of dietary polyphenols. *Functional Foods – Phytochemicals and Health Promoting Potential*: IntechOpen.

Sung, H., Ferlay, J., Siegel, R.L., Laversanne, M., Soerjomataram, I., Jemal, A. and Bray, F. 2021. Global cancer statistics 2020: GLOBOCAN estimates of incidence and mortality worldwide for 36 cancers in 185 countries. *CA: Cancer J Clin.* 71: 209–249.

Tao, J.-h., Zhao, M., Wang, D.-g., Yang, C., Du, L.-Y., Qiu, W.-q. and Jiang, S. 2016. Biotransformation and metabolic profile of catalpol with human intestinal microflora by ultra-performance liquid chromatography coupled with quadrupole time-of-flight mass spectrometry. *J Chromatogr B.* 1009: 163–169.

Teng, H. and Chen, L. 2019. Polyphenols and bioavailability: An update. *Crit Rev Food Sci Nutr.* 59: 2040–2051.

Tresserra-Rimbau, A., Lamuela-Raventos, R.M. and Moreno, J.J. 2018. Polyphenols, food and pharma. Current knowledge and directions for future research. *Biochem Pharmacol.* 156: 186–195.

Tsao, R. 2010. Chemistry and biochemistry of dietary polyphenols. *Nutrients* 2, 1231–1246.

Tomás-Barberán, F.A., and Clifford, M.N. 2000. Flavanones, chalcones and dihydrochalcones – nature, occurrence and dietary burden. *J. Sci. Food Agric.* 80: 1073–1080.

Ulevitch, R. and Tobias, P. 1995. Receptor-dependent mechanisms of cell stimulation by bacterial endotoxin. *Ann Rev immunol.* 13: 437–457.

Van Buiten, C.B. and Elias, R.J. 2021. Gliadin sequestration as a novel therapy for Celiac disease: A prospective application for polyphenols. *Int J MolSci .* 22: 595.

Van Giau, V., Wu, S.Y., Jamerlan, A., An, S.S.A., Kim, S.Y. and Hulme, J. 2018. Gut microbiota and their neuroinflammatory implications in Alzheimer's disease. *Nutrients* 10:1765.

Verma, A.K. and Pratap, R. 2010. The biological potential of flavones. *Nat Prod Rep.* 27: 1571–1593.

Vogt, N.M., Romano, K.A., Darst, B.F., Engelman, C.D., Johnson, S.C., Carlsson, C.M., Asthana, S., Blennow, K., Zetterberg, H., Bendlin, B.B. and Rey, F.E. 2018. The gut microbiota-derived metabolite trimethylamine N-oxide is elevated in Alzheimer's disease. *Alzheimer's Res and Ther.* 10: 124.

Wadsworth, T.L., McDonald, T.L. and Koop, D.R. 2001.Effects of Ginkgo biloba extract (EGb 761) and quercetin on lipopolysaccharide-induced signaling pathways involved in the release of tumor necrosis factor-α. *Biochem Pharmacol.* 62: 963–974.

Wan, J.Y., Zhang, Y.Z., Yuan, J.B., Yang, F.Q., Chen, Y., Zhou, L.D. and Zhang, Q.H. 2017. Biotransformation and metabolic profile of anemoside B4 with rat small and large intestine microflora by ultra-performance liquid chromatography–quadrupole time-of-flight tandem mass spectrometry. *Biomed Chromatogr.* 31: e3873.

Ware, C.F. 2005. Network communications: Lymphotoxins, LIGHT, and TNF. *Annu. Rev. Immunol.* 23: 787–819.

Weng, Y.-P., Hung, P.-F., Ku, W.-Y., Chang, C.-Y., Wu, B.-H., Wu, M.-H., Yao, J.-Y., Yang, J.-R. and Lee, C.-H. 2018. The inhibitory activity of gallic acid against DNA methylation: application of gallic acid on epigenetic therapy of human cancers. *Oncotarget* 9: 361.

Willett, W.C.1994. Diet and health: What should we eat?*Science* 264: 532–537.

Wojtunik-Kulesza, K., Oniszczuk, A., Oniszczuk, T., Combrzyński, M., Nowakowska, D. and Matwijczuk, A. 2020. Influence of in vitro digestion on composition, bioaccessibility and antioxidant activity of food polyphenols – a non-systematic review. *Nutrients* 12: 1401.

Wu, M., Luo, Q., Nie, R., Yang, X., Tang, Z. and Chen, H. 2021. Potential implications of polyphenols on aging considering oxidative stress, inflammation, autophagy, and gut microbiota. *Crit Rev Food Sci Nutr* 61: 2175–2193.

Xiang, Q., Guo, W., Tang, X., Cui, S., Zhang, F., Liu, X., Zhao, J., Zhang, H., Mao, B. and Chen, W. 2021. Capsaicin – the spicy ingredient of chili peppers: A review of the gastrointestinal effects and mechanisms. *Trends Food Sci Technol.* 116: 755–765.

Xu, Y., Wang, Z., You, W., Zhang, X., Li, S., Barish, P.A., Vernon, M.M., Du, X., Li, G. and Pan, J. 2010. Antidepressant-like effect of trans-resveratrol: involvement of serotonin and noradrenaline system. *Eur Neuropsychopharmacol.* 20: 405–413.

Yang, H., Tuo, X., Wang, L., Tundis, R., Portillo, M.P., Simal-Gandara, J., Yu, Y., Zou, L., Xiao, J. and Deng, J. 2021a. Bioactive procyanidins from dietary sources: The relationship between bioactivity and polymerization degree. *Trends Food Sci Technol.* 111: 114–127.

Yang, S., Zhang, Y., Li, W., You, B., Yu, J., Huang, X. and Yang, R. 2021b. Gut microbiota composition affects Procyanidin A2-attenuated atherosclerosis in ApoE–/–Mice by modulating the bioavailability of its microbial metabolites. *J Agric Food Chem.*

Young, I. and Woodside, J. 2001. Antioxidants in health and disease. *J Clin Pathol.* 54: 176–186.

Zeng, Z., Centner, C., Gollhofer, A. and Konig, D. 2021. Effects of dietary strategies on exercise-induced oxidative stress: A narrative review of human studies. *Antioxidants* (Basel) 10 .

Zhang, A. H., Ma, Z. M., Kong, L., Gao, H. l., Sun, H., Wang, X. Q., Yu, J. B., Han, Y., Yan, G. l. and Wang, X. J. 2020. High-throughput lipidomics analysis to discover lipid biomarkers and profiles as potential targets for evaluating efficacy of Kai-Xin-San against APP/PS1 transgenic mice based on UPLC–Q/TOF–MS. *Biomed Chromatogr.* 34: e4724.

Zhang, G., Gong, T., Kano, Y. and Yuan, D. 2014. Screening for in vitro metabolites of kakkalide and irisolidone in human and rat intestinal bacteria by ultra-high performance liquid chromatography/quadrupole time-of-flight mass spectrometry. *J Chromatogr B.* 947: 117–124.

Zhang, H. and Tsao, R. 2016. Dietary polyphenols, oxidative stress and antioxidant and anti-inflammatory effects. *Curr Opini Food Sci.* 8: 33–42.

Zhao, L., Chen, S., Liu, T., Wang, X., Huang, H. and Liu, W. 2019. Callistephin enhances the protective effects of isoflurane on microglial injury through downregulation of inflammation and apoptosis. *Mol Med Rep.* 20: 802–812.

Zheng, J., Cheng, J., Zheng, S., Feng, Q. and Xiao, X. 2018. Curcumin, a polyphenolic curcuminoid with its protective effects and molecular mechanisms in diabetes and diabetic cardiomyopathy. *Front Pharmacol.* 9: 472–472.

Zhou, Y., Jiang, Z., Lu, H., Xu, Z., Tong, R., Shi, J. and Jia, G. 2019. Recent advances of natural polyphenols activators for Keap1-Nrf2 signaling pathway. *Chem Biodivers.* 16: e1900400.

Zuiter, A.S. 2014. *Proanthocyanidin: Chemistry and Biology: From Phenolic Compounds to Proanthocyanidins.* Elsevier.

3 Dietary Polyphenols in Aging, Neurological, and Cognitive Disorders

*Lalit Mohan Nainwal and *Poonam Arora

*E-mail: lalit12331@yahoo.com; poonamarora96@gmail.com

CONTENTS

3.1 INTRODUCTION

Neurodegenerative disorders (NDs), collectively referred to as group of diseases, belongs to heterogeneous and multifactorial pathologies that affect different parts and structure of the brain ,but ultimately leads to debilitating and life-threatening conditions. All these disorders are generally chronic and progressive, having complex pathogenesis and mechanisms, leading to neuronal cell death that causes nervous system dysfunction (Brettschneider et al., 2015).

There are more than 600 CNS disorders that come under the umbrella of neurological disorders; the most common are Parkinson's disease (PD), Alzheimer's disease (AD), schizophrenia, epilepsy, prion disease, encephalitis, stroke, and Huntington's disease (WHO, 2021). Most of these diseases share common pathological and molecular signaling pathways such as neuro-inflammation, decreased level of antioxidant enzymes, up-regulated oxidative-stress and apoptosis of neurons in the brain.

3.1.1 PARKINSON'S DISEASE (PD)

The term "Parkinson's disease" refers to a group of neurodegenerative disorders of the extrapyramidal system, characterized by the loss of dopaminergic function with consequent loss of motor function, resulting in development of the disease (Schrag et al., 2015). PD affects approximately 1 percent of the population by the age of 65, increasing to 4–5 percent by the age of 85 years. In 2020, 9.4 million people globally suffered with PD (Kouli, et al., 2018), with an increasing trend, over the past 30 years. The percentage of sufferers aged above 80 years increased in the population (Tunc et al., 2020).

DOI: 10.1201/9781003251538-3

Clinically, PD is characterized by four cardinal signs of physical disabilities, akinesia, bradykinesia, ataxia, tremor, rigidity, and postural instability at rest. In large part, these symptoms result from the degeneration of dopaminergic neurons in substantia nigra (SNpc) of the midbrain by 70 percent, with concomitant loss of their axons that project to the striatum along the nigrostriatal pathway. This results in deficiency of the neurotransmitter dopamine in brain areas receiving dopamine, particularly, basal ganglia regions including neostriatum, the subthalamic nucleus (STN), the external and internal pallidal segments, post-commissural putamen and the substantia nigra with its pars reticulata and pars compacta (ThomasWichmann, 2019). The progressive loss of dopamine in the striatum of individuals with PD leads to increased activity in the GPi/SNpr pathways and concomitant dysfunction of gamma aminobutyric acid (GABA) (Salin et al., 2002).This inhibits thalamus from activating the functions of motor activity and frontal cortex in PD patients.

In addition, abnormalities in other brain areas, including midbrain, brainstem, the cerebral cortex, and elements of the peripheral nervous system also contribute in pathogenesis of PD (ThomasWichmann, 2019). Binding of presynaptically released dopamine to the postsynaptic membrane activates dopamine receptors. However, monoamine oxidase (MAO) and catechol-O-methyl transferase break down the dopamine, rendering it inactive dopamine. The MAO inhibitor impedes the action of MAO enzymes and prevents the breakdown of dopamine, thereby allowing more dopamine to remain in the synapse and binding to the postsynaptic membrane (Riederer and Laux, 2011). In severe PD, the depletion of dopamine alters the function of basal ganglia neurotransmitters, such as GABA, glutamate, and serotonin (Gasparini et al., 2013).

In patients with progressive PD, the protein inclusions, Lewy bodies, are seen inside nerve cells. These bodies are abnormal aggregates and consist of insoluble fibrillary aggregates containing misfolded proteins, including α-synuclein (α-syn) (Spillantini et al., 1997). Other factors, recognized as neuroinflammation, and altered mitochondrial function, are also reported to result in vascular leakiness in PD development (Williams-Gray et al., 2016). Risk factors for the development of PD include oxidative stress, production of free radicals, and a number of environmental toxins (Zhou et al., 2008). Although environmental risk factors for PD have received significant interest, genetic factors underlying development of PD cannot be ignored. Several genes such as synuclein, UCHL1 (a ubiquitin carboxy-terminal hydrolase L1), parkin, DJ1 (a parkin associated protein involved with oxidative stress), and PINK1 (a putative serine threonine kinase) have shown a possible role in early-onset PD.

Considering complex and unresolved molecular cellular and molecular mechanisms involved in the path genesis of PD, there, are no specific laboratory tests or radio-imaging measures that confirm the PD. At the clinical desk, the diagnosis of disease is made by only analyzing the motor characteristics (Marino et al., 2020). The primary goal in the management of PD aims to control the symptoms to improve the mobility and life expectancy ratio and overall quality of the treated patients. The treatment consists of single or combination of dopaminergic agents with levodopa, COMT inhibitors, anticholinergics drugs, and MAO-B inhibitors to maximize clinical outcomes (Marino et al., 2020).

To date, treatment of PD focuses primarily on controlling motor signs with the use of dopaminergic medications. Lately, finding the effects of non-motor pathways on multiple body systems in PD (Todorova et al., 2014) has posed a challenge for developing novel therapeutic approaches that could treat multiple symptoms of the illness.

3.1.2 ALZHEIMER DISEASE (AD)

Alzheimer disease (AD) is an age-related neurodegenerative disorder associated with dementia and cognitive impairment. As per report of the Centers for Disease Control and Prevention (CDC), approximately, 5.8 million Americans were diagnosed with Alzheimer's disease in year 2020, and this number is expected double every five years (Kuman et al., 2018). AD affects the cortical and hippocampus regions of the brain, thereby disrupting cholinergic neuronal transmission. Formation of neurofibrillary tangles, composed mainly of modified tau protein, are the peculiar features of AD. The hyper-phosphorylation of tau protein is essential for disease pathology. In Alzheimer's disease, extracellular β-amyloid plaques are formed due to abnormal action of amyloid precursor protein (APP). The deposition of β-amyloid in the brain of the AD patient occurs at an earlier clinical stage, known as *mild neurocognitive disorder*. Positron Emission Tomography (PET) of an Alzheimer brain presents acetylcholine (ACh), dopamine, β-amyloid, and tau proteins as important clinical biomarkers for the diagnosis of AD (Zafari et al., 2015). In general, two types of Alzheimer's disease, sporadic and familiar, are known; 95 percent of sporadic and 5 percent cases of familiar are present in Alzheimer patients. The cases identified by gene mutation in Alzheimer's disease show alterations in the gene expression of APP, presenilin 1 and 2, and tau protein. mutation in APP leads to production of $A\beta_{40}$ and $A\beta_{42}$ peptides, and this $A\beta_{42}$ forms protofibrils and fibrils. The precursor protein and tau protein-encoding genes are involved in the sporadic cases. Two types of Aβ are seen in AD, and $A\beta_{1-42}$ is more toxic for AD progression due to its long peptide. Diminished activity of Cyclooxygenase (COX) in early-onset sporadic AD patients under the age of 60 has been (Reddy and Reddy, 2011) largely reported.

Binding of heme-α to Aβ interferes with the function of COX. Increased expression of Aβ in the cerebral cortex of transgenic mice has shown to attenuate the function of COX mediated via the amyloid precursor protein (APP)/ Presenilin proteins 1 (PSEN1) mechanism.

Increasing evidence suggests that intercellular accumulation of β-amyloid can damage OXPHOS (oxidative phosphorylation) and increase ROS production in mitochondria. Increased oxidative stress alters mitochondrial membrane potential and disrupts complex IV function (cyto-oxidase) and Adenosine Triphosphate (ATP) production (Reddy and Reddy, 2011). Altered mitochondrial redox potential causes tau-induced neurotoxicity in experimental animals (De Castro et al., 2011). Excessive generation of ROS causes oxidative stress and cell death in Alzheimer's disease.

In the human central nervous system, cholinergic synapses are densely present in the thalamus, striatum, limbic system, and neocortex, demonstrating the crucial role of cholinergic transmission in several functions of the brain, including dementia, Alzheimers, and age-related cognitive diseases. The primary cause for functional loss of cholinergic neurons in the forebrain is stated to be neurofibrillary degeneration in the basal forebrain, resulting in presynaptic cholinergic denervation in the nucleus basalis of Meynert (Md Torequl Islam, 2017). Growing pieces of evidence report that nearly 100,000 cholinergic neurons located in the nucleus basalis of Alzheimer patients are damaged in AD.

The most likely expected treatment strategies include AChE inhibitors, Aβ inhibitors/vaccination, secretase inhibitors, Cu-Zn chelators, non-steroidal anti-inflammatory drugs (NSAIDs) (Cai and Tammineni, 2017). To date, only six drugs are approved by the US Food and Drug Administration (FDA), including five anticholinergics and lately approved monoclonal antibody (Christopher H. van Dyck, 2018; Prins and Scheltens, 2013). These drugs primarily slow down progression of cognitive symptoms in AD patients, Therefore, much effort is being directed towards the novel pharmacological therapies that can inhibit the molecular signaling pathways and thereby block progression of clinical symptoms.

3.1.3 DEPRESSION

Depression is considered as psychotic illness, affecting approximately, 3.8 percent of the population worldwide, including 5.7 percent of adults older than 60 years. Studies report that more than 350 million people of all ages suffer from depression, and over 700,000 people commit suicide every year. Suicide is the fourth leading cause of death in 15–29-year-olds (Evans-Lacko et al., 2018). Depression, as described by the World Health Organization (WHO), is a mood disorder characterized by specific symptoms including sadness, loss of interest and pleasure (anhedonia), lack of appetite, low self-esteem or self-worth, sleep disturbance, feelings of tiredness, and poor concentration (WHO, 2021). Individuals living with depression describe varying degrees of helplessness and hopelessness, accompanied by thoughts of death. Depression can be recurrent or long-lasting and results in substantial impairment in one's ability to function.

This major depressive disorder, characterized by the presence of depressed mood or loss of interest (or both) for at least two weeks, is believed to be influenced by genetic and biological factors, while minor depression is considered as result of life events. With increasing understanding of pathophysiology of disease, depression is recognized as a heterogeneous disorder with a complex phenomenon and multiple etiologies, including dysregulation of the hypothalamic-pituitary-adrenal (HPA) axis with hypothesis of biogenic amines and inclusion of genetic and environmental factors (Jesulola et al., 2018; Schmidt et al., 2011). In addition, other biological factors, such as neurogenesis, disrupted inflammatory cytokines secretion, elevated levels of corticotrophin-releasing factor (CRF) contribute to development of depression (Hepgul et al., 2013).

Alterations in monoaminergic molecule levels are associated with abnormal behavioral symptoms of depression, such as low mood, vigilance, reduced motivation, fatigue, and psychomotor agitation or retardation while changes in cerebral 5-HT levels are responsible for changes in behavioral and somatic functions (including body temperature, appetite, sleep, sex, pain response, and body circadian rhythm) that are seen in depression. Dopamine abnormalities have been linked to impaired motivation, concentration and aggression, while dopamine transmission may improve cognitive outcomes including decision making and motivation. Increased activity of monoamine oxidase (MAO) enzyme is measured (Ramesh et al., 2017). This leads to reduction in the biogenic amines, thereby resulting in the decreased transmission of neurotransmitters, including glutamate, NE and histamine and primarily, 5HT. Abnormal neurotransmitter serotonin receptors (5-HT1 and 5-HT2) or presynaptic β2- adrenoceptors, G-protein and protein receptors transport protein function, contribute to lower monoamine neurotransmitter levels.

Stress has been considered as another precipitating factor of depression in depressed individuals. Stress-response system in mammals include prefrontal cortex, amygdala, hippocampus and nucleus accumbens of the brain. Further, depressive symptoms may be produced via peripheral inflammatory cytokines crossing the blood–brain barrier (BBB) to induce neuroinflammation. These include interleukins IL-1ß, IL-2, IL-4, IL-6, IL-8, IL-10 and interferon gamma, CRP, TNF-α, and chemokine monocyte chemoattractant protein-1 (McCusker et al., 2013; Strawbridge et al., 2018).

Though well known treatments are available for depression, more than 75 percent of the world population receives no treatment. There is no clinically effective treatment for mild, moderate, and severe depression. People are often misdiagnosed and randomly given antidepressants for psychological problems such as behavioral activation, cognitive behavioral therapy and interpersonal psychotherapy. Drugs most commonly prescribed are selective serotonin reuptake inhibitors (SSRIs) and tricyclic antidepressants (TCAs).

3.1.4 SCHIZOPHRENIA

Schizophrenia affects approximately 0.32 percent of the world population. This rate is 1 in 222 people among adults. The disease is not as common as many other psychotic disorders. Schizophrenia is classified as psychotic or positive symptoms, including delusions and hallucinations, where patients feel aloof regarding reality, while negative symptoms are characterized by feelings of social withdrawal, impaired motivation, a drop in spontaneous speech. Cognitive symptoms comprise disorganized speech, attention, and thought, eventually impairing the person's capability to communicate (Dennison et al., 2021). Environmental factors contributing to schizophrenia-like clinical symptoms include minority ethnicity, economic adversity, social isolation, childhood trauma (Levit et al., 2021).

Pathophysiology of schizophrenia is described mainly as disruption in the normal transmission of neurotransmitters, including dopamine, serotonin, and glutamate with addition of aspartate, glycine, and gamma-aminobutyric acid (GABA) (Jia et al., 2021). Abnormal activity of dopamine receptors, specifically D2, is an important part of the neurochemical imbalance in schizophrenia (Yüksel et al., 2021). The prognosis for patients with schizophrenia is usually found to be unpredictable. Approximately, 20 percent of patient's experience favorable treatment results. The patients experience many psychotic episodes, long-term symptoms, and a meager response to antipsychotics. Diagnosis of schizophrenia is based on assessing patient-specific signs and symptoms, as described in the Diagnostic and Statistical Manual of Mental Disorders: persistence of two or more active-phase symptoms that last for a minimum period of one month: delusions, hallucinations, disorganized speech, catatonic behavior.

The major goals of treatment for schizophrenia are based on both pharmacological and non-pharmacological therapies. The aim is to prevent relapse and increase adaptive functioning (Campana et al., 2021). Pharmacological therapy starts with prescribing first- or second-generation antipsychotics in acute phases or use of both as combination therapy. However, the adverse effects can contribute to the increased risk of cardiovascular mortality observed in schizophrenia patients (De Hert et al., 2012).

3.2 BIOAVAILABILITY OF DIETARY POLYPHENOLS

Therapeutic potential and health benefits of polyphenols has been extensively studied by researchers seeking to understand their detailed mechanism of action (Peterson et al., 2012). Bioavailability of dietary polyphenols have been extensively studied in the past using *in vitro* and *in vivo* studies on animals and humans. A very slight change in molecular structure of different polyphenols completely alter and give a great difference in their metabolic fate (Figueira et al., 2017).

The absorption and biological activity of polyphenols were always questionable because of their fast absorption, extensive metabolism, and rapid excretion rate (Alminger et al., 2014). However, various human clinical trials have found that polyphenols express a high volume of distribution and absorption and remain present for longer durations in different organs and tissues, thus exerting their therapeutic effects. Glycosides of polyphenolic compounds are less absorbed as compared to their aglycone part due to low lipophilic character. Polyphenolic glycosides are hydrolysed in the gastro-intestinal tract and converts into their more absorbable aglycone part. Duodenum and proximal jejunum part of the small intestine are the major sites of absorption of polyphenols (Alminger et al., 2014). Enterocytes present in the small intestine plays an active role in absorption of polyphenols. In their aglycone form, polyphenols have easier entry into the circulation through passive diffusion (Tian et al., 2009). Absorption of glycosylated polyphenols are absorbed through epithelial cells of the small intestine through ATP-dependent active transport channels like sodium-dependent glucose transporter, (SGLT1). Hydrolysis of glycosides are also done by cytosolic β-glucosidase (CBG), and enzyme lactase phloridzin hydrolase (LPH), depending on the glycoside (Day et al., 2000; Gee et al., 2000). Highly hydrolysis-resistible polymeric flavanols exhibit poor to low absorption (Serra et al., 2010). Unabsorbed fractions of compounds get absorbed in the large intestine, however, intestinal microbiota convert these polyphenols into simpler low molecular weight phenols, acids and oxaloacetates (Borges et al., 2013).

After absorption, dietary polyphenols faced extensive phase I and phase II metabolism known as biotransformation reactions and are converted into highly hydrophilic compounds. These metabolites could be easily excreted from the body with the help of phase III transporters, such as P-glycoprotein (P-gp), solute carrier family (SLC) transporters, multidrug resistance-associated proteins (MRP2, BCRP); this helps in detoxification/elimination of undesired compounds through urine, bile and faeces. Metabolites also enter into enterohepatic recirculation through bile and get reabsorbed through the duodenum, thus showing biphasic absorption (Döring and Petzinger, 2014; Xu et al., 2005; Yoshida et al., 2013). Therefore, bioavailability of polyphenols is a multistage and complex process and currently not fully understood, involving absorption, metabolism, conjugation, de-conjugation, catabolism, and excretion. Due to the high volume of distribution, the exact amount of polyphenol deposited in different organs after absorption has not yet been fully studied, thus it is very dubious to give a negative comment on the exact bioavailability of absorbed polyphenols.

The presence of polyphenols in the central nervous system is also an interesting and attractive issue for researchers. The blood–brain barrier penetration, concentration of several polyphenolics in the brain has been extensively studied through in vitro, in vivo animals and in humans during clinical studies. Neurotherapeutic applications of several polyphenols and crude extracts of plants have been studied in the past by several researchers.

3.3   POLYPHENOLS BEYOND BLOOD–BRAIN BARRIER

The exchange of molecular species between blood and neuronal tissues of the brain is regulated by three layers of barriers constituted of (a) *blood–brain barrier* is a complex dynamic and powerful thick lining of cerebrovascular endothelial cells that separate blood and interstitial fluid of brain and is selective for toxicants and pathogens thus protect CNS from these substances; (b) the *choroid plexusepithelium* responsible for exchange and transport of inorganic ions, amino acids, vitamins, hormones, and small peptides to cerebrospinal fluid; and (c) the arachnoid epithelium.

Various efflux pumps like P-gp, BCRP, MRP2 are also present in the BBB that efflux out small polar and macromolecules (Campos-Bedolla et al., 2014). Besides all of these restrictions where it is impossible for polyphenols like polar molecules to cross such barriers, the presence of polyphenols in the brain is amazing and has attained attention of researchers across the world in last few years. Membrane transport modulation capacity of polyphenols also has been studied.

The BBB penetration and permeability of polyphenols has been extensively studied by Youdim et al. (2004a, 2004b, 2003). They found that lipophilic conjugates of polyphenol such as methylated conjugates could easily penetrate and attain higher concentration in the brain than their polar sulphate and glucuronide conjugates. However, it is still a mystery how polyphenols acquire entrance inside the brain and how they interact with the BBB. Therefore, all receptors and transporters that are present in the epithelial membranes must be considered while studying the penetration and interactions of polyphenolic compounds with BBB and other CNS barriers (Ziberna et al., 2014). Still, there are several strong evidences from in vitro, in vivo studies that polyphenols effectively reach the brain and exert neuroprotection (listed in Table 3.1).

3.4   EVIDENCE-BASED UTILITY OF DIETARY POLYPHENOLICS IN CLINICAL TRIALS

This section provides an overview and emphasizes the clinical potentials of dietary polyphenolics in neuroprotection, enhancement in memory and cognitive functions of on human subjects. This section also gives clear evidence about the potential application of dietary polyphenolics in the treatment of various neurological diseases/disorders (Tables 3.2 and 3.3).

Most of the clinical studies discussed in Tables 3.2 and 3.3 support the use of flavonoids in the management and prevention of cognitive disorders. Acute improvement in the memory functions also has been established with consumption of flavonoids over a short span of time. However, long-term and consistent consumption of flavonoids was required to produce more specific improvement

TABLE 3.1
Blood-brain barrier(BBB) penetrability, neuroprotective potential of several dietary polyphenols and their interaction different types of active drug transporters present at BBB

Compound	Role	Interaction with efflux pumps	Study model	Reference
Apigenin	BBB penetrative, neuroprotective (Yang et al., 2014)	P-gp, and BCRP inhibitor	In vitro Rat BMEC, in vitro CEM/ADR5000 and CCRF-CEM leukemia cells and in vitro BCRP positive MDA-MB-231 cancer cells	(Yang et al., 2014; Saeed et al., 2015)
Catechin/ Epicatechin	BBB penetrative, neuroprotective (Bitu et al., 2015; Hussein et al., 2015; Faria et al., 2011)	P-gp activator, MRP1/MRP2 substrate, MRP2 substrate	In vitro study using NIH-3T3-G185 cells, MDCKII/MRP1 cells & MDCKII/MRP2 cells and Caco-2 cells	(Wang et al., 2012; Vaidyanathan and Walle, 2001; Hong et al., 2003)
Curcumin	BBB penetrative, neuroprotective (Wu et al., 2015; Tsai et al., 2000)	P-gp and BCRP inhibitor	In vitro studies using rat brain capillaries and MCF-7 cells	(Wang et al., 2014; Shukla et al., 2009)
Fisetin	BBB penetrative, neuroprotective (Lapchak, 2013)	MRPs inhibitor	In-vitro studies using Caco-2 cells	(Yoshimura et al., 2013)
Genistein	BBB penetrative, neuroprotective (Huang and Zhang, 2010; Yu et al., 2009)	P-gp, MRP2, BCRP inhibitor	MCF7/BC19-3, rats, Transgenic-rats in-vitro using K562/BCRP and MCF7/BC19-3 cells and in vivo studies on rats	(Castro and Altenberg, 1997; Imai et al., 2004)
Hesperitin	BBB penetrative, neuroprotective (Youdim et al., 2003)	BCRP and P-gp inhibitor	In vitro study using primary mouse brain microvascular endothelial cells (BMEC), BCRP-overexpressing epithelial breast cancer cell line MCF/MR and BCRP transfected human myelogenous leukemia cell line K562/BCRP	(Cooray et al., 2004; Mitsunaga et al. 2000)
Kaempferol	BBB penetrative, neuroprotective (Yang et al., 2014; Lagoa et al., 2009)	MRPs and P-gp, BCRP inhibitor	In vitro study using primary mouse brain microvascular endothelial cells (BMEC), human glioblastoma cell line T98G, and wild-type MDCK (Madin-Darby canine kidney) cells having drug resistant BCRP1 gene expression	(Mitsunaga et al. 2000; Nakatsuma et al., 2010; An et al., 2011)
Myricetin	BBB penetrative, neuroprotective (Lei et al., 2012)	P-gp and BCRP inhibitor	In vitro study using multidrug resistant MCF-7/ADR cells and human embryonic kidney (HEK293) cells with ABCG2 expression	(Choi et al., 2011; Tan et al., 2013)
Naringenin	BBB penetrative, neuroprotective (Sarkar et al., 2012; Zbarsky et al., 2005)	P-gp and MRPs inhibitor	In vitro study using primary mouse brain microvascular endothelial cells and cells and human embryonic kidney (HEK293) cells with ABCG2 expression	(Mitsunaga et al. 2000; Tan et al., 2013)

(continued)

TABLE 3.1 (Continued)
Blood-brain barrier(BBB) penetrability, neuroprotective potential of several dietary polyphenols and their interaction different types of active drug transporters present at BBB

Compound	Role	Interaction with efflux pumps	Study model	Reference
Quercetin	BBB penetrative, neuroprotective (Youdim et al., 2004; Lei et al., 2015)	P-gp, BCRP and MRPs inhibitor	In vitro study using primary mouse brain microvascular endothelial cells, multi target resistant receptor (MRP)-1, 2, 3, 4, 5 over-expressed human embryonic kidney cells like HEK293/MRP1, HEK/MRP4, HEK/MRP5 cells	[97, 98, 109] (Mitsunaga et al. 2000; Cooray et al., 2004; Wu et al., 2005)
Resveratrol	BBB penetrative, neuroprotective (Vingtdeux et al., 2010; Narayanan et al., 2015)	BCRP and P-gp inhibitors	In vitro study using multidrug resistant MCF-7/ADR cells and human embryonic kidney (HEK293) cells with ABCG2 expression	(Cooray et al., 2004; Choi et al., 2009)
Rutin	BBB penetrative, neuroprotective (Yang et al., 2014; An et al., 2011)	P-gp inhibitor	In vitro study using primary mouse brain microvascular endothelial cells (BMEC)	(Yang et al., 2014)
Ferulic acid	BBB penetrative, neuroprotective	--	In vitro studies using SH-SY5Y neuroblastoma cells and differentiated rat pheochromocytoma (PC12) cells	(Catino et al., 2016; Yin et al., 2019)
Caffeic acid	BBB penetrative, neuroprotective	--	PC12 cells	(Jeong et al., 2011)
Luteolin	BBB penetrative, neuroprotective	--	In vivo rat model and in in vitro studies using rat pheochromocytoma PC12 cells	(Woo et al., 2013; Guo et al., 2013; Zhang et al., 2017)
Epigallocatechin-3-gallate (EGCG)	BBB penetrative, neuroprotective	--	In vivo mice model and during in vitro study using human neuroblastoma SH-SY5Y cells	(Pervin et al., 2017)
Daidzein	BBB penetrative, neuroprotective	--	In vitro study using human derived Hypothelamic GnRH-neurons	(Morelli et al., 2021; Johnson et al., 2020)
Equol	BBB penetrative, neuroprotective	--	In vitro studies using murine microglial BV2 cells and neuroblastoma SH-SY5Y Cells	• (Johnson et al., 2020)
Genistein	BBB penetrative, neuroprotective	--	In vivo studies in rats	(Tung-Hu Tsai et al., 2005; Duan et al., 2021)
Nobiletin	BBB penetrative, neuroprotective	--	In vitro studies using primary cortical neurons	(Lee et al., 2018)
Gallic acid	BBB penetrative, neuroprotective	--	In vitro study using BBB kit (RBT-24) comprises of brain pericytes, capillary endothelial cells and astrocytes.	(Sun et al., 2014)

TABLE 3.2

Dietary polyphenolic compounds in clinical trials against cognition

Disease/ Indication	Intervention/ Treatment	Study size	Study design	Result	Year
Cognitive function	Consumption of flavonoid-rich foods	1091 participants of mean age around 70 years	CSS	No significant improvement in cognitive function with consumption of flavonoid rich foods	2011 (Butchart et al., 2011)
Cognitive function	Consumption of equol	152 participants of mean age of 69.2 years	CSS	Soy consumption improved the cognitive performance	2017 (Igase et al., 2017)
Cognitive function	Consumption of tofu	517 participants of age between 50-98 years	CSS	High tofu intake decrease memory function in elder people.	2015 (Xu et al., 2015)
Cognitive function	Consumption of chocolate	968 participants of age between 28-98 years	CSS	Consumption of chocolate improve cognition power	2016 (Crichton et al., 2016)
Dementia and AD	Consumption of flavonoids for 9.6 years	968 participants of age between 28-98 years	CS	No correlation between consumption of flavonoids and AD	2010 (Devore et al., 2010)
ADRD andAD	Consumption of flavonoids for 19.7 years	2801 participants of age above 50 years	CS	An association between higher long-term flavonoid intake and lower risks of ADRD and AD	2020 (Shishtar et al., 2020)
Cognitive function	Consumption of flavonoids for 4 year	16010 participants of age greater than 70 years	CS	An association between higher flavonoid intake and decreased rates of cognitive decline	2012 (Devore et al., 2012)
Cognitive function	Consumption of cocoa polyphenols for a period of 1 year	55 participants of age between 56-75 years with Amnesia	CS	Consumption of cocoa polyphenols decreased the progression of dementia	2019 (Calabrò et al., 2019)
Cognitive function	Consumption of caffeine and chocolate for a period of 4 years	531 participants of age above 65 years	CS	Cognitive impairment was resolved with daily intake of chocolate/ caffeine (greater than 75 mg)	2016 (Moreira et al., 2016)
Cognitive function	Consumption of fruits and vegetables for a period of 13 years	2533 participants of age between 45-60 years	CS	Correlation of fruit and vegetable intake with verbal memory scores, and a negative association with executive functioning scores	2011 (Péneau et al., 2011)

(continued)

TABLE 3.2 (Continued)
Dietary polyphenolic compounds in clinical trials against cognition

Disease/ Indication	Intervention/ Treatment	Study size	Study design	Result	Year
AD	Consumption of strawberry for a period of 6.7 years	925 participants of age between 58-98 years	CS	Associations of strawberries and foods rich in anthocyanidins ,and total flavonoids consumption with a decreased risk of AD	2019 (Agarwal et al., 2019)
Cognitive function	Consumption of fruit and vegetable for a period of 5 years	2613 participants of age between 43-70 years	CS	No significant correlation between intake of and improvement in cognitive function	2011 (Nooyens et al., 2011)
Cognitive function	Consumption of soy protein rich in isoflavones for 2.5 years	350 participants of age between 45-92 years	RCT, DB, PC	No improvement in cognitive functions.	2012 (Henderson et al., 2012)
Cognitive function	Isoflavone (80 mg/ day) for 4 months	38 participants of age between 50-65 years	RCT, DB, PC	Improvement in cognitive functions.	2010 (Santos-Galduróz et al., 2010)
Cognitive function	Consumption of single dose of ECGC (135 mg or 270 mg)	30 participants of age between 18-30 years	RCT, DB, PC, CO	No significant changes in cognitive function	2012 (Wightman et al., 2012)
Cognitive function	Consumption of 2 g of green tea as a source of catechins (227 mg/day) and theanine (42 mg/ day)	12 participants of mean age 88 years	PS	Improved cognitive function	2014 (Ide et al., 2014)
Cognitive function	Consumption of 2 g of green tea powder as a source of catechin (220.2 mg/ day) for 1 year	33 participants of mean age of 84.8 years	RCT, DB, PC	Cognitive function was not improved	2015 (Ide et al., 2015)
Cognitive function	Consumption of catechin capsules (336.4 mg/day) as supplement for 12 weeks	52 participants of age between 50-69 years	RCT, DB, PC	Long term consumption improves the cognitive functions	2020 (Baba et al., 2020)
Cognitive function	Consumption of cocoa flavanols (994 mg /day)	30 participants of age between 18-35 years	RCT, DB, PC, CO	Acute enhancements in cognitive behavior	2010 (Scholey et al., 2010)
Cognitive function	Consumption of 35 g of dark chocolate as a source of cocoa flavanols (720 mg/day)	30 participants of age between 18-25 years	RCT, single-blinded, CO	Acute enhancements in cognitive behavior and memory.	2011 (Field et al., 2011)

TABLE 3.2 (Continued)
Dietary polyphenolic compounds in clinical trials against cognition

Disease/ Indication	Intervention/ Treatment	Study size	Study design	Result	Year
Cognitive function	Consumption of 35 g of dark chocolate as a source of cocoa flavanols (900 mg/day) and (-) epicatechin (138 mg/ day) with or without aerobic exercise for 3 months	37 participants of age between 50-69 years	RCT, DB	Perfections in dentate gyrus function	2014 (Brickman et al., 2014)
Cognitive function	Consumption of cocoa flavanols (993 mg/ day), for 2 weeks	90 participants of mean age 69.55 years	RCT, DB	Improvement in verbal fluency and train making test A and B	2015 (Mastroiacovo et al., 2015)
Cognitive function	Consumption of cocoa flavanols (990 mg/ day), for 2 weeks	90 participants of mean age 71.17 years	RCT, DB	Improvement in verbal fluency and train making test A and B	2012 (Desideri et al., 2012)
Cognitive function	Consumption of cocoa flavanols (250 and 500 mg/ day) for 1 month	78 participants of age between 40-65 years	RCT, DB, PC	No improvement in cognitive function	2013 (Pase et al., 2013)
Cognitive function	Consumption of 240 ml of orange juice as a source of flavonoids (272 mg/day)	24 participants of age between 30-65 years	RCT, DB, PC, CO	Acute enhancement in cognitive function	2016 (Alharbi et al., 2016)
Cognitive function	Consumption of 230 ml of purple grape juice as a source of anthocyanins (32 mg/day)	20 participants of age between 18-35 years	RCT, DB, PC, CO	Acute improvements in aspects of cognition	2017 (Haskell-Ramsay et al., 2017)
Cognitive function	Consumption of 500 ml of citrus juice as a source of flavonoids (70.5 mg/day)	24 participants of age between 18-30 years	RCT, SB, PC, CO	Acutely improved performance on the Digit Symbol Substitution Test, no improvement in other cognitive tests.	2016 (Lamport et al., 2016)
Cognitive function	Consumption of 60 ml of montmorency tart cherry concentrate, as source of cyanidine-3-glycoside	27 participants of age between 45-60 years	RCT, DB, PC, CO	No significant changes in cognitive function	2016 (Keane et al., 2016)

(*continued*)

TABLE 3.2 (Continued)
Dietary polyphenolic compounds in clinical trials against cognition

Disease/ Indication	Intervention/ Treatment	Study size	Study design	Result	Year
Cognition function	Consumption of 24 g of freeze- dried blueberry extract as a source of phenolics (864 mg/day) and anthocyanin (460.8 mg/day) for 3 months	37 participants of age between 60-75 years	RCT, DB, PC	Improvements in executive function	2018 (Miller et al., 2018)
Cognition function	Consumption of 30 ml of blueberry concentrate as a source of anthocyanidins (387 mg/day) for 3 months	26 participants of age above 65 years	RCT, DB, PC	Improvements in working memory and brain activation	2017 (Bowtell et al., 2017)
Memory function	Consumption of 444-621 ml of wild blueberry juice per day for 3 months	9 participants of mean age between 76.2 years	Preliminary study	Improvements in paired associate learning and wordlist recall	2010 (Krikorian et al., 2010)
Dementia using blood oxygen level-dependent signal	Consumption of 25 g of blueberry powder as a source of gallic acid (417 mg/day), and cyaniding-3-glucoside (269 mg/day) for 4 months	16 participants of age between 68-92 years with MCI	RCT, DB, PC	Improvements in brain activation	2018 (Boespflug eta l., 2018)
Cognitive function	Consumption of 25 g of blueberry powder as a source of gallic acid (417 mg/day), and cyaniding-3-glucoside (269 mg/day) for 6 months	76 participants of age between 62-80 years with MCI	RCT, DB, PC	Improvements in memory discrimination	2018 (McNamara et al., 2018)
Cognitive function	Consumption of 355 ml of concord grape juice as a source of polyphenols (777 mg/day) for 3 months	25 participants of age between 40-50 years	RCT, DB, PC, CO	Improvements in immediate spatial memory	2016 (Lamport et al., 2016)
Memory function	Consumption of 444-621 ml of concord grape juice as a source of polyphenols per day for 3 months	12 participants of mean age of 78.2 years	RCT, DB, PC	Improvements in verbal learning, no significant Changes in verbal and spatial recall	2010 (Krikorian, et al., 2010)

TABLE 3.2 (Continued)
Dietary polyphenolic compounds in clinical trials against cognition

Disease/ Indication	Intervention/ Treatment	Study size	Study design	Result	Year
Memory function and brain activation	Consumption of 355-621 ml of concord grape juice per day as a source of polyphenols for 4 months	21 participants of age between 68-90 years with MCI	RCT, DB, PC	Improvements in neurocognitive function	2012 (Krikorian et al., 2012)
Cognitive function	Consumption of 500 ml of orange juice as a source of flavanones (305 mg/day) for 2 months	37 participants of age between 60-81 years	RCT,DB,CO	Improvements in global cognitive function	2015 (Kean et al., 2015)
Memory function	Consumption of 8 ounces of pomegranate juice per day for 2 months	28 participants of mean age of 62.6 years with decline in memory function	RCT,DB,PC	Improvements in memory performance and functional brain activation	2013 (Bookheimer et al., 2013)

* CS = Cohort study; CSS = cross-sectional study; CO = cross-over study; DB = double-blind; RCT = randomized controlled trial; PS = pilot study; PC = placebo controlled; MCI = mild cognitive impairment.

in overall cognitive functions. Short-term consumption of flavonoids did not improve complete cognitive functions but progressively enhanced other associated tasks like frontal lobe functions, information integration, sustained attention, category fluency, speed, verbal memory, incidental learning and the recalling power of the brain. This proved that to obtain major observable effects in human beings, long-term consumption of polyphenols are essential. Polyphenols exert perceptible effects on memory, locomotor and executive function by improving cerebral blood flow, reducing oxidative-stress, thus suppressing neuro-inflammation and restoring hippocampus connectivity (Joris et al., 2018). Despite low oral bioavailability (about 1 percent) of resveratrol, repeated administration of it accelerated its bioavailability and ameliorated age-related cognitive decline (Ayaz et al., 2019). In addition, these benefits may be also relevant for younger adults. The ongoing study on polyphenol consumption in older adults at risk for dementia/AD is significant, while it is possible that the window of opportunity to ameliorate the age-related cognitive deterioration by nutritious diet emerges early in the life course. However, it is quite evident from the above studies that combinations of polyphenols give synergistic response and increases bioavailability of other polyphenols, for example, caffeine and piperine increases the bioavailability of ECGC and curcumin, respectively.

3.5 CONCLUSION AND FUTURE PERSPECTIVE

Randomized controlled trials (RCTs) provide a stronger confirmation on the pivotal connotation of polyphenol consumption with cognitive function; however, the sample sizes are quite small, with less than a hundred participants in most studies, and the outcome of short-term supplementation in clinical studies might not be comparable with long-term consumption. Therefore, future studies on large populations, predominantly with patients on mild cognitive impairment (MCI) stage, with

TABLE 3.3

Pure Dietary Polyphenolics Tested for their Cognitive Functions in Clinical Trials

Disease/Indication	Intervention/ Treatment	Study size	Study design	Result	Year
Health promoting effects	Consumption of Lapidated curcumin formulation Longvida (400 mg/day) that contain 80 mg of curcumin for a period of 4 weeks	38 participants of age between 40-60 years	RCT, PC	Positive correction between consumption of curcumin and decrease in plasma concentration of amyloidβ-protein	2012 (DiSilvestro et al., 2012)
Cognitive function	Consumption of Lapidated curcumin formulation Longvida (400 mg/day) that contain 80 mg of curcumin for a period of 4 weeks	60 participants of age between 60-85 years	RCT, DB, PC	Improved attention and memory power	2015(Cox et al., 2015)
Working memory	Consumption of 1 g of turmeric	48 participants of age above 60 years	RCT, DB, PC	Working memory was improved	2014(Lee et al., 2014)
Cognitive function	Consumption of formulation of curcumin contain curcumin (180 mg/day) for a period of 18 months	40 participants of age between 51-84 years	RCT, DB, PC	Attention power and memory was improved	2018 (Small et al., 2018)
Cognitive function	Consumption of curcumin formulation Biocurcumax (1500 mg/day) that contain 88 percent of curcumin for a period of 1 year	96 participants of age between 40-90 years	RCT, DB, PC	No positive change and improvement in the cognitive function	2016 (Rainey-Smith et al., 2016)
Cognitive function	Consumption of resveratrol (250 mg/ day and 500 mg/ day) OD	22 participants of age between 18-25 years	RCT, DB, PC, CO	No positive change and improvement in the cognitive function	2010 (Kennedy et al., 2010)
Cognitive function	Consumption of resveratrol with or without piperine (20 mg), OD	23 participants of age between 19-34 years	RCT, DB, PC, CO	No positive change and improvement in the cognitive function	2014 (Wightman et al., 2014)
Cognitive function	Consumption of resveratrol supplement (70, 150 and 300 mg/day), OD	36 participants of age between 40-80 years	RCT, DB, PC	Multi-tasking ability of the candidate was improved	2016(Wong et al., 2016)

TABLE 3.3 (Continued)
Pure Dietary Polyphenolics Tested for their Cognitive Functions in Clinical Trials

Disease/Indication	Intervention/ Treatment	Study size	Study design	Result	Year
Cognitive function	Consumption of resveratrol (500 mg/day) for 28 days	60 participants of age between 18-30 years	RCT, DB, PC	Brain function was improved in 3-Back task	2015 (Wightman et al., 2015)
Cognitive function	Consumption of 200 ml of liquid juice that contains 150 mg of resveratrol for 6 months	37 participants of age between 68-83 years	RCT, DB, PC	No positive change and improvement in the cognitive function	2018 (Moran et al., 2018)
Cognitive function	Consumption of resveratrol (1 g/day) for a period of 3 months	32 participants of age between 65-93 years	RCT (Phase IIa), DB, PC	Psychomotor activity was improved but no effect on other cognitive functions	2018 (Anton et al., 2018)
Memory performance	Consumption of resveratrol (200 mg/day) for a period 26 weeks	46 participants of age between 50-75 years	RCT, DB, PC	Memory power and functional activity in hippocampus was improved	2014 (Witte et al., 2014)
Verbal memory performance	Consumption of resveratrol (200 mg/day) for a period 26 weeks	53 participants of age between 60-79 years	RCT, DB, PC	No positive change and improvement in the verbal memory performance	2018 (Neil J. Vickers, 2018)
Memory performance	Consumption of resveratrol (200 mg/day) for a period 26 weeks	40 participants of age between 50-80 years	RCT, DB, PC	Memory power and functional activity in hippocampus was improved performance	2017 (Köbe et al., 2017)
Cognitive function	Consumption of resveratrol (150 mg/day) for a period 14 weeks	80 participants of age between 45-85 years	RCT, DB, PC	Memory power, verbal memory and functional activity in hippocampus was improved	2017 (Evans et al., 2017)
Cognitive function	Consumption of resveratrol (150 mg/day) for a period 14 weeks	80 participants of age between 45-85 years	RCT, DB, PC	Cognitive functions were improved	2016 (Evans et al., 2016)
Cognitive function	Consumption of antioxidant mainly in the form of flavonoids and lignan for a period 5 years	2613 participants of age between 43-77 years	CS	Positive correlation. Cognitive function was improved in terms of fluency, speed and recognition power.	2015 (Nooyens et al., 2015)
Cognitive function	Consumption of polyphenol for a period of 3 years	652 participants of age above 65 years	CS	No statistical improvement in cognitive function	2015 (Rabassa et al., 2015)
Cognitive function	Consumption of total or class-specific polyphenolic compounds for a period of 13 years	2574 participants of mean age 66 years	CS	Verbal memory was improved but there was no improvement in executive function	2012 (Kesse-Guyot et al., 2012)

(*continued*)

TABLE 3.3 (Continued)
Pure Dietary Polyphenolics Tested for their Cognitive Functions in Clinical Trials

Disease/Indication	Intervention/ Treatment	Study size	Study design	Result	Year
Dementia and AD	Consumption of polyphenols for a period of 11.7 years	1329 participants of mean age 75.8 years	CS	Polyphenol intake decreases the risk of dementia and AD	2018 (Lefèvre-Arbogast et al., 2018)
Cognitive function	Consumption of a 72 g of grape formulation consists of several flavonoids and resveratrol total 371.25 mg of flavonoids per day for 6 months	10 participants of age between 66-82 years	PS	Cognitive function was improved.	2017 (Lee et al., 2017)

* CS = Cohort study; CSS = cross-sectional study; CO = cross-over study; DB = double-blind; RCT = randomized controlled trial; PS = pilot study; PC = placebo controlled; MCI = mild cognitive impairment.

extended intervention time and longer follow-up after the supplementation period should establish relative benefits of consumption of polyphenols on cognitive dysfunctions and other related diseases.

In the unavailability of curative treatment, dietary polyphenols are important adaptable and indispensable factors for cognitive aging and dementia. Several polyphenols publicized their persuasive neuroprotective, and principally flavonoids curcumin, and resveratrol, have pleiotropic pharmacological activities capable of being ideal aspirants to reserve or increase cognitive performances that pave the way for onset of dementia and AD. Undoubtedly, disagreements in study design partially bring about erratic findings on the same compound. Noteworthy: it is critical to determine the effective dose and period of supplementation, boost the bioavailability of the polyphenols, and formulate a standardized preparation to confirm the level of polyphenols sufficient to offer protective action. However, research on the action of polyphenols on cognitive disorders is at an early stage. More extensive work is needed in this field to lay down a concrete basis for future investigation.

REFERENCES

Agarwal, P., Holland, T.M., Wang, Y., Bennett, D.A., and Morris, M.C. 2019. Association of strawberries and anthocyanidin intake with Alzheimer's dementia risk. *Nutrients* 11: 3060.

Alharbi, M.H., Lamport, D.J., Dodd, G.F., Saunders, C., Harkness, L., Butler, L.T., and Spencer, J.P. 2016. Flavonoid-rich orange juice is associated with acute improvements in cognitive function in healthy middle-aged males. *Eur J Nutr* 55: 2021–2029.

Alminger, M., Aura, A.M., Bohn, T., Dufour, C., El, S.N., Gomes, A., Karakaya, S.İ.B.E.L., Martínez-Cuesta, M.C., Mcdougall, G.J., Requena, T., and Santos, C.N. 2014. Studies of intestinal permeability of 36 flavonoids using Caco-2 cell monolayer model. In vitro models for studying secondary plant metabolite digestion and bioaccessibility. *Compr Rev Food Sci* 13: 413–436.

An, G., Gallegos, J., and Morris, M.E. 2011. The bioflavonoid kaempferol is an ABCG2 substrate and inhibits ABCG2-mediated quercetin efflux. *Drug Metab Dispos* 39: 426–432.

Anton, S.D., Ebner, N., Dzierzewski, J.M., Zlatar, Z.Z., Gurka, M.J., Dotson, V.M., Kirton, J., Mankowski, R.T., Marsiske, M., and Manini, T.M. 2018. Effects of 90 days of resveratrol supplementation on cognitive function in elders: A pilot study. *J Altern Complement Med* 24: 725–732.

Ayaz, M., Sadiq, A., Junaid, M., Ullah, F., Ovais, M., Ullah, I., Ahmed, J., and Shahid, M. 2019. Flavonoids as prospective neuroprotectants and their therapeutic propensity in aging associated neurological disorders. *Front Aging Neurosci* 11: 155.

Baba, Y., Inagaki, S., Nakagawa, S., Kaneko, T., Kobayashi, M., and Takihara, T. 2020. Effect of daily intake of green tea catechins on cognitive function in middle-aged and older subjects: A randomized, placebo-controlled study. *Molecules* 25: 4265.

Bitu Pinto, N., da Silva Alexandre, B., Neves, K.R.T., Silva, A.H., Leal, L.K.A., and Viana, G.S. 2015. Neuroprotective properties of the standardized extract from Camellia sinensis (green tea) and its main bioactive components, epicatechin and epigallocatechin gallate, in the 6-OHDA model of Parkinson's disease. *Evid Based Complementary Altern Med* 2015: 161092.

Boespflug, E.L., Eliassen, J.C., Dudley, J.A., Shidler, M.D., Kalt, W., Summer, S.S., Stein, A.L., Stover, A.N., and Krikorian, R. 2018. Enhanced neural activation with blueberry supplementation in mild cognitive impairment. *Nutr Neurosci* 21: 297–305.

Bookheimer, S.Y., Renner, B.A., Ekstrom, A., Li, Z., Henning, S.M., Brown, J.A., Jones, M., Moody, T., and Small, G.W. 2013. Pomegranate juice augments memory and FMRI activity in middle-aged and older adults with mild memory complaints. *Evid Based Complementary Altern Med*: 946298.

Borges, G., Lean, M.E., Roberts, S.A., and Crozier, A. 2013. Bioavailability of dietary (poly) phenols: A study with ileostomists to discriminate between absorption in small and large intestine. *Food Funct* 4: 754–762.

Bowtell, J.L., Aboo-Bakkar, Z., Conway, M.E., Adlam, A.L.R., and Fulford, J. 2017. Enhanced task-related brain activation and resting perfusion in healthy older adults after chronic blueberry supplementation. *Appl Physiol Nutr Metab* 42: 773–779.

Brettschneider, J., Tredici, K.D., Lee, V.M.Y., and Trojanowski, J.Q. 2015. Spreading of pathology in neurodegenerative diseases: A focus on human studies. *Nat Rev Neurosci* 16: 109–120.

Brickman, A.M., Khan, U.A., Provenzano, F.A., Yeung, L.K., Suzuki, W., Schroeter, H., Wall, M., Sloan, R.P., and Small, S.A. 2014. Enhancing dentate gyrus function with dietary flavanols improves cognition in older adults. *Nat Neurosci* 17: 1798–1803.

Butchart, C., Kyle, J., McNeill, G., Corley, J., Gow, A.J., Starr, J.M., and Deary, I.J. 2011. Flavonoid intake in relation to cognitive function in later life in the Lothian Birth Cohort 1936. *Br J Nutr* 106: 141–148.

Cai, Q., and Tammineni, P. 2017. Mitochondrial aspects of synaptic dysfunction in Alzheimer's disease. *J Alzheimer's Dis* 57: 1087–1103.

Calabrò, R.S., De Cola, M.C., Gervasi, G., Portaro, S., Naro, A., Accorinti, M., Manuli, A., Marra, A., De Luca, R., and Bramanti, P. 2019. The efficacy of cocoa polyphenols in the treatment of mild cognitive impairment: A retrospective study. *Medicina* 55: 156.

Campana, M., Falkai, P., Siskind, D., Hasan, A., and Wagner, E. 2021. Characteristics and definitions of ultra-treatment-resistant schizophrenia – A systematic review and meta-analysis. *Schizophr Res* 228: 218–226.

Campos-Bedolla, P., Walter, F.R., Veszelka, S., and Deli, M.A. 2014. Role of the blood-brain barrier in the nutrition of the central nervous system. *Arch Med Res* 45: 610–638.

Castro, A.F., and Altenberg, G.A. 1997. Inhibition of drug transport by genistein in multidrug-resistant cells expressing P-glycoprotein. *Biochem Pharmacol* 53: 89–93.

Catino, S., Paciello, F., Miceli, F., Rolesi, R., Troiani, D., Calabrese, V., Santangelo, R., and Mancuso, C. 2016. Ferulic acid regulates the Nrf2/heme oxygenase-1 system and counteracts trimethyltin-induced neuronal damage in the human neuroblastoma cell line SH-SY5Y. *Front Pharmacol* 6: 305.

Choi, J.S., Choi, B.C., and Kang, K.W. 2009. Effect of resveratrol on the pharmacokinetics of oral and intravenous nicardipine in rats: Possible role of P-glycoprotein inhibition by resveratrol. *Pharmazie* 64: 49–52.

Choi, S.J., Shin, S.C., and Choi, J.S. 2011. Effects of myricetin on the bioavailability of doxorubicin for oral drug delivery in rats: Possible role of CYP3A4 and P-glycoprotein inhibition by myricetin. *Arch Pharm Res* 34: 309–315.

Cooray, H.C., Janvilisri, T., van Veen, H.W., Hladky, S.B., and Barrand, M.A. 2004. Interaction of the breast cancer resistance protein with plant polyphenols. *Biochem Biophys Res Commun* 317: 269–275.

Cox, K.H., Pipingas, A., and Scholey, A.B. 2015. Investigation of the effects of solid lipid curcumin on cognition and mood in a healthy older population. *J Psychopharmacol* 29: 642–651.

Crichton, G.E., Elias, M.F., and Alkerwi, A.A. 2016. Chocolate intake is associated with better cognitive function: The Maine-Syracuse Longitudinal Study. *Appetite* 100: 126–132.

Day, A.J., Cañada, F.J., Díaz, J.C., Kroon, P.A., Mclauchlan, R., Faulds, C.B., Plumb, G.W., Morgan, M.R., and Williamson, G. 2000. Dietary flavonoid and isoflavone glycosides are hydrolysed by the lactase site of lactase phlorizin hydrolase. *FEBS Letters* 468: 166–170.

De Castro, I.P., Martins, L.M., and Loh, S.H.Y. 2011. Mitochondrial quality control and Parkinson's disease: A pathway unfolds. *Mol Neurobiol* 43: 80–86.

De Hert, M., Detraux, J., Van Winkel, R., Yu, W., and Correll, C.U. 2012. Metabolic and cardiovascular adverse effects associated with antipsychotic drugs. *Nat Rev Endocrinol* 8: 114–126.

Dennison, C.A., Legge, S.E., Hubbard, L., Lynham, A.J., Zammit, S., Holmans, P., Cardno, A.G., Owen, M.J., O'Donovan, M.C., and Walters, J.T. 2021. Risk factors, clinical features, and polygenic risk scores in schizophrenia and schizoaffective disorder depressive-type. *Schizophr Bull* 47: 1375–1384.

Desideri, G., Kwik-Uribe, C., Grassi, D., Necozione, S., Ghiadoni, L., Mastroiacovo, D., Raffaele, A., Ferri, L., Bocale, R., Lechiara, M.C., and Marini, C. 2012. Benefits in cognitive function, blood pressure, and insulin resistance through cocoa flavanol consumption in elderly subjects with mild cognitive impairment: The Cocoa, Cognition, and Aging (CoCoA) study. *Hypertension* 60: 794–801.

Devore, E.E., Grodstein, F., van Rooij, F.J., Hofman, A., Stampfer, M.J., Witteman, J.C., and Breteler, M.M. 2010. Dietary antioxidants and long-term risk of dementia. *Arch Neurol* 67: 819–825.

Devore, E.E., Kang, J.H., Breteler, M.M., and Grodstein, F. 2012. Dietary intakes of berries and flavonoids in relation to cognitive decline. *Ann Neurol* 72: 135–143.

DiSilvestro, R.A., Joseph, E., Zhao, S., and Bomser, J. 2012. Diverse effects of a low dose supplement of lipidated curcumin in healthy middle aged people. *Nutrition* 11: 1–8.

Döring, B., and Petzinger, E. 2014. Phase 0 and phase III transport in various organs: combined concept of phases in xenobiotic transport and metabolism. *Drug Metab Rev* 46: 261–282.

Duan, X., Li, Y., Xu, F., and Ding, H. 2021. Study on the neuroprotective effects of Genistein on Alzheimer's disease. *Brain Behav* 11: e02100.

Evans, H.M., Howe, P.R., and Wong, R.H. 2017. Effects of resveratrol on cognitive performance, mood and cerebrovascular function in post-menopausal women; a 14-week randomised placebo-controlled intervention trial. *Nutrients* 9: 27.

Evans, H.M., Howe, P.R.C., and Wong, R.H.X. 2016. Clinical evaluation of effects of chronic resveratrol supplementation on cerebrovascular function, cognition, mood, physical function and general well-being in postmenopausal women –rationale and study design. *Nutrients* 8: 150.

Evans-Lacko, S., Aguilar-Gaxiola, S., Al-Hamzawi, A., Alonso, J., Benjet, C., Bruffaerts, R., Chiu, W.T., Florescu, S., de Girolamo, G., Gureje, O., and Haro, J.M. 2018. Socio-economic variations in the mental health treatment gap for people with anxiety, mood, and substance use disorders: results from the WHO World Mental Health (WMH) surveys. *Psychol Med* 48: 1560–1571.

Faria, A., Pestana, D., Teixeira, D., Couraud, P.O., Romero, I., Weksler, B., de Freitas, V., Mateus, N., and Calhau, C. 2011. Insights into the putative catechin and epicatechin transport across blood-brain barrier. *Food Funct* 2: 39–44.

Field, D.T., Williams, C.M., and Butler, L.T. 2011. Consumption of cocoa flavanols results in an acute improvement in visual and cognitive functions. *Physiol Behav* 103: 255–260.

Figueira, I., Menezes, R., Macedo, D., Costa, I., and Nunes dos Santos, C. 2017. Polyphenols beyond barriers: A glimpse into the brain. *Curr Neuropharmacol* 15: 562–594.

Gasparini, F., Di Paolo, T., and Gomez-Mancilla, B. 2013. Metabotropic glutamate receptors for Parkinson's disease therapy *Parkinsons Dis* 2013: 196028.

Gee, J.M., DuPont, M.S., Day, A.J., Plumb, G.W., Williamson, G., and Johnson, I.T. 2000. Intestinal transport of quercetin glycosides in rats involves both deglycosylation and interaction with the hexose transport pathway. *J Nutr* 130: 2765–2771.

Guo, D.J., Li, F., Yu, P.H.F., and Chan, S.W. 2013. Neuroprotective effects of luteolin against apoptosis induced by 6-hydroxydopamine on rat pheochromocytoma PC12 cells. *Pharm Biol* 51: 190–196.

Haskell-Ramsay, C.F., Stuart, R.C., Okello, E.J., and Watson, A.W. 2017. Cognitive and mood improvements following acute supplementation with purple grape juice in healthy young adults. *Eur J Nutr* 56: 2621–2631.

Henderson, V.W., John, J.S., Hodis, H.N., Kono, N., McCleary, C.A., Franke, A.A., and Mack, W.J. 2012. Long-term soy isoflavone supplementation and cognition in women: a randomized, controlled trial. *Neurology* 78:1841–1848.

Hepgul, N., Cattaneo, A., Zunszain, P.A., and Pariante, C.M. 2013. Depression pathogenesis and treatment: What can we learn from blood mRNA expression? *BMC Med* 11: 1–13.

Hong, J., Lambert, J.D., Lee, S.H., Sinko, P.J., and Yang, C.S. 2003. Involvement of multidrug resistance-associated proteins in regulating cellular levels of (−)-epigallocatechin-3-gallate and its methyl metabolites. *Biochem Biophys Res Commun* 310: 222–227.

Huang, Y.H., and Zhang, Q.H. 2010. Genistein reduced the neural apoptosis in the brain of ovariectomised rats by modulating mitochondrial oxidative stress. *Br J Nutr* 104: 1297–1303.

Hussein, S.S.S., Kamarudin, M.N.A. and Abdul Kadir, H., 2015. (+)-Catechin attenuates NF-κB activation through regulation of Akt, MAPK, and AMPK signaling pathways in LPS-induced BV-2 microglial cells. *Am J Chin Med* 43: 927–952.

Ide, K., Yamada, H., Takuma, N., Kawasaki, Y., Harada, S., Nakase, J., Ukawa, Y., and Sagesaka, Y.M. 2015. Effects of green tea consumption on cognitive dysfunction in an elderly population: A randomized placebo-controlled study. *Nutrition* 15: 1–9.

Ide, K., Yamada, H., Takuma, N., Park, M., Wakamiya, N., Nakase, J., Ukawa, Y., and Sagesaka, Y.M. 2014. Green tea consumption affects cognitive dysfunction in the elderly: a pilot study. *Nutrients* 6: 4032–4042.

Igase, M., Igase, K., Tabara, Y., Ohyagi, Y., and Kohara, K. 2017. Cross-sectional study of equal producer status and cognitive impairment in older adults. *Geriatr Gerontol Int* 17: 2103–2108.

Imai, Y., Tsukahara, S., Asada, S., and Sugimoto, Y. 2004. Phytoestrogens/flavonoids reverse breast cancer resistance protein/ABCG2-mediated multidrug resistance. *Cancer Res* 64: 4346–4352.

Islam, M.T. 2017. Oxidative stress and mitochondrial dysfunction-linked neurodegenerative disorders. *Neurol Res* 39: 73–82.

Jeong, C.H., Jeong, H.R., Choi, G.N., Kim, D.O., Lee, U.K., and Heo, H.J. 2011. Neuroprotective and anti-oxidant effects of caffeic acid isolated from Erigeron annuus leaf. *Chin Med* 6: 1–9.

Jesulola, E., Micalos, P., and Baguley, I.J. 2018. Understanding the pathophysiology of depression: From monoamines to the neurogenesis hypothesis model – Are we there yet? *Behav Brain Res* 341: 79–90.

Jia, L., Li, S., Dai, W., Guo, L., Xu, Z., Scott, A.M., Zhang, Z., Ren, J., Zhang, Q., Dexheimer, T.S., and Chung-Davidson, Y.W. 2021. Convergent olfactory trace amine-associated receptors detect biogenic polyamines with distinct motifs via a conserved binding site. *J Biol Chem* 297: 101268. https://doi.org/10.1016/j.jbc.2021.101268.

Johnson, S.L., Park, H.Y., Vattem, D.A., Grammas, P., Ma, H., and Seeram, N.P. 2020. Equol, a blood–brain barrier permeable gut microbial metabolite of dietary isoflavone daidzein, exhibits neuroprotective effects against neurotoxins induced toxicity in human neuroblastoma SH-SY5Y cells and Caenorhabditis elegans. *Plant Foods Hum Nutr* 75: 512–517.

Joris, P.J., Mensink, R.P., Adam, T.C., and Liu, T.T. 2018. Cerebral blood flow measurements in adults: a review on the effects of dietary factors and exercise. *Nutrients* 10: 530.

Kean, R.J., Lamport, D.J., Dodd, G.F., Freeman, J.E., Williams, C.M., Ellis, J.A., Butler, L.T., and Spencer, J.P. 2015. Chronic consumption of flavanone-rich orange juice is associated with cognitive benefits: An 8-wk, randomized, double-blind, placebo-controlled trial in healthy older adults. *Am J Clin Nutr* 101: 506–514.

Keane, K.M., Haskell-Ramsay, C.F., Veasey, R.C., and Howatson, G. 2016. Montmorency Tart cherries (Prunus cerasus L.) modulate vascular function acutely, in the absence of improvement in cognitive performance. *Br J Nutr* 116: 1935–1944.

Kennedy, D.O., Wightman, E.L., Reay, J.L., Lietz, G., Okello, E.J., Wilde, A., and Haskell, C.F. 2010. Effects of resveratrol on cerebral blood flow variables and cognitive performance in humans: A double-blind, placebo-controlled, crossover investigation. *Am J Clin Nutr* 91: 1590–1597.

Kesse-Guyot, E., Fezeu, L., Andreeva, V.A., Touvier, M., Scalbert, A., Hercberg, S., and Galan, P. 2012. Total and specific polyphenol intakes in midlife are associated with cognitive function measured 13 years later. *J Nutr* 142: 76–83.

Köbe, T., Witte, A.V., Schnelle, A., Tesky, V.A., Pantel, J., Schuchardt, J.P., Hahn, A., Bohlken, J., Grittner, U., and Flöel, A. 2017. Impact of resveratrol on glucose control, hippocampal structure and connectivity, and memory performance in patients with mild cognitive impairment. *Front Neurosci* 11: 105.

Kouli, A., Torsney, K.M., and Kuan, W.L. 2018. Parkinson's disease: Etiology, neuropathology, and pathogenesis. *Exon Publications* 3–26. 10.15586/codonpublications.parkinsonsdisease.2018.ch1.

Krikorian, R., Boespflug, E.L., Fleck, D.E., Stein, A.L., Wightman, J.D., Shidler, M.D., and Sadat-Hossieny, S. 2012. Concord grape juice supplementation and neurocognitive function in human aging. *J Agric Food Chem* 60: 5736–5742.

Krikorian, R., Nash, T.A., Shidler, M.D., Shukitt-Hale, B., and Joseph, J.A. 2010. Concord grape juice supplementation improves memory function in older adults with mild cognitive impairment. *Br J Nutr* 103: 730–734.

Krikorian, R., Shidler, M.D., Nash, T.A., Kalt, W., Vinqvist-Tymchuk, M.R., Shukitt-Hale, B., and Joseph, J.A. 2010. Blueberry supplementation improves memory in older adults. *J Agric Food Chem* 58: 3996–4000.

Kumar, K., Kumar, A., Keegan, R.M., and Deshmukh, R. 2018. Recent advances in the neurobiology and neuropharmacology of Alzheimer's disease. *Biomed Pharmacother* 98: 297–307.

Lagoa, R., Lopez-Sanchez, C., Samhan-Arias, A.K., Gañan, C.M., Garcia-Martinez, V., and Gutierrez-Merino, C. 2009. Kaempferol protects against rat striatal degeneration induced by 3-nitropropionic acid. *J Neurochem* 111: 473–487.

Lamport, D.J., Lawton, C.L., Merat, N., Jamson, H., Myrissa, K., Hofman, D., Chadwick, H.K., Quadt, F., Wightman, J.D., and Dye, L. 2016. Concord grape juice, cognitive function, and driving performance: A 12-wk, placebo-controlled, randomized crossover trial in mothers of preteen children. *Am J Clin Nutr* 103: 775–783.

Lamport, D.J., Pal, D., Macready, A.L., Barbosa-Boucas, S., Fletcher, J.M., Williams, C.M., Spencer, J.P., and Butler, L.T. 2016. The effects of flavanone-rich citrus juice on cognitive function and cerebral blood flow: An acute, randomised, placebo-controlled cross-over trial in healthy, young adults. *Br J Nutr* 116: 2160–2168.

Lapchak, P.A. 2013. Drug-like property profiling of novel neuroprotective compounds to treat acute ischemic stroke: Guidelines to develop pleiotropic molecules. *Transl Stroke Res* 4: 328–342.

Lee, J., Torosyan, N., and Silverman, D.H. 2017. Examining the impact of grape consumption on brain metabolism and cognitive function in patients with mild decline in cognition: A double-blinded placebo-controlled pilot study. *Exp Gerontol* 87: 121–128.

Lee, J.H., Amarsanaa, K., Wu, J., Jeon, S.C., Cui, Y., Jung, S.C., Park, D.B., Kim, S.J., Han, S.H., Kim, H.W., and Eun, S.Y. 2018. Nobiletin attenuates neurotoxic mitochondrial calcium overload through K+ influx and ΔΨm across mitochondrial inner membrane. *Korean J Physiol Pharmacol* 22: 311–319.

Lee, M.S., Wahlqvist, M.L., Chou, Y.C., Fang, W.H., Lee, J.T., Kuan, J.C., Liu, H.Y., Lu, T.M., Xiu, L., Hsu, C.C., and Andrews, Z.B. 2014. Turmeric improves post-prandial working memory in pre-diabetes independent of insulin. *Asia Pac J Clin Nutr* 23: 581–591.

Lefèvre-Arbogast, S., Gaudout, D., Bensalem, J., Letenneur, L., Dartigues, J.F., Hejblum, B.P., Féart, C., Delcourt, C., and Samieri, C. 2018. Pattern of polyphenol intake and the long-term risk of dementia in older persons. *Neurology* 90: e1979-e1988.

Lei, X., Chao, H., Zhang, Z., Lv, J., Li, S., Wei, H., Xue, R., Li, F., and Li, Z. 2015. Neuroprotective effects of quercetin in a mouse model of brain ischemic/reperfusion injury via anti-apoptotic mechanisms based on the Akt pathway. *Mol Med Rep* 12: 3688–3696.

Lei, Y., Chen, J., Zhang, W., Fu, W., Wu, G., Wei, H., Wang, Q., and Ruan, J. 2012. In vivo investigation on the potential of galangin, kaempferol and myricetin for protection of D-galactose-induced cognitive impairment. *Food Chem* 135: 2702–2707.

Levit, J., Valderrama, J., Georgakopoulos, P., Hansen, S.K., Salisu, M., Valderrama, J., Georgakopoulos, P., Fanous, A., Bigdeli, T., Knowles, J., Pato, C., Pato, M., Pato, C.N., and Pato, M.T. 2021. Childhood trauma and psychotic symptomatology in ethnic minorities with schizophrenia. *Schizophr Bull Open* 2: 68.

Marino, B.L., de Souza, L.R., Sousa, K., Ferreira, J.V., Padilha, E.C., da Silva, C.H., Taft, C.A., and Hage-Melim, L.I. 2020. Parkinson's disease: A review from pathophysiology to treatment. *Mini Rev Med Chem* 20: 754–767.

Mastroiacovo, D., Kwik-Uribe, C., Grassi, D., Necozione, S., Raffaele, A., Pistacchio, L., Righetti, R., Bocale, R., Lechiara, M.C., Marini, C., and Ferri, C. 2015. Cocoa flavanol consumption improves cognitive function, blood pressure control, and metabolic profile in elderly subjects: the Cocoa, Cognition, and Aging (CoCoA) Study – a randomized controlled trial. *Am J Clin Nutr* 101: 538–548.

McCusker, R.H., and Kelley, K.W. 2013. Immune–neural connections: How the immune system's response to infectious agents influences behavior. *J Exp Biol* 216: 84–98.

McNamara, R.K., Kalt, W., Shidler, M.D., McDonald, J., Summer, S.S., Stein, A.L., Stover, A.N., and Krikorian, R. 2018. Cognitive response to fish oil, blueberry, and combined supplementation in older adults with subjective cognitive impairment. *Neurobiol Aging* 64: 147–156.

Miller, M.G., Hamilton, D.A., Joseph, J.A., and Shukitt-Hale, B. 2018. Dietary blueberry improves cognition among older adults in a randomized, double-blind, placebo-controlled trial. *Eur J Nutr* 57: 1169–1180.

Mitsunaga, Y., Takanaga, H., Matsuo, H., Naito, M., Tsuruo, T., Ohtani, H., and Sawada, Y. 2000. Effect of bioflavonoids on vincristine transport across blood-brain barrier. *Eur J Pharmacol* 395: 193–201.

Moran, C., Scotto di Palumbo, A., Bramham, J., Moran, A., Rooney, B., De Vito, G., and Egan, B. 2018. Effects of a six-month multi-ingredient nutrition supplement intervention of omega-3 polyunsaturated fatty acids, vitamin D, resveratrol, and whey protein on cognitive function in older adults: a randomised, double-blind, controlled trial. *J Prev Alzheimer's Dis* 5: 175–183.

Moreira, A., Diógenes, M.J., De Mendonça, A., Lunet, N., and Barros, H. 2016. Chocolate consumption is associated with a lower risk of cognitive decline. *J Alzheimer's Dis* 53: 85–93.

Morelli, S., Piscioneri, A., Guarnieri, G., Morelli, A., Drioli, E., and De Bartolo, L. 2021. Anti-neuroinflammatory effect of daidzein in human hypothalamic GnRH neurons in an in vitro membrane-based model. *BioFactors* 47: 93–111.

Nakatsuma, A., Fukami, T., Suzuki, T., Furuishi, T., Tomono, K., and Hidaka, S. 2010. Effects of kaempferol on the mechanisms of drug resistance in the human glioblastoma cell line T98G. *Pharmazie* 65: 379–383.

Narayanan, S.V., Dave, K.R., Saul, I., and Perez-Pinzon, M.A. 2015. Resveratrol preconditioning protects against cerebral ischemic injury via nuclear erythroid 2–related factor 2. *Stroke* 46: 1626–1632.

Nooyens, A.C., Bueno-de-Mesquita, H.B., van Boxtel, M.P., van Gelder, B.M., Verhagen, H., and Verschuren, W.M. 2011. Fruit and vegetable intake and cognitive decline in middle-aged men and women: The Doetinchem Cohort Study. *Br J Nutr* 106: 752–761.

Nooyens, A.C., Milder, I.E., Van Gelder, B.M., Bueno-de-Mesquita, H.B., Van Boxtel, M.P., and Verschuren, W.M. 2015. Diet and cognitive decline at middle age: The role of antioxidants. *Br J Nutr* 113: 1410–1417.

Pase, M.P., Scholey, A.B., Pipingas, A., Kras, M., Nolidin, K., Gibbs, A., Wesnes, K., and Stough, C. 2013. Cocoa polyphenols enhance positive mood states but not cognitive performance: A randomized, placebo-controlled trial. *J Psychopharmacol* 27: 451–458.

Péneau, S., Galan, P., Jeandel, C., Ferry, M., Andreeva, V., Hercberg, S., Kesse-Guyot, E., and SU. VI. MAX 2 Research Group. 2011. Fruit and vegetable intake and cognitive function in the SU. VI. MAX 2 prospective study. *Am J Clin Nutr* 94: 1295–1303.

Pervin, M., Unno, K., Nakagawa, A., Takahashi, Y., Iguchi, K., Yamamoto, H., Hoshino, M., Hara, A., Takagaki, A., Nanjo, F., and Minami, A. 2017. Blood brain barrier permeability of (−)-epigallocatechin gallate, its proliferation-enhancing activity of human neuroblastoma SH-SY5Y cells, and its preventive effect on age-related cognitive dysfunction in mice. *Biochem Biophys Rep* 9: 180–186.

Peterson, J.J., Dwyer, J.T., Jacques, P.F., and McCullough, M.L. 2012. Associations between flavonoids and cardiovascular disease incidence or mortality in European and US populations. *Nutr Rev* 70: 491–508.

Prins, N.D., and Scheltens, P. 2013. Treating Alzheimer's disease with monoclonal antibodies: Current status and outlook for the future. *Alzheimers Res Ther* 5: 1–6.

Rabassa, M., Cherubini, A., Zamora-Ros, R., Urpi-Sarda, M., Bandinelli, S., Ferrucci, L., and Andres-Lacueva, C. 2015. Low levels of a urinary biomarker of dietary polyphenol are associated with substantial cognitive decline over a 3-year period in older adults: The Invecchiare in Chianti Study. *J Am Geriatr Soc* 63: 938–946.

Rainey-Smith, S.R., Brown, B.M., Sohrabi, H.R., Shah, T., Goozee, K.G., Gupta, V.B., and Martins, R.N. 2016. Curcumin and cognition: A randomised, placebo-controlled, double-blind study of community-dwelling older adults. *Br J Nutr* 115: 2106–2113.

Ramesh, M., Dokurugu, Y.M., Thompson, M.D., and Soliman, M.E. 2017. Therapeutic, molecular and computational aspects of novel monoamine oxidase (MAO) inhibitors. *Comb Chem High Throughput Screen* 20: 492–509.

Reddy, H.P., and P. Reddy, T. 2011. Mitochondria as a therapeutic target for aging and neurodegenerative diseases. *Curr Alzheimer Res* 8: 393–409.

Riederer, P., and Laux, G. 2011. MAO-inhibitors in Parkinson's Disease. *Exp Neurobiol* 20: 1.

Saeed, M., Kadioglu, O., Khalid, H., Sugimoto, Y., and Efferth, T. 2015. Activity of the dietary flavonoid, apigenin, against multidrug-resistant tumor cells as determined by pharmacogenomics and molecular docking. *J Nutr Biochem* 26: 44–56.

Salin, P., Manrique, C., Forni, C., and Kerkerian-Le Goff, L. 2002. High-frequency stimulation of the subthalamic nucleus selectively reverses dopamine denervation-induced cellular defects in the output structures of the basal ganglia in the rat. *J Neurosci* 22: 5137–5148.

Santos-Galduróz, R.F., Galduróz, J.C.F., Facco, R.L., Hachul, H., and Tufik, S. 2010. Effects of isoflavone on the learning and memory of women in menopause: A double-blind placebo-controlled study. *Braz J Med Biol Res* 43: 1123–1126.

Sarkar, A., Angeline, M.S., Anand, K., Ambasta, R.K., and Kumar, P. 2012. Naringenin and quercetin reverse the effect of hypobaric hypoxia and elicit neuroprotective response in the murine model. *Brain Res* 1481: 59–70.

Schmidt, H.D., Shelton, R.C., and Duman, R.S. 2011. Functional biomarkers of depression: Diagnosis, treatment, and pathophysiology. *Neuropsychopharmacology* 36: 2375–2394.

Scholey, A.B., French, S.J., Morris, P.J., Kennedy, D.O., Milne, A.L., and Haskell, C.F. 2010. Consumption of cocoa flavanols results in acute improvements in mood and cognitive performance during sustained mental effort. *J Psychopharmacol* 24: 1505–1514.

Schrag, A., Horsfall, L., Walters, K., Noyce, A., and Petersen, I. 2015. Prediagnostic presentations of Parkinson's disease in primary care: A case-control study. *Lancet Neurol* 14: 57–64.

Serra, A., Macia, A., Romero, M.P., Valls, J., Bladé, C., Arola, L., and Motilva, M.J. 2010. Bioavailability of procyanidin dimers and trimers and matrix food effects on in vitro and in vivo models. *Br J Nutr* 103: 944–952.

Shishtar, E., Rogers, G.T., Blumberg, J.B., Au, R., and Jacques, P.F. 2020. Long-term dietary flavonoid intake and risk of Alzheimer disease and related dementias in the Framingham Offspring Cohort. *Am J Clin Nutr* 112: 343–353.

Shukla, S., Zaher, H., Hartz, A., Bauer, B., Ware, J.A., and Ambudkar, S.V. 2009. Curcumin inhibits the activity of ABCG2/BCRP1, a multidrug resistance-linked ABC drug transporter in mice. *Pharm Res* 26: 480–487.

Small, G.W., Siddarth, P., Li, Z., Miller, K.J., Ercoli, L., Emerson, N.D., Martinez, J., Wong, K.P., Liu, J., Merrill, D.A., and Chen, S.T. 2018. Memory and brain amyloid and tau effects of a bioavailable form of curcumin in non-demented adults: A double-blind, placebo-controlled 18-month trial. *Am J Geriatr Psychiatry* 26: 266–277.

Spillantini, M.G., Schmidt, M.L., Lee, V.M.Y., Trojanowski, J.Q., Jakes, R., and Goedert, M. 1997. α-Synuclein in Lewy bodies. *Nature* 388: 839–840.

Strawbridge, R., Young, A.H., and Cleare, A.J. 2018. Biomarkers for depression: Recent insights, current challenges and future prospects. *Focus* 16: 194–209.

Sun, J., Li, Y.Z., Ding, Y.H., Wang, J., Geng, J., Yang, H., Ren, J., Tang, J.Y., and Gao, J. 2014. Neuroprotective effects of gallic acid against hypoxia/reoxygenation-induced mitochondrial dysfunctions in vitro and cerebral ischemia/reperfusion injury in vivo. *Brain Res* 1589: 126–139.

Syed, H.S.S., Kamarudin, M.N.A., and Abdul Kadir, H. 2015. (+)-Catechin attenuates NF-κB activation through regulation of Akt, MAPK, and AMPK signaling pathways in LPS-induced BV-2 microglial cells. *Am. J. Chin. Med.* 43: 927–952.

Tan, K.W., Li, Y., Paxton, J.W., Birch, N.P., and Scheepens, A. 2013. Identification of novel dietary phytochemicals inhibiting the efflux transporter breast cancer resistance protein (BCRP/ABCG2). *Food Chem* 138: 2267–2274.

Tian, X.J., Yang, X.W., Yang, X., and Wang, K. 2009. Studies of intestinal permeability of 36 flavonoids using Caco-2 cell monolayer model. *Int J Pharm* 367: 58–64.

Todorova, A., Jenner, P., and Chaudhuri, K.R. 2014. Non-motor Parkinson's: Integral to motor Parkinson's, yet often neglected. *Pract Neurol* 14: 310–322.

Tsai, T., and Chen, Y. 2000. Determination of unbound hesperetin in rat blood and brain by microdialysis coupled to microbore liquid chromatography. *J Food Drug Anal* 8: 331–336.

Tsai, T.H. 2005. Concurrent measurement of unbound genistein in the blood, brain and bile of anesthetized rats using microdialysis and its pharmacokinetic application. *J Chromatogr A* 1073: 317–322.

Tunc, H.C., Sakar, C.O., Apaydin, H., Serbes, G., Gunduz, A., Tutuncu, M., and Gurgen, F. 2020. Estimation of Parkinson's disease severity using speech features and extreme gradient boosting. *Med Biol Eng Comput* 58: 2757–2773.

Vaidyanathan, J.B., and Walle, T. 2001. Transport and metabolism of the tea flavonoid (–)-epicatechin by the human intestinal cell line Caco-2. *Pharm Res* 18: 1420–1425.

Van Dyck, Christopher H. 2018. Anti-amyloid-β monoclonal antibodies for Alzheimer's disease: Pitfalls and promise. *Biol Psychiatry* 83: 311–319.

Vickers, N.J. 2017. Animal communication: When I'm calling you, will you answer too? *Curr Biol* 27: R713-R715.

Vingtdeux, V., Giliberto, L., Zhao, H., Chandakkar, P., Wu, Q., Simon, J.E., Janle, E.M., Lobo, J., Ferruzzi, M.G., Davies, P., and Marambaud, P. 2010. AMP-activated protein kinase signaling activation by resveratrol modulates amyloid-β peptide metabolism. *J Biol Chem* 285: 9100–9113.

Wang, J., Ferruzzi, M.G., Ho, L., Blount, J., Janle, E.M., Gong, B., Pan, Y., Gowda, G.N., Raftery, D., Arrieta-Cruz, I., and Sharma, V. 2012. Brain-targeted proanthocyanidin metabolites for Alzheimer's disease treatment. *J Neurosci* 32: 5144–5150.

Wang, S., Chen, R., Zhong, Z., Shi, Z., Chen, M., and Wang, Y. 2014. Epigallocatechin-3-gallate potentiates the effect of curcumin in inducing growth inhibition and apoptosis of resistant breast cancer cells. *Am J Chin Med* 42: 1279–1300.

WHO Fact Sheet 2021. www.who.int/news-room/fact-sheets/detail/depression (assessed on 28.1.2022).

Wichmann, T. 2019. Changing views of the pathophysiology of Parkinsonism. *Mov Disord* 34: 1130–1143.

Wightman, E.L., Haskell, C.F., Forster, J.S., Veasey, R.C., and Kennedy, D.O. 2012. Epigallocatechin gallate, cerebral blood flow parameters, cognitive performance and mood in healthy humans: A double-blind, placebo-controlled, crossover investigation. *Hum Psychopharmacol* 27: 177–186.

Wightman, E.L., Haskell-Ramsay, C.F., Reay, J.L., Williamson, G., Dew, T., Zhang, W., and Kennedy, D.O. 2015. The effects of chronic trans-resveratrol supplementation on aspects of cognitive function, mood, sleep, health and cerebral blood flow in healthy, young humans. *Br J Nutr* 114: 1427–1437.

Wightman, E.L., Reay, J.L., Haskell, C.F., Williamson, G., Dew, T.P., and Kennedy, D.O. 2014. Effects of resveratrol alone or in combination with piperine on cerebral blood flow parameters and cognitive performance in human subjects: a randomised, double-blind, placebo-controlled, cross-over investigation. *Br J Nutr* 112: 203–213.

Williams-Gray, C.H., Wijeyekoon, R., Yarnall, A.J., Lawson, R.A., Breen, D.P., Evans, J.R., Cummins, G.A., Duncan, G.W., Khoo, T.K., Burn, D.J., and Barker, R.A. 2016. Serum immune markers and disease progression in an incident Parkinson's disease cohort (ICICLE-PD). *Mov Disord* 31: 995–1003.

Witte, A.V., Kerti, L., Margulies, D.S., and Flöel, A. 2014. Effects of resveratrol on memory performance, hippocampal functional connectivity, and glucose metabolism in healthy older adults. *J Neurosci* 34: 7862–7870.

Wong, R.H., Raederstorff, D., and Howe, P.R. 2016. Acute resveratrol consumption improves neurovascular coupling capacity in adults with type 2 diabetes mellitus. *Nutrients* 8: 425.

Woo, T.T., Li, S.Y., Lai, W.W., Wong, D., and Lo, A.C. 2013. Neuroprotective effects of lutein in a rat model of retinal detachment. *Graefes Arch Clin Exp Ophthalmol* 251: 41–51.

Wu, C.P., Calcagno, A.M., Hladky, S.B., Ambudkar, S.V., and Barrand, M.A. 2005. Modulatory effects of plant phenols on human multidrug-resistance proteins 1, 4 and 5 (ABCC1, 4 and 5). *J FEBS* 272: 4725–4740.

Wu, J.X., Zhang, L.Y., Chen, Y.L., Yu, S.S., Zhao, Y., and Zhao, J. 2015. Curcumin pretreatment and post-treatment both improve the antioxidative ability of neurons with oxygen-glucose deprivation. *Neural Regen Res* 10: 481.

Xu, C., Li, C.Y.T., and Kong, A.N.T. 2005. Induction of phase I, II and III drug metabolism/transport by xenobiotics. *Arch Pharm Res* 28: 249–268.

Xu, X., Xiao, S., Rahardjo, T.B., and Hogervorst, E. 2015. Tofu intake is associated with poor cognitive performance among community-dwelling elderly in China. *JPAD-J Prev Alzheim* 43: 669–675.

Yang, Y., Bai, L., Li, X., Xiong, J., Xu, P., Guo, C., and Xue, M. 2014. Transport of active flavonoids, based on cytotoxicity and lipophilicity: An evaluation using the blood–brain barrier cell and Caco-2 cell models. *Toxicol in Vitro* 28: 388–396.

Yin, C.L., Lu, R.G., Zhu, J.F., Huang, H.M., Liu, X., Li, Q.F., Mo, Y.Y., Zhu, H.J., Chin, B., Wu, J.X., and Liu, X.W. 2019. The study of neuroprotective effect of ferulic acid based on cell metabolomics. *Eur J Pharmacol* 864: 172694.

Yoshida, K., Maeda, K., and Sugiyama, Y. 2013. Hepatic and intestinal drug transporters: Prediction of pharmacokinetic effects caused by drug-drug interactions and genetic polymorphisms. *Annu Rev Pharmacol Toxicol* 53: 581–612.

Yoshimura, K., Inui, N., Karayama, M., Inoue, Y., Enomoto, N., Fujisawa, T., Nakamura, Y., Takeuchi, K., Sugimura, H. and Suda, T., 2017. Successful crizotinib monotherapy in EGFR-mutant lung adenocarcinoma with acquired MET amplification after erlotinib therapy. *Respir Med Case Rep* 20: 160–163.

Yoshimura, S., Kawano, K., Matsumura, R., Sugihara, N., and Furuno, K. 2009. Inhibitory effect of flavonoids on the efflux of *N*-Acetyl 5-Aminosalicylic Acid intracellularly formed in Caco-2 cells. *J Biomed Biotechnol* 2009.

Youdim, K.A., Dobbie, M.S., Kuhnle, G., Proteggente, A.R., Abbott, N.J., and Rice-Evans, C. 2003. Interaction between flavonoids and the blood–brain barrier: In vitro studies. *J Neurochem* 85: 180–192.

Youdim, K.A., Qaiser, M.Z., Begley, D.J., Rice-Evans, C.A., and Abbott, N.J. 2004. Flavonoid permeability across an in situ model of the blood–brain barrier. *Free Radic Biol Med* 36: 592–604.

Youdim, K.A., Shukitt-Hale, B., and Joseph, J.A. 2004. Flavonoids and the brain: interactions at the blood–brain barrier and their physiological effects on the central nervous system. *Free Radic Biol Med* 37: 1683–1693.

Yu, H.L., Li, L., Zhang, X.H., Xiang, L., Zhang, J., Feng, J.F., and Xiao, R. 2009. Neuroprotective effects of genistein and folic acid on apoptosis of rat cultured cortical neurons induced by β-amyloid 31–35. *Br J Nutr* 102: 655–662.

Yüksel, A., and Bahadır-Yılmaz, E. 2021. The effect of mindfulness-based psychosocial skills training on functioning and insight level in patients with schizophrenia. *Community Ment Health* J 57: 365–371.

Zafari, S., Backes, C., Meese, E., and Keller, A. 2015. Circulating biomarker panels in Alzheimer's disease. *Gerontology* 61: 497–503.

Zbarsky, V., Datla, K.P., Parkar, S., Rai, D.K., Aruoma, O.I., and Dexter, D.T. 2005. Neuroprotective properties of the natural phenolic antioxidants curcumin and naringenin but not quercetin and fisetin in a 6-OHDA model of Parkinson's disease. *Free Radic Res* 39: 1119–1125.

Zhang, J.X., Xing, J.G., Wang, L.L., Jiang, H.L., Guo, S.L., and Liu, R. 2017. Luteolin inhibits fibrillary β-Amyloid1–40-induced inflammation in a human blood-brain barrier model by suppressing the p38 MAPK-mediated NF-κB signaling pathways. *Molecules* 22: 334.

Zhou, C., Huang, Y., and Przedborski, S. 2008. Oxidative stress in Parkinson's disease: A mechanism of pathogenic and therapeutic significance. *Ann N Y Acad Sci* 1147: 93–104.

Ziberna, L., Fornasaro, S., Čvorović, J., Tramer, F. and Passamonti, S., 2014. Bioavailability of flavonoids: the role of cell membrane transporters. In *Polyphenols in Human Health and Disease,* 489–511: Academic Press.

4 Dietary Polyphenols in Cancer

Ashish Shah, Vaishali Patel, and Ghanshyam Parmar
*E-mail: shah.ashishpharmacy@gmail.com; vaishalipatel48@yahoo.com;
ghanstaurus22@gmail.com

CONTENTS

DOI: 10.1201/9781003251538-4

4.1   INTRODUCTION

Cancer is a group of diseases characterized by uncontrollable growth of abnormal cells that have the ability to spread in various parts of the body. Cancer in body cells normally grows and multiplies as per the need of the body. Sometimes this normal response process breaks down and, instead of normal new cells, damaged or abnormal cells grow and multiply and may result in formation of tumor. Tumor may be benign or metastatic. Tumor cells have the ability to spread to, or invade nearby tissues. Benign tumor does not spread or invade nearby tissues, while malignant tumor can spread. The more than a hundred types of cancers, and classification or nomenclature are based on organ or tissues from where cancer starts to grow. The major classes of cancer include carcinoma, sarcoma, and lymphoma (Shewach and Kuchta, 2009).

Cancer is the leading cause of death. According to Global Cancer statistics, there were about 19.3 million new cancer cases and 10 million cancer deaths reported in the year 2020. The pharmacological treatment of cancer includes hormone therapy, immunotherapy, chemotherapy or a combination of all. Among these, the major obstacle to chemotherapy is Multi Drug Resistance (MDR). Continuous doses of chemotherapeutic agents reduce efflux outside the cell, which hampers permeation across the cell membranes. To understand permeation of cell across the cell, various nanocarriers are used in the treatment of cancer (Sung et al., 2021).

Many years ago, phytomedicines were widely explored for the treatment of different types of cancer, and which provide alternative therapy to that of conventional treatment. The major advantage of using phytochemicals as an anticancer agent is that they do not show any deleterious effect during the course of treatment. Phytochemicals are found in fruits, vegetables, plants, dietary fibres and natural products that can regulate many molecular processes. Plant-based phytochemicals like oils, gums, terpins, saponins, minerals, glycosides, flavonoids, lignins, podophyllotoxins, polyphenolic compounds, vinca alkaloids, and taxol constitute primary and secondary metabolites that play a major role in inhibition. Phytochemicals do not show any toxic effect on normal cells. The phytochemicals act by inhibiting various pathways like CDC2, CDK2, CDK4 kinases, topoisomerase, mTOR, TNK, PI3K, Akt, Bcl-2, Bcl-1, cytokines and interleukins.

4.1.1   Oxidative stress in cancer

Oxidative stress considered as the situation in which reactive oxygen species (ROS) formation exceed cellular antioxidant capacity. ROS producing continuously in aerobic cells as by-products of normal oxygen metabolism. Reactive oxygen species are generated as a response of endogenous and exogenous stimuli. Reactive oxygen species such as super oxide radical, hydroxy radical, singlet oxygen and hydrogen peroxide. Free radicals are highly reactive chemicals that have potential to harm cells. Free radicals are generated when atom or molecules gain or lose an electron. Although free radical generation is a normal response of the human body and plays an important role in various biological process, at high concentration they have proven hazardous to the body (Fiaschi and Chiarugi, 2012; Hayes et al., 2020). Free radicals have the ability to damage various components such as DNA, proteins and cell membranes. The damage caused by free radicals, especially with respect to DNA, may play role in development of cancer. Various types of carcinogens mediate their effect through formation of ROS during their metabolism. DNA damage via oxidative

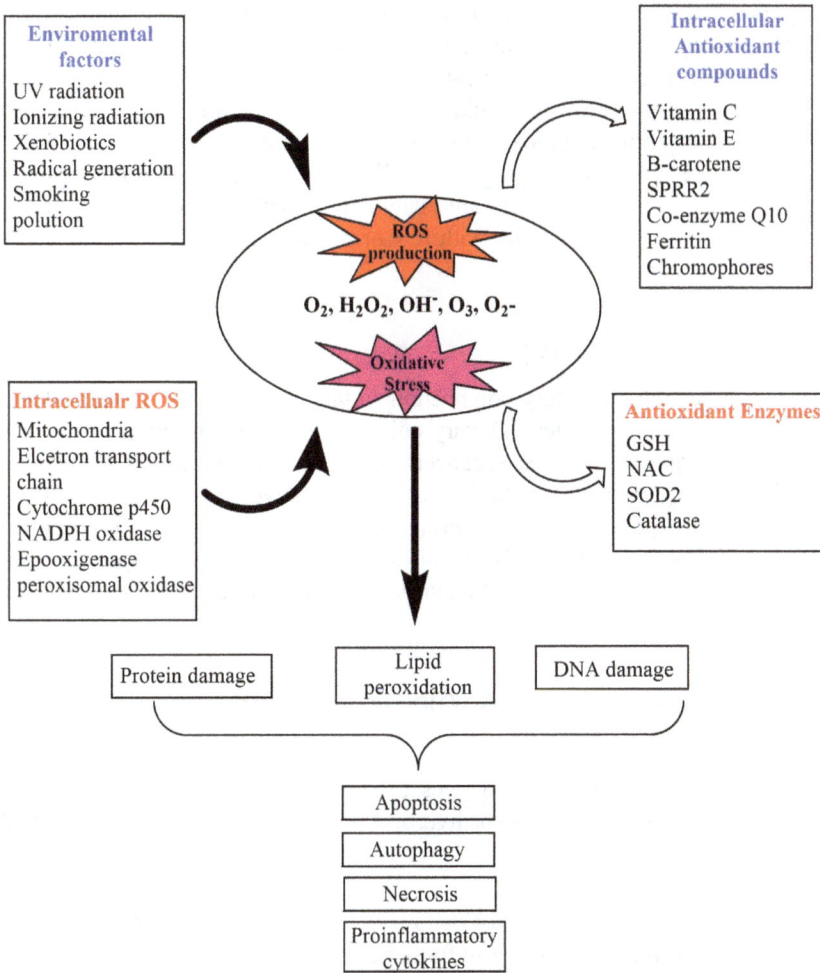

FIGURE 4.1 Oxidative stress-generation and its biological effect. Environmental factors and Intracellular ROS increase the oxidative stress. Antioxidant enzymes and intracellular antioxidant compounds decrease the oxidative stress.

stress cause mutations and therefore play important role in initiation and progression of carcinogenesis (Figure 4.1) (Gonzalez-Hunt et al., 2018).

Therefore, in case of disease conditions where ROS levels are elevated, antioxidants provide an important role in the balance of ROS. Oxidative stress is one of the main components involved in the development and growth of cancer cells. Elevated ROS levels cause down signaling of various antioxidant enzymes that result in abnormal activity of different molecular targets like NF-KB and Nrf2. Over-activation of these targets cause cancer cell proliferation, angiogenesis, metastasis, and suppression of cell death process (Kurutas, 2016).

4.1.2 WHAT ARE ANTIOXIDANTS?

Antioxidants are the chemicals that have the ability to interact with free radicals and have the potential to neutralize their effect. Antioxidants are also known as "free radical scavengers." The antioxidant synthesized directly by the body mechanism are known as endogenous antioxidants, and those that are taken from external sources are known as exogenous antioxidants. Exogenous antioxidants are also referred to as dietary antioxidants, which are obtained from various sources like fruits, vegetables, and grains. The antioxidant may obtain either from natural resources or synthetic products. The synthetic antioxidants are useful to improve health but, at the same time, adverse effects from the synthetic compounds constitute a serious issue. Natural products can provide solutions for the said purpose. There are many natural herbs/phytochemicals that have been reported for potential antioxidant activity with few or no side effects (Bouayed and Bohn, 2010; Dias et al., 2020).

4.2   DIETARY POLYPHENOLICS

The potential health benefits of dietary polyphenols in human health fetched tremendous attrition amongst various discipline researchers. Dietary polyphenols can be useful in a variety of clinical situations like neuro-degenerative disease, cancers, and cardiovascular diseases. Polyphenols have strong antioxidant activity, which helps to fight against oxidative stress caused by ROS (Williams et al., 2004). Chemically, polyphenols are a group of natural compounds with phenolic structural features (Figure 4.2). The majority of polyphenolics are found in nature as glycoside with different sugar units. For simplification, polyphenols can be classified based on the chemical structure of aglycone (Tsao, 2010).

4.2.1   Classification of Polyphenolics (Zhou et al., 2016)

4.2.1.1   Phenolic acids
Phenolic acid can be divided into major two types; benzoic acid and cinnamic acid derivatives. They are often found in bound form, which can be hydrolyzed upon enzymatic or chemical hydrolysis.

4.2.1.2   Flavonoids
The general structure of flavonoids has C6-C3-C5 backbone in which two C6 in ring A and Ring B are phenolic in nature. Flavonoids can be also further divided into sub-groups such as anthocyanins, flavan-3-ols, flavones, flavanones and flavanols. In the natural products most of these subclasses exist as glycoside.

4.2.1.3   Polyphenoleic amides
Capsaicinoids and avenanthramides are the two major polyphenolic amides which have major signification as first one is found in chili peppers and second one found in oats. Capsaicinoids like capsaicin has strong antioxidant and anti-inflammatory properties.

4.2.1.4   Other polyphenols
Several non-flavonoids, polyphenols found in foods, are considered as important for human health. These non-flavonoids include resveratrol, ellagic acid, curcumin, and Rosamarinic acid.

4.2.2   Potential Health Effects of Polyphenols

In the plant, the naturally occurring polyphenols are stored either in glycoside or non-glycoside conjugate form. The bioavailability of flavonoids in humans is dependent on the type of polyphenolics. The preliminary mechanism of action polyphenols was thought to be antioxidant, but this fact is not considered now, because of the inadequate concentration of polyphenolics in a tissue and due to this antioxidant effect of polyphenols is not enough for scavenging of free radicals from the tissue. Apart

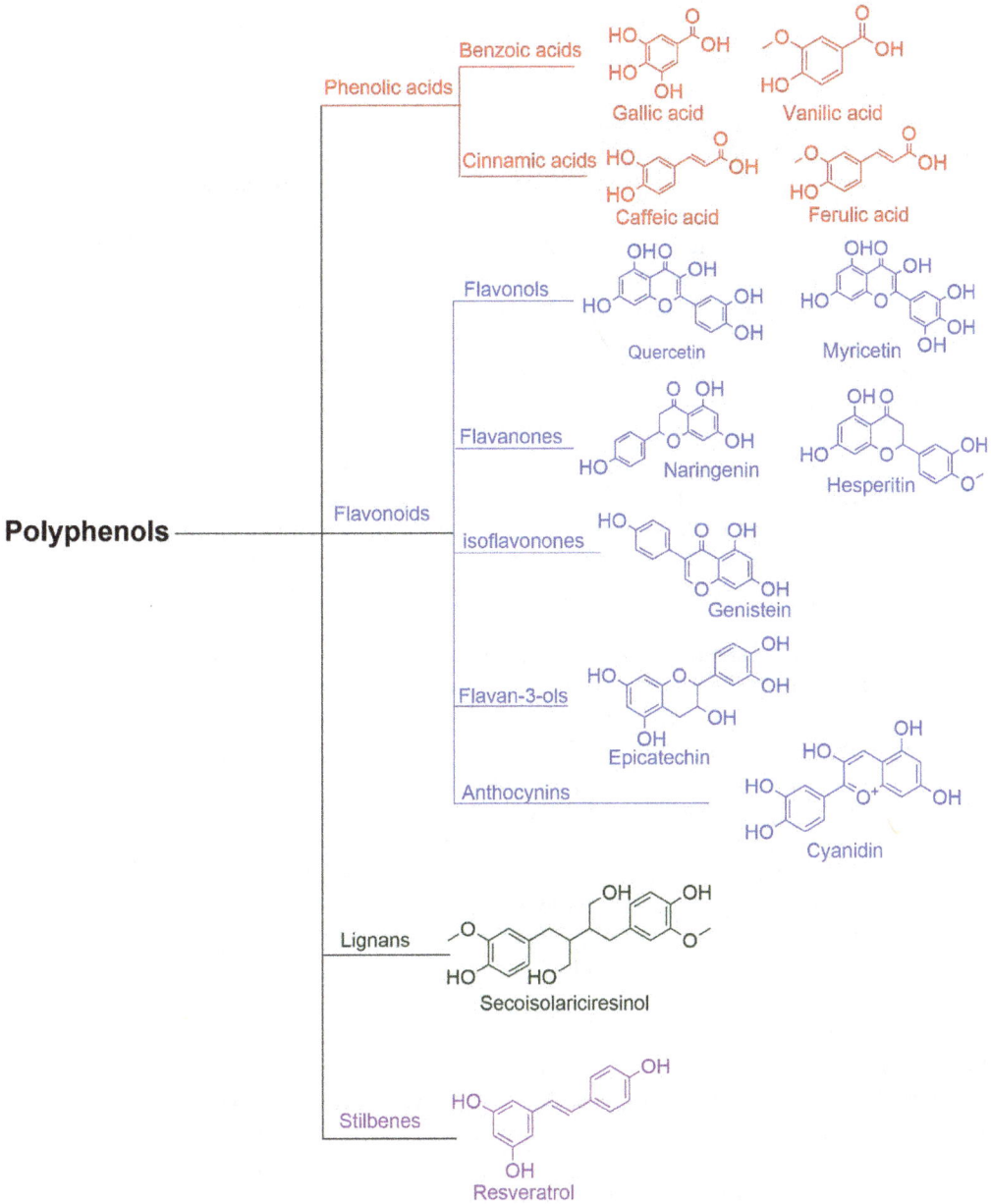

FIGURE 4.2 Classification of polyphenols with examples.

from antioxidant effect, certain biological effects have been also identified. This effect includes regulation of transcription factors and fat metabolism, diverse effect on inters and intra cellular signaling pathways, and modulation of various inflammatory mediators like TNF-α, IL-1β and IL-6. These facts are supported by a variety of clinical evidence. For example, some flavonoids havea potential effect on glucoregulation through downstream signaling that increase insulin secretion. The other beneficial effects include reduction of inflammation and oxidative stress and promotion of β-cell proliferation. Phlorizin found in apple is the competitive inhibitor of SGLT-1 and SGLT-2 and may be useful to treat hyperglycemia (Fraga et al., 2019).

4.2.3 ANTIOXIDANT ACTIVITY OF POLYPHENOLS

Antioxidants are chemicals that have the ability to interact with free radicals and have potential to neutralize their effect. Antioxidants are also known as "free radical scavengers." The antioxidants synthesized directly by the body mechanism are known as endogenous antioxidants, and those taken from external sources are known as exogenous antioxidants. Exogenous antioxidants are also referred as dietary antioxidants and are obtained from various sources like fruits, vegetables, and grains. The antioxidant may obtain either from natural resources or synthetic products. The synthetic antioxidants are useful to improve health but, at the same time, adverse effects of the synthetic compounds is a serious issue. Natural products can provide solutions for the said purpose. There are many natural herbs/phytochemicals that have been reported for potential antioxidant activity with few or no side effects (Bouayed and Bohn, 2010; Dias et al., 2020).

In order to fight against reactive oxygen species, cells develop different mechanisms to maintain redox homeostasis. This mechanism could provide direct and indirect antioxidant activity. Based on the mechanism, classification of antioxidants can be done. The direct antioxidant administered in high doses neutralize the deleterious effect of ROS (Tan et al., 2018). The indirect antioxidant enhances the antioxidant capacity of specific genes of antioxidant proteins through various transcription factors and, due to these physiological effects, last longer than direct antioxidants. The key transcription factor, nuclear factor-like 2 (Nrf2), is a basic zipper-containing transcription factor that activates many other enzymes that play important roles in cell defense by improving removal of ROS.

Polyphenols and carotenoids are two main phytochemicals that contribute most of the antioxidant properties of foods/plants. Natural polyphenolic compounds found in the human diet has antioxidant activity due to radical scavenging properties. Dietary polyphenols are classified in five different groups: flavonoids, stilbenes, tannins, phenolic acids, and coumarins (Yang, 2009). Total phenolic content and antioxidant activity have direct relationships. The literature data suggests that phytochemical extract that have higher phenolic content possesses higher antioxidant activity (Song et al., 2010). Carotenoids are a group of phytochemicals responsible for such as red, yellow, and orange colors in food. Various fruits and vegetables are the major sources of carotenoids. The various carotenoids, like α-carotene, β-carotene, cryptoxanthin, lycopene, and lutein, are found commonly in diets that have antioxidant activity (Stahl and Sies, 2003).

Some examples of antioxidant polyphenols are depicted as follows:

4.2.3.1 Catechin

Catechins are monomers of flavonols with a variety of similar compounds that include catechin, epigallocatechin, epicatechingallate (EGC) and epigallocatechingallate (EGCG). EGCG found in green tea as major component of polyphenols. EGCG reported for variety of biological activities, including antioxidant and anticancer (Beecher, 2003). Sahin et al. (Sahin et al., 2010) reported mode of action of EGCG to fight against oxidative stress. In that study it was noted that oxidative biomarkers are strongly correlated with hepatic nuclear transcription factors. In another study Romeo et al. (Romeo et al., 2009) reported that EGCG increase HO-1 activity in rat neurons, which is associated with Nrf2 activation and translocation. The resulting effect provides protection against oxidative stress. In the various studies it was found that catechin in combination with curcumin gives synergistic activity against human colon adenocarcinoma and larynx carcinoma.

4.2.3.2 Lycopene

Lycopene is a type of carotenoid that has powerful antioxidant activity and provides protection against free-radical induced damage to cells. The singlet oxygen-quenching capacity of lycopene is two-fold higher than β-carotene and ten-fold higher than α-tocopherol. Lycopene has the ability to regulate redox-sensitive molecular targets such as mitogen-activated protein kinas and Nrf2 (Palozza et al., 2012). The supplementation of lycopene in the form of tomato powder increases the expression of Nrf2, HO-1, gluthione, and other antioxidant enzymes. Lycopene has the ability to stimulate various transcription factors that induce antioxidant activity; in addition, lycopene metabolites also increase intracellular GSH levels and decrease ROS levels. The combination of lycopene with β-carotene could inhibit cell proliferation and arrest the cell cycle (Linnewiel et al., 2009). Lycopene along with β-carotene could induce apoptosis in human breast cancer. Additionally, lycopene is also useful against various types of cancer, such as colon, tongue and gastric (Jiang et al., 2018).

4.2.3.3 Curcumin

Curcumin, a polyphenolic found in turmeric has been reported for anti-inflammatory, antioxidant and anticancer activity. The antioxidant activity of curcumin is due to modulation of antioxidant enzyme by modulating Nrf2 (He et al., 2012). It has been reported that curcumin increases ARE binding od Nrf2 and induce expression of GST and NQO1. It also has been reported that the liver and lung of mice treated with dietary curcumin reduced the oxidative stress (Tapia et al., 2013). Sahin et al. reported that Nrf2 level linearly increases as the supplemental level of curcumin increases (Sahin et al., 2012). The anticancer potential of curcumin has been tested against a variety of cancer cell lines that include cervical, colorectal, and breast (Danafar et al., 2018; de Matos et al., 2018; Jalili-Nik et al., 2018). The anticancer potential of curcumin is due to its ability to inhibit cancer cell proliferation and to induce cancer cell death.

4.2.3.4 Resveratrol

Resveratrol, a polyphenolic compound found in peanuts, cranberries, and grapes, is reported for diverse biological actions (Calabrese et al., 2008). Resveratrol provides an antioxidant effect by reducing oxidative stress and preventing the action of free radicals. The antioxidant activity is due to increasing Nrf2 expression and its accumulation in the nucleus and, due to this, the deleterious effect of free radicals is reduced or nullified. The treatment of resveratrol can increase expression of Nrf2 targeted genes such as CAT, SOD, NQO1, HO-1 and γ-GCS. Furthermore, research data suggest that resveratrol increases Nrf2 expression that upregulates antioxidant/phase II enzymes that reduce the expression of heat shock protein (HSP) (Palsamy and Subramanian, 2011). Resveratrol has potent cytotoxic effect on cancer cells. The evidence is supported by transglutaminase type II inhibition in cholangiocarcinoma and gallbladder carcinoma (Sundar Dhilip Kumar et al., 2018).

4.2.3.5 Genistein

Genistein is type of isoflavone found in soybean that has structural similarity with mammalian estradiol. Genistein reported for antioxidant and anticancer activity. Genestein induces the expression of antioxidant enzymes manganese superoxide dismutase and catalase, which may associate with AMPK-activated protein kinase and phosphate. The anticancer activity of genistein observe in a variety of cancers, such as breast, ovary, prostate, urinary bladder, colon, liver, and stomach. In several studies it has been demonstrated that dietary soy supplementation reduces the inflammatory process related to prostate carcinogenesis (Kim et al., 2014; Mukund et al., 2017). Apart from this, many of the phytoantioxidants have been tested for anticancer activity, and some of those molecules are currently used in clinical practice. Still, detailed studies and preclinical investigations are needed to clarify their potential chemopreventive and anticancer activities.

4.2.4 ADME PROFILE OF POLYPHENOLS

ADME profile of polyphenols are dependent on their classes. There are substantial variations observed as per the specific class of polyphenols. Isoflavones and phenolic acids, such as gallic acid and caffeine, have more absorption compared to flavanones, catechins, and quercetin glycosides, while proanthocyanidines, anthocyanins, and galloylated tea catechins are the least-absorbed polyphenols. In polyphenolics, as the molecular weight increases the absorption rate decreases. In general, the absorption of polyphenols around 10–20 percent in the small intestine as most of the polyphenols present in the form of glycoside, esters, and polymers. However, the aglycon part of the polyphenols can be absorbed in the small intestine but, as most of the polyphenols are available in conjugated form, they must be hydrolyzed by intestinal enzymes or gut microbiota. The poor bioavailability in the small intestine allows polyphenols to enter into the large intestine, where it encounters colonic microflora. For example, hydrolyzation of quercetin-3-O-rhamnoside and quercetin-3-O-rhamnoglucoside by gut flora to the readily available form, quercetin. The absorption of polyphenols depends upon the amount ingested. Large doses of polyphenols are metabolized in the liver, while small doses catalyzed by intestinal mucosa with the liver playing a secondary role to further modify the polyphenols conjugate from the small intestine. The conversion into phytosomes may improve the efficacy without compromising safety. In the phytosome technology, polyphenol molecules bind with phospholipid or phosphatidylcholine by intermolecular bonding (Luca et al., 2020; Williamson, 2017).

4.3   POLYPHENOLS IN CANCER

There is ample evidence that suggests that dietary polyphenols could reduce the risk of cancers. Phenolic compounds have the ability to act against carcinogenesis by increasing the cell defense system, inhibition of anti-inflammatory and anti-cellular growth signal pathways, and activation of antioxidant enzymes. The natural polyphenols present in the diet give anticancer activity through a different mechanism that includes modulation of cell-cycle signaling, removal of anticancer agents, induction of antioxidant enzymes, apoptosis, and cell-cycle arrest. Resveratrol, epigallocatechin, and curcumin are the most extensively studied polyphenols (Zhang et al., 2000).The inhibitor effect of polyphenols in the cancer cell is due to two major reasons: (1) Modification of redox status, and (2) interference with basic cellular functions.

4.3.1   MODIFICATION OF REDOX STATUS

Reactive oxygen species have important rolse in regulation of cell replication and signal trans-duction activities. Severe oxidative stress and lack of antioxidant defense may cause damage to lipid, protein, and DNA. Due to this, oxidative damage is considered as one of the major factors in development and progression of cancer cells. Polyphenols due to their antioxidant properties may provide a chemopreventive effect. For example, tea polyphenols have the ability to inhibit carcinogen-induced DAN damage and oxidative stress. In addition, they also minimize the harmful effects of UV light on skin. EGCG a constituent of tea polyphenol has the ability to inhibit produc-tion of hydrogen peroxide by 60-80 percent. Resveratrol and tannins also possess hydrogen per-oxide inhibition activity (Barbaste et al., 2002; Higdon and Frei, 2003; Saija et al., 2000).

4.3.2   INTERFERENCE WITH CANCER CELL FUNCTIONS

4.3.2.1   Cell cycle

Proliferation of cells is controlled by the cell cycle, governed by activation of CDK complexes. Overexpression of CDKs, inactivation of tumor suppressor genes like p53 and retinoblastoma (Rb) results in dysregulation of the cell cycle. Due to this, CDKs, Rb, and p53 are considered important hallmarks of cancer. Down regulation of CDKs results in cell-cycle arrest that can occur either at G1/S or G2/M phase, ultimately leads to cell apoptosis. Polyphenols have shown the cell-cycle arrest effect on a different cell line that includes gastric, prostate, breast, pancreatic, hepatoma, endometrial, and lymphoma. Arrest particularly observed in the G1/S or G2/M phase in the presence of a mutated p53 gene. Quercetin-induced G1 phase arrest leads to apoptosis. Quercetin is also reported to arrest G2/M phase in the breast cancer cell line and the S phase in colorectal tumors. Several studies reported on the G1 phase arrest of human breast and prostate cancer induced by EGCG. EGCG have the direct ability to inhibit CKKs and to induce expression of p21 and p27. The ability of resveratrol to induce S-phase arrest and provide protective effect on paclitaxel-induced apoptosis of neuroblastoma cells (Masuda et al., 2001; Ramos et al., 2005).

4.3.2.2 Apoptosis

Excessive proliferation and reduction of apoptosis are two major characteristics of malignant cells. Oxidative stress, DNA damage, ion fluctuations, and inhibitory effects on cytokines – all these factors increase the apoptosis process. Much literature data suggest that polyphenols inhibit the cancer-cell growth by inducing apoptosis in different cell lines. The induction of apoptosis is caspase-3 dependent in many cases. The other possible mechanism for induction of apoptosis is due to their ability to stimulate H_2O_2 generation by cancer cells. For example, resveratrol increases the apoptosis effect in different leukemia β-cell lines. Resveratrol-induced cell death is also observed in breast carcinoma cell lines. In another example of EGCG-treated LNCaP cells, it was observed that apoptosis can also be observed after cell-cycle arrest, due to inhibition of NK-κB and activation of the p53 gene (Chen et al., 2005).

4.3.2.3 Invasion and metastasis

Cancer cells have the ability to migrate in surrounding tissues via blood vessels or lymphatic vessels and create secondary lesions. Secretion of several protein and down regulation of kinases play an important role in this process. Polyphenols were also found to suppress malignant cell migration, invasion and metastasis. Polyphenols have been shown to modulate expression of number of adhesion molecules. EGCG upregulates β1-integrin subunit while kaempferol, chrysin, apigenin, and luteolin inhibit intracellular adhesion molecule-1 expression in A549 alveolar epithelial cancer cells. In mouse melanoma animal model metastasis was 73 percent reduced upon intravenous administration of pterostilbene and quercetin. The anticancer activity of green tea catechins was reported by Liu et al. and found that green tea catechins suppress *in vitro* melanoma cell migration through inhibition of FAK activity (Pilorget et al., 2003).

4.3.2.4 Angiogenesis

Angiogenesis is considered one of the important processes for tumor growth, providing nutrient supply. Various growth factors, such as vascular endothelial growth factor (VEGF), transforming growth factor, platelet-derived growth factors, and modulation of metalloproteinases are considered as key regulators of the angiogenesis process. Flavonoids have been reported to affect the cancer-cell invasion process. For example, silymarin was found to reduce MMP secretion and vascular endothelial tube formation. Luteolin reported for inhibition of tumor growth and angiogenesis in murine xenograftmodel. Green tea polyphenols also found inhibitors of angiogenesis process by inhibiting VEGF and MMP-2 and 9 (Annabi et al., 2002; Jiang et al., 2000).

4.4   USE OF POLYPHENOLS IN VARIOUS TYPES OF CANCER

4.4.1   PROSTATE CANCER

Prostate cancer is the second most-commonly diagnosed cancer in males. Lifestyle and diet choice are the main reasons for development of this cancer type. Dietary components include several types of vegetables, and fruits can lower the incidence of this cancer. Scientific data suggest that various polyphenols present in the various dietary components have chemopreventive roles in prostate cancer. The mechanism for anticancer activity of polyphenols includes DNA methylation, epigenetic alterations, and down regulation of cell-signaling pathways that ultimately lead to inhibition of cell proliferation, induction of apoptosis, and cell-cycle arrest. Various epidemiological studies indicate that diet components, such as green tea red wine, tomatoes, turmeric, and cruciferous, and alliaceous vegetables can have the ability to reduce the incidence of prostate cancer (Costa et al., 2017; Deb et al., 2019; Nasir et al., 2020; Pejcic et al., 2019).

4.4.2 COLON CANCER

Colorectal cancer is the third most-common cancer type with a high rate of mortality. Multiple factors play a role in development and progression of this type of cancer. Eating habits can particularly affect the intestinal microbiota and could be one of the major reasons for the development of this cancer as intestinal microbiota have a defined role in homeostasis. Dietary polyphenols have favorable effects on intestinal microbiota. They also have the capability to reduce free-radical concentration and inflammation. Upon biotransformation, polyphenols are converted into simple aromatic carboxylic acids (phenolic acids) and these metabolites are more bioactive than precursor polyphenols. However, only 5–10 percent of total intake of polyphenol can be absorbed in the small intestine. EGCG gives inhibitory action on DNA methyltransferase that prevents abnormal methylation of DNA strands. For this inhibitory action, methylation of EGCG by catechol O-methyl transferase plays a very important role. Djulis, polyphenols reach crop, have also the ability to prevent colorectal cancer. Lee et al. reported that polyphenols present in this plant can reduce oxidative stress, increase apoptosis, and decrease the proliferation of oncogenic cells in mice. Another study, carried out by Hu et al., demonstrates the action of ginkgetinin prevention of VEGF-mediated angiogenesis in combination with resveratrol (Hu et al., 2019; Li et al., 2019; Vivarelli et al., 2019; Zhao and Zhang, 2020).

4.4.3 BREAST CANCER

Breast cancer is second most-frequent type of cancer in women throughout the world. A variety of reasons, like age, obesity, unhealthy diet, smoking, alcohol consumption, and occupational exposure can increase the risk of breast cancer. The effect of polyphenols in the prevention of breast cancer is controversial. At high concentrations polyphenols obtained from food or as a food supplement may inhibit the growth of breast cancer cells but, at lower concentrations, they can even stimulate the growth of breast cancer cells. Anticancer properties of resveratrol against breast cancer are well established. Resveratrol and its analogue are able to bind with the estrogen receptor and helpful in hormone anticancer therapy. Resveratrol can modulate ERα and p53 expression in estrogen receptor positive breast cancer. Apart from resveratrol, other polyphenols like curcumin, genistein, quercetin, silibinin, and luteolin are also widely investigated. Luteolin is the recent compound studied for breast cancer therapy. Antioxidants, anti-inflammatory, antiestrogenic and reduction in DNA alterations are the main mechanisms for the anticancer potential of luteolin. Apart from this, other natural polyphenols like EGCG and oleuropein have been also investigated (Ahmed et al., 2019; Horgan et al., 2019; Shaikh et al., 2019).

4.4.4 LUNG CANCER

Lung cancer is the second most common type of cancer with high mortality rate. Small-cell and non-small cell lung carcinoma are the two-types of lung cancer. A balanced diet of vegetables and fruit with consumption of natural products can be very useful to fight against lung carcinoma. The anticancer properties of some natural polyphenols due to antiproliferative, antimigratory, antimetastatic. and antiangiogenic properties. Resveratrol, curcumin and EGCG are the most studied compounds against lung-cancer treatment. Dieckol is another polyphenolic compound which was found to be active against non-small cell lung carcinoma (NSCLC) (Zhang et al., 2019; Zhou et al., 2019).

4.4.5 BLADDER CANCER

EGCG, curcumin and resveratrol exert antioxidant effects and also give a genotoxic effect for cancer cells. Resveratrol gives anticancer activity against bladder cancer cells by modulating TP53 gene. This gene responds to a stress signal that includes cell-cycle arrest, apoptosis, and DAN repair. Curcumin is also considered as a promising candidate in bladder cancer therapy as it modulates various signaling pathways such as PI3K, Akt, mTOR, and VEGF involved in progression and malignancy of bladder cancer (Almeida et al., 2019; Ashrafizadeh et al., 2020).

4.4.6 SKIN CANCER

One of the most common types of cancer worldwide is skin cancer, mainly occurring due to mutation in proto-oncogenes and tumor-suppressor genes. The mutation leads to excessive proliferation of skin cells. Exposure to UV radiation is considered one of the major reasons for development of skin cancer. Various literature data suggest that curcumin, resveratrol, coumarin, and epigallocatechin have anticancer effects due to their various pharmacological properties, such as antioxidant, anti-inflammatory, antiproliferative, and chemopreventive. Rodriguez_luna et al. reported the role of fucoxanthin in combination with rosmarinic acid in prevention of skin carcinoma. In another study Shin et al. reported the role of quercetin in skin cancer due to suppression of multiple targets (Heenatigala Palliyage et al., 2019; Sajadimajd et al., 2020).

4.4.7 PANCREATIC CANCER

One of the deadliest cancers, pancreatic cancer is caused due to poor prognosis, and that is the reason for diagnosis at an advance stage. Current chemotherapy is not very effective and unable to significantly improve life expectancy. Use of epigallocatechin in combination with gemcitabine have been found safe and effective. This combination is useful in blocking migration of pancreatic cancer. The combination inhibits the Akt pathway, and this pathway plays an important role in growth of pancreatic cancer cells (Wei et al., 2019).

4.4.8 LEUKEMIA

The effectiveness of chemotherapeutic agents is very low, as leukemia has an unfavorable prognosis and chances of reoccurrence are likely. Recent findings suggest a potential role of resveratrol in adjuvant therapy. The effectiveness is due to antioxidant and anti-inflammatory properties of resveratrol. Resveratrol is also helpful in lowering drug resistance. The role of curcumin was also established recently in myeloid leukemia therapy (Ivanova et al., 2019; Siedlecka-Kroplewska et al., 2019).

4.5 POLYPHENOLICS IN CANCER TARGETED DRUG DELIVERY SYSTEMS

The nanodelivery system of polyphenolic compounds is widely used as an anticancer agent. The polyphenolic compounds perform a key role in the suppression of reactive oxygen species. The ROS activity in the cancerous cells is higher and polyphenolic compounds inhibit ROS activity along with oxidative stress regulation. The polylactic-co-glycolic acid (PLGA) nanoparticles consist of different amounts of lactic acid and glycolic acid. They are biocompatible, biodegradable and hydrophobic in nature. Nanoformulations of polyphenols have been prepared using a wide variety of techniques that use carriers such as liposomes, nanocapsules, dendrimers, nanosheets, micelles, and so forth (Table 4.1 here).

Polyphenol nano formulations perform their anticancer activity via several cellular mechanisms: cell-cycle arrest at different phases of cancer, activation of caspase enzymes, reduction of tumor vascularization, reducing tumor cell invasion and metastasis. Apoptosis is the major indication of anticancer therapy. Many polyphenolic drugs exert their action through apoptosis, which is a programmed cell death for removal of unwanted cells (Abbaszadeh et al., 2019). For higher bioavailability, better antineoplastic activity, and to improve passive targeting to the cells, polyphenolicnano formulations are widely used.

Nanoparticles are ultrafine microscopic particles with a size less than 100 nm. They are further classified as nanospheres or nanocapsules. Nanosphere constitute the matrix system where drug is dispersed in the polymer matrix, and nanocapsules are the reservoir-based system in which a drug is enclosed in the inner cavity of polymeric membrane. Nanoparticles are obtained from natural sources, and their surfaces can be modified with polyethylene glycol to improve cellular uptake. Several polymers such as poly(lactide-co-glycolide), (PLGA), polylactide (PLA), polyglycolide, polycaprolactone (PCL), chitosan, PLGA-Polyethylene glycol have been developed for passive and ligand targeted delivery of active ingredient (Zhou et al., 2021). Various phyto antioxidants are used in combination with different polymers. Several findings suggest that the quercetin has proven its anticancer activity against C6 gliomacells. The PLGA poly lactic glycolic acid (PLGA) is used as a carrier for loading quercetin. Curcumin-loaded nanoparticles demonstrated a sustained release property and enhanced cellular bioavailability of curcumin, and further decreased cell viability in in vitro studies. The method of preparing nanoparticles include solvent evaporation method, nanoprecipitation, salting out, supercritical fluid, and dialysis.

4.5.1 LIPOSOMES

Liposomes are lipid vesicles which consist of lipid bilayer as an outer membrane and aqueous vehicle enclosed in the inner membrane. The internal core serves as a carrier for drugs, protein, vaccine and enzymes to be delivered to the targeted site. The liposomal delivery of betulinic acid enhances the penetration across the cell membrane and shows anticancer activity. Icariin-loaded liposomes formulated by anti-solvent evaporation technique showed potent cytotoxic activity against ovarian cancer. Liposomes of curcumin were found to be active against breast cancer cells, adenocarcinoma, and other tumor cells. Curcumin-loaded liposomes act by targeting MDF-7 and SKRB3 cells (Shindikar et al., 2016). Further, the in vivo and in vitro effects of curcumin on proliferation, apoptosis, signaling was carried out on human pancreatic cells. The curcumin was found to suppress its binding and decreased expression of this protein complex regulated gene products including cyclooxygenase-linked immunoassay, both of which can be implicated in tumor growth/invasiveness. The liposomes containing resveratrol and curcumin were developed in order to increase the bio availability of curcumin. The study was conducted on mice by taking serum and prostate tissue. The concentration of combination was found to be increased as compared to the curcumin alone. Quercetin liposomes were investigated to have an antiproliferative effect in numerous tumor cells like breast, leukaemia, colon, ovary, squamous cell, endometrial cells and non-small lung (Van Tran et al., 2019). The inhibition of tyrosine kinase was shown by the parenteral preparation of quercetin. The combination of quercetin and vincristine into the liposomal formulation exhibited effectiveness against the tumor growth of breast cells. The resveratrol in the pegylated state with and without glycine was evaluated for its anticancer activity. The results showed increased drug-loading efficiency without glycine.

4.5.2 SOLID LIPID NANOPARTICLES

Solid lipid nanoparticles consist of the solid and lipid phase in the formulation. The particle size of solid lipid nanoparticles ranges from 10 to 1000 nm. They are stabilized by salts, sterols, surfactants, phospholipids, and so forth. The nanoparticles fabricated from biocompatible polymer can incorporate hydrophilic and lipophilic drugs in their structure. The trans-resveratrol coated with chitosan has shown effectiveness for target cells HT-29 in colon cancer when combined with folic acid. The combination of piperine and quercetin shows cytotoxic effect against caco-2 and C26 cells. The myricetin loaded nanoparticles showed a cytotoxic effect against adenocarcinoma cancer cells. The encapsulation efficiency directly inhibits the growth of cancer cells. Sesamol is the phenol component of the sesame oil and it is produced due to hydrolysis of sesamolin during thermal oxidation. Sesamol is the phenolic component of sesame oil whose antioxidant activity is attributed to the presence of benzodioxole group in its ring structure (Bhardwaj et al., 2017). Solid lipid nanoparticles of sesame oil prepared using lecithin and polysorbate 80 showed hepatoprotective effect.

4.5.3 Nano micelles

If the solvent is polar, the hydrophilic portion of a molecule orient towards the outer surface to maximize contact with a polar solvent while the hydrophobic parts are clustered in the core region to minimize contact with polar solvent. This kind of molecular arrangement is referred to as normal micelles. Alternatively, the assembly is formed in the non-polar solvent which is known as reverse nano micelles. Insoluble drugs are loaded in the normal nano micelles while the soluble drugs are loaded in the reverse nano micelles. Here hydrophobic part of copolymer provides the core which allows encapsulation of drug whereas hydrophilic portion of polymer forms shell of the micelles. This property prevents its uptake by reticuloendothelial system and thereby prolongs its circulation into the blood stream. Polymer derives nano micelles, shows greater stability, and lowers critical micellar concentration value as compared to surfactant-based nanoparticles. Micelles of crosslinked N-isopropylacrylamide with N-Vinyl-2-pyrrolidone and PEG monoacrylate encapsulate curcumin. These nanoparticles demonstrated better efficacy comparable to that of pure curcumin against human pancreatic cancer. Marked improvement in the in vivo antitumor efficacy with two times decrease in the tumor growth was shown by folate conjugated micelles as compared to the non-targeted micelles. Folate-anchored pluronic-P105 and L-101 were investigated as micellar carriers for the delivery in the human breast cancer cells (Arora et al., 2010). It showed enhanced internalization of the entrapped therapeutic ingredient. Chitosan quercetinnano micelles, which demonstrated a strong release profile in the gastrointestinal environment, and has shown rapid cellular uptake as compared to free drugs (Hu et al., 2013). The polymeric micelles of luteolin with PEG-PCL by self-assembly method showed increased bioavailability in the C-26 colon carcinoma cells.

4.5.4 Dendrimers

Dendrimer is a nano-sized repetitively branched, tree like structure that comprises a central core, interior branches and exterior multifunctional surface groups. The role of branches is to accumulate as many as possible molecules like proteins, amino acids and so forth. The functional groups like carboxyl, amine or hydroxyl terminated can be easily incorporated with the help of surface functional groups (Abderrezak et al., 2012). Dendrosome is an amphipathic and biodegradable that has been used in the delivery of genes into cell lines. Poly (glycerol-succinic acid) dendrimers exhibited potential carrier for camptothecin. Increased cytotoxicity was observed in the human breast adenocarcinoma, colorectal carcinoma and glioblastoma. The Ursolic acid and folic acid conjugated with polyamidoamine (PAMAM) dendrimer can target HepG2 cells and increases cellular uptake. Dendrimers formulated by using gallic acid with polyamidoamine exhibited enhanced cytotoxic effect on MCF-7 breast cancer cells (Shen et al., 2018).

4.5.5 Gold nanoparticles

Gold is one of the noble metals, which is known for its unique optical properties and this popular phenomenon is known as localized surface plasmon resonance. Gold can easily make bonds with thiol and amine groups so that it would be preferable for gold to be chosen as carrier. Gold nanoparticles can be easily tagged with ligands for the site targeting. Biocompatibility, surface

TABLE 4.1
Various polyphenolic nanoformulations used in the cancer therapy (Abbaszadeh et al., 2019)

Phytoconstituent category	Phytoconstituent used	Nanoformulations	Targeted cells	Anticancer activity
Flavonoid	Baicalein	Targeted ligands of folate and hyaluronic acid	Human lung cancer dells	Decreases cell viability
	Chrysin	Nanosuspension	Human hepatocellular carcinoma	Decreases cell growth
	ECEG	Nanoparticles	Human melanoma	Increases apoptosis, Bax, Caspase-9
	Quercetin	Liposomes	Human breast cancer cells	Increased apoptosis
Flavonolignan	Silibinin	Pegylatednanoliposomes	Human hepatocellular carcinoma	Decreases cell viability
Stilbene	Resveratrol	Pegylatednanoliposomes	Human head and neck cancer	Anticancer effect at high concentration
Naphthoquinone	Plumbagin	Silver nano particles	Human skin cells	Increased apoptosis by production of free radicals

property, unique size and shape characteristics with internalization have shown capability of different gold nanoparticles in retinal pigment epithelium cells (Roy et al.). The resveratrol coated gold nanoparticles showed better effect inn diabetic retinopathy with the size of 20 nm. Curcuma wenyujin loaded gold nanoparticles showed good anticancer activity against breast cancer. Gold nanoparticles from ethanolic extract of *Plectranthus amboinicus* showed their anticancer effect in adenocarcinoma of lung cancer (Stakleff et al., 2012). Polyphenolic nanoformulations used in the cancer therapy are summarized in Table 4.1.

4.6   CLINICAL EVIDENCE OF POLYPHENOLS

Exogenous antioxidants have been shown to prevent the types of free radical damage that have been associated with cancer development. Therefore, researchers have investigated whether taking dietary antioxidant supplements can help lower the risk of developing or dying from cancer in humans. Many observational studies, including case–control studies and cohort studies, have been conducted to investigate (see Table 4.2) whether the use of dietary antioxidant supplements is associated with reduced risks of cancer in humans. Overall, these studies have yielded mixed results. Because observational studies cannot adequately control for biases that might influence study outcomes, the results of any individual observational study must be viewed with caution.

4.7   CONCLUSION

There is plenty of evidence suggesting that a balanced diet of fruits and vegetables due to presence of polyphenols can reduce the possible incidence of several types of cancers. Antioxidant and anti-inflammatory properties of polyphenols are the main pharmacological actions that contributing in anticancer properties. Apart from this there are multiple mechanisms that modulate the various

TABLE 4.2
Phytoantioxidants in Clinical Trial

Sl. No.	Trial name	Dose	Types of cancer pateint	Results
1	Linxian	β-carotene (15 mg)+ α-tocopherol (30 mg)+ selenium (50 μg) daily	Oesophageal and gastric	Initial: No effect on risk of developing either cancer; decreased risk of dying from gastric cancer only Later: no effect on risk of dying from gastric cancer Later: No effect on risk of dying from gastric cancer
2	α-tocopherol β-carotene cancer prevention study	β-carotene (20 mg)+ α-tocopherol (50 mg) daily	Middle-aged male smokers	Initial: increased incidence of lung cancer for those who took beta-carotene supplements Later: No effect of either supplement on incidence of urothelial, pancreatic, colorectal, renal cell, or upper aerodigestive tract cancers
3	Carotene and retinol efficacy trial	15 mg beta-carotene and 25,000 International Units	People at high risk of lung cancer because of a history of smoking or exposure to asbestos	Initial: increased risk of lung cancer and increased death from all causes – trial ended early. Later: higher risks of lung cancer and all-cause mortality persisted; no effect on risk of prostate cancer
4	Physicians' Health Study I (PHS I), United States	Beta-carotene supplementation (50 mg every other day for 12 years)	Male physicians	No effect on cancer incidence, cancer mortality, or all-cause mortality in either smokers or non-smokers
5	Women's Health Study (WHS), United States	Beta-carotene supplementation (50 mg every other day), vitamin E supplementation (600	Women ages 45 and older	Initial: no benefit or harm associated with 2 years of beta-carotene supplementation Later: No benefit or harm associated with 2 years of vitamin E supplementation
6	Supplémentation en Vitamines et MinérauxAntioxydants (SU.VI.MAX) Study, France	Daily supplementation with vitamin C (120 mg), vitamin E (30 mg), beta-carotene (6 mg), and the minerals selenium (100 μg) and zinc (20 mg) for a median of 7.5 years	Men and women	Initial: lower total cancer and prostate cancer incidence and all-cause mortality among men only; increased incidence of skin cancer among women only Later: No evidence of protective effects in men or harmful effects in women within 5 years of ending supplementation
7	Heart Outcomes Prevention Evaluation–The Ongoing Outcomes (HOPE–TOO) Study, International	Daily supplementation with alpha-tocopherol (400 IU) for a median of 7 years	People diagnosed with cardiovascular disease or diabetes	No effect on cancer incidence, death from cancer, or the incidence of major cardiovascular events

(*continued*)

TABLE 4.2 (Continued)
Phytoantioxidants in Clinical Trial

Sl. No.	Trial name	Dose	Types of cancer pateint	Results
8	Selenium and Vitamin E Cancer Prevention Trial (SELECT), United States	Daily supplementation with selenium (200 µg), vitamin E (400 IU), or both	Men ages 50 and older	Initial: no reduction in incidence of prostate or other cancers—trial stopped early Later: more prostate cancer cases among those who took vitamin E alone
9	Physicians' Health Study II (PHS II), United States	400 IU vitamin E every other day, 500 mg vitamin C every day, or a combination of the two	Male physicians ages 50 years and older	No reduction in incidence of prostate cancer or other cancers

molecular pathways, which ultimately decrease cancer cell survival, cancer cell proliferation and increase in apoptosis process. Resveratrol, epigallocatechin and curcumin are the most extensively studied polyphenols. Apart from having promising roles in cancer prevention, one of the major limitations is poor bioavailability upon administration of pure compound and that limits their use. The efficacy of polyphenols also effects by other constituents present in the diet. The efficacy and bioavailability can be improved by combinations with other phytochemical chemical and anticancer drugs or formulating in nanotechnology-based drug delivery system. Although, most of the polyphenols are found safe in terms of toxicity profile, some isoflavones like genistein and daidzein have hormone related adverse effect. Therefore, safety concern should also be considered while using of polyphenols. An other context is about limited clinical trials indicating anticancer activity of polyphenols. Larger randomized clinical trials should be carried out to generate more evidence justifying the anticancer potential of polyphenols.

REFERENCES

Abbaszadeh, H., Keikhaei, B., Mottaghi, S., 2019. A review of molecular mechanisms involved in anticancer and antiangiogenic effects of natural polyphenolic compounds. *Phytother Res* 33: 2002–2014.

Abderrezak, A., Bourassa, P., Mandeville, J.S., Sedaghat-Herati, R., Tajmir-Riahi, H.A., 2012. Dendrimers bind antioxidant polyphenols and cisplatin drug. *PloS one* 7: e33102.

Ahmed, S., Khan, H., Fratantonio, D., Hasan, M.M., Sharifi, S., Fathi, N., Ullah, H., Rastrelli, L., 2019. Apoptosis induced by luteolin in breast cancer: Mechanistic and therapeutic perspectives. *Phytomedicine* 59: 152883.

Almeida, T.C., Guerra, C.C.C., De Assis, B.L.G., de Oliveira Aguiar Soares, R.D., Garcia, C.C.M., Lima, A.A., da Silva, G.N., 2019. Antiproliferative and toxicogenomic effects of resveratrol in bladder cancer cells with different TP53 status. *Environ Mol Mutagen* 60: 740–751.

Annabi, B., Lachambre, M.P., Bousquet-Gagnon, N., Page, M., Gingras, D., Beliveau, R., 2002. Green tea polyphenol (-)-epigallocatechin 3-gallate inhibits MMP-2 secretion and MT1-MMP-driven migration in glioblastoma cells. *Biochim Biophys Acta* 1542: 209–220.

Antioxidants and Cancer Prevention, www.cancer.gov/about-cancer/causes-prevention/risk/diet/antioxidants-fact-sheet, Last Accessed on 26–01–2022.

Arora, R., Malhotra, P., Chawla, R., Gupta, D., Juneja, M., Kumar, R., Sharma, A., Baliga, M., Sharma, R., Tripathi, R., 2010. Herbal drugs for oncology: current status and future directions in cancer chemoprevention. *Herbal Medicine: A Cancer Chemopreventive and Therapeutic Perspective.* Jaypee Brothers Medical Publishers, Delhi: 3–41.

Ashrafizadeh, M., Yaribeygi, H., Sahebkar, A., 2020. Therapeutic effects of curcumin against bladder cancer: A review of possible molecular pathways. *Anticancer Agents Med Chem* 20: 667–677.

Barbaste, M., Berke, B., Dumas, M., Soulet, S., Delaunay, J.C., Castagnino, C., Arnaudinaud, V., Cheze, C., Vercauteren, J., 2002. Dietary antioxidants, peroxidation and cardiovascular risks. *J Nutr Health Aging* 6: 209–223.

Beecher, G.R., 2003. Overview of dietary flavonoids: Nomenclature, occurrence and intake. *J Nutr* 133: 3248S–3254S.

Bhardwaj, R., Sanyal, S.N., Vaiphei, K., Kakkar, V., Deol, P.K., Kaur, I.P., Kaur, T., 2017. Sesamol induces apoptosis by altering expression of Bcl-2 and Bax proteins and modifies skin tumor development in Balb/c mice. *Anticancer Agents Med Chem* 17: 726–733.

Bouayed, J., Bohn, T., 2010. Exogenous antioxidants – double-edged swords in cellular redox state: Health beneficial effects at physiologic doses versus deleterious effects at high doses. *Oxid Med Cell Longev* 3: 228–237.

Calabrese, V., Cornelius, C., Mancuso, C., Pennisi, G., Calafato, S., Bellia, F., Bates, T.E., Giuffrida Stella, A.M., Schapira, T., Dinkova Kostova, A.T., Rizzarelli, E., 2008. Cellular stress response: A novel target for chemoprevention and nutritional neuroprotection in aging, neurodegenerative disorders and longevity. *Neurochem Res* 33: 2444–2471.

Chen, D., Daniel, K.G., Chen, M.S., Kuhn, D.J., Landis-Piwowar, K.R., Dou, Q.P., 2005. Dietary flavonoids as proteasome inhibitors and apoptosis inducers in human leukemia cells. *Biochem Pharmacol* 69: 1421–1432.

Costa, C., Tsatsakis, A., Mamoulakis, C., Teodoro, M., Briguglio, G., Caruso, E., Tsoukalas, D., Margina, D., Dardiotis, E., Kouretas, D., Fenga, C., 2017. Current evidence on the effect of dietary polyphenols intake on chronic diseases. *Food Chem Toxicol* 110: 286–299.

Danafar, H., Sharafi, A., Kheiri, S., Kheiri Manjili, H., 2018. Co-delivery of Sulforaphane and Curcumin with PEGylated iron oxide-gold core shell nanoparticles for delivery to breast cancer cell line. *Iran J Pharm Res* 17: 480–494.

de Matos, R.P.A., Calmon, M.F., Amantino, C.F., Villa, L.L., Primo, F.L., Tedesco, A.C., Rahal, P., 2018. Effect of Curcumin-Nanoemulsion associated with photodynamic therapy in cervical carcinoma cell lines. *Biomed Res Int* 2018: 4057959.

Deb, G., Shankar, E., Thakur, V.S., Ponsky, L.E., Bodner, D.R., Fu, P., Gupta, S., 2019. Green tea-induced epigenetic reactivation of tissue inhibitor of matrix metalloproteinase-3 suppresses prostate cancer progression through histone-modifying enzymes. *Mol Carcinog* 58: 1194–1207.

Dias, T.R., Martin-Hidalgo, D., Silva, B.M., Oliveira, P.F., Alves, M.G., 2020. Endogenous and exogenous antioxidants as a tool to ameliorate male infertility induced by reactive oxygen species. *Antioxid Redox Signal*, 33: 11, 767–785

Fiaschi, T., Chiarugi, P., 2012. Oxidative stress, tumor microenvironment, and metabolic reprogramming: A diabolic liaison. *Int J Cell Biol*: 762825.

Fraga, C.G., Croft, K.D., Kennedy, D.O., Tomas-Barberan, F.A., 2019. The effects of polyphenols and other bioactives on human health. *Food Funct* 10: 514–528.

Gonzalez-Hunt, C.P., Wadhwa, M., Sanders, L.H., 2018. DNA damage by oxidative stress: Measurement strategies for two genomes. *Curr. Opin. Toxicol.* 7: 87–94.

Hayes, J.D., Dinkova-Kostova, A.T., Tew, K.D., 2020. Oxidative stress in cancer. *Cancer Cell* 38: 167–197.

He, H.J., Wang, G.Y., Gao, Y., Ling, W.H., Yu, Z.W., Jin, T.R., 2012. Curcumin attenuates Nrf2 signaling defect, oxidative stress in muscle and glucose intolerance in high fat diet-fed mice. *World J Diabetes* 3: 94–104.

Heenatigala Palliyage, G., Singh, S., Ashby, C.R., Jr., Tiwari, A.K., Chauhan, H., 2019. Pharmaceutical topical delivery of poorly soluble polyphenols: Potential role in prevention and treatment of melanoma. *AAPS PharmSciTech* 20: 250.

Higdon, J.V., Frei, B., 2003. Tea catechins and polyphenols: health effects, metabolism, and antioxidant functions. *Crit Rev Food Sci Nutr* 43: 89–143.

Horgan, X.J., Tatum, H., Brannan, E., Paull, D.H., Rhodes, L.V., 2019. Resveratrol analogues surprisingly effective against triple-negative breast cancer, independent of ERalpha. *Oncol Rep* 41: 3517–3526.

Hu, B., Ting, Y., Zeng, X., Huang, Q., 2013. Bioactive peptides/chitosan nanoparticles enhance cellular antioxidant activity of (-)-epigallocatechin-3-gallate. *J Agric Food Chem* 61: 875–881.

Hu, W.H., Chan, G.K., Duan, R., Wang, H.Y., Kong, X.P., Dong, T.T., Tsim, K.W., 2019. Synergy of ginkgetin and resveratrol in suppressing VEGF-induced angiogenesis: A therapy in treating colorectal cancer. *Cancers (Basel)* 11.

Ivanova, D., Zhelev, Z., Semkova, S., Aoki, I., Bakalova, R., 2019. Resveratrol modulates the redox-status and cytotoxicity of anticancer drugs by sensitizing leukemic lymphocytes and protecting normal lymphocytes. *Anticancer Res* 39: 3745–3755.

Jalili-Nik, M., Soltani, A., Moussavi, S., Ghayour-Mobarhan, M., Ferns, G.A., Hassanian, S.M., Avan, A., 2018. Current status and future prospective of Curcumin as a potential therapeutic agent in the treatment of colorectal cancer. *J Cell Physiol* 233: 6337–6345.

Jiang, C., Agarwal, R., Lu, J., 2000. Anti-angiogenic potential of a cancer chemopreventive flavonoid antioxidant, silymarin: Inhibition of key attributes of vascular endothelial cells and angiogenic cytokine secretion by cancer epithelial cells. *Biochem Biophys Res Commun* 276: 371–378.

Jiang, L.N., Liu, Y.B., Li, B.H., 2018. Lycopene exerts anti-inflammatory effect to inhibit prostate cancer progression. *Asian J Androl* 21: 80–85.

Kim, S.H., Kim, C.W., Jeon, S.Y., Go, R.E., Hwang, K.A., Choi, K.C., 2014. Chemopreventive and chemotherapeutic effects of genistein, a soy isoflavone, upon cancer development and progression in preclinical animal models. *Lab Anim Res* 30: 143–150.

Kurutas, E.B., 2016. The importance of antioxidants which play the role in cellular response against oxidative/nitrosative stress: urrent state. *Nutr J* 15: 71.

Li, Y., Gao, X., Lou, Y., 2019. Interactions of tea polyphenols with intestinal microbiota and their implication for cellular signal conditioning mechanism. *J Food Biochem* 43: e12953.

Linnewiel, K., Ernst, H., Caris-Veyrat, C., Ben-Dor, A., Kampf, A., Salman, H., Danilenko, M., Levy, J., Sharoni, Y., 2009. Structure activity relationship of carotenoid derivatives in activation of the electrophile/antioxidant response element transcription system. *Free Radic Biol Med* 47: 659–667.

Luca, S.V., Macovei, I., Bujor, A., Miron, A., Skalicka-Wozniak, K., Aprotosoaie, A.C., Trifan, A., 2020. Bioactivity of dietary polyphenols: The role of metabolites. *Crit Rev Food Sci Nutr* 60: 626–659.

Masuda, M., Suzui, M., Weinstein, I.B., 2001. Effects of epigallocatechin-3-gallate on growth, epidermal growth factor receptor signaling pathways, gene expression, and chemosensitivity in human head and neck squamous cell carcinoma cell lines. *Clin Cancer Res* 7: 4220–4229.

Mukund, V., Mukund, D., Sharma, V., Mannarapu, M., Alam, A., 2017. Genistein: Its role in metabolic diseases and cancer. *Crit Rev Oncol Hematol* 119: 13–22.

Nasir, A., Bullo, M.M.H., Ahmed, Z., Imtiaz, A., Yaqoob, E., Jadoon, M., Ahmed, H., Afreen, A., Yaqoob, S., 2020. Nutrigenomics: Epigenetics and cancer prevention: A comprehensive review. *Crit Rev Food Sci Nutr* 60: 1375–1387.

Palozza, P., Catalano, A., Simone, R., Cittadini, A., 2012. Lycopene as a guardian of redox signalling. *Acta Biochim Pol* 59: 21–25.

Palsamy, P., Subramanian, S., 2011. Resveratrol protects diabetic kidney by attenuating hyperglycemia-mediated oxidative stress and renal inflammatory cytokines via Nrf2-Keap1 signaling. *Biochim Biophys Acta* 1812: 719–731.

Pejcic, T., Tosti, T., Dzamic, Z., Gasic, U., Vuksanovic, A., Dolicanin, Z., Tesic, Z., 2019. The Polyphenols as potential agents in prevention and therapy of prostate diseases. *Molecules* 24.

Pilorget, A., Berthet, V., Luis, J., Moghrabi, A., Annabi, B., Beliveau, R., 2003. Medulloblastoma cell invasion is inhibited by green tea (-)epigallocatechin-3-gallate. *J Cell Biochem* 90: 745–755.

Ramos, S., Alia, M., Bravo, L., Goya, L., 2005. Comparative effects of food-derived polyphenols on the viability and apoptosis of a human hepatoma cell line (HepG2). *J Agric Food Chem* 53: 1271–1280.

Romeo, L., Intrieri, M., D'Agata, V., Mangano, N.G., Oriani, G., Ontario, M.L., Scapagnini, G., 2009. The major green tea polyphenol, (-)-epigallocatechin-3-gallate, induces heme oxygenase in rat neurons and acts as an effective neuroprotective agent against oxidative stress. *J Am Coll Nutr* 28: Suppl: 492S-499S.

Roy, A.D., Varghese, K.K., Ghosh, S., Paul, P., Mitra, I., 2022. Metallic nanoparticles from natural products as a potential anti-cancer. *Int J Pharm Sci Rev Res* 67 (1): 171–177. doi:10.47583/ijpsrr.2021.v67i01.028

Sahin, K., Orhan, C., Tuzcu, M., Ali, S., Sahin, N., Hayirli, A., 2010. Epigallocatechin-3-gallate prevents lipid peroxidation and enhances antioxidant defense system via modulating hepatic nuclear transcription factors in heat-stressed quails. *Poult Sci* 89: 2251–2258.

Sahin, K., Orhan, C., Tuzcu, Z., Tuzcu, M., Sahin, N., 2012. Curcumin ameliorates heat stress via inhibition of oxidative stress and modulation of Nrf2/HO-1 pathway in quail. *Food Chem Toxicol* 50: 4035–4041.

Saija, A., Tomaino, A., Trombetta, D., De Pasquale, A., Uccella, N., Barbuzzi, T., Paolino, D., Bonina, F., 2000. In vitro and in vivo evaluation of caffeic and ferulic acids as topical photoprotective agents. *Int J Pharm* 199: 39–47.

Sajadimajd, S., Bahramsoltani, R., Iranpanah, A., Kumar Patra, J., Das, G., Gouda, S., Rahimi, R., Rezaeiamiri, E., Cao, H., Giampieri, F., Battino, M., Tundis, R., Campos, M.G., Farzaei, M.H., Xiao, J., 2020. Advances on natural polyphenols as anticancer agents for skin cancer. *Pharmacol Res* 151: 104584.

Shaikh, A.A., Braakhuis, A.J., Bishop, K.S., 2019. The Mediterranean Diet and breast cancer: A personalised approach. *Healthcare* (Basel) 7.

Shen, Z., Li, B., Liu, Y., Zheng, G., Guo, Y., Zhao, R., Jiang, K., Fan, L., Shao, J., 2018. A self-assembly nanodrug delivery system based on amphiphilic low generations of PAMAM dendrimers-ursolic acid conjugate modified by lactobionic acid for HCC targeting therapy. *Nanomedicine* 14: 227–236.

Shewach, D.S., Kuchta, R.D., 2009. Introduction to cancer chemotherapeutics. *Chem Rev* 109: 2859–2861.

Shindikar, A., Singh, A., Nobre, M., Kirolikar, S., 2016. Curcumin and resveratrol as promising natural remedies with nanomedicine approach for the effective treatment of triple negative breast cancer. *J Oncol* 2016.

Siedlecka-Kroplewska, K., Wozniak, M., Kmiec, Z., 2019. The wine polyphenol resveratrol modulates autophagy and induces apoptosis in MOLT-4 and HL-60 human leukemia cells. *J Physiol Pharmacol* 70.

Song, F.L., Gan, R.Y., Zhang, Y., Xiao, Q., Kuang, L., Li, H.B., 2010. Total phenolic contents and antioxidant capacities of selected Chinese medicinal plants. *Int J Mol Sci* 11: 2362–2372.

Stahl, W., Sies, H., 2003. Antioxidant activity of carotenoids. *Mol Aspects Med* 24: 345–351.

Stakleff, K.S., Sloan, T., Blanco, D., Marcanthony, S., Booth, T.D., Bishayee, A., 2012. Resveratrol exerts differential effects in vitro and in vivo against ovarian cancer cells. *Asian Pac J Cancer Prev* 13: 1333–1340.

Sundar Dhilip Kumar, S., Houreld, N.N., Abrahamse, H., 2018. Therapeutic potential and recent advances of Curcumin in the treatment of aging-associated diseases. *Molecules* 23.

Sung, H., Ferlay, J., Siegel, R.L., Laversanne, M., Soerjomataram, I., Jemal, A., Bray, F., 2021. Global Cancer Statistics 2020: GLOBOCAN Estimates of Incidence and Mortality Worldwide for 36 Cancers in 185 Countries. *CA Cancer J Clin* 71: 209–249.

Tan, B.L., Norhaizan, M.E., Liew, W.P., Sulaiman Rahman, H., 2018. Antioxidant and oxidative stress: A mutual interplay in age-related diseases. *Frontiers in Pharmacology* 9: 1162.

Tapia, E., Zatarain-Barron, Z.L., Hernandez-Pando, R., Zarco-Marquez, G., Molina-Jijon, E., Cristobal-Garcia, M., Santamaria, J., Pedraza-Chaverri, J., 2013. Curcumin reverses glomerular hemodynamic alterations and oxidant stress in 5/6 nephrectomized rats. *Phytomedicine* 20: 359–366.

Tsao, R., 2010. Chemistry and biochemistry of dietary polyphenols. *Nutrients* 2: 1231–1246.

Van Tran, V., Moon, J.-Y., Lee, Y.-C., 2019. Liposomes for delivery of antioxidants in cosmeceuticals: Challenges and development strategies. *J Control Release* 300: 114–140.

Vivarelli, S., Salemi, R., Candido, S., Falzone, L., Santagati, M., Stefani, S., Torino, F., Banna, G.L., Tonini, G., Libra, M., 2019. Gut microbiota and cancer: From pathogenesis to therapy. *Cancers* (Basel) 11 .

Wei, R., Penso, N.E.C., Hackman, R.M., Wang, Y., Mackenzie, G.G., 2019. Epigallocatechin-3-Gallate (EGCG) suppresses pancreatic cancer cell growth, invasion, and migration partly through the inhibition of Akt Pathway and Epithelial-Mesenchymal Transition: Enhanced efficacy when combined with Gemcitabine. *Nutrients* 11.

Williams, R.J., Spencer, J.P., Rice-Evans, C., 2004. Flavonoids: antioxidants or signalling molecules? *Free Radic Biol Med* 36: 838–849.

Williamson, G., 2017. The role of polyphenols in modern nutrition. *Nutr Bull* 42: 226–235.

Yang, J., 2009. Brazil nuts and associated health benefits: A review. *LWT – Food Science and Technology* 42: 1573–1580.

Zhang, L., Xie, J., Gan, R., Wu, Z., Luo, H., Chen, X., Lu, Y., Wu, L., Zheng, D., 2019. Synergistic inhibition of lung cancer cells by EGCG and NF-kappaB inhibitor BAY11-7082. *J Cancer* 10: 6543–6556.

Zhang, M., Zhang, J.P., Ji, H.T., Wang, J.S., Qian, D.H., 2000. Effect of six flavonoids on proliferation of hepatic stellate cells in vitro. *Acta Pharmacol Sin* 21: 253–256.

Zhao, Y., Zhang, X., 2020. Interactions of tea polyphenols with intestinal microbiota and their implication for anti-obesity. *J Sci Food Agric* 100: 897–903.

Zhou, H., Zheng, B., McClements, D.J., 2021. Encapsulation of lipophilic polyphenols in plant-based nanoemulsions: impact of carrier oil on lipid digestion and curcumin, resveratrol and quercetin bioaccessibility. *Food Funct* 12: 3420–3432.

Zhou, Q., Pan, H., Li, J., 2019. Molecular insights into potential contributions of natural polyphenols to lung cancer treatment. *Cancers (Basel)* 11.

Zhou, Y., Zheng, J., Li, Y., Xu, D.P., Li, S., Chen, Y.M., Li, H.B., 2016. Natural polyphenols for prevention and treatment of cancer. *Nutrients* 8.

5 Dietary Polyphenols in Cardiovascular Diseases (CVDs)

Johra Khan and Mithun Rudrapal
E-mail: j.khan@mu.edu.sa; rsmrpal@gmail.com

CONTENTS

5.1 INTRODUCTION

Cardiovascular disease (CVD) is not only the leading cause of death in developing countries, but also in developed countries. It is found responsible for one third of deaths across the globe (Del Rio et al., 2010). Cardiovascular disease is a term used for heart and blood vessels related disease condition (Del Rio et al., 2013, Kumar and Pandey, 2013). Most CVD is found to be caused by atherosclerosis, which is an inflammatory disease resulting from intrinsic activation of different factors like coagulation factors, chemotactic factors, vasoactive mediators, and complement pathways (Grosso et al., 2014). An increasing risk of CVD is found related with sedentary lifestyles, high cholesterol levels, obesity, hypertension, smoking, a high level of triglycerides, diabetes, and genetic factors that are activated due to bad diet and lifestyle (Zamora-Ros et al., 2013). Some specific mutations in low-density lipoprotein (LDL-R) receptors cause disruption in absorption of LDL-cholesterol, its metabolism, and homeostasis (Owen et al., 2009). The macrophages uptake LDL-C in subendothelial space and promotes the formation of plaques in blood vessels causing weakness and narrowing of vessel walls (Godos et al., 2017a). These plaques ultimately rupture in vessels, leading to vascular obstruction development.

Nowadays a lot of researchers are focusing on abundant dietary compounds to find the solution for different disease conditions, including CVD (Gaziano et al., 2006). Plant polyphenols are secondary metabolites found in various fruits, cereals, and vegetables (Verhaar et al., 2002, Wilkins et al., 2017). Fruits and vegetables having significant concentration of polyphenols are bitter, have a specific taste, and color. Beverages like a cup of coffee or a glass of red wine contains around 100 mg of polyphenols; similarly, chocolates and legumes also help in dietary polyphenol supply (Scalbert et al., 2005, Grosso et al., 2017a).

Different epidemiological meta-analysis studies at the end of twentieth century suggest that continuous use of dietary polyphenols increase immunity against various diseases such as cancer, CVD, osteoporosis, neurodegenerative disease, and diabetes (Pandey and Rizvi, 2009). Due to

DOI: 10.1201/9781003251538-5

medicinal benefits of plant-based polyphenols, research and scientific studies are increasing (Cory et al., 2018). The studies until now have grouped polyphenols into two broad classes: (a) flavonoid and (b) nonflavonoid (Abbas et al., 2017). Flavonoids are structurally made of rings of oxygenated heterocycle and two rings of phenols (Kinoshita et al., 2005). The different types of flavonoids are identified on the basis of oxidation of the heterocyclic pyran ring (Singla et al., 2019, Rudrapal et al., 2020). The three classifications on the basis of these structural modifications are flavonols, anthrocyanins, and flavanols (Cheynier et al., 2006, Tanaka et al., 2008). Until now, around 4,000 have been identified in plants and this is continuously increasing with data from current research (Khan et al., 2021). The non-flavonoids include different phenolic acids like benzoic and hydroxycinnamic acids, with many other naturally occurring nonflavonoids, such as gallotannins, lignans, ellagitannins, and stilbene oligomers (Andrés-Lacueva et al., 2010, Boccellino and D'Angelo, 2020). As per the US Department of Health and Human Services, each individual should consume at least five portions of fruits and vegetables every day, which is equates to 80 grams, as this helps to maintain body mass index between 18.5 and 24.9 kg/m^2 (Guenther et al., 2006, Blanck et al., 2008). The *in vivo* effect of polyphenols depends on several factors, like method and grade of extraction and type of food (Wijesinghe and Jeon, 2012). It also depends on intestinal absorption, target tissues, metabolism, and their biological action (Tsao, 2010).

Flavonols like kaempferol, isorhamnetin, myricetin, and quercetin are found in foods such as spices, cocoa, and berries in high concentration (Erdman Jr et al., 2007, Miean and Mohamed, 2001). Citrus fruits contain large amount of different types of flavanones, the highest being naringenin and hesperetin (Erlund et al., 2002, Wilcox et al., 1999). Black and green teas contain flavan-3-ols, including gallocatechin, epicatechingallate, procyanidin, epicatechin, epigallocatechin, epigallocatechingallate, and epicatechingallate (Nakai et al., 2005, Nawrot-Hadzik et al., 2021). Green tea is reported to contain fourteen different polyphenolic compounds in flavan-3-ol and 4 compounds of flavonol and phenolic acid class (Díaz-Mula et al., 2019, Dai et al., 2018). Anthocyanin classes chiefly contain anthocyanidin and anthocyanin, especially in red and purple fruits and vegetables (Mattioli et al., 2020, Martinotti et al., 2021). Soy products like tofu, roasted soy nuts, and miso contain isoflavones, which resemble estrogen in structure and are also known as phytoestrogen (Coward et al., 1993). Lignans is another class of polyphenols, which are categorized by 1, 4 diarylbutane structure found in high concentrations in flax and sesame seed (Tangney and Rasmussen, 2013). The basic polyphenols in phenolic acid are hydroxybenzoic and hydroxycinnamic acid, which are found in significant amount in walnuts, blueberries, coffee, and plums (Abbas et al., 2017). Tyrosol is another subclass of polyphenols specially found in oils like olive oil (Tangney and Rasmussen, 2013).

Many recent studies on the co-relation of dietary polyphenol consumption and cardiovascular diseases in many countries reported a linear association between increased consumption of flavonoids and low risk of CVD (Adriouch et al., 2018, Bazzano et al., 2005). Similar study results with flavan-3-ol and anthocyanin with CVD are also reported. In this chapter we summarized the relation between polyphenol rich food and different cardiovascular diseases, vascular endothelium health, and hypertension risk.

5.2   EFFECT OF POLYPHENOL ON HYPERTENSION

A study on polyphenol consumption and hypertension including 15 cross-section and 7 prospective cohort studies found no correlation between total flavonoids with risk of hypertension, whereas another group consuming anthocyanin was reportedly associated with decreases in hypertension risk (Billingsley and Carbone, 2018, Sosnowska et al., 2017). Another meta-analysis based study including more than 200,000 volunteers with more than 45,000 cases of hypertension patients showed high phytoestrogens and phenolic acid consumption reduces the risk of hypertension (Godos et al., 2018, Grassi et al., 2008).

Five different studies using total polyphenol, and some based on individual classes of polyphenols on approximately 2,500 individuals, included patients with type 2 diabetes who reported significant decreases in systolic and diastolic blood pressure, especially in combined or total polyphenols, whereas the single flavonoids, or phenolic acid, have low effect on CVD (Godos et al., 2019). Another study was conducted on females aged between 18 to 75 years and focused on intake of total flavonoids with their subclasses with their effect on in association systolic blood pressure (Lye et al., 2009, Durazzo et al., 2019). A pan-European study of an adolescent population joined the HELENA (Healthy Lifestyle in Europe by Nutrition in Adolescence) study in which polyphenol dietary intake assessed on the basis of 24-hour consumption for 2 days was evaluated on blood pressure (Moreno et al., 2008, Ruiz et al., 2006). Results of total phenolic intake in diet showed a significant effect on high blood pressure, but polyphenol subclass intakes have no association with high diastolic blood pressure (Ruiz et al., 2006). This study collected data from more than 8,000 participants. HAPIEE (Health, Alcohol and Psychosocial factors In Eastern Europe) is another recent study conducted on more than 8,800 Polish individuals, including both male and female, using a questioner based on intake of dietary polyphenols (Grosso et al., 2017b). This study estimated total and subclasses of polyphenol using the phenol explorer database, which showed linear inverse correlation between total polyphenol, and a significant inverse association was reported between phenolic acid and blood pressure (Marventano et al., 2016).

Some studies focused on single classes, or single compounds, of polyphenols. A Brazilian cohort-based study focused on the association between coffee polyphenols, including phenolic acids and hypertension (Miranda et al., 2017, Carnauba et al., 2021a, O'Keefe et al., 2013, Carnauba et al., 2021b). This study found an inverse correlation between phenolic acid and hypertension. MEAL (Men's Eating and Living), a cohort based study conducted on more than 1900 southern Italian adults, assessed a food-based questioner using a phenol explorer database with a median intake of 522.2 mg/day, found to significantly reduce the hypertension, whereas some subclasses of phenolic acid like hydroxyphenylacetic acids is also inversely associated with hypertension. Another study, known as KNHANES, surveyed a South Korean population of approximately 1,800 men and 2,900 women, and studied a 24-hour dietary recall of flavonols intake using different databases, including USDA (United States Data base) flavonoid content of food, the Japan Functional Food Factor Database, and the Rural Development Administration Food Functional Composition (Yang et al., 2012, Lee et al., 2017). This study result also reported an inverse association between high BP (systolic blood pressure ≥ 130 mm Hg, and diastolic blood pressure ≥ 85 mm Hg) in women participants but did not show the same association in male participants.

5.2.1 Effect of Phytoestrogens on Hypertension

To study the effect of phytoestrogen, several researchers focused on the dietary intake of polyphenols on the occurrence of hypertension. Eight different studies investigated the association of phyto-estrogen on hypertension on different populations (Woo et al., 2019). One study based on Caucasian and African-American populations recruited males and females between 18 to 30 years of age for a multivariate study of dietary isoflavone intake, and the participants were followed for up to 20 years (Zhang et al., 2005). This study reported that the highest quartile of isoflavone intake can significantly reduce the systolic blood pressure by 4.4 nm Hg. A second study was conducted on a Mediterranean population, using a MEAL cohort (Miyake et al., 2005). It reported that even the highest quartile of total phytoestrogen has no significant effect on isoflavones and lignans individually – however, compounds like biochanin A and pinoresinol show an inverse effect with hypertension. The PREDIMED (Prevención con Dieta Mediterránea) trial is another study conducted on an

TABLE 5.1

Different studies investigating the relation between dietary polyphenol intake and hypertension

Polyphenol Classes/ subclasses	Cohort Name	Population size	Hypertension correlation	References
Flavonoids	NHS I, USA	46,672	Self- reported, inversely associated with hypertension	(Cassidy et al., 2011)
Flavonoids (and subclasses)	HPFS, USA	5,629	Self-reported, inversely associated with hypertension	(Grosso et al., 2018)
Total polyphenols, flavonoids (and subclasses), phenolic acids (and subclasses), stilbenes, lignans, other	HAPIEE, Poland	2725	Self-reported, inversely associated with hypertension	(Jennings et al., 2012)
Flavonoids (and subclasses)	E3N, France	657	Significantly associated with hypertension	(Grosso et al., 2017b)
Flavonoids (and subclasses)	HELENA, multicenter Europe	9350	Self-reported, inversely associated with hypertension	(Wisnuwardani et al., 2020)
Total polyphenols, flavonoids, phenolic acids, stilbenes, lignans, others (and subclasses)	TLGS, Iran	550	Significantly associated with hypertension	(Sohrab et al., 2013)
Phenolic acids (and subclasses and individual polyphenols)	ISA-Capital Study, Brazil	1936	Significantly associated with hypertension	(Miranda et al., 2016)
Flavanols	KNHANES, Korea	4745	Significantly associated with hypertension	(Miranda et al., 2017)
Isoflavones	CARDIA, USA	3142	Significantly associated with hypertension	(Godos et al., 2017b)
Total phytoestrogens, isoflavones (and individual polyphenols), lignans (and individual polyphenols	MEAL, Italy	1936	Significantly associated with hypertension	(Yang et al., 2012)
Lignans	PREDIMED, Spain	7169	Significantly associated with hypertension	(Richardson et al., 2016)
Isoflavones	JPHC II, Japan	4165	Significantly associated with hypertension	(Creus-Cuadros et al., 2017)
Isoflavones, lignans	MR Cohort, South Korea	5509	Significantly associated with hypertension	(Nozue et al., 2017)

elderly population in the Mediterranean area (Godos et al., 2019). This study focused on finding the relation between the intake of lignin and blood pressure. The result of the study showed no significant effect on blood pressure alone, but when intake of lignin is consumed with low-fat dairy products, it shows a significant reduction on blood pressure (de Kleijn et al., 2002).

Research on soy products as a source of phytoestrogens was carried out on an Asian population in JPHC-II (Japan Public Health Center), including more than 900 males and 3200 females (Kreijkamp-Kaspers et al., 2004). This study was based on a food questioner's survey using Japanese STFC (standard tables of food consumption). The study included individuals with systolic blood pressure of >130 mm Hg or diastolic blood pressure >85 mm Hg (Witkowska et al., 2018). The result of the study shows that consumption of soy products lowers the risk of high blood pressure, especially in women, but no effect was observed in men. A fourth study was conducted on Korean adults in age group more than 40 years in association with soy isoflavone and high blood pressure with systolic blood pressure of >130 mm Hg or diastolic blood pressure >85 mm Hg (Jayalath et al., 2014). The study found an inverse relation between consumption of soy isoflavone and high blood pressure, in both males and females, and a linear association in females only. To study the effect of phytoestrogens in females with post-menopause, a fifth study was conducted on more than 900 females with post-menopause conditions based on the Framingham Offspring Study (Chang et al., 2007). This study was conducted to find the correlation between phytoestrogen intake and the cardiovascular risk factor, assessed using a self-administered questioner and literature reviews (Choi et al., 2015). This study reported no significant effect on systolic and diastolic blood pressure by intake of isoflavones and lignin (Soleimani et al., 2018). Another study was conducted on females living in the Netherlands with post-menopausal conditions, and it assessed dietary intake of phytoestrogen by questioner (Cotterchio et al., 2008). The participants were enrolled for one year and the result was assessed in year 2019, finding no association between systolic and diastolic blood pressure and consumption of phytoestrogen, whereas a significant effect of lignans were found on lowering the prevalence of hypertension (Lowcock et al., 2013).

A recent study on Polish females with post-menopause was conducted to find the effect of dietary lignans using 24-hour recall and a Dutch lignin database (WOBASZ and WOBASZ II) (Buja et al., 2020). The study found no relation between total polyphenols and individual lignin and hypertension (Muirhead, 2010).

5.3 POLYPHENOLS AS INHIBITORS OF LOW-DENSITY LIPOPROTEINS (LDL)

LDL oxidation is found to play an important role in acute anthrogenesis. Many studies reported the Mediterranean diet to be effective in reducing cardiovascular diseases. This diet contains 68% total dietary antioxidant from beverages and 20% from vegetables and fruits and a much less amount is contributed by cereals (Saura-Calixto and Goñi, 2006). The Mediterranean diet receives 1171 mg of gallic acid per day, which can be estimated by Folin Ciocalteau method (Saura-Calixto and Goñi, 2006, Wahrburg et al., 2002). Polyphenols in fruit and vegetables like onions are not found to have enough effect on LDL. Polyphenols like quercentin, apigenin, kaempferol, and myricetin from fruits, beverages, and vegetables when consumed by humans increases the amount of plasma flavonoid in blood and shows a significant effect in LDL reduction (Cook and Samman, 1996). Lycopene from dietary supplements was found to increase blood serum lycopene up to two-fold and reduces LDL oxidation significantly (Guven et al., 2019).

5.3.1 POLYPHENOLS FROM FRUITS AND VEGETABLES

Some studies focused on fruit-based polyphenols and their effect on LDL oxidation. Some researchers conducted studies using freeze-dried strawberry consumption on females to see its effect on lipid peroxidation in blood. The results showed direct correlation between consumption of strawberries and reduction in lipid peroxidation and cholesterol level in blood. A similar study was conducted on consumption of kiwifruit for eight weeks on LDL oxidation (Chang and Liu, 2009). The result of the study support the inverse correlation of kiwifruit polyphenol and reduction of LDL oxidation. Consumption of pomegranate juice was found to inhibit LDL oxidation in carotid artery stenosis patients, and also in healthy mice (Duttaroy, 2013). LDL oxidation reduction enhances the serum paraoxanase-1, reduces NADH (Nicotinamide Adenine Dinucleotide Hydrogen) -oxidase, and increased activity of glutathione redutase (Rezaei et al., 2020). Consuming 200 ml of cranberry vinegar/juice or grape juice 910 ml/day) twice a day for ten weeks can reduce LDL oxidation and thiobarbituric acid substances. Flavonoids like catechin, quercetin, isorhamnetin, epicatechin, and kaempferol present in the skin of almonds was found to be effective in reducing LDL oxidation up to 18 percent and significantly increases upto 52 percent of vitamin E (Rezaei et al., 2020).

Flavonoids in cocoa products were found to reduce LDL oxidation in human serum. A chocolate product with cocoa powder significantly decreases LDL oxidation, whereas another study showed that high consumption of flavonoids from chocolate did not show any significant increase in LDL

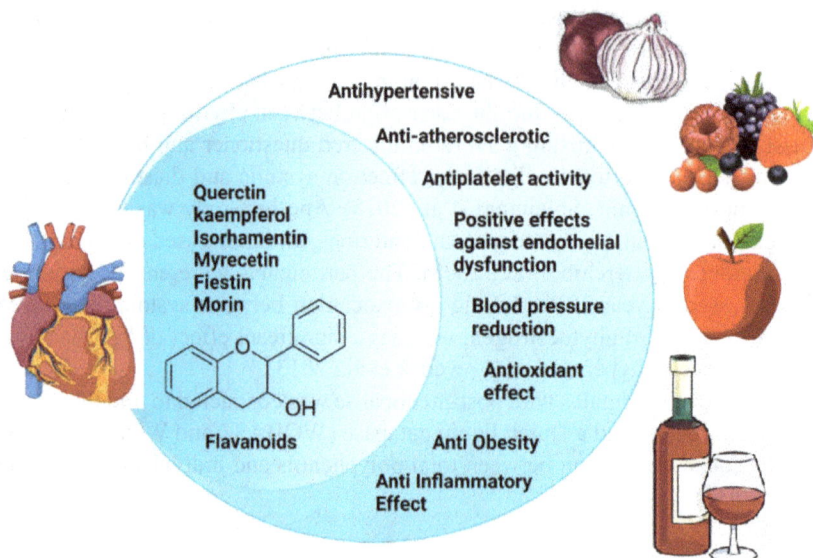

FIGURE 5.1   Different polyphenols and their effect on different cardiovascular conditions from different sources.

oxidation inhibition (Bakuradze et al., 2019). Polyphenols present in green tea, normal tea, and coffee are also reported to significantly reduce LDL oxidation and also increases ascorbic acid in plasma of patients having deficiency of ascorbic acid (Sadek, 2012). Whereas some studies noticed variation in these data and related it to plasma sampling and dose levels with individual absorption capacities that may vary from person to person. Onions and black tea have high flavonoid content but they fail to reduce LDL (Low Density Lipoproteins) oxidation or any lipid peroxidation markers in human blood (Alim et al., 2020). Some studies on green tea flavonoids found them to reduce serum Malondialdehyde-LDL concentration in serum. After drinking coffee the chances of LDL oxidative modification increases and around eight cups of coffee a day increases total cholesterol, HDL (High Density Lipoproteins), apolipoproteins in serum and decreases LDL to HDL ration. Dealcoholized red wine reduces atherosclerosis without LDL oxidation reduction in arterial wall (Wang et al., 2005). Limited consumption of red wine, beer, and stout were found to be useful in LDL oxidation, whereas the consumption of three or more drinks was reported to be pro-oxidant due to ethanol metabolism.

Research conducted by Benito et al., on intake of flavonoids reported that excess intake of polyphenols is not found beneficial when consumed without a proportional increase of other vitamins (Duthie et al., 2000, Benito et al., 2002)). Polyphenols at certain concentrations act as prooxidants and can start LDL oxidation, as reported with oleuropein and hydroxytryrosol. Oxidation conjugates of quercetin as inhibitors of LDL and the metabolites of flavonoids – like sulfates, quercetin, and glucuronides – were found to be more effective in comparison to glucoside and aglycone (Gentile et al., 2018).

5.3.2   Effect of polyphenols on vasodilation

Impaired functioning of endothelia is one of the initial symptoms in atherogenesis, which is caused by low accessibility of nitric oxide, resulting in vasodilation. Researchers studied the effect of polyphenols on vasodilated arteries and reported a positive effect on these arteries (Siasos et al.,

2013). Some polyphenols from plants like maize, aubergine (eggplant), and cranberry are found to be highly vasorelaxant (Faggio et al., 2017). *In vitro* studies on fruits, nuts, spices, tea, vegetables, and whole grains finds significant reductions in the markers related to endothelial dysfunction, whereas the high-fat diet increases the marker levels (Das et al., 2021). Fruit research in flavonoids show improved vasodilation by increasing synthase expression and activity by searching the peroxynitrite resulting radicals protecting the cofactor tetrahydrobiopterin, which is important for NO synthase activity, which can be done by reducing the stress induced redox gene (Pascual-Teresa et al., 2010).

Reductions of NADPH (Nicotinamide Adenine Dinucleotide Phosphate) and endothelium-dependent vasodilation superoxide anions were seen with ferulic acid and citrus extract (Wiggers et al., 2008). The aqueous extract from fruits of *Berberis vulgaris* were found effective in endothelium independent vasodilation (Akbar, 2020). A study on human consumption of fruits reported no significant effect on endothelium-independent vasodilation. Polyphenol intake from fruits and vegetables increases NO synthesis by increasing Ca^{+2}flux within a short time frame and, with time, in increases NO synthase (Özgen et al., 2012).

5.4 CONCLUSION

Dietary polyphenols are important health-promoting components present in our diet; however, the primary mechanism related to activity of these polyphenols in regard to anti-inflammatory, cell modulating, anti-platelet, and antioxidant action are not clear. Various pharmacological studies are focusing on receptor, cell signaling and expression at gene level in cardiovascular diseases and intake of dietary polyphenols. Further studies are required to explain the association between different subclasses of polyphenol and decreased risk of hypertension and other related cardiovascular diseases.

REFERENCES

Abbas, M., Saeed, F., Anjum, F.M., Afzaal, M., Tufail, T., Bashir, M.S., Ishtiaq, A., Hussain, S. and Suleria, H.A.R. 2017. Natural polyphenols: An overview. *Int J Food Prop* 20: 1689–1699.

Adriouch, S., Lampuré, A., Nechba, A., Baudry, J., Assmann, K., Kesse-Guyot, E., Hercberg, S., Scalbert, A., Touvier, M. and Fezeu, L.K. 2018. Prospective association between total and specific dietary polyphenol intakes and cardiovascular disease risk in the Nutrinet-Sante French cohort. *Nutrients* 10: 1587.

Akbar, S. 2020. *Berberis vulgaris L.(Berberidaceae). Handbook of 200 Medicinal Plants.* Springer.

Alim, A., Li, T., Nisar, T., Ren, D., Liu, Y. and Yang, X. 2020. Consumption of two whole kiwifruit (Actinide chinensis) per day improves lipid homeostasis, fatty acid metabolism and gut microbiota in healthy rats. *Int J Biol Macromol* 156: 186–195.

Andrés-Lacueva, C., Medina-Remon, A., Llorach, R., Urpi-Sarda, M., Khan, N., Chiva-Blanch, G., Zamora-Ros, R., Rotches-Ribalta, M. and Lamuela-Raventos, R.M. 2010. Phenolic compounds: chemistry and occurrence in fruits and vegetables. Fruit and vegetable phytochemicals: Chemistry, nutritional value and stability. *Wiley* 1: 384.

Bakuradze, T., Tausend, A., Galan, J., Groh, I.A.M., Berry, D., Tur, J.A., Marko, D. and Richling, E. 2019. Antioxidative activity and health benefits of anthocyanin-rich fruit juice in healthy volunteers. *Free Radic Res* 53: 1045–1055.

Bazzano, L.A., Joint, F. and Organization, W.H. 2005. Dietary intake of fruit and vegetables and risk of diabetes mellitus and cardiovascular diseases. *World Health Organization.*

Benito, S., Lopez, D., Saiz, M.P., Buxaderas, S., Sanchez, J., Puig-Parellada, P. and Mitjavila, M.T., 2002. A flavonoid-rich diet increases nitric oxide production in rat aorta. *British journal of pharmacology*, 135(4), pp.910–916.

Billingsley, H.E. and Carbone, S. 2018. The antioxidant potential of the Mediterranean diet in patients at high cardiovascular risk: an in-depth review of the PREDIMED. *Nutr Diabetes* 8: 1–8.

Blanck, H.M., Gillespie, C., Kimmons, J.E., Seymour, J.D. and Serdula, M.K. 2008. Trends in fruit and vegetable consumption among US men and women, 1994–2005. *Prev Chronic Dis* 5: 1–10

Boccellino, M. and D'angelo, S. 2020. Anti-obesity effects of polyphenol intake: Current status and future possibilities. *Int J Mol Sci* 21: 5642.

Buja, A., Pierbon, M., Lago, L., Grotto, G. and Baldo, V. 2020. Breast cancer primary prevention and diet: An umbrella review. *Int J Environ Res Public Health* 17: 4731.

Carnauba, R.A., Hassimotto, N.M. and Lajolo, F.M. 2021a. Estimated dietary polyphenol intake and major food sources of the Brazilian population. *British J Nutr* 126: 441–448.

Carnauba, R.A., Sarti, F.M., Hassimotto, N.M. and Lajolo, F.M. 2021b. Assessment of dietary intake of bioactive food compounds according to income level in the Brazilian population. *British J Nut* 1–8.

Cassidy, A., O'reilly, É.J., Kay, C., Sampson, L., Franz, M., Forman, J., Curhan, G. and Rimm, E.B. 2011. Habitual intake of flavonoid subclasses and incident hypertension in adults. *Am J Clin Nutr* 93: 338–347.

Chang, A.Y., Abdullah, S.M., Jain, T., Stanek, H.G., Das, S.R., Mcguire, D.K., Auchus, R.J. and De Lemos, J.A. 2007. Associations among androgens, estrogens, and natriuretic peptides in young women: Observations from the Dallas Heart Study. *J Am Coll Cardiol* 49: 109–116.

Chang, W.-H. and Liu, J.-F. 2009. Effects of kiwifruit consumption on serum lipid profiles and antioxidative status in hyperlipidemic subjects. *Int J Food Sci Nutr* 60: 709–716.

Cheynier, V., Duenas-Paton, M., Salas, E., Maury, C., Souquet, J.-M., Sarni-Manchado, P. and Fulcrand, H. 2006. Structure and properties of wine pigments and tannins. *Am J Enol Vitic* 57: 298–305.

Choi, Y., Chang, Y., Kim, B.-K., Kang, D., Kwon, M.-J., Kim, C.-W., Jeong, C., Ahn, Y., Park, H.-Y. and Ryu, S. 2015. Menopausal stages and serum lipid and lipoprotein abnormalities in middle-aged women. *Maturitas* 80: 399–405.

Cook, N.C. and Samman, S. 1996. Flavonoids – chemistry, metabolism, cardioprotective effects, and dietary sources. *J Nutr Biochem* 7: 66–76.

Cory, H., Passarelli, S., Szeto, J., Tamez, M. and Mattei, J. 2018. The role of polyphenols in human health and food systems: A mini-review. *Front Nutr* 5: 87.

Cotterchio, M., Boucher, B.A., Kreiger, N., Mills, C.A. and Thompson, L.U. 2008. Dietary phytoestrogen intake – lignans and isoflavones – and breast cancer risk (Canada). *CCC,* 19: 259–272.

Coward, L., Barnes, N.C., Setchell, K.D. and Barnes, S. 1993. Genistein, daidzein, and their. beta.-glycoside conjugates: antitumor isoflavones in soybean foods from American and Asian diets. *J Agric Food Chem* 41: 1961–1967.

Creus-Cuadros, A., Tresserra-Rimbau, A., Quifer-Rada, P., Martínez-González, M. A., Corella, D., Salas-Salvadó, J., Fitó, M., Estruch, R., Gómez-Gracia, E. and Lapetra, J. 2017. Associations between both lignan and yogurt consumption and cardiovascular risk parameters in an elderly population: Observations from a cross-sectional approach in the PREDIMED study. *J Acad Nutr Diet* 117: 609–622.

Dai, W., Tan, J., Lu, M., Zhu, Y., Li, P., Peng, Q., Guo, L., Zhang, Y., Xie, D. and HU, Z. 2018. Metabolomics investigation reveals that 8-CN-ethyl-2-pyrrolidinone-substituted flavan-3-ols are potential marker compounds of stored white teas. *J Agric Food Chem* 66: 7209–7218.

Das, M., Devi, K.P., Belwal, T., Devkota, H.P., Tewari, D., Sahebnasagh, A., Nabavi, S.F., Khayat Kashani, H.R., Rasekhian, M. and Xu, S. 2021. Harnessing polyphenol power by targeting eNOS for vascular diseases. *Crit Rev Food Sci Nutr* 1–26.

De Kleijn, M.J., Van Der Schouw, Y.T., Wilson, P.W., Grobbee, D.E. and Jacques, P.F. 2002. Dietary intake of phytoestrogens is associated with a favorable metabolic cardiovascular risk profile in postmenopausal US women: The Framingham study. *J Nutr* 132: 276–282.

Del Rio, D., Costa, L., Lean, M. and Crozier, A. 2010. Polyphenols and health: What compounds are involved? *Nutr Metabo Cardio Dis* 20: 1–6.

Del Rio, D., Rodriguez-Mateos, A., Spencer, J.P., Tognolini, M., Borges, G. and Crozier, A. 2013. Dietary (poly) phenolics in human health: structures, bioavailability, and evidence of protective effects against chronic diseases. *Antioxid. Redox Signal* 18: 1818–1892.

Díaz-Mula, H.M., Tomás-Barberán, F.A. and García-Villalba, R. 2019. Pomegranate fruit and juice (cv. Mollar), rich in ellagitannins and anthocyanins, also provide a significant content of a wide range of proanthocyanidins. *J Agric Food Chem* 67: 9160–9167.

Durazzo, A., Lucarini, M., Souto, E.B., Cicala, C., Caiazzo, E., Izzo, A.A., Novellino, E. and Santini, A. 2019. Polyphenols: A concise overview on the chemistry, occurrence, and human health. *Phytother Res* 33: 2221–2243.

Duthie, G.G., Duthie, S.J. and Kyle, J.A. 2000. Plant polyphenols in cancer and heart disease: implications as nutritional antioxidants. *Nutr. Res. Rev.,* 13: 79–106.

Duttaroy, A.K. 2013. Cardioprotective properties of kiwifruit. *Adv Food Nutr Res* 68: 273–282.

Erdman Jr, J.W., Balentine, D., Arab, L., Beecher, G., Dwyer, J.T., Folts, J., Harnly, J., Hollman, P., Keen, C.L. and Mazza, G. 2007. Flavonoids and heart health: proceedings of the ILSI North America flavonoids workshop, May 31–June 1, 2005, Washington, DC. *J Nutr* 137: 718S–737S.

Erlund, I., Silaste, M., Alfthan, G., Rantala, M., Kesäniemi, Y. and Aro, A. 2002. Plasma concentrations of the flavonoids hesperetin, naringenin and quercetin in human subjects following their habitual diets, and diets high or low in fruit and vegetables. *Eur J Clin Nutr* 56: 891–898.

Faggio, C., Sureda, A., Morabito, S., Sanches-Silva, A., Mocan, A., Nabavi, S.F. and Nabavi, S.M. 2017. Flavonoids and platelet aggregation: A brief review. *Eur J Pharmacol* 807: 91–101.

Gaziano, T., Reddy, K.S., Paccaud, F., Horton, S. and Chaturvedi, V. 2006. Cardiovascular disease. *N Engl J Med* 60–72.

Gentile, D., Fornai, M., Pellegrini, C., Colucci, R., Blandizzi, C. and Antonioli, L. 2018. Dietary flavonoids as a potential intervention to improve redox balance in obesity and related co-morbidities: A review. *Nutr Res Rev* 31: 239–247.

Godos, J., Bergante, S., Satriano, A., Pluchinotta, F.R. and Marranzano, M. 2018. Dietary phytoestrogen intake is inversely associated with hypertension in a cohort of adults living in the Mediterranean area. *Molecules* 23: 368.

Godos, J., Marventano, S., Mistretta, A., Galvano, F. and Grosso, G. 2017a. Dietary sources of polyphenols in the Mediterranean healthy Eating, Aging and Lifestyle (MEAL) study cohort. *Int J Food Sci Nutr* 68: 750–756.

Godos, J., Sinatra, D., Blanco, I., Mulè, S., La Verde, M. and Marranzano, M. 2017b. Association between dietary phenolic acids and hypertension in a Mediterranean cohort. *Nutrients* 9: 1069.

Godos, J., Vitale, M., Micek, A., Ray, S., Martini, D., Del Rio, D., Riccardi, G., Galvano, F. and Grosso, G. 2019. Dietary polyphenol intake, blood pressure, and hypertension: A systematic review and meta-analysis of observational studies. *Antioxidants* 8: 152.

Grassi, D., Desideri, G., Necozione, S., Lippi, C., Casale, R., Properzi, G., Blumberg, J. B. and Ferri, C. 2008. Blood pressure is reduced and insulin sensitivity increased in glucose-intolerant, hypertensive subjects after 15 days of consuming high-polyphenol dark chocolate. *J Nutr* 138: 1671–1676.

Grosso, G., Bella, F., Godos, J., Sciacca, S., Del Rio, D., Ray, S., Galvano, F. and Giovannucci, E.L. 2017a. Possible role of diet in cancer: systematic review and multiple meta-analyses of dietary patterns, lifestyle factors, and cancer risk. *Nutr Rev* 75: 405–419.

Grosso, G., Stepaniak, U., Micek, A., Kozela, M., Stefler, D., Bobak, M. and Pajak, A. 2018. Dietary polyphenol intake and risk of hypertension in the Polish arm of the HAPIEE study. *Eur J Nutr* 57: 1535–1544.

Grosso, G., Stepaniak, U., Micek, A., Stefler, D., Bobak, M. and Pająk, A. 2017b. Dietary polyphenols are inversely associated with metabolic syndrome in Polish adults of the HAPIEE study. *Eur J Nutr* 56: 1409–1420.

Grosso, G., Stepaniak, U., Topor-Mądry, R., Szafraniec, K. and Pająk, A. 2014. Estimated dietary intake and major food sources of polyphenols in the Polish arm of the HAPIEE study. *Nutrition* 30: 1398–1403.

Guenther, P.M., Dodd, K.W., Reedy, J. and Krebs-Smith, S. M. 2006. Most Americans eat much less than recommended amounts of fruits and vegetables. *J Am Diet Assoc* 106: 1371–1379.

Guven, H., Arici, A. and Simsek, O. 2019. Flavonoids in our foods: A short review. *J Basic Clin Health Sci* 3: 96–106.

Jayalath, V.H., De Souza, R.J., Sievenpiper, J.L., Ha, V., Chiavaroli, L., Mirrahimi, A., Di Buono, M., Bernstein, A.M., Leiter, L.A. and Kris-Etherton, P.M. 2014. Effect of dietary pulses on blood pressure: A systematic review and meta-analysis of controlled feeding trials. *Am J Hypertens* 27: 56–64.

Jennings, A., Welch, A.A., Fairweather-Tait, S.J., Kay, C., Minihane, A.-M., Chowienczyk, P., Jiang, B., Cecelja, M., Spector, T. and Macgregor, A. 2012. Higher anthocyanin intake is associated with lower arterial stiffness and central blood pressure in women. *Am J Clin Nutr* 96: 781–788.

Khan, J., Deb, P.K., Priya, S., Medina, K.D., Devi, R., Walode, S.G. and Rudrapal, M. 2021. Dietary flavonoids: Cardioprotective potential with antioxidant effects and their pharmacokinetic, toxicological and therapeutic concerns. *Molecules* 26: 4021.

Kinoshita, T., Lepp, Z. and Chuman, H. 2005. Construction of a novel database for flavonoids. *J Med Investiga* 52: 291–292.

Kreijkamp-Kaspers, S., Kok, L., Bots, M.L., Grobbee, D.E. and Van Der Schouw, Y.T. 2004. Dietary phytoestrogens and vascular function in postmenopausal women: A cross-sectional study. *J Hyper* 22: 1381–1388.

Kumar, S. and Pandey, A.K. 2013. Chemistry and biological activities of flavonoids: An overview. *The sci. world J.* 2013.

Lee, H., Han, E., Ji Kwon, N., Kim, Y., Kim, S., Kim, H. and Min, S.G. 2017. Korean Rural Development Administration's web based food and nutrient database management and validation system (NutriManager) – A report. *J Food Compos Anal* 62: 231–238.

Lowcock, E.C., Cotterchio, M. and Boucher, B.A. 2013. Consumption of flaxseed, a rich source of lignans, is associated with reduced breast cancer risk. *CCC* 24: 813–816.

Lye, H.-S., Kuan, C.-Y., Ewe, J.-A., Fung, W.-Y. and Liong, M.-T. 2009. The improvement of hypertension by probiotics: Effects on cholesterol, diabetes, renin, and phytoestrogens. *Int J Mol Sci* 10: 3755–3775.

Martinotti, S., Bonsignore, G., Patrone, M. and Ranzato, E. 2021. Mediterranean Diet Polyphenols: Anthocyanins and their Implications for Health. *Mini-Rev Med Chem* 21: 1692–1700.

Marventano, S., Salomone, F., Godos, J., Pluchinotta, F., Del Rio, D., Mistretta, A. and Grosso, G. 2016. Coffee and tea consumption in relation with non-alcoholic fatty liver and metabolic syndrome: A systematic review and meta-analysis of observational studies. *Clin Nutr* 35: 1269–1281.

Mattioli, R., Francioso, A., Mosca, L. and Silva, P. 2020. Anthocyanins: A comprehensive review of their chemical properties and health effects on cardiovascular and neurodegenerative diseases. *Molecules* 25: 3809.

Miean, K.H. and Mohamed, S. 2001. Flavonoid (myricetin, quercetin, kaempferol, luteolin, and apigenin) content of edible tropical plants. *J Agric Food Chem* 49: 3106–3112.

Miranda, A.M., Steluti, J., Fisberg, R.M. and Marchioni, D.M. 2016. Association between polyphenol intake and hypertension in adults and older adults: A population-based study in Brazil. *PloS one* 11: e0165791.

Miranda, A.M., Steluti, J., Fisberg, R.M. and Marchioni, D.M. 2017. Association between coffee consumption and its polyphenols with cardiovascular risk factors: A population-based study. *Nutrients* 9: 276.

Miyake, Y., Sasaki, S., Ohya, Y., Miyamoto, S., Matsunaga, I., Yoshida, T., Hirota, Y. and Oda, H. 2005. Soy, isoflavones, and prevalence of allergic rhinitis in Japanese women: The Osaka Maternal and Child Health Study. *J Allergy Clin Immunol* 115: 1176–1183.

Moreno, L., De Henauw, S., Gonzalez-Gross, M., Kersting, M., Molnar, D., Gottrand, F., Barrios, L., Sjöström, M., Manios, Y. and Gilbert, C. 2008. Design and implementation of the healthy lifestyle in Europe by nutrition in adolescence cross-sectional study. *Int J Obes* 32: S4-S11.

Muirhead, N. 2010. The impact of whole grains and their constituents on markers of vascular health and metabolism – An ex vivio and in vitro investigation. University of Surrey, UK.

Nakai, M., Fukui, Y., Asami, S., Toyoda-Ono, Y., Iwashita, T., Shibata, H., Mitsunaga, T., Hashimoto, F. and Kiso, Y. 2005. Inhibitory effects of oolong tea polyphenols on pancreatic lipase in vitro. *J Agric Food Chem* 53: 4593–4598.

Nawrot-Hadzik, I., Matkowski, A., Hadzik, J., Dobrowolska-Czopor, B., Olchowy, C., Dominiak, M. and Kubasiewicz-Ross, P. 2021. Proanthocyanidins and flavan-3-ols in the prevention and treatment of periodontitis – Antibacterial effects. *Nutrients* 13: 165.

Nozue, M., Shimazu, T., Sasazuki, S., Charvat, H., Mori, N., Mutoh, M., Sawada, N., Iwasaki, M., Yamaji, T. and Inoue, M. 2017. Fermented soy product intake is inversely associated with the development of high blood pressure: The Japan Public Health Center–Based Prospective Study. *J Nutr* 147: 1749–1756.

O'keefe, J.H., Bhatti, S.K., Patil, H.R., Dinicolantonio, J.J., Lucan, S.C. and Lavie, C.J. 2013. Effects of habitual coffee consumption on cardiometabolic disease, cardiovascular health, and all-cause mortality. *J Am Coll Cardiol* 62: 1043–1051.

Owen, D.M., Huang, H., Ye, J. and Gale Jr., M. 2009. Apolipoprotein E on hepatitis C virion facilitates infection through interaction with low-density lipoprotein receptor. *Virology* 394: 99–108.

Özgen, M., Saraçoğlu, O. and Geçer, E.N. 2012. Antioxidant capacity and chemical properties of selected barberry (Berberis vulgaris L.) fruits. *Hortic Environ Biotechnol* 53: 447–451.

Pandey, K.B. and Rizvi, S.I. 2009. Plant polyphenols as dietary antioxidants in human health and disease. *Oxid. Med. Cell. Longev.* 2: 270–278.

Pascual-Teresa, D., Moreno, D.A. and García-Viguera, C. 2010. Flavanols and anthocyanins in cardiovascular health: a review of current evidence. *Int J Mol Sci* 11: 1679–1703.

Rezaei, N., Zaherijamil, Z., Moradkhani, S., Saidijam, M., Oshaghi, E.A. and Tavilani, H. 2020. Kiwifruit supplementation increases the activity of the Paraoxonase enzyme and decreases oxidized low-density lipoprotein in high-fat diet fed hamsters. *Avicenna J Med Biochem* 8: 58–63.

Richardson, S.I., Steffen, L.M., Swett, K., Smith, C., Burke, L., Zhou, X., Shikany, J.M. and Rodriguez, C.J. 2016. Dietary total isoflavone intake is associated with lower systolic blood pressure: The Coronary Artery Risk Development in Young Adults (CARDIA) study. *J Clin Hypertens* 18: 778–783.

Rudrapal, M., Khairnar, S.J., Bin Dukhyil, A.A.A., Khan, J., Alaidarous, M., Palai, S., Deb, P.K., Bhattacharjee, S. and Devi, R. 2020. Dietary polyphenols and their role in oxidative stress-induced human diseases: Insights into protective effects, antioxidant potentials and mechanism (s) of action. *Front Pharmacol* 283.

Ruiz, J.R., Ortega, F.B., Gutierrez, A., Meusel, D., Sjöström, M. and Castillo, M.J. 2006. Health-related fitness assessment in childhood and adolescence: a European approach based on the AVENA, EYHS and HELENA studies. *J Public Health* 14: 269–277.

Sadek, M.A. 2012. Impact of Actinidia deliciosa (Kiwi fruit) consumption on oxidative stress status in carcinogenesis Mona A. *Biol Sci* 2012: 117–127.

Saura-Calixto, F. and Goñi, I. 2006. Antioxidant capacity of the Spanish Mediterranean diet. *Food Chem* 94: 442–447.

Scalbert, A., Manach, C., Morand, C., Rémésy, C. and Jiménez, L. 2005. Dietary polyphenols and the prevention of diseases. *Crit Rev Food Sci Nutr* 45: 287–306.

Siasos, G., Tousoulis, D., Tsigkou, V., Kokkou, E., Oikonomou, E., Vavuranakis, M., Basdra, E., Papavassiliou, A. and Stefanadis, C. 2013. Flavonoids in atherosclerosis: an overview of their mechanisms of action. *Curr Med Chem* 20: 2641–2660.

Singla, R.K., Dubey, A.K., Garg, A., Sharma, R.K., Fiorino, M., Ameen, S.M., Haddad, M.A. and Al-Hiary, M. 2019. Natural polyphenols: Chemical classification, definition of classes, subcategories, and structures. *J AOAC Int* 5: 1397–1400.

Sohrab, G., Hosseinpour-Niazi, S., Hejazi, J., Yuzbashian, E., Mirmiran, P. and Azizi, F. 2013. Dietary polyphenols and metabolic syndrome among Iranian adults. *Int J Food Sci Nutra* 64: 661–667.

Soleimani, A., Pourmoghaddas, A., Sadeghi, M., Roohafza, H., Talaei, M., Dianatkhah, M., Oveisgharan, S., Soleimani, M. and Sarrafzadegan, N. 2018. Risk and age of cardiovascular event in women with metabolic syndrome: menopause age in focus. *Metab Syndr Relat Diso* 16: 127–134.

Sosnowska, B., Penson, P. and Banach, M. 2017. The role of nutraceuticals in the prevention of cardiovascular disease. *Cardiovas Drug Ther* 7: S21.

Tanaka, Y., Sasaki, N. and Ohmiya, A. 2008. Biosynthesis of plant pigments: anthocyanins, betalains and carotenoids. *Plant J* 54: 733–749.

Tangney, C.C. and Rasmussen, H.E. 2013. Polyphenols, inflammation, and cardiovascular disease. *Curr Atheroscler Rep* 15: 1–10.

Tsao, R. 2010. Chemistry and biochemistry of dietary polyphenols. *Nutrients* 2: 1231–1246.

Verhaar, M., Stroes, E. and Rabelink, T. 2002. Folates and cardiovascular disease. *Curr Atheroscler Rep* 22: 6–13.

Wahrburg, U., Kratz, M. and Cullen, P. 2002. Mediterranean diet, olive oil and health. *Eur J Lipid Sci Technol* 104: 698–705.

Wang, Z., Zou, J., Cao, K., Hsieh, T.-C., Huang, Y. and Wu, J.M. 2005. Dealcoholized red wine containing known amounts of resveratrol suppresses atherosclerosis in hypercholesterolemic rabbits without affecting plasma lipid levels. *Int J Mol Med* 16: 533–540.

Wiggers, G.A., PeçAnha, F.M., Briones, A.M., Perez-Giron, J.V., Miguel, M., Vassallo, D.V., Cachofeiro, V., Alonso, M.J. and Salaices, M. 2008. Low mercury concentrations cause oxidative stress and endothelial dysfunction in conductance and resistance arteries. *Am J Physiol Heart Circ* 295: H1033-H1043.

Wijesinghe, W. and Jeon, Y.-J. 2012. Enzyme-assistant extraction (EAE) of bioactive components: A useful approach for recovery of industrially important metabolites from seaweeds: A review. *Fitoterapia* 83: 6–12.

Wilcox, L.J., Borradaile, N.M. and Huff, M.W. 1999. Antiatherogenic properties of naringenin, a citrus flavonoid. *Cardiovasc Drug Rev* 17: 160–178.

Wilkins, E., Wilson, L., Wickramasinghe, K., Bhatnagar, P., Leal, J., Luengo-Fernandez, R., Burns, R., Rayner, M. and Townsend, N. 2017. European cardiovascular disease statistics. Department for Health, University of Bath

Wisnuwardani, R.W., De Henauw, S., Forsner, M., Gottrand, F., Huybrechts, I., Knaze, V., Kersting, M., Donne, C.L., Manios, Y. and Marcos, A. 2020. Polyphenol intake and metabolic syndrome risk in European adolescents: The HELENA study. *Eur J Nutr* 59: 801–812.

Witkowska, A.M., Waśkiewicz, A., Zujko, M.E., Szcześniewska, D., Stepaniak, U., Pająk, A. and Drygas, W. 2018. Are total and individual dietary lignans related to cardiovascular disease and its risk factors in postmenopausal women? A nationwide study. *Nutrients* 10: 865.

Woo, H.W., Kim, M.K., Lee, Y.-H., Shin, D.H., Shin, M.-H. and Choi, B.Y. 2019. Habitual consumption of soy protein and isoflavones and risk of metabolic syndrome in adults≥ 40 years old: A prospective analysis of the Korean Multi-Rural Communities Cohort Study (MRCohort). *Eur J Nutr* 58: 2835–2850.

Yang, Y.J., Kim, Y.J., Yang, Y.K., Kim, J.Y. and Kwon, O. 2012. Dietary flavan-3-ols intake and metabolic syndrome risk in Korean adults. *Nutr Res Pract* 6: 68–77.

Zamora-Ros, R., Knaze, V., Luján-Barroso, L., Romieu, I., Scalbert, A., Slimani, N., Hjartåker, A., Engeset, D., Skeie, G. and Overvad, K. 2013. Differences in dietary intakes, food sources and determinants of total flavonoids between Mediterranean and non-Mediterranean countries participating in the European Prospective Investigation into Cancer and Nutrition (EPIC) study. *Br J Nutr* 109: 1498–1507.

Zhang, X., Shu, X.-O., Li, H., Yang, G., Li, Q., Gao, Y.-T. and Zheng, W. 2005. Prospective cohort study of soy food consumption and risk of bone fracture among postmenopausal women. *Arch Intern Med* 165: 1890–1895.

6 Dietary Polyphenols in Arthritis and Inflammatory Disorders

Lalit Mohan Nainwal and Poonam Arora
E-mail: lalit12331@yahoo.com; poonamarora96@gmail.com

CONTENTS

DOI: 10.1201/9781003251538-6

6.1 INTRODUCTION

Arthritis is a chronic, degenerative disorder generally characterized with swelling, pain, stiffness, and tenderness of one or more joints (Nigrovic and Raychaudhuri, 2018). Generally, it is grouped as age-associated disease but at present it is more often found in young adults due to lack of exercise. Now arthritis has emerged as life-style associated disease(Testa et al., 2021; Amoako and George, 2014; Antony et al., 2016; Quicke et al., 2021).

Arthritis is majorly classified as osteoarthritis, rheumatoid arthritis, and gout. However, there are some other kinds of arthritis associated with other diseases, including septic arthritis, lupus arthritis, psoriatic arthritis, juvenile idiopathic arthritis, ankylosing spondylitis, and Still's disease (Nigrovic and Raychaudhuri, 2018).

Osteoarthritis (OA) is the most common type of arthritis in elderly people and affects more than 240 million globally (Peat and Thomas, 2021). Osteoarthritis is classified as primary osteoarthritis and secondary osteoarthritis on the basis of the underlying cause. Primary arthritis is localized or generalized, and mostly happens in postmenopausal women. The cause of secondary osteoarthritis is trauma, other inflammatory arthritis, obesity and Paget's disease. Rheumatoid arthritis (RA) is an autoimmune disease characterized by the hyper-responsiveness of the body's own immune system that attacks one's own body tissues (Peat and Thomas, 2021; Hunter et al., 2020; Wolf and Bruce,

2003). Unlike osteoarthritis, rheumatoid arthritis is not age related and attacks not only joints but also other parts of the body. When uric acid crystals are deposited in the synovial joints, it produces an inflammatory response and progressively causes bone joint erosion. This type of arthritis is known as gout or gouty arthritis.

The World Health Organisation (WHO) compiled all joint-related diseases under a single heading of musculoskeletal conditions (WHO, 2021). According to a recent WHO fact sheet 2021, approximately 1.71 billion people globally suffer from arthritis or other associated conditions (Baková and Adriana, 2021). According to the WHO, the global burden of osteoarthritis percentage has progressively increased by 120 percent in 2017 in respect to year 1990. The global prevalence of RA is around 1 percent of the population and mostly affects women (WHO, 2021; Baková and Adriana, 2021).

There is no complete cure for this progressive disease, however; only symptomatic treatment is available and clinically practiced, such as the use of non-steroidal anti-inflammatory drugs (NSAIDS), corticosteroids, diseases modifying anti-rheumatic drugs, and antibodies. Chronic use of NSAIDS, corticosteroids, produces serious side effects and interferes with functioning of other organs, therefore there is a consistent and huge demand for new anti-arthritic agents that are economical, safe, less toxic and/or delay progression of arthritis (WHO, 2021).

6.2   DIETARY POLYPHENOLS USED IN TREATMENT OF ARTHRITIS

Plants, fruits, and vegetables are indispensable parts of our diet and are an abundant source of dietary polyphenols. As the name itself explains, polyphenols are a group of naturally occurring phytochemicals comprised of several hydroxyl groups on an aromatic ring. Polyphenols could be classified in different ways, as depicted in Figure 6.1. The structures of polyphenols are represented in Figure 6.2.

In this chapter, we summarize potential dietary polyphenols that have been studied recently for their potential anti-arthritic activities. This chapter summarizes only recent developments related to the discussed polyphenols and provides an update on previously available data.

6.2.1   Quercetin

Quercetin is a flavanoid chemically known as 3, 3′, 4′, 5, 7-pentahydroxy flavone showing a wide spectrum of pharmacological and therapeutic potential, such as antiproliferative, immunosuppressive,

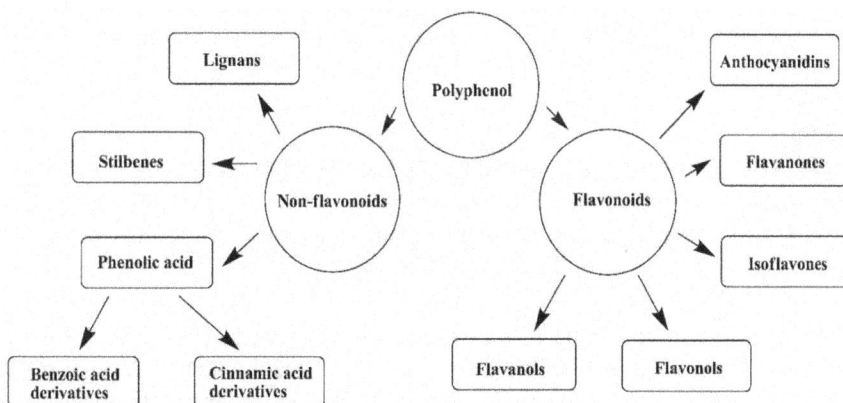

FIGURE 6.1   Classification of dietary polyphenolic compounds.

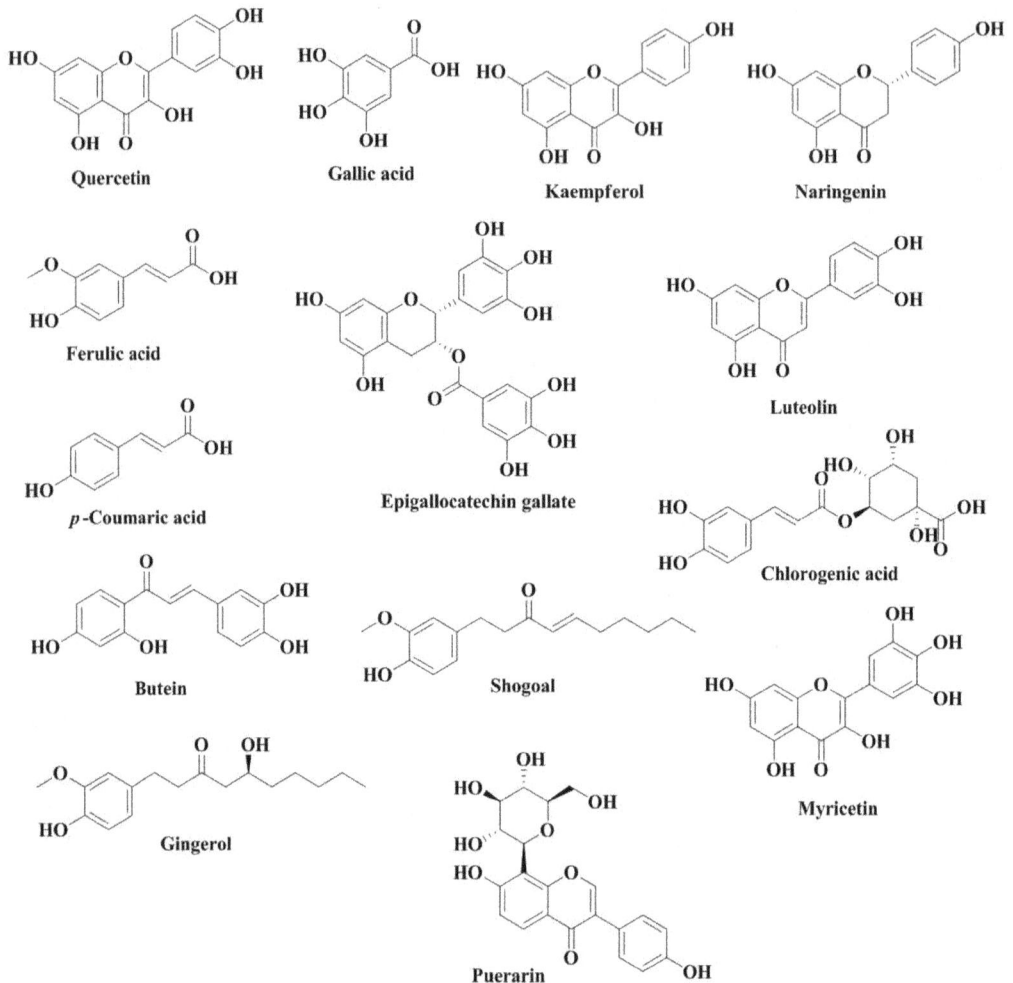

FIGURE 6.2 Chemical structure of some dietary polyphenols.

pro-apoptotic, antioxidant, and anti-inflammatory activities. Naturally, it is found as quercetin-3-O-β-glucoside, which on hydrolysis yields free aglycone quercetin (Sharma et al., 2018; Wong et al., 2020).

6.2.1.1 Source

Red onion (45 mg/100 g), apples (4.01 mg/100 g), fennel leaves (46.8 mg/100 g), oregano (42.0 mg/100 g), chili pepper (32.6 mg/100 g), dill (79.0 mg/100 g), cherries, (16.4 mg/100 g) and broccoli (13.7 mg/100 g) are some common sources of quercetin in our daily diet (Hollman and Katan, 1997).

6.2.1.2 Chemistry

When quercetin reacts with free radicals, it provides a proton and itself becomes a free radical, but the free radical, or unpaired electrons, thus formed is delocalized throughout the ring and keeps the molecule at a low potential energy level (Mariani et al., 2008), thus preventing further reaction of the quercetin free radical and helping in maintaining stability of the quercetin molecule.

The most important structural features in the quercetin molecule are (1) ortho-dihydroxyl group on ring B, (2) conjugated 2, 3-olefin with 4-oxo group, and (3) hydroxyl group at 3 and 5 positions (Alexandra Bentz, 2017). Electron donating functional groups support the resonance in the quercetin molecule that thus increases its therapeutic potential (Saccol et al., 2019).

6.2.1.3 Pharmacology

Kim et al. (2019) reported the myeloperoxidase (MPO), ectonucleoside triphosphate diphosphohydrolase (E-NTPDase), and ectoadenosine deaminase (E-ADA) modulating the potential of quercetin in complete Freund adjuvant (CFA) induced rat model of arthritis at different dose levels. Regular administration of quercetin at a dose of 50 mg/kg b.w. for 45 days significantly increased the latency period against thermal hyperalgesia and reduced the arthritis score and paw oedema in treated-group rats as compared to control group rats. Quercetin treatment increased the diminished E-ADA activity and decreased the elevated E-NTPDase, MPO, and cytokine levels in the arthritic rats. This study stated the attenuating role of quercetin in arthritis through the immune system via the purinergic system and by decreasing the levels of IFN-gamma and IL-4 in serum. Quercetinhas the ability to prevent the IL-17 mediated osteoclastogenesis in rheumatoid arthritis (RA). Quercetin significantly decreased the IL-17 and Th-17 cell provoked osteoclast formation, osteoclast differentiation, and RANKL (receptor activator of nuclear factor kappa-B ligand) protein production during *in vitro* studies. It also decreases the Th-17 differentiation, IL-17 production but has no effect on T-regulatory cells. Quercetin suppressed the IL-17-stimulated the phosphorylation of mTOR, ERK, and IκB-α in rheumatoid arthritis fibroblasts-like synoviocytes (RA FLS). However, it enhanced the IL-7 mediated phosphorylation of AMPK in RA fibroblasts-like synoviocytes (Haleagrahara et al., 2017).

Administration of quercetin alone was found more efficacious than methotrexate treatment as well as in combination with methotrexate in the management of collagen-induced arthritis (CIA). Decreased levels of TNF-α, IL-1β, IL-17, and monocyte chemotactic protein (MCP)-1 by quercetin was reported in the collagen-induced arthritis in C57BL/6 mice model. Quercetin treatment delivered the lowest degree of joint inflammation and highest degree of protection at a dose of 30 mg/kg b.w. administered orally for a period of 49 days. Methotrexate also reduced the circulating levels of TNF-α, IL-1β and IL-17 but had not shown superior anti-inflammatory and protective effect in C57BL/6 mice. No significant improvement in joint destruction was observed when quercetin and methotrexate were given in combination. However, unexpected body-weight loss and death was reported when quercetin and methotrexates were given in combination to CIA challenged C57BL/6 mice (Jeyadevi et al., 2013). Quercetin loaded quantum dots amplified the levels of superoxide dismutase (SOD), glutathione peroxidase (GPX), glutathione (GSH) and catalase (CAT) enzyme levels thus stifled cartilage degradation by the reducing of ROS generation (Haleagrahara et al., 2018).

Quercetin (30 mg/kg, b.w.) monotherapy significantly decreased the c-reactive protein (CRP), IL-6 levels along with mRNA expression of matrix metalloproteinases (MMP)-3, MMP-9, TNF-α and monocyte chemoattractant protein CCL2in C57/BL6 mice (Li et al., 2021). Intraperitoneal administration of quercetin (100 mg/kg, b.w.) repressed cytokines production (IL-1β and TNF-α) and apoptosis of articular cartilage in osteoarthritic rats. In vivo studies in rats and in vitro studies confirmed that quercetin protects IL-1β-induced rat chondrocyte injuries by suppressing IRAK1, NLRP3, caspase-3, iNOS and COX-2 expression in osteoarthritic rats (Yuan et al., 2020). Neutrophil-inactivating effects of quercetin in rheumatoid arthritis-challenged mice was reported recently (Ventura-Martínez et al., 2021). Quercetin impedes the neutrophil infiltration and endorses the apoptosis of activated neutrophils, thus reduces the levels of circulatory inflammatory cytokines in plasma. Quercetin treatment prevents neutrophil extracellular trap formation and supresses autophagy. Simultaneous administration of quercetin with diclofenac decreases the antinociceptive activity of diclofenac in uric acid-induced arthritic gout pain in rats (Sullivan et al., 1951). However, exact mechanism of action of quercetin in arthritis is still not known.

6.2.1.4   Pharmacokinetics and Toxicity

In mouse, LD_{50} values of quercetin after subcutaneous and oral administration were 100 mg/kg b.w. and 160 mg/kg b.w., respectively (Gugler et al., 1975). In humans, after giving 4g of quercetin orally, free quercetin as well as its conjugates were not detected in blood in first 24 hours; 53 percent of the administered dose was eliminated through faeces within 72 hours (Moon et al., 2008). When administered intravenously, it exhibits biphasic profile with half-lives of 8.8 mins and 2.4 hours; and more than 98 percent plasma binding (Hollman et al., 1996). In nine hours, 0.65 percent and 7.4 percent of quercetin and its conjugate were excreted through urine (Graefe et al., 1999). These reports suggest that quercetin has a slow elimination rate and thus could accumulate in tissues and plasma (Hollman et al., 1997). 4-Hydroxy-3-methoxyphenylacetic acid, 3,4-dihydroxyphenylacetic acid, and meta hydroxyphenylacetic acid are metabolites of quercetin identified in the urine samples (Gross et al., 1996).

6.2.2   GALLIC ACID

Gallic acid is a phenolic acid molecule, chemically known as 3,4,5-trihydroxybenzoic acid.

6.2.2.1   Source

Clove (458.19 mg/100 g), chestnut (479.78 mg/100 g), chicory (25.84 mg/100 g), and black tea (4.63 mg/100 g) are some common source of quercetin in our daily diet.

6.2.2.2   Chemistry

In gallic acid a carboxylic acid group and three hydroxyl groups at position 3, 4 and 5 are grafted to a benzene ring. The presence of three hydroxyl groups in *ortho*-position to each other is the most important feature for the best free-radical scavenging activity (Badhani et al., 2015; Gao et al., 2019).

6.2.2.3   Pharmacology

Gallic acid exhibited an apoptosis-inducing characteristic against RA FLS by modulating the expression of pro-inflammatory genes in RA FLS during in vitro studies. Gallic acid at 0.1 μM and 1 μM significantly augmented the caspase-3 activity and enhanced the expression of proapoptotic markers like p53 and Bax in RA FLS. Simultaneously, a significant decrease in the expression of anti-apoptotic proteins Bcl-2 and pAkt were also reported in dose- and time-dependent manners in gallic acid-treated RA FLS. Gallic acid also showed promising anti-inflammatory activity by decreasing the elevated mRNA expression of MMP-9, IL-1β, COX-2, IL-6, CCL-7/MCP-3C and CL-2/MCP-1 in RA FLS cells (Yoon et al., 2013; Alam et al., 2014).

In TNF-α stimulated RA FLS cells, gallic acid exhibited a similar kind of pro-anti-inflammatory activity and suppresses the mRNA expression of MMP-9, IL-1β, COX-2 and IL-6 (Chung et al., 2010). Gallic acid had shown protective action against advanced glycation end products (AGEs) activated chondrocytes during in vitro studies. Gallic acid (25 μM) treatment decreases the release and production of PEG2 reactive oxygen species, SOD, GSH and expression of iNOS and COX-2 enzyme were suppressed in AGEs challenged chondrocytes (Wen et al., 2015). Gallic acid reduces gouty arthritis by blocking NLRP3 inflammasome activation and pyroptosis by increasing NRF2 signalling. It dampens the monosodium urate (MSU) crystal-induced NLRP3 activation and IL-1β release and sluggish the leukocyte infiltration in the joints of MSU challenged mice. Expression of caspase 1 and p20 proteins were suppressed by the gallic acid. Administration of gallic acid also decreases the neutrophil infiltration in the MSU-induced gouty arthritis in mice (Lin et al., 2020).

6.2.2.4 Pharmacokinetics and Toxicity

In healthy humans gallic acid exhibited maximum plasma concentration (C_{max}), maximum time (t_{max}), area under the plasma concentration-time curve ($AUC_{(0-12h)}$), half-life ($t_{1/2}$), elimination rate constant (k), total amount of GA collected in urine (A_e) and renal clearance (CL_r) as 1.83 µmol/L, 1.27 h, 4.29 µmol/h/L, 1.19 h, 0.58 h^{-1}, 35.9 µmol, and 8.4 L/h, respectively (Shahrzad et al., 2001).

During acute and subacute toxicity studies of 28 days, LD_{50} of gallic acid was reported greater than 2000 mg/kg in albino mice (Variya et al., 2019). While in subchronic toxicity profiling dose of 119 mg/kg/day for male rats and 128 mg/kg/day, for female rats was found safe (Niho et al., 2001).

6.2.3 Kaempferol

Kaempferol is chemically 3,4',5,7-tetrahydroxyflavone belonging to the flavonolcategory of flavonoid.

6.2.3.1 Source

Spinach (55 mg/100 g), dill (40 mg/100 g), kale (47 mg/100 g), chives (12.5 mg/100 g), cherries (5.14 mg/100 g) and broccoli (7.2 mg/100 g) are some common sources of quercetin in our daily diet.

6.2.3.2 Chemistry

Chemically, kaempferol is less reactive and more stable than quercetin because it has one less hydroxyl group than quercetin (Dabeek and Melissa, 2019; Kumar and Panday, 2013; Sharma et al., 2018). Aglycone of flavonols are lipophilic in nature and are fat soluble, thus show a large volume of distribution (Jiang et al., 2015). Glycoside form of flavonols are water soule. Generally mono saccharides such as xylose, glucose, rhamnose are linked by a glycosidic linkage (Xiao et al., 2021).

6.2.3.3 Pharmacology

By regulating the XIST/miR-130a/STAT3 axis, kaempferol showed protective effects on arthritic chondrocytes. In C28/I2 cells, kaempferol reduced proinflammatory cytokine production and extra-cellular matrix breakdown. Kaempferol also improved X-inactive specific transcript (XIST) expression and increased miR-130a expression. XIST interacted with miR-130a, and STAT3 was shown to be a miR-130a target. In C28/I2 cells, knocking down XIST expression reduced pro-inflammatory cytokine production and extracellular matrix breakdown. The effects of XIST knockdown were reversed when signal transducer and activator of transcription 3 (STAT3) was overexpressed (Xiao et al., 2021). Kaempferol suppresses the proliferation of RA FLS and synthesis of MMP-1, MMP-3, COX-2, and PGE2 in rheumatoid arthritis synovial fibroblasts triggered by IL-1β. IL-1β induced activation of NF-κB and phosphorylation of JNK, ERK-1/2 and p38 proteins were suppressed by kaempferol (Yoon et al., 2013). Inhibition of Src, Syk, IRAK1, and IRAK4 during in vitro and in vivo studies confirmed that kaempferol directly suppresses the activation of NF-κβ (p65 and p50) and AP-1 (c-Jun and c-Fos) thus elicited its promising anti-inflammatory action. The mRNA expression of iNOS, TNF-α, and COX-2 encoding genes in lipopolysaccharide- (LPS-) and sodium nitroprusside sensitized RAW264.7 cells was downregulated by kaempferol (Kim et al., 2015; Kadioglu et al., 2015). In CIA mice model, kaempferol significantly dampens the TNF-α facilitated mitogen-activated protein kinase (MAPK) activation without disturbing the expression of TNF-α receptors. Suppressed MAPK activation leads to reduced migration, invasionof RA FLSs (Pan et al., 2018). In collagen-induced arthritis mice, kaempferol treatment reduced the incidence and severity of arthritis. During in vitro and in vivo experiments, kaempferol slowed down the osteoclast differentiation and cause suppression of other osteoclast biomarkers such as MMP-9, MMP-2, tartrate-resistant acid phosphatase and integrin-3. It decreased the synovial fluid basic fibroblast growth factor (bFGF) -induced bFGF-FGF receptor-3 (FGFR-3) phosphorylation in animal and in vitro experiments (Lee et al., 2018). Kaempferol decreased $CD3^+$ T cell invasion, NF-κB signaling and gene expression of IL-6, IL-17A and TNF-α in psoriatic skin. Kaempferol mediated anti-inflammatory and anti-psoriasis activity is due to enhanced augmented expression of $CD4^+$ forkhead box protein 3 (FoxP3)$^+$regulatory T cell (T-reg) in the skin, spleen and lymph nodes. In vitro studies confirmed that the kaempferol inhibits T-cells proliferation and mTOR signalling in psoriasis. Kaempferol (50 mg/kg and 100 mg/kg) treated psoriatic mice exhibited significantly low migration and percentage of IL-17A$^+$CD4$^+$ T cells in lymph nodes and spleen (Liu et al., 2019).

6.2.3.4 Pharmacokinetics and Toxicity

Pharmacokinetics and pharmacodynamics profiling of kaempferol was studied in both animals and in humans. In rats, after oral administration, kaempferol (100 mg/kg) exhibited extensively rapid absorption with faced first pass metabolism and had shown $AUC_{0-\infty}$ value of 0.76 ± 0.1 µg/ml, C_{max} value of 253.9 ± 48 µg/ml, T_{max} value of 1.5 ± 0.8 h, and percent bioavailability (F) about 2.7 percent. Low bioavailbility of kaempferol is due to its glucuronidation and other metabolic conversions in the liver and the gut. On i.v. administration (10 mg/kg) in rats, kaempferol exhibited $AUC_{0-\infty}$ value of 3.2 ± 0.3 µg/ml, V_d value of 12.7 ± 0.5 L/kg, $T_{1/2}$ value of 4.1 ± 0.2 h, Cl_{total} value of 3.2 ± 0.1 (L/hr/kg), Cl_{renal} value of 2.3 ± 0.16 (L/hr/kg) and $Cl_{non-renal}$ value of 0.8 ± 0.1 (L/hr/kg). The AUC value was increases proportionally on increasing the dose in both IV and oral administration (Barve et al., 2009).

Absorption of kaempferol from endive was studied in human subjects to understand its pharmacokinetic profile. Only 0.1 µM maximum plasma concentration of kaempferol was detected after 5.8 hour of orals administration (9 mg), represent its absorption form distal part of small intestine. A 1.9 percent of the dose was excreted in urine within 24 hours. It showed $AUC_{(0-24)}$ value of 0.757 ± 0.15 h/mmol/L (DuPont et al., 2004).

6.2.4 NARINGENIN

Naringenin is a flavonoid belonging to the subclass of flavanones. It is structurally (2S)-5,7-dihydroxy-2-(p-hydroxyphenyl)-4-chromanone, having benzo- γ-pyrone as structural core.

6.2.4.1 Source

Rosemary (55.1 mg/100 g), grapefruit juice (37.76 mg/100 mL) are common and rich source of naringenin in our diet.

6.2.4.2 Pharmacology

Naringenin gives protection from LPS and Con-A induced inflammation and organ injury in C57BL/6 and BALB/c mice model. LPS-stimulated release of IL-6 and TNF-α from macrophages and T cells was shunted after naringenin treatment during in vivo and in vitro (RAW 264.7 cell) studies. Intracellular cytokine degradation was upregulated by naringenin through transcription factor EB and lysosome-dependent mechanisms. However, naringenin did not alter the expression and stability of other mediators or factors involved in toll like receptor-3 (TLR-3) signalling cascade. Thus naringenin regulate its anti-inflammatory action by controlling post-translational mechanism (Jin et al., 2017).

Oral administration of naringenin (150 mg/Kg) sluggish theTiO_2-induced mechanical hyperalgesia, edema, and leukocyte recruitment in mice with 96 percent, 77 percent and 74 percent respectively. Naringenin improves the bone resorption by suppressing mRNA expression of receptor activator of nuclear factor κ B (RANK) and amplifying mRNA expression osteoprotegerin (OPG). Upregulation of NADPH oxidase 2 subunits (gp91phox and NBT), thiobarbituric acid reactive substances, mRNA expression of TNF-α, IL-6, IL-33, pro-IL-1β and activation of NF-κB are account for the potent in vivo anti-arthritic and anti-inflammatory activity of naringenin (Manchope et al., 2018). In monosodium iodoacetate induced osteoarthritis in rats, naringenin (40 mg/kg) decreased the pain and expression of MMP-3 thus regulated synthesis of MMP-3 and controled NF-κB mediated pathophysiology of osteoarthritis. During in vitro studies, transcriptional expression of various biomarkers (MMP-13, MMP-3, MMP-1, IL-1β, IkB-α, NF-κB-p65) and proteolytic enzymes specifically, a disintegrin and metalloproteinase with thrombospondin motifs (ADAMTS-4, and ADAMTS-5) were diminished in IL-1β-treated chondrocytes after naringenin treatment (Wang et al., 2017).

Naringenin exhibits its antioxidant, antinociceptive, and anti-inflammatory responses by triggering NO-cGMP-PKG-K_{ATP}channel signaling that stimulate Nrf2/HO-1 (heme oxygenase-1) pathway. Naringenin (50 mg/kg b.w. orally) decreased the neutrophil trafficking, TNF-α production, suppressed preproendothelin-1, gp91phox, IL-33, preproET-1 and COX-2 mRNA expression and upregulated the expression of Nrf2/HO-1 in superoxide anion donor (KO_2)-induced inflammation and algesia in mice (Manchope et al., 2016; Bussmann et al., 2019). Naringenin treatment enhanced the GSH levels and halts the lipid peroxidation and superoxide (O_2^-) production in murin model of pain (Manchope et al., 2016). Naringenin (40 mg/kg b.w. orally) ameliorates the inflammation in CIF RA rats by modifying Bax/Bcl-2 balance. A significant reduction in the concentration of TNF-α, IL-1β, and IL-6 in serum was noticed in naringenin-treated rats. Naringenin stimulates the apoptosis of synovial tissues through mitochondrial-dependent pathway by increasing Bax and reducing Bc2 expression (Wang et al., 2015). Naringenin suppresses the maturation of dendritic cells and thus could be explored for the treatment of arthritis (Pannu et al., 2019; Li et al., 2015). Naringenin suppressed the carrageenan-induced rat paw oedema and pain at a dose of 200 mg /kg b.w. orally (Chung et al., 2019).

6.2.4.3   Pharmacokinetics and Toxicity

A complete, broad, and in-depth pharmacokinetic study of naringein and naringenin was done by Yang et al. using rats, dogs, and humans at different dose levels. In a single orally administered doses of naringenin (480 mg) in humans, different pharmacokinetics parameters were reported as $t_{1/2}$ (3.37 h), T_{max} (13.1 h), C_{max} (96.8 ng/mL^{-1}), AUC_{0-t} (615 h. ng mL^{-1}), $AUC_{0-\infty}$ (805 h. ng mL^{-1}), V_d (55.8 L/kg) and Cl (6.32 L/h/kg). Similarly, single orally administered doses of naringin (480 mg) in humans different pharmacokinetics parameters were reported as $t_{1/2}$ (2.01 h), T_{max} (1.66 h), C_{max} (5.69 ng/mL^{-1}), AUC_{0-t} (19.4 h.ng mL^{-1}), $AUC_{0-\infty}$ (21.2 h.ng mL^{-1}), V_d (1,360 L/kg) and Cl (477 L/h/kg) (Bai et al., 2020).

6.2.5   Ferulic Acid

Ferulic acid is chemically known as 4-hydroxy-3-methoxycinnamic acid. It belongs to phenolic acid class.

6.2.5.1   Source

Wheat flour (72.21 mg/100 g), refined wheat flour (14.11 mg/100 mL), dark chocolate (24 mg/100 g) and dates (11.83 mg/100 mL) are some common and rich sources of ferulic acid in our diets.

6.2.5.2   Pharmacology

In CFA-induced arthritis in rats, ferulic acid (100 mg/Kg, b.w.; orally) and ferulic acid ethyl ester (100 mg/Kg, b.w.; orally) treatment significantly decreased release of TNF-α, and expression of Janus kinase 2 (JAK2), IL-1β, NO and IL-6 in RA rats. Decreased body weight, edema, C-reactive protein and albunin/globulin ratio was quite improved in the ferulic acid-treated group (Zhu et al., 2020; Cunha et al., 2016). Ferullic acid (30 mg/Kg, b.w.; orally) reduces joint inflammation, release of NO, TNF-α, IL-1β, and suppresses the mRNA expression of NF-κB p65, caspase-1, NLRP3 inflammasomes in MSU-induced gouty arthritis in rats. Antioxidant enzyme SOD and catalase

(CAT) levels was uplifted after ferulic acid treatment (Doss et al., 2016). Ferulic acid quashes osteoclastdifferentiation and bone destruction by blocking RANKL-dependent NF-κB signalling cascade (Doss et al., 2018). A significant decrease in the level of downstream mediators like nuclear factor of activated T-cells 1 (NFATc1), NK-κBp-65, cathepsin K, c-Fos and MMP-9 was noticed in RAW 264.7 cells after ferulic acid treatment (Das et al., 2014).

Ferulic acid (50 mg/Kg, b.w.; orally) is capable of reversing radiation induced inflammatory response by suppressing IKKα/β and IkBα and NK-κB phosphorylation and gene expression of COX-2 and iNOS-2 in murine model. A significant decrease in TNF-α, IL-1β levels along with increase in GSH, SOD and CAT enzyme levels in serum was reported in after ferulic acid treatment (Das et al., 2014). Ferulic acid (20 mg/Kg/day, orally) recovers inflammatory erosive arthritis in TNF-Tg mice. Enhanced expression of muscle genes h1-calponinl and sMYH in lymphatic smooth muscle cells and complete shutdown of TNF-mediated NO by lymphatic endothelial cells in ferulic acid-treated TNF-Tg mice was observed. Ferulic acid treatment improved lymphatic flow and con-traction and restricted bone and cartilage erosion in TNF-Tg mice (Liang et al., 2016).TNF-α and IL-1β levels were diminished in ferulic acid-treated, collagen-induced arthritic rats in dose- and time-dependent manners (Zhu et al., 2014). Ferulic acid repressed IL-17 mediated inflammatory signalling in RA FLS by blocking IL-17 receptor type A and STAT-3 signalling pathway. It also suppressed expression of TLR-3, IL-23, granulocyte-macrophage colony stimulating factor (GM-CSF) and cysteine-rich angiogenic inducer 61 (Cyr61) in arthritic rat model. Thus it is evident that it blocks IL-17/IL-17RA/STAT-3 signalling cascade in arthritis (Ganesan and Rasool, 2019).

6.2.5.3 Pharmacokinetics and Toxicity

Ferulic acid was found to be non-toxic to platelets, RBC, and WBC up to a concentration of 300 μg/ml. However, a different study reported it as toxic to human cells at a concentration of 40 mg/L. The sub-maximal dose that causes death of 50 percent of the population (LD_{50}) for ferulic acid in mice was reported 866 ± 28 mg/kg when administered through i.v. route (Li et al., 2021). This dose leads to death within 6 hours with tremors and spasticity. Ferulic acid could aggravate cancer and renal damage in diabetics and rats with renal complication. No evidence about tumurogenicity of ferulic acid is reported to date. Short $T_{1/2}$ of 1.77 hour describes fast absorption and distribution of ferulic acid in rats. Bioavailability of ferulic acid is quite enhanced when it is given with clopidogrel. Clopidogrel increased AUC, C_{max} and T_{max} of ferulic acid in rats (Li et al., 2017).

6.2.6 Epigallocatechin-3-gallate

Epigallocatechin gallate, is an ester of gallic acid and epigallocatechin and commonly also termed as epigallocatechin-3-gallate.

6.2.6.1 Source

Apples (10-43 mg/100 g), green tea (10-80 mg/100 mL), dark chocolate (46-61 mg/100 g) and black tea (6-50 mg/100 mL) are some common and rich sources of Epigallocatechin-3-gallate in our diet.

6.2.6.2 Pharmacology

Chondroprotective and cartilage-preserving property of EGCG has been quite well studied in the past two decades by various researchers and scientists by mean of in vitro and in vivo studies (Salahuddin Ahmed, 2010). During in vitro studies, when IL-1β-stimulated OA human chondrocytes were treated with EGCG, a substantial reduction in COX-2 expression and iNOS expression results in corresponding decreases in iNOS, PGE2, and NO release was observed (Ahmed at al., 2002; Singh et al., 2002; Panda et al., 2021). IL-1β-induced phosphorylation and proteolysis of IκBα was inhibited by EGCG thus impede overexpression or expression of NF-κB. IL-1β-induced upregulation of p46 isoform of c-Jun-N-terminal kinase, MMP-1, MMP-13, ADAMTS-1, ADAMTS-4, and ADAMTS-5 was suppressed by EGCG (Singh et al., 2003; Ahmed et al., 2004; Adcocks et al., 2002; Vankemmelbeke et al., 2003; Morinobu et al., 2008.

Oncostatin M stimulated upregulation of cFos and MCP-1/CCL2 was suppressed by EGCG (20 mg/Kg, i.p.) (Lin et al., 2008). LPS-induced bone resorption was halted after EGCG treatment through decrease in IL-1β and TNF-α levels. In RA, EGCG works by downregulating RANTES/CCL5, Gro-α/CXCL1, ENA-78/CXCL5, protein kinase Cδ and NF-κB pathways. Serum IgE level was significantly decreased after EGCG treatment in CIA murine model (Haqqi et al., 1999).

In adjuvant-induced arthritis model, EGCG (100 mg/Kg, i.p.) boosts the production of soluble gp130 protein thus suppresses IL-6 mediated downs stream and trans-signalling pathways (Ahmed et al., 2008; Ahmed, 2010). Shunted MMP-2 activity was also accounted for anti-arthritic activity of EGCG (Ahmed et al., 2008).

Long-term oral administration of ECGC in Pristane-induced arthritis (PIA) reactivates the chlorinating MPO activity and suppressed acute and chronic inflammation and pain in RA (Leichsenring et al., 2016). EGCG (10 mg/kg) supresses the serum type-II collagen-specific immunoglobulin (Ig) IgG2a antibodies in mice RA model (Min et al., 2015). Diminished levels of B cells, marginal zone B cells, CD4 T cells and CD8 T cells were quantified in ECGC-treated animals. A significant increase in indoleamine-2,3-dioxygenase (IDO) expression through sitimulation of CD11b+ dendritic cells (DC) along with increased levels of CD4+ Foxp3+ regulatory T cells (Tregs) were observed in ECGC treated groups. ECGC ameliorates NRf-2 mediated anti-oxidant pathway by increasing Nrf-2 and heme oxygenase-1 (HO-1) levels in RA mice. Release of interleukin (IL)-1β, IL-8, chemokine (C-C motif) ligand 2 (CCL2) and transforming growth-factor beta was suppressed by THP-1 cells by EGCG in a concentration-dependent manner in a calcium pyrophosphate induced in vitro model of RA (Oliviero et al., 2013).

6.2.6.3 Pharmacokinetics and Toxicity

In rats, oral administration of EGCG at a dose of 500 mg/kg/day for 13 weeks was found non-toxic and safe (Chow et al., 2001). In humans, acute as well as long-term safety of ECGC has been well established (Lin et al., 2007; Lambert et al., 2003; Dang et al., 2013). Absorption of ECGC is quite increased under fasting conditions; this is might be due to reduced glucuronidation process (Isbrucker et al., 2006). In humans, 800 mg/day dose of EGCG was found safe and non-toxic (Younes et al., 2018). However, circulating and synovial concentrations of EGCG after administration has not been yet studied even in vitro and in preclinical studies.

6.2.7 Luteolin

6.2.7.1 Source

Oregano (1028.80 mg/100 g), thyme (51 mg/100 mL), lemon (1.5 mg/100 g), rosemary (4 0 mg/100 g), peppermint (11.33 mg/100 g), broccoli (7.45 mg/100 g) and carrot (3.75 mg/100 mL) are some common and rich source of luteolin in our diet (Kumar and Pandey 2013).

6.2.7.2 Pharmacology

In Freund's complete adjuvant (FCA)-induced arthritis rat model, luteolin (10mg/kg and 20mg/kg, b.w.) treated rats had shown decreased arthritic score and reduced level of several inflammatory cytokines like TNF-α, IL-6, IL-1β and IL-17. Luteolin stopped the infiltration of inflammatory cells and thus decreases the inflammatory response, synovial hyperplasia and destruction of bone in arthritic rats. Several pivotal inflammatory markers (P2X4, NLRP1, ASC and Caspase-1p10) and their expression at transcription level were significantly suppressed in luteolin-treated arthritic rats (Shi et al., 2015). Luteoin in combination with natural, endogenous fatty acid molecule

N-palmitoylethanolamine suppressed the erosion of hind-paw arthritis in CII-challenged mice by reducing elevated nitrotyrosine and malondialdehyde (MDA) levels, thus decreasing the oxidative and nitrosative damage. Elevated levels of key proinflammatory cytokines and chemokines such as TNF-α, IL-6 and IL-1β were also suppressed by the luteolin and *N*-palmitoylethanolamine combined treatment. Neutrophil infiltration and MPO activity were also suppressed by this combined treatment (Impellizzeri et al., 2013). Elevated uric acid level and joint-inflammation were suppressed by luteolin when administered in monosodium urate crystal-induced arthritic rats. At a dose of 50 mg/kg, b.w. luteolin suppressed the production and release of TNF-α, IL-6, IL-1β, lysosomal and other oxidative enzymes (lipid peroxidase and myeloperoxidase) in arthritic rats (Lodhi et al., 2020).

Luteolin nanocomposite containing polyethylene glycol-mesoporous silica (PEG-MSN) as nano-carrier was synthesized and evaluated for its protective role in the treatment RA by Pang et al. (2021). The toxicity profile of designed nanocomposite was found similar to that of luteolin. Serum level of major inflammatory markers such as TNF-α, IL-2, IL-6, IL-1β and IL-17 was significantly reduced in treated RA rats. These nanocomposite increases the existence time, decreases the clearance rate of luteolin and offers target drug delivery capability (Pang et al., 2021). Luteolin exhibited its chondroprotective effects by inhibiting release of MMP-3 from articular chondrocytes during in vitro studies as well as in vivo studies (Kang et al., 2014). Gene expression and release of of ADAMTS-4, ADAMTS-5,MMP-3, MMP-1 and MMP-13 were suppressed by luteolin. This confirmed the MMP-3 inhibitory potential of luteolin. In an another study, IL-1β-induced synthesis and release of NO, MMP-9, MMP-8, MMP-2, TNF-α and PGE2 was inhibited by luteolin treatment in OA rats. Gene expression of metalloproteinase MMP-1, MMP-3, MMP-13 and other pro-inflammatory markers COX-2, and iNOS were also suppressed in luteolin treated OA challenged-animals (Fei et al., 2019).

6.2.7.3 Pharmacokinetics and Toxicity

Luteolin and luteolin 7-*O*-β-glucoside had reached peak plasma concentration 3.08 nmol/ml and 14.1 nmol/ml after 15 minutes and 30 minutes respectively. Luteolin was excreted 9 percent and 31.3 percent through urine and faeces, respectively, confirming that it is metabolized majorly through other routes as well (Wang et al., 2021). In liver luteolin and its glycosides converted into its glucuronide and methylated conjugates and were detected in human and rat plasma. Luteolin was also distributed in organs like kidney and liver thus showing its biological activity in diseases related to these organs (Shimoi et al., 1998). The percent oral bioavailability of luteolin and luteolin-7-*O*-glucoside in rats was found to be about 26 percent and 10 percent, respectively (Shimoi et al., 1998; Deng et al., 2017).

6.2.8 Cumaric Acid

Chemically, *p*-coumaric acid is 3-(4-hydroxyphenyl)-2-propenoic acid. It is a hydroxyl derivative of cinnamic acid.

6.2.8.1 Source

Oregano (5.75 mg/100 g), common sage (4.95 mg/100 mg), clove (8.49 mg/100 g), rosemary (3.67 mg/100 g), peanut (6.46 mg/100 g), and olive (1.43 mg/100 g) are some common and rich sources of *p*-coumaric acid in our diet.

6.2.8.2 Chemistry

p-Coumaric acid is unstable at high temperatures and therefore its percentage quite decreases during preparation of wines. In wines, and during laboratory research, it reacts with ethanol and this creates a huge problem in its detection as well as in predicting its properties (Salameh et al., 2008).

6.2.8.3 Pharmacology

p-Coumaric acid inhibits the paw oedema and release and expression of various proinflammatorycytokines and chemokines such as TNF-α, IL-6, IL-1β and MCP-1 in serum and synovial joints. *p*-Coumaric acid also suppresses the transcriptional expression of iNOS, COX-2R and osteoclastogenic factors RANKL and TRAP, in adjuvant induced arthritic rats (Neog et al., 2017). Diminutions of various transcription factors NF-jB-p65, p-NF-jB-p65, NFATc-1, and c-Fos, JNK, p-JNK and ERK1/2 and MAP kinase pathway was abolished by p-coumaric acid. Various lysosomal enzymes such as acid phosphatase, N-acetyl glucosaminidase, β-galactosidase and cathepsin D were significantly deceased in surrounding tissues and in serum after *p*-coumaric acid treatment in arthritic rats. ALT, AST, ALP levels were also decreased in treated rats (Zhu et al., 2018; Pragasam et al., 2013).

6.2.8.4 Pharmacokinetics and Toxicity

p-Coumaric acid has a much higher absorption efficiency than gallic acid in rats (Kim et al., 2021). After oral administration of coumaric acid (258 mg) different pharmacokinetic parameters documented from human subjects were $t_{1/2}$ (0.9 h), T_{max} (0.5 h), C_{max} (21.95 ng/mL^{-1}), AUC_{0-t} (20.55 h.ng mL^{-1}), and $AUC_{0-\infty}$ (20.55 h.ng mL^{-1}).

6.2.9 CHLOROGENIC ACID

Chlorogenic acid belongs to the phenolic category and is a hydroxycinnamic acid derivative. Chemically, it is 5-*O*-caffeoylquinic acid (5-CQA) having caffeic acid moiety bound to a quinic acid fragment.

6.2.9.1 Source

Carrot (purple) (18.79 mg/100 g), carrot (red) (1.35 mg/100 mg), tomato (0.19 mg/100 g), apple (1.16 mg/100 g) and green coffee (86.45-61.67 mg/g) are some common and rich sources of chlorogenic acid in our diet.

6.2.9.2 Chemistry

CQAs are quite highly sensitive and unstable in the presence of light, heat and change in pH (Jaiswal et al., 2012; Johnston et al., 2003; Dawidowicz and Typre, 2017; Li et al., 2015; Gil and Wianowska, 2017). Heating of 5-O-caffeoylquinic acid (5-CQA) causes isomerization and transformation into 3-O-caffeoylquinic acid and 4-O-caffeoylquinic acid in the presence of water. It also undergoes esterification, hydrolysis, and addition reaction (addition of water molecule to double bond) when it comes in contact with water. It also is isomerized into other CQAs during extraction from plant, and undergoes trans-cis isomerization in the presence of light (Jaiswal et al., 2012; Johnston et al., 2003; Dawidowicz and Typre, 2017; Li et al., 2015; Gil and Wianowska, 2017.

6.2.9.3 Pharmacology

Chlorogenic acid exhibited its chondroprotective effects by inhibiting release of MMP-3 from articular chondrocytes during in vitro studies as well as in vivo studies of inarthritic rabbits (Chen et al., 2011). Cartilage gene expression of MMP-3, MMP-1 and MMP-13 were suppressed, while expression of metalloproteinase-1 (TIMP-1) was enhanced by chlorogenic acid in treated animals. IL-1β-induced expression of nuclear factor kappa-B (NF-κB) and the degradation of inhibitor of κB (IκB)-α were stifled by chlorogenic acid. Chlorogenic acid inhibits TNF-α expression if B cell activating factor (BAFF) in MH7A cells and their apoptosis in concentration dependent manner. Molecular and mechanistic studies proved that cholrogenic acid reduces the expression of BAFF by inhibiting TNF-α signalling as well as by inhibiting DNA-binding of NF-κB to the BAFF (Fu et al., 2019). Similarly, it also averts IL-1β-stimulated inflammatory responses in human SW-1353 chondrocytes (Liu et al., 2017). Chlorogenic acid exhibited its anti-inflammatory effect in collagen-induced rheumatoid arthritis by modulating dimethylarginine dimethylaminohydrolase (DDAH)/ asymmetric (N^G, N^G) dimethylarginine (ADMA)/ cortistatin (CST)pathway. Elevated mRNA expression of TNF-α, IL-1β and IL-6 in tissues were suppressed by chlorogenic acid in treated animals. Expression of DDAH-1, DDAH-2 and CST were significantly enhanced in the tissues of chlorogenic acid treated arthritic animals (Zhang et al., 2020).

At a dose of 40 mg/kg, b.w. chlorogenic acid competently suppressed CD80/86 cell count as well as modulate and balance total (CD3) and differentiated T cell (CD4 and CD8) counts in adjuvant induced-arthritic rats. It did not modulate CD28 cell count. It suppressed Th1 cytokines level whereas Th2 cytokines levels were upregulated by the chlorogenic acid in dose-dependent manner (Chauhan et al., 2012; Naveed et al., 2018). Reduction of the mRNA expressions of inflammatory factors IL-1β, COX-2, TNF-α and IL-6; and autophagy factor mTOR were significantly reduced in synovium of footpad joint of chlorogenic acid treated CFA rats. The m-RNA expression of several autophagy factors anti-rat autophagy related gene (ATG)-5, ATG12, ATG7, adenosine monophosphate (AMP)-activated protein kinase (AMPK), Beclin1 and anti-rat light chain 3-II (LC3-II) were found upregulated in synovium of footpad joints of chlorogenic acid treated CFA rats (Lang-Fang et al., 2021). Cholrogenic acid also exhibited attractive anti-edematogenic, antinociceptive and antipyretic activity in rats at a dose of 100 mg/kg (Dos et al., 2006).

6.2.9.4 Pharmacokinetics and Toxicity

Pharmacological parameters reported after oral administration of chlorogeic acid (50 mg/kg) in rats are $t_{1/2}$ (1.70 h), T_{max} (0.48 h), C_{max} (0.55 mg/L), AUC_{0-t} (1.50 h.mg/L), and $AUC_{0-\infty}$ (1.61 h.mg/L), V_d (97.5 L/kg), K (0.41 h^{-1}) and CL (39 L/(h.kg)). The absorption of chlorogenic acid from *Solanum lyratum* whole plant extract was found significantly low. Clearance rate of chlorogenic acid absorbed from extract had shown a quite high renal clearance rate compared to crude chlorogenic acid. This result shows that the presence of other phytocompounds might alter pharmacokinetics (absorption, distribution, excretion, metabolism) of chlorogenic acid (Qi et al., 2011).

In humans, when chlorogenic acid (1000 mg) was administered in ileostomy subjects, 33 ± 17 percent of chlorogenic acid was absorbed while 11 percent of chlorogenic acid was excreted through urine (Olthof et al., 2001). It is quite clear from previous studies on animals and humans that absorption of chlorogenic acid occurs in the small intestine, and it is extensively metabolized in other compounds after absorption. Further research is required in this area.

6.2.10 Butein

Butein is a polyphenolic compound chemically known as 2',3,4,4'-tetrahydroxychalcone. It is primarily found in the plant *Butea monoserpma* (Padmavathi et al., 2017).

6.2.10.1 Source

Broad bean pods contain 0.08 mg/100g of butein.

6.2.10.2 Pharmacology

In OA rat model, butein supressed the IL-1β mediated invigorating IL-6 expression. It enhanced the phosphorylation of AMPKα$^{Thr-172}$, and its mediated downward phosphorylation of tuberous sclerosis complex 2 on Ser1387 primes (TSC2$^{Ser-1387}$) and Unc-51 Like Autophagy Activating Kinase 1 on Ser317 primes (ULK1$^{Ser-317}$). Strong inhibition of mTOR$^{Ser-2448}$ and its downstream target ribosomal protein S6 kinase beta-1 (p70S6K) was also reported with butein treatment that amplify autophagy flux and supress IL-6 expression in OA chondrocytes. Thus, butein activates autophagy through AMPK/TSC2/ULK1/mTOR Pathway (Ansari et al., 2018).

Butein has shown significantly higher anti-inflammatory activity than luteolin in RAW264.7 cells. The anti-inflammatory action of butein is mediated through of activation and upregulation of Heme Oxygenase-1 (HO-1) as well as inhibition of NFκB activation, in concentration-dependent manner. It also restricts the synthesis and production of NO by supressing expression of inducible nitric oxide synthase (iNOS) in a dose-dependent manner. LPS stimulated translocation of NFκB and NFκB gene activity was more strongly abolished by butein than luteolin (Sung et al., 2015). Butein supressed the ulcerative colitis inflammation score by more than 50 percent by downregulating mRNA expression and activation of MMP-9, IL-6, interferon (IFN)-γ and IL-1β in mice. IL-6-upregulated activation of STAT3 was inhibited by butein in Colo 205 cells (Lee et al., 2015). Butein inhibited mast cell activation and its mediated production and release of pro-inflammatory cytokines like NFκB, TNF-α, IL-8, IκB and IL-6 (Raseed et al., 2010).

6.2.11 Shogaol

6-Shogaol is (4E)-1-(4-hydroxy-3-methoxyphenyl)dec-4-en-3-one.

6.2.11.1 Source
Ginger is the principal source of shogaol.

6.2.11.2 Pharmacology
6-Shogaol potentially ameliorates intense inflammation and other manifestations occurring in gouty arthritis. In MSU crystal-induced gouty arthritis in mice, 6-shogaol suppressed the gouty inflammation, elevated β-glucuronidase, lactate dehydrogenase, TNF-α levels in serum and in synovial joint tissues (Sabina et al., 2010). In CFA-induced arthritic rats, it significantly lowered the migration of neutrophils, leukocytes, macrophages and monocytes in the synovial joint. It also decreased the expression of soluble vascular cell adhesion molecule-1 (VCAM-1) in synovial fluid and joint tissues (Levy et al., 2006).

6.2.11.3 Pharmacokinetics and Toxicity
Pharmacological parameters reported (Zick et al., 2008) after oral administration of 6-Shogaol (2000 mg) in healthy humans were $t_{1/2}$ (120.4 minutes), T_{max} (65.6 minutes), C_{max} (0.15 μg/mL), and AUC_{0-24} (10.9 μg/mL). No free 6-shogaol level was detected in blood plasma, however 6-shogaol sulphate and glucuronide conjugates were also not detected even at 2000 mg dose. This study was conducted at different dose levels. However, no sign of toxicity was observed in the participants (Yu et al., 2011).

6.2.12 Myricetin

Myricetin belongs to class of flavonoid, chemically it is 3,3',4',5,5',7-hexahydroxyflavone.

6.2.12.1 Source
Garlic (1600 mg/ 100 g), lemon (500 mg/ 100 g), lotus root (600 mg/ 100 g) pepper and green chilli (1200 mg/ 100 g) are common food sources of myricetin in our food (Taheri et al., 2020).

6.2.12.2 Pharmacology
Cathepsin K exhibits potent collagenase activity and is overexpressed in osteoclast and synovial fibroblasts in arthritis. Myricetin directly blocked the cathepsin K activity with IC_{50} value of 585.3 μmol/L. In CIA mice model, daily oral administration of myricetin at a dose of 25 mg/kg, b.w. significantly lowered the level of C-terminal telopeptide degradation product of type I collagen

(CTX-I), deoxypyridinoline and cartilage oligomeric matrix protein that directly confirmed the cartilage protection (Yuan et al., 2015). In rheumatoid arthritis, myricetin treatment stopped the production and release of inflammatory mediators from the synovial fibroblasts, thus protecting destruction of cartilages and bones. Myricetin suppressed IL-1β-stimulated production of MMPs, mitogen-activated protein kinases (MAPKs) and inflammatory cytokines from SW982 synovial cells. IL-6 and MMP-1 production was significantly diminished by myricetin in IL-1β- stimulated SW982 synovial cells. The phosphorylation of Jun NH_2-terminal kinase (JNK) and p38-MAPK were also inhibited by myricetin at a concentration of 10μM (Lee et al., 2010). In IL-1β-stimulated SW1353 cells myricetin downregulated the production and expression of PEG2, NO, expression of MMP-1, MMP-3, MMP-13, ERK, JNK and p38 kinases. It also suppressed expression and downstream signalling pathway of TNF-α and activator protein-1 (AP-1) a transcription factor, thus blocked expression of c-Fos and c-Jun and of other cytokines. Downregulation of STAT-1/JNAK2 signalling pathway was also reported by myricetin (Wang et al., 2016). A glycoside of myricetin (myricetin-3-O-β-dxylofuranosyl-(1 → 2)-α-l-rhamnopyranoside) potentially inhibited the TNF-α with IC_{50} value of 1.11 μM. In murin model of CFA, it significantly reduces the level of TNF-α, C-reactive protein, and fibrinogen at a dose of 5 mg/kg (Ticona et al., 2021)

6.2.13 6-GINGEROL

6-Gingerol is chemically (5S)-5-hydroxy-1-(4-hydroxy-3-methoxyphenyl)decan-3-one.

6.2.13.1 Source
Ginger is the principal source of 6-gingerol.

6.2.13.2 Pharmacology
6-Gengerol selectively inhibits osteoclast differentiation by suppressing IL-1-induced overexpression of receptor activator of nuclear factor kappa-B ligand (RANKL) in osteoblast. It also decreases the activity of cyclooxygenase and PGE synthase enzymes, thus diminished IL-1-induced production of prostaglandin E2 (PGE2) in osteoclasts, thus exerted its anti-inflammatory and anti-arthritis potential (Hwang et al., 2018). 6-Gingerol is also studied for its anti-neutrophil migration property in lupus and other inflammatory diseases. It inhibits neutrophil oxidative burst, formation of H_2O_2 in neutrophils, reactive oxygen species formation in nueutrophil that leads to neutrophil extracellular trap (NET) release or NETosis. It inhibits cAMP-specific PDE activity thus shunted pro-inflammatory cytokines production and release. A dramatic decrease in plasma NET levels was observed with decrease in both both cell-free DNA and myeloperoxidase-DNA complexes. It had shown potent PDE enzyme and protein kinase-A (PKA) enzyme inhibition activity. Beside it, various other immunogenic and inflammatory markers were also significantly suppressed by 6-gingerol such as anti-β2GPI, total IgG and anti-dsDNA (Ali et al., 2021). 6-Gingerol at a concentration of 10 μM ameliorate TNF-α stimulated proliferation, maturation and differentiation ofosteoblast-like MG-63 cells. Treatment with 6-gingerol enhanced the ALP enzyme level and bone mineralization. TNF-α mediated inflammatory response and IL-6 production was completely baselined by 6-gingerol (Fan et al., 2015). 6-Gingerol suppressed the inflammatory response and damage in dextran sulfate sodium (DDS)-induced ulcerative colitis in mice and maintained Th17/Treg balance. Additionally, a sharp decline mRNA expression of IL-6, IL-10 and IL-17 were noticed in 6-gingererol-treated animals as compared to the control group animals. Serum level of IL-6, IL-10 and IL-17 also declined in the treated mice group. Upregulated Th17 count and downregulated Treg cell count in ulcerative colitis challenged mice were found to be reversed by 6-gingerol treatment. At the molecular level, 6-gingerol supressed DDS augmented overexpression of RORγT and upregulated the FOXP3 gene expression, thus modulating systemic and localized Th17/Treg balance (Sheng et al., 2020).

6.2.13.3 Pharmacokinetics and Toxicity

Pharmacological parameters reported after oral administration of 6-gingerol (2000 mg) in healthy humans (Zick et al., 2008) were $t_{1/2}$ (110 minutes), T_{max} (65.6 minutes), C_{max} (0.85 µg/mL), and AUC_{0-24} (65.6 µg/mL). No free 6-gingerol level was detected in blood plasma, however 6-gingerol sulphate conjugates was detected in plasma while its glucuronide conjugate was detected in bile, however it is not clear which conjugate is formed first. This study was conducted at different dose levels. It was quite surprising that no conjugate metabolite was detected below 1000 mg dose (Yu et al., 2011).

6.2.14 Puerarin

Puerarin belongs to a family of isoflavone and is chemically known as 8-(β-D-Glucopyranosyl)-4',7-dihydroxyisoflavone. It is 8-C-glucoside of daidzein.

6.2.14.1 Source

It is an active constituent of root of pueraria plant (*Radix puerariae*).

6.2.14.2 Pharmacology

In OArat model puerarin (100 mg/kg) protected bone and joint degradation due to its anti-matrix-degrading potential and anti-inflammatory action. Puerarin-treated animals had shown a significant decrease expression of COX-2, ADAMTS-5, MMP-3, MMP-13 in cartilage tissues. Production of inflammatory markers like TNF-α, IL-6 and IL-1β; and cartilage degradation biomarkers CTX-I, CTX-II, COMP, and PIINP levels were also found suppressed in puerarin-treated animals (Ma et al., 2021).

Puerarin mitigates clinical and pathological manifestations of OA by upregulating AMP-activated protein kinase/proliferator-activated receptor-γ coactivator-1 signalling pathway. Phosphorylation of AMPKα and total peroxisome proliferator-activated receptor-gamma coactivator-1 alpha (PGC-1α) content was enhanced by puerarin, but it blocked the acetylation of PGC-1α (Yang et al., 2018).

To fully understand the OA protective mechanism of action of puerarin, Liang et al. (2016) have studied the methylation level at the promoter region of genes that are closely associated with OA. Expression level of tissue inhibitor of metalloproteinase-3 (TIMP-3) and MMP-9 at mRNA level were significantly altered by puerarin in OA chondrocytes. It enhanced TIMP-3 expression while supressing MMP-9 expression in OA chondrocytes. Similarly, it upregulated the methylation of TIMP-3 and decreased the methylation of MMP-9 at gene level, thus exerting its cartilage and bone protective role (Liang et al., 2016).

Puerarin alleviates oxidative stress and inflammation collagen antibody-induced arthritis in mice by regulating toll-like receptor 4 (TLR4)/nuclear factor-κB (NF-κB) signaling pathway. Puerarin treatment enhanced the phosphorylation of Janus kinase 2, while it downregulates the mRNA expression of phosphorylated signal transducer and activator of transcription 3 (STAT3) protein in arthritic mice (Wang et al., 2016).

6.3 DIETARY POLYPHENOLS UNDER CLINICAL TRIALS FOR THE TREATMENT OF ARTHRITIS AND INFLAMMATORY DISEASES

Despite most polyphenol enriched food and natural products for the treatment of various diseases having a long history of tradition and ethano-medicinal use, there is quite insufficient scientific data and studies that could bring these natural phyto-compounds to clinical stages. Many scientists and researchers look at these molecules as potential sources of novel multi-targeting drug molecules. At present some polyphenols are going to be tested for their clinical efficacy for the treatment of arthritis and other inflammatory diseases (Table 6.1).

TABLE 6.1

Dietary Polyphenols under Clinical Trials for the Treatment of Arthritis (Clinical trial 2022)

Clinical Trial Number	Title	Condition	Status	Study Type
NCT02251678	Evaluate the Effect of Elimune Capsules	Plaque psoriasis	Phase 1	Interventional
NCT02905799	Resveratrol in Knee Osteoarthritis	Knee osteoarthritis	Phase 3	Interventional
NCT04661267	Randomized Trial of Regenexx Stem Cell Support Formula	Knee osteoarthritis	--	Interventional
NCT02744703	Effect of Caffeic Acid Phenethyl Ester as a Matrix Metalloproteinase Inhibitor: Randomized Controlled Clinical Trial	Matrix Metalloproteinase Inhibitors	--	Interventional
NCT04638387	PB125, Osteoarthritis, Pain, Mobility, and Energetics	Osteoarthritis, Knee Muscle Weakness Pain, Joint	--	Interventional
NCT05038410	Study to Investigate the Mechanism of Action of an Oral Enzyme Treatment With Bromelain, Trypsin and Rutoside Versus Placebo in Subjects With Osteoarthritis	Osteoarthritis	--	Interventional

In a randomized, double-blind clinical trial with placebo-control on 50 women suffering with rheumatoid arthritis was conducted where quercetin was supplemented for 8 weeks at dose of 500 mg/day orally. Quercetin supplement significantly reduced the clinical symptoms of rheumatoid arthritis in patients evaluated by means of Disease Activity Score 28 (DAS-28) and health assessment questionnaire. Significant decrease in swollen joint counts ($p < 0.005$) and circulating levels of TNF-α ($p < 0.005$) in clinical candidates with after quercetin supplementation was observed in comparison to a placebo control group (Javadi et al., 2017).

6.4 CONCLUSION AND FUTURE PERSPECTIVE

In summary, several dietary polyphenols have been found to promise anti-inflammatory, anti-arthritic, and bone restorative potential that intercede at all stages of arthritis progression and restore bone density. These compounds are scientifically recognized at preclinical and some clinical studies, and also against other diseases. These compounds are capable of inhibiting the initiation, promotion, and progression of several types of arthritis and other associated pathological manifestation. However, successful clinical use of these compounds could only be proven after human clinical trials. The rapid absorption, metabolism, and large volume of distribution of these compounds put a serious question in front of scientists and researchers about their mechanism of action and therapeutic potency. It is quite interesting that most of the dietary polyphenols are non-toxic and are safe to consume up to 2 g dose in humans. In most cases concentration of polyphenol in affected area, synovial fluid and synovial joints are still not studied, therefore a doses and tissue concentration ratio are not completely clarified. A wide number of published in vitro, in vivo research and clinical studies ascertained the anti-arthritic and anti-inflammatory role of polyphenols. Various regulatory cellular and molecular regulatory mechanisms through which polyphenols work are now well understood. All these findings revealed the promising role of polyphenols in prevention and therapy of arthritis and other inflammatory diseases. There are so many questions still unanswered about their clinical setting. Metabolomics studies of polyphenols should be further explored to learn their active metabolites. The concept of bioavailability and its impact on therapeutic efficacy and effectiveness should also be revisited.

REFERENCES

Adcocks, C., Collin, P., and Buttle, D.J. 2002. Catechins from green tea (Camellia sinensis) inhibit bovine and human cartilage proteoglycan and type II collagen degradation in vitro. *J Nutr* 132: 341–346.

Ahmed, S. 2010. Green tea polyphenol epigallocatechin 3-gallate in arthritis: progress and promise. *Arthritis Res Ther* 12: 1–9.

Ahmed, S., Marotte, H., Kwan, K., Ruth, J.H., Campbell, P.L., Rabquer, B.J., Pakozdi, A., and Koch, A.E. 2008. Epigallocatechin-3-gallate inhibits IL-6 synthesis and suppresses transsignaling by enhancing soluble gp130 production. *Proc. Natl. Acad. Sci. U.S.A.* 105: 14692–14697.

Ahmed, S., Rahman, A., Hasnain, A., Lalonde, M., Goldberg, V.M., and Haqqi, T.M. 2002. Green tea polyphenol epigallocatechin-3-gallate inhibits the IL-1β-induced activity and expression of cyclooxygenase-2 and nitric oxide synthase-2 in human chondrocytes. *Free Radic Biol Med* 33: 1097–1105.

Ahmed, S., Wang, N., Lalonde, M., Goldberg, V.M. and Haqqi, T.M. 2004. Green tea polyphenol epigallocatechin-3-gallate (EGCG) differentially inhibits interleukin-1β-induced expression of matrix metalloproteinase-1 and-13 in human chondrocytes. *J. Pharmacol. Exp. Ther.* 308: 767–773.

Alam, M., Rahman, M.M., and Khalil, M. 2014. Nutraceuticals in arthritis management: A contemporary prospect of dietary phytochemicals. *Open Nutraceuticals J* 7. https://doi.org/10.2174/187639600140 7010021.

Ali, R.A., Gandhi, A.A., Dai, L., Weiner, J., Estes, S.K., Yalavarthi, S., Gockman, K., Sun, D., and Knight, J.S. 2021. Antineutrophil properties of natural gingerols in models of lupus. *JCI insight* 6: e138385. https://doi.org/10.1172/jci.insight.138385.

Amoako, A.O., and Pujalte, G.G.A. 2014. Osteoarthritis in young, active, and athletic individuals. *Clin Med Insights: Arthritis Musculoskelet Disord* 7: 14386. https://doi.org/10.4137/CMAMD.S14386.

Ansari, M.Y., Ahmad, N., and Haqqi, T.M. 2018. Butein activates autophagy through AMPK/TSC2/ULK1/ mTOR pathway to inhibit IL-6 expression in IL-1β stimulated human chondrocytes. *Cell Physiol Biochem* 49: 932–946.

Antony, B., Jones, G., Jin, X., and Ding, C. 2016. Do early life factors affect the development of knee osteoarthritis in later life: a narrative review. *Arthritis Res Ther* 18: 1–8.

Apaza Ticona, L., Souto Pérez, B., Martín Alejano, V., and Slowing, K. 2021. Anti-inflammatory and anti-arthritic activities of Glycosylated Flavonoids from Syzygium jambos in edematogenic agent-induced paw edema in mice. *Rev Bras Farmacogn* 31: 429–441.

Badhani, B., Sharma, N., and Kakkar, R. 2015. Gallic acid: A versatile antioxidant with promising therapeutic and industrial applications. *RSC Adv* 5: 27540–27557.

Bai, Y., Peng, W., Yang, C., Zou, W., Liu, M., Wu, H., Fan, L., Li, P., Zeng, X., and Su, W. 2020. Pharmacokinetics and metabolism of naringin and active metabolite naringenin in rats, dogs, humans, and the differences between species. *Front Pharmacol* 11: 364.

Baková, Z., and Kolesárová, A. 2021. Bioflavonoid quercetin-food sources, bioavailability, absorbtion and effect on animal cells. *J Microbiol Biotechnol Food Sci 2021*: 426–433.

Barve, A., Chen, C., Hebbar, V., Desiderio, J., Saw, C.L.L., and Kong, A.N. 2009. Metabolism, oral bioavailability and pharmacokinetics of chemopreventive kaempferol in rats. *Biopharm Drug Dispos* 30: 356–365.

Bentz, A.B. 2017. A review of quercetin: Chemistry, antioxident properties, and bioavailability. *J Young Investig.* www.jyi.org/2009-april/2017/10/15/a-review-of-quercetin-chemistry-antioxidant-properties-and-bioavailability.

Bussmann, A.J., Borghi, S.M., Zaninelli, T.H., Dos Santos, T.S., Guazelli, C.F., Fattori, V., Domiciano, T.P., Pinho-Ribeiro, F.A., Ruiz-Miyazawa, K.W., Casella, A., and Vignoli, J.A. 2019. The citrus flavanone naringenin attenuates zymosan-induced mouse joint inflammation: Induction of Nrf2 expression in recruited CD45+ hematopoietic cells. *Inflammopharmacology*, 27: 1229–1242.

Chauhan, P.S., Satti, N.K., Sharma, P., Sharma, V.K., Suri, K.A., and Bani, S. 2012. Differential effects of chlorogenic acid on various immunological parameters relevant to rheumatoid arthritis. *Phytother Res* 26: 1156–1165.

Chen, W.P., Tang, J.L., Bao, J.P., Hu, P.F., Shi, Z.L., and Wu, L.D. 2011. Anti-arthritic effects of chlorogenic acid in interleukin-1β-induced rabbit chondrocytes and a rabbit osteoarthritis model. *Int Immunopharmacology* 11: 23–28.

Chow, H.S., Cai, Y., Alberts, D.S., Hakim, I., Dorr, R., Shahi, F., Crowell, J.A., Yang, C.S., and Hara, Y. 2001. Phase I pharmacokinetic study of tea polyphenols following single-dose administration of epigallocatechin gallate and polyphenon E. *Cancer Epidemiol Biomark Prev* 10: 53–58.

Chung, S.J., Kim, T.Y., Kwon, Y.J., Park, Y.B., Lee, S.K., and Park, M.C. 2010. Gallic acid diminishes cellular proliferation and pro-inflammatory gene expressions in fibroblast like synoviocytes from patients with rheumatoid arthritis. *Arthritis Rheumatol* 62: 27.

Chung, T.W., Li, S., Lin, C.C., and Tsai, S.W. 2019. Antinociceptive and anti-inflammatory effects of the citrus flavanone naringenin. *Tzu-Chi Med J* 31: 81.

Cunha, F.V.M., Gomes, B.D.S., Neto, B.D.S., Ferreira, A.R., de Sousa, D.P. and Oliveira, F.D.A. 2016. Ferulic acid ethyl ester diminished Complete Freund's Adjuvant-induced incapacitation through antioxidant and anti-inflammatory activity. *Naunyn-Schmiedeb Arch Pharmacol* 389: 117–130.

Dabeek, W.M., and Marra, M.V. 2019. Dietary quercetin and kaempferol: Bioavailability and potential cardiovascular-related bioactivity in humans. *Nutrients* 11: 2288.

Dang, T., Honda, M., Shiraishi, M., Qiu, X., Hotta, T., Tsusaki, T., Matsuyama, Y., Shimasaki, Y. and Oshima, Y., 2013. Pharmacokinetic study of catechin (Epigallocatechin Gallate) after intraperitoneal and oral administration to Yellowtail Seriola quinqueradiata. *Aquaculture Science* 61: 205–206.

Das, U., Manna, K., Sinha, M., Datta, S., Das, D.K., Chakraborty, A., Ghosh, M., Saha, K.D., and Dey, S. 2014. Role of ferulic acid in the amelioration of ionizing radiation induced inflammation: a murine model. *PloS one* 9: e97599.

Dawidowicz, A.L., and Typre, R. 2017. Transformation of chlorogenic acids during the coffee beans roasting process. *Eur Food Res Technol* 243: 379–390.

Deng, C., Gao, C., Tian, X., Chao, B., Wang, F., Zhang, Y., Zou, J., and Liu, D. 2017. Pharmacokinetics, tissue distribution and excretion of luteolin and its major metabolites in rats: Metabolites predominate in blood, tissues and are mainly excreted via bile. *J Funct Foods* 35: 332–340.

Dos Santos, M.D., Almeida, M.C., Lopes, N.P., and De Souza, G.E.P. 2006. Evaluation of the anti-inflammatory, analgesic and antipyretic activities of the natural polyphenol chlorogenic acid. *Biol Pharm Bull* 29: 2236–2240.

Doss, H.M., Dey, C., Sudandiradoss, C., and Rasool, M.K. 2016. Targeting inflammatory mediators with ferulic acid, a dietary polyphenol, for the suppression of monosodium urate crystal-induced inflammation in rats. *Life Sci* 148: 201–210.

Doss, H.M., Samarpita, S., Ganesan, R., and Rasool, M. 2018. Ferulic acid, a dietary polyphenol suppresses osteoclast differentiation and bone erosion via the inhibition of RANKL dependent NF-κB signalling pathway. *Life Sciences*, 207, 284–295.

DuPont, M.S., Day, A.J., Bennett, R.N., Mellon, F.A., and Kroon, P.A. 2004. Absorption of kaempferol from endive, a source of kaempferol-3-glucuronide, in humans. *Eur J Clin Nutr* 58: 947–954.

EFSA Panel on Food Additives and Nutrient Sources added to Food (ANS), Younes, M., Aggett, P., Aguilar, F., Crebelli, R., Dusemund, B., Filipič, M., Frutos, M.J., Galtier, P., Gott, D., and Gundert-Remy, U. 2018. Scientific opinion on the safety of green tea catechins. *EFSA J* 16: e05239.

Fan, J.Z., Yang, X. and Bi, Z.G., 2015. The effects of 6-gingerol on proliferation, differentiation, and maturation of osteoblast-like MG-63 cells. *Braz J Med Biol Res* 48: 637–643.

Fei, J., Liang, B., Jiang, C., Ni, H., and Wang, L. 2019. Luteolin inhibits IL-1β-induced inflammation in rat chondrocytes and attenuates osteoarthritis progression in a rat model. *Biomed Pharmacother* 109: 586–1592.

Fu, X., Lyu, X., Liu, H., Zhong, D., Xu, Z., He, F., and Huang, G. 2019. Chlorogenic acid inhibits BAFF expression in collagen-induced arthritis and human synoviocyte MH7A cells by modulating the activation of the NF-κB signaling pathway. *J Immunol Res* 2019. https://doi.org/10.1155/2019/8042097.

Ganesan, R., and Rasool, M. 2019. Ferulic acid inhibits interleukin 17-dependent expression of nodal pathogenic mediators in fibroblast-like synoviocytes of rheumatoid arthritis. *J Cell Biochem* 120: 1878–1893.

Gao, J., Hu, J., Hu, D., and Yang, X. 2019. A role of gallic acid in oxidative damage diseases: A comprehensive review. *Nat Prod Commun* 14: 1934578X19874174.

Gil, M., and Wianowska, D. 2017. Chlorogenic acids – their properties, occurrence and analysis. *Annales Universitatis Mariae Curie-Sklodowska, sectio AA–Chemia* 72: 61.

Graefe, E.U., Derendorf, H., and Veit, M. 1999. Pharmacokinetics and bioavailability of the flavonol quercetin in humans. *Int J Clin Pharmacol Ther* 37: 219–233.

Gross, M., Pfeiffer, M., Martini, M., Campbell, D., Slavin, J., and Potter, J. 1996. The quantitation of metabolites of quercetin flavonols in human urine. *Cancer Epidemiol Biomark Prev* 5: 711–720.

Gugler, R., Leschik, M., and Dengler, H.J. 1975. Disposition of quercetin in man after single oral and intravenous doses. *Eur J Clin Pharmacol* 9: 229–234.

Haleagrahara, N., Hodgson, K., Miranda-Hernandez, S., Hughes, S., Kulur, A.B., and Ketheesan, N. 2018. Flavonoid quercetin–methotrexate combination inhibits inflammatory mediators and matrix metalloproteinase expression, providing protection to joints in collagen-induced arthritis. *Inflammopharmacology* 26: 1219–1232.

Haleagrahara, N., Miranda-Hernandez, S., Alim, M.A., Hayes, L., Bird, G., and Ketheesan, N. 2017. Therapeutic effect of quercetin in collagen-induced arthritis. *Biomed Pharmacother* 90: 38–46.

Haqqi, T.M., Anthony, D.D., Gupta, S., Ahmad, N., Lee, M.S., Kumar, G.K., and Mukhtar, H. 1999. Prevention of collagen-induced arthritis in mice by a polyphenolic fraction from green tea. *Proc. Natl. Acad. Sci. U.S.A.* 96: 4524–4529.

Hollman, P.C., Gaag, M.V., Mengelers, M.J., Van Trijp, J.M., De Vries, J.H., and Katan, M.B. 1996. Absorption and disposition kinetics of the dietary antioxidant quercetin in man. *Free Radic Biol Med* 21: 703–707.

Hollman, P.C., Van Trijp, J.M., Mengelers, M.J., De Vries, J.H., and Katan, M.B. 1997. Bioavailability of the dietary antioxidant flavonol quercetin in man. *Cancer Lett* 114: 139–140.

Hollman, P.C.H., and Katan, M.B. 1997. Absorption, metabolism and health effects of dietary flavonoids in man. *Biomed Pharmacother* 51: 305–310. https://clinicaltrials.gov/ (accessed on 17, January, 2022).

Hunter, D.J., March, L., and Chew, M. 2020. Osteoarthritis in 2020 and beyond: A Lancet Commission. *The Lancet* 396: 1711–1712.

Hwang, Y.H., Kim, T., Kim, R., and Ha, H. 2018. The natural product 6-gingerol inhibits inflammation-associated osteoclast differentiation via reduction of prostaglandin E2 levels. *Int J Mol Sci* 19: 2068.

Impellizzeri, D., Esposito, E., Di Paola, R., Ahmad, A., Campolo, M., Peli, A., Morittu, V.M., Britti, D., and Cuzzocrea, S. 2013. Palmitoylethanolamide and luteolin ameliorate development of arthritis caused by injection of collagen type II in mice. *Arthritis Res Ther* 15: 1–14.

Isbrucker, R.A., Bausch, J., Edwards, J.A., and Wolz, E. 2006. Safety studies on epigallocatechin gallate (EGCG) preparations. Part 1: Genotoxicity. *Food Chem Toxicol* 44: 626–635.

Jaiswal, R., Matei, M.F., Golon, A., Witt, M., and Kuhnert, N. 2012. Understanding the fate of chlorogenic acids in coffee roasting using mass spectrometry based targeted and non-targeted analytical strategies. *Food Funct* 3: 976–984.

Javadi, F., Ahmadzadeh, A., Eghtesadi, S., Aryaeian, N., Zabihiyeganeh, M., Rahimi Foroushani, A., and Jazayeri, S. 2017. The effect of quercetin on inflammatory factors and clinical symptoms in women with rheumatoid arthritis: A double-blind, randomized controlled trial. *J Am Coll Nutr* 36: 9–15.

Jeyadevi, R., Sivasudha, T., Rameshkumar, A., Ananth, D.A., Aseervatham, G.S.B., Kumaresan, K., Kumar, L.D., Jagadeeswari, S., and Renganathan, R. 2013. Enhancement of anti-arthritic effect of quercetin using thioglycolic acid-capped cadmium telluride quantum dots as nanocarrier in adjuvant induced arthritic Wistar rats. *Colloids Surf. B: Biointerfaces* 112: 255–263.

Jiang, H., Engelhardt, U.H., Thräne, C., Maiwald, B., and Stark, J. 2015. Determination of flavonol glycosides in green tea, oolong tea and black tea by UHPLC compared to HPLC. *Food Chem* 183: 30–35.

Jin, L., Zeng, W., Zhang, F., Zhang, C., and Liang, W. 2017. Naringenin ameliorates acute inflammation by regulating intracellular cytokine degradation. *J Immunol* 199: 3466–3477.

Johnston, K.L., Clifford, M.N., and Morgan, L.M. 2003. Coffee acutely modifies gastrointestinal hormone secretion and glucose tolerance in humans: glycemic effects of chlorogenic acid and caffeine. *Am J Clin Nutr* 78: 728–733.

Kadioglu, O., Nass, J., Saeed, M.E., Schuler, B., and Efferth, T. 2015. Kaempferol is an anti-inflammatory compound with activity towards NF-κB pathway proteins. *Anticancer Res* 35: 2645–2650.

Kang, B.J., Ryu, J., Lee, C.J., and Hwang, S.C. 2014. Luteolin inhibits the activity, secretion and gene expression of MMP-3 in cultured articular chondrocytes and production of MMP-3 in the rat knee. *Biomol Ther* 22: 239.

Kim, H., Choi, Y., An, Y., Jung, Y.R., Lee, J.Y., Lee, H.J., Jeong, J., Kim, Z., and Kim, K. 2020. Development of p-Coumaric acid analysis in human plasma and its clinical application to PK/PD Study. *J Clin Med* 10: 108.

Kim, H.R., Kim, B.M., Won, J.Y., Lee, K.A., Ko, H.M., Kang, Y.S., Lee, S.H., and Kim, K.W. 2019. Quercetin, a plant polyphenol, has potential for the prevention of bone destruction in rheumatoid arthritis. *J Med Food* 22: 152–161.

Kim, S.H., Park, J.G., Lee, J., Yang, W.S., Park, G.W., Kim, H.G., Yi, Y.S., Baek, K.S., Sung, N.Y., Hossen, M.J., and Lee, M.N. 2015. The dietary flavonoid Kaempferol mediates anti-inflammatory responses via the Src, Syk, IRAK1, and IRAK4 molecular targets. *Mediat Inflamm* 2015. https://doi.org/10.1155/2015/904142.

Kumar, S. and Pandey, A.K., 2013. Chemistry and biological activities of flavonoids: an overview. Sci. World J. 2013. https://doi.org/10.1155/2013/162750.

Lambert, J.D., Lee, M.J., Lu, H., Meng, X., Hong, J.J.J., Seril, D.N., Sturgill, M.G., and Yang, C.S. 2003. Epigallocatechin-3-gallate is absorbed but extensively glucuronidated following oral administration to mice. *J Nutr* 133: 4172–4177.

Lan-Fang, C., Dan, X., Jun, S., Fu-Yong, Q., Xiao-Wan, W., and Liang, X. 2021. Anti-inflammatory effect of Eucommia Chlorogenic Acid (ECA) on adjuvant-induced arthritis rats and its effect on autophagy related pathway. *Food Sci Technol.* 42: e74521, 2022.

Lee, C.J., Moon, S.J., Jeong, J.H., Lee, S., Lee, M.H., Yoo, S.M., Lee, H.S., Kang, H.C., Lee, J.Y., Lee, W.S., and Lee, H.J. 2018. Kaempferol targeting on the fibroblast growth factor receptor 3-ribosomal S6 kinase 2 signaling axis prevents the development of rheumatoid arthritis. *Cell Death Dis* 9: 1–21.

Lee, S.D., Choe, J.W., Lee, B.J., Kang, M.H., Joo, M.K., Kim, J.H., Yeon, J.E., Park, J.J., Kim, J.S., and Bak, Y.T. 2015. Butein effects in colitis and interleukin-6/signal transducer and activator of transcription 3 expression. *World J Gastroenterol* 21: 465.

Lee, Y.S., and Choi, E.M. 2010. Myricetin inhibits IL-1β-induced inflammatory mediators in SW982 human synovial sarcoma cells. *Int Immunopharmacol* 10: 812–814.

Leichsenring, A., Bäcker, I., Furtmüller, P.G., Obinger, C., Lange, F., and Flemmig, J. 2016. Long-Term Effects of (−)-Epigallocatechin Gallate (EGCG) on Pristane-Induced Arthritis (PIA) in Female Dark Agouti Rats. *PLoS One* 11: e0152518.

Levy, A.S., Simon, O., Shelly, J., and Gardener, M. 2006. 6-Shogaol reduced chronic inflammatory response in the knees of rats treated with complete Freund's Adjuvant. *BMC Pharmacol* 6: 1–8.

Li, D., Rui, Y.X., Guo, S.D., Luan, F., Liu, R., and Zeng, N. 2021. Ferulic acid: A review of its pharmacology, pharmacokinetics and derivatives. *Life Sci* 284: 119921.

Li, J., Bai, Y., Bai, Y., Zhu, R., Liu, W., Cao, J., An, M., Tan, Z. and Chang, Y.X., 2017. Pharmacokinetics of caffeic acid, ferulic acid, formononetin, cryptotanshinone, and tanshinone IIA after oral administration of Naoxintong capsule in rat by HPLC-MS/MS. *Evid-based Complement Altern Med* . https://doi.org/10.1155/2017/9057238.

Li, W., Wang, Y., Tang, Y., Lu, H., Qi, Y., Li, G., He, H., Lu, F., Yang, Y., and Sun, H. 2021. Quercetin alleviates osteoarthritis progression in rats by suppressing inflammation and apoptosis via inhibition of IRAK1/NLRP3 signaling. *J Inflamm Res* 14: 3393.

Li, Y.J., Zhang, C.F., Ding, G., Huang, W.Z., Wang, Z.Z., Bi, Y.A., and Xiao, W. 2015. Investigating the thermal stability of six caffeoylquinic acids employing rapid-resolution liquid chromatography with quadrupole time-of-flight tandem mass spectrometry. *Eur Food Res Technol* 240: 1225–1234.

Li, Y.R., Chen, D.Y., Chu, C.L., Li, S., Chen, Y.K., Wu, C.L., and Lin, C.C. 2015. Naringenin inhibits dendritic cell maturation and has therapeutic effects in a murine model of collagen-induced arthritis. *J Nutr Biochem* 26: 1467–1478.

Liang, Q., Ju, Y., Chen, Y., Wang, W., Li, J., Zhang, L., Xu, H., Wood, R.W., Schwarz, E., Boyce, B.F. and Wang, Y., 2016. Lymphatic endothelial cells efferent to inflamed joints produce iNOS and inhibit lymphatic vessel contraction and drainage in TNF-induced arthritis in mice. *Arthritis Res Ther* 18: 1–14.

Liang, Y., Chen, S., Yang, Y., Lan, C., Zhang, G., Ji, Z. and Lin, H., 2016. Effect of puerarin on TIMP3, MMP-9 expression and methylation in chondrocytes of rat osteoarthritis. *Int J Clin Exp Med* 9: 17952–17957.

Lin, L.C., Wang, M.N., Tseng, T.Y., Sung, J.S., and Tsai, T.H. 2007. Pharmacokinetics of (−)-epigallocatechin-3-gallate in conscious and freely moving rats and its brain regional distribution. *J Agric Food Chem* 55: 1517–1524.

Lin, S.K., Chang, H.H., Chen, Y.J., Wang, C.C., Galson, D.L., Hong, C.Y., and Kok, S.H. 2008. Epigallocatechin-3-gallate diminishes CCL2 expression in human osteoblastic cells via up-regulation of phosphatidylinositol 3-kinase/Akt/Raf-1 interaction: a potential therapeutic benefit for arthritis. *Arthritis Rheumatol* 58: 3145–3156.

Lin, Y., Luo, T., Weng, A., Huang, X., Yao, Y., Fu, Z., Li, Y., Liu, A., Li, X., Chen, D., and Pan, H. 2020. Gallic acid alleviates gouty arthritis by inhibiting NLRP3 inflammasome activation and pyroptosis through enhancing Nrf2 signaling. *Front immunol* 11: 3197.

Liu, C., Liu, H., Lu, C., Deng, J., Yan, Y., Chen, H., Wang, Y., Liang, C.L., Wei, J., Han, L., and Dai, Z. 2019. Kaempferol attenuates imiquimod-induced psoriatic skin inflammation in a mouse model. *Clin Exp Immunol* 198: 403–415.

Liu, C.C., Zhang, Y., Dai, B.L., Ma, Y.J., Zhang, Q., Wang, Y., and Yang, H. 2017. Chlorogenic acid prevents inflammatory responses in IL-1β-stimulated human SW-1353 chondrocytes, a model for osteoarthritis. *Mol Med Rep* 16: 1369–1375.

Lodhi, S., Vadnere, G.P., Patil, K.D., and Patil, T.P. 2020. Protective effects of luteolin on injury induced inflammation through reduction of tissue uric acid and pro-inflammatory cytokines in rats. *J Tradit Complement Med* 10: 60–69.

Ma, T.W., Wen, Y.J., Song, X.P., Hu, H.L., Li, Y., Bai, H., Zhao, M.C., and Gao, L. 2021. Puerarin inhibits the development of osteoarthritis through antiinflammatory and antimatrix-degrading pathways in osteoarthritis-induced rat model. *PhytotheR Res* 35: 2579–2593.

Manchope, M.F., Artero, N.A., Fattori, V., Mizokami, S.S., Pitol, D.L., Issa, J.P., Fukada, S.Y., Cunha, T.M., Alves-Filho, J.C., Cunha, F.Q., and Casagrande, R. 2018. Naringenin mitigates titanium dioxide (TiO2)-induced chronic arthritis in mice: Role of oxidative stress, cytokines, and NFκB. *Inflamm Res* 67: 997–1012.

Manchope, M.F., Calixto-Campos, C., Coelho-Silva, L., Zarpelon, A.C., Pinho-Ribeiro, F.A., Georgetti, S.R., Baracat, M.M., Casagrande, R., and Verri Jr, W.A. 2016. Naringenin inhibits superoxide anion-induced inflammatory pain: role of oxidative stress, cytokines, Nrf-2 and the NO− cGMP− PKG− KATPChannel signaling pathway. *PloS one* 11: e0153015.

Mariani, C., Braca, A., Vitalini, S., De Tommasi, N., Visioli, F., and Fico, G. 2008. Flavonoid characterization and in vitro antioxidant activity of Aconitum anthora L.(Ranunculaceae). *Phytochemistry* 69: 1220–1226.

Mereles, D., and Hunstein, W. 2011. Epigallocatechin-3-gallate (EGCG) for clinical trials: more pitfalls than promises? *Int J Mol Sci* 12: 5592–5603.

Min, S.Y., Yan, M., Kim, S.B., Ravikumar, S., Kwon, S.R., Vanarsa, K., Kim, H.Y., Davis, L.S., and Mohan, C. 2015. Green tea epigallocatechin-3-gallate suppresses autoimmune arthritis through indoleamine-2, 3-dioxygenase expressing dendritic cells and the nuclear factor, erythroid 2-like 2 antioxidant pathway. *J Inflamm* 12: 1–15.

Moon, Y.J., Wang, L., DiCenzo, R., and Morris, M.E. 2008. Quercetin pharmacokinetics in humans. *Biopharm Drug Dispos* 29: 205–217.

Morinobu, A., Biao, W., Tanaka, S., Horiuchi, M., Jun, L., Tsuji, G., Sakai, Y., Kurosaka, M., and Kumagai, S. 2008. (−)-Epigallocatechin-3-gallate suppresses osteoclast differentiation and ameliorates experimental arthritis in mice. *Arthritis Rheumatol* 58: 2012–2018.

Naveed, M., Hejazi, V., Abbas, M., Kamboh, A.A., Khan, G.J., Shumzaid, M., Ahmad, F., Babazadeh, D., FangFang, X., Modarresi-Ghazani, F., and WenHua, L. 2018. Chlorogenic acid (CGA): A pharmacological review and call for further research. *Biomed Pharmacother* 97: 67–74.

Neog, M.K., Joshua Pragasam, S., Krishnan, M., and Rasool, M. 2017. p-Coumaric acid, a dietary polyphenol ameliorates inflammation and curtails cartilage and bone erosion in the rheumatoid arthritis rat model. *BioFactors* 43: 698–717.

Nigrovic, P.A., et al. 2018. Genetics and the classification of arthritis in adults and children. *Arthritis Rheumatol* 70: 7–17.

Niho, N., Shibutani, M., Tamura, T., Toyoda, K., Uneyama, C., Takahashi, N., and Hirose, M. 2001. Subchronic toxicity study of gallic acid by oral administration in F344 rats. *Food Chem Toxicol* 39: 1063–1070.

Oliviero, F., Sfriso, P., Scanu, A., Fiocco, U., Spinella, P., and Punzi, L. 2013. Epigallocatechin-3-gallate reduces inflammation induced by calcium pyrophosphate crystals in vitro. *Front Pharmacol* 4: 51.

Olthof, M.R., Hollman, P.C., and Katan, M.B. 2001. Chlorogenic acid and caffeic acid are absorbed in humans. *J Nut* 131: 66–71.

Padmavathi, G., Roy, N.K., Bordoloi, D., Arfuso, F., Mishra, S., Sethi, G., Bishayee, A., and Kunnumakkara, A.B. 2017. Butein in health and disease: A comprehensive review. *Phytomedicine* 25: 118–127.

Pan, D., Li, N., Liu, Y., Xu, Q., Liu, Q., You, Y., Wei, Z., Jiang, Y., Liu, M., Guo, T., and Cai, X. 2018. Kaempferol inhibits the migration and invasion of rheumatoid arthritis fibroblast-like synoviocytes by blocking activation of the MAPK pathway. *Int Immunopharmacol* 55: 174–182.

Panda, S.P., Panigrahy, U.P., Mallick, S.P., Prasanth, D.S.N.B.K., and Raghavendra, M. 2021. Screening assessment of trimethoxy flavonoid and-(-)-epigallocatechin-3-gallate against formalin-induced arthritis in Swiss albino rats and binding properties on NF-κB-MMP9 proteins. *Future J Pharm Sci* 7: 1–13.

Pang, J., Yang, F., Zhang, Z., Yang, W., Li, Y., and Xu, H. 2021. The role of luteolin nanocomposites in rheumatoid arthritis treatment. *Materials Express* 11: 303–309.

Pannu, A., Goyal, R.K., Ojha, S. and Nandave, M., 2019. Naringenin: A promising flavonoid for herbal treatment of rheumatoid arthritis and associated inflammatory disorders. In *Bioactive Food as Dietary Interventions for Arthritis and Related Inflammatory Diseases*, pp. 343–354. Academic Press, 2019.

Peat, G., and Thomas, M.J. 2021. Osteoarthritis year in review 2020: epidemiology & therapy. Osteoarthr Cartil 29: 180–189. https://doi.org/10.1016/j.joca.2020.10.007.

Pragasam, S.J., Murunikkara, V., Sabina, E.P., and Rasool, M. 2013. Ameliorative effect of p-coumaric acid, a common dietary phenol, on adjuvant-induced arthritis in rats. *Rheumatol Int* 33: 325–334.

Qi, W., et al. 2011. Comparative pharmacokinetics of chlorogenic acid after oral administration in rats. *Journal of Pharm Anal* 1: 30.

Quicke, J.G., Conaghan, P.G., Corp, N., and Peat, G. 2021. Osteoarthritis year in review 2021: Epidemiology and therapy. *Osteoarthr Cartil*. 10.1016/j.joca.2021.10.003

Rasheed, Z., Akhtar, N., Khan, A., Khan, K.A., and Haqqi, T.M. 2010. Butrin, isobutrin, and butein from medicinal plant Butea monosperma selectively inhibit nuclear factor-κB in activated human mast cells: Suppression of tumor necrosis factor-α, interleukin (IL)-6, and IL-8. *J Pharmacol Exp Ther* 333: 354–363.

Sabina, E.P., Rasool, M., Mathew, L., EzilRani, P., and Indu, H. 2010. 6-Shogaol inhibits monosodium urate crystal-induced inflammation-An in vivo and in vitro study. *Food Chem Toxicol* 48: 229–235.

Saccol, R.D.S.P., da Silveira, K.L., Adefegha, S.A., Manzoni, A.G., da Silveira, L.L., Coelho, A.P.V., Castilhos, L.G., Abdalla, F.H., Becker, L.V., Martins, N.M.B., and Oliveira, J.S. 2019. Effect of quercetin on E-NTPDase/E-ADA activities and cytokine secretion of complete Freund adjuvant–induced arthritic rats. *Cell Biochem Funct* 37: 474–485.

Salameh, D., Brandam, C., Medawar, W., Lteif, R., and Strehaiano, P. 2008. Highlight on the problems generated by p-coumaric acid analysis in wine fermentations. *Food Chem* 107: 1661–1667.

Shahrzad, S., Aoyagi, K., Winter, A., Koyama, A., and Bitsch, I. 2001. Pharmacokinetics of gallic acid and its relative bioavailability from tea in healthy humans. *J Nutr*, 131: 1207–1210.

Sharma, A., Kashyap, D., Sak, K., Tuli, H.S., and Sharma, A.K. 2018. Therapeutic charm of quercetin and its derivatives: a review of research and patents. *Pharm Pat Anal* 7: 15–32.

Sharma, A., Sharma, P., Tuli, H.S., and Sharma, A.K. 2018. Phytochemical and pharmacological roperties of flavonols. In eLS.; American Cancer Society; Wiley: Hoboken, NJ. pp. 1–12. ISBN 978–0–470–01590–2.

Sheng, Y., Wu, T., Dai, Y., Ji, K., Zhong, Y., and Xue, Y. 2020. The effect of 6-gingerol on inflammatory response and Th17/Treg balance in DSS-induced ulcerative colitis mice. *Ann Transl Med* 8: 442. https://doi.org/10.21037/atm.2020.03.141.

Shi, F., Zhou, D., Ji, Z., Xu, Z., and Yang, H. 2015. Anti-arthritic activity of luteolin in Freund's complete adjuvant-induced arthritis in rats by suppressing P2X4 pathway. *Chem Biol Interact* 226: 82–87.

Shimoi, K., Okada, H., Furugori, M., Goda, T., Takase, S., Suzuki, M., Hara, Y., Yamamoto, H., and Kinae, N. 1998. Intestinal absorption of luteolin and luteolin 7-O-β-glucoside in rats and humans. *FEBS Lett* 438: 220–224.

Singh, R., Ahmed, S., Islam, N., Goldberg, V.M., and Haqqi, T.M. 2002. Epigallocatechin-3-gallate inhibits interleukin-1β–induced expression of nitric oxide synthase and production of nitric oxide in human chondrocytes: suppression of nuclear factor κB activation by degradation of the inhibitor of nuclear factor κB. *Arthritis Rheum* 46: 2079–2086.

Singh, R., Ahmed, S., Malemud, C.J., Goldberg, V.M., and Haqqi, T.M. 2003. Epigallocatechin-3-gallate selectively inhibits interleukin-1 β-induced activation of mitogen activated protein kinase subgroup c-Jun N-terminal kinase in human osteoarthritis chondrocytes. *J Orthop Res* 21: 102–109.

Sullivan, M., Follis Jr, R.H., and Hilgartner, M. 1951. Toxicology of podophyllin. *Proc Soc Exp Biol Med* 77: 269–272.

Sung, J., and Lee, J. 2015. Anti-inflammatory activity of butein and luteolin through suppression of NF κ B activation and induction of heme oxygenase-1. *J Med Food* 18: 557–564.

Taheri, Y., Suleria, H.A.R., Martins, N., Sytar, O., Beyatli, A., Yeskaliyeva, B., Seitimova, G., Salehi, B., Semwal, P., Painuli, S., and Kumar, A. 2020. Myricetin bioactive effects: Moving from preclinical evidence to potential clinical applications. *BMC Complement Med Ther* 20: 1–14.

Testa, D., Calvacchi, S., Petrelli, F., Giannini, D., Bilia, S., Alunno, A., and Puxeddu, I. 2021. One year in review 2021: Pathogenesis of rheumatoid arthritis. *Clin Exp Rheumatol* 39: 445–452.

Vankemmelbeke, M.N., Jones, G.C., Fowles, C., Ilic, M.Z., Handley, C.J., Day, A.J., Knight, C.G., Mort, J.S., and Buttle, D.J. 2003. Selective inhibition of ADAMTS-1, -4 and -5 by catechin gallate esters. *Eur J Biochem* 270: 2394–2403.

Variya, B.C., Bakrania, A.K., Madan, P., and Patel, S.S. 2019. Acute and 28-days repeated dose sub-acute toxicity study of gallic acid in albino mice. *Regul Toxicol Pharmacol* 101: 71–78.

Ventura-Martínez, R., Déciga-Campos, M., Bustamante-Marquina, A., Ángeles-López, G.E., Aviles-Herrera, J., González-Trujano, M.E., and Navarrete-Vázquez, G. 2021. Quercetin decreases the antinociceptive effect of diclofenac in an arthritic gout-pain model in rats. *J Pharm Pharmacol* 73: 1310–1318.

Wang, C., Wang, W., Jin, X., Shen, J., Hu, W. and Jiang, T., 2016. Puerarin attenuates inflammation and oxidation in mice with collagen antibody-induced arthritis via TLR4/NF-κB signaling. *Mol Med Rep* 14: 1365–1370.

Wang, C.C., Guo, L., Tian, F.D., An, N., Luo, L., Hao, R.H., Wang, B., and Zhou, Z.H. 2017. Naringenin regulates production of matrix metalloproteinases in the knee-joint and primary cultured articular chondrocytes and alleviates pain in rat osteoarthritis model. *Braz J Med Biol Res* 50. https://doi.org/10.1590/1414-431X20165714.

Wang, J., Wei, T., Gao, J., He, H., Chang, X., and Yan, T. 2015. Effects of Naringenin on inflammation in complete freund's adjuvant-induced arthritis by regulating Bax/Bcl-2 balance. *Inflammation* 38: 245–251.

Wang, P., Li, S.S., and Wang, X.H. 2016. Myricetin exerts anti-osteoarthritic effects in IL-1 beta stimulated SW1353 cells via regulating matrix metalloproteinases and modulating JNK/P38MAPK/Ap-1/c-Fos and JAK/STAT Signalling. *Int J Pharmacol* 12: 440–450.

Wang, Z., Zeng, M., Wang, Z., Qin, F., Chen, J., and He, Z. 2021. Dietary luteolin: A narrative review focusing on its pharmacokinetic properties and effects on glycolipid metabolism. *J Agric Food Chem* 69: 1441–1454.

Wen, L., Qu, T.B., Zhai, K., Ding, J., Hai, Y., and Zhou, J.L. 2015. Gallic acid can play a chondroprotective role against age-induced osteoarthritis progression. *J Orthop Sci* 20: 734–741.

Wong, S.K., Chin, K.Y., and Ima-Nirwana, S. 2020. Quercetin as an agent for protecting the bone: A review of the current evidence. *Int J Mol Sci* 21: 6448.

Woolf, A.D., and Pfleger, B. 2003. Burden of major musculoskeletal conditions. *Bull World Health Organ* 81: 646–656.

World Health Organisation (WHO) Musculoskeletal conditions fact sheet 2021. www.who.int/news-room/fact-sheets/detail/musculoskeletal-conditions. Accessed on 17–01–2021.

Xiao, J., Muzashvili, T.S., and Georgiev, M.I. 2014. Advances in the biotechnological glycosylation of valuable flavonoids. *Biotechnol Adv* 32: 1145–1156.

Xiao, Y., Liu, L., Zheng, Y., Liu, W., and Xu, Y. 2021. Kaempferol attenuates the effects of XIST/miR-130a/STAT3 on inflammation and extracellular matrix degradation in osteoarthritis. *Future Med Chem* 13: 1451–1464.

Yang, M., Luo, Y., Liu, T., Zhong, X., Yan, J., Huang, Q., Tao, J., He, Q., Guo, M., and Hu, Y. 2018. The effect of puerarin on carotid intima-media thickness in patients with active rheumatoid arthritis: a randomized controlled trial. *Clin Ther* 40: 1752–1764.

Yoon, C.H., Chung, S.J., Lee, S.W., Park, Y.B., Lee, S.K., and Park, M.C. 2013. Gallic acid, a natural polyphenolic acid, induces apoptosis and inhibits proinflammatory gene expressions in rheumatoid arthritis fibroblast-like synoviocytes. *Jt Bone Spine* 80: 274–279.

Yoon, H.Y., Lee, E.G., Lee, H., Cho, I.J., Choi, Y.J., Sung, M.S., Yoo, H.G., and Yoo, W.H. 2013. Kaempferol inhibits IL-1β-induced proliferation of rheumatoid arthritis synovial fibroblasts and the production of COX-2, PGE2 and MMPs. *Int J Mol Med* 32: 971–977.

Yu, Y., Zick, S., Li, X., Zou, P., Wright, B., and Sun, D. 2011. Examination of the pharmacokinetics of active ingredients of ginger in humans. *AAPS J* 13: 417–426.

Yuan, K., Zhu, Q., Lu, Q., Jiang, H., Zhu, M., Li, X., Huang, G., and Xu, A. 2020. Quercetin alleviates rheumatoid arthritis by inhibiting neutrophil inflammatory activities. *J Nutr Biochem* 84: 108454.

Yuan, X., Liu, Y., Hua, X., Deng, X., Sun, P., Yu, C., Chen, L., Yu, S., Liu, S. and Pang, H., 2015. Myricetin ameliorates the symptoms of collagen-induced arthritis in mice by inhibiting cathepsin K activity. *Immunopharmacol Immunotoxicol* 37: 513–519.

Zhang, M., Yi, X., Xie, S., and Zhan, Q. 2020. Study on the mechanism of chlorogenic acid inhibiting collagen-induced rheumatoid arthritis through dimethylarginine dimethylaminohydrolase/asymmetric dimethylarginine/cortistatin signaling pathway. *J Biomater Tissue Eng* 10: 1385–1391.

Zhu, H., Liang, Q.H., Xiong, X.G., Chen, J., Wu, D., Wang, Y., Yang, B., Zhang, Y., Zhang, Y. and Huang, X., 2014. Anti-inflammatory effects of the bioactive compound ferulic acid contained in Oldenlandia diffusa on collagen-induced arthritis in rats. *Evid-based Complement Altern Med*. https://doi.org/10.1155/2014/ 573801.

Zhu, H., Liang, Q.H., Xiong, X.G., Wang, Y., Zhang, Z.H., Sun, M.J., Lu, X., and Wu, D. 2018. Anti-inflammatory effects of p-coumaric acid, a natural compound of Oldenlandia diffusa, on arthritis model rats. *Evid.-Based Complement Altern Med*. https://doi.org/10.1155/2018/5198594

Zhu, L., Zhang, Z., Xia, N., Zhang, W., Wei, Y., Huang, J., Ren, Z., Meng, F., and Yang, L. 2020. Anti-arthritic activity of ferulic acid in complete Freund's Adjuvant (CFA)-induced arthritis in rats: JAK2 inhibition. *Inflammopharmacology* 28: 463–473.

Zick, S.M., Djuric, Z., Ruffin, M.T., Litzinger, A.J., Normolle, D.P., Alrawi, S., Feng, M.R., and Brenner, D.E. 2008. Pharmacokinetics of 6-gingerol, 8-gingerol, 10-gingerol, and 6-shogaol and conjugate metabolites in healthy human subjects. *Cancer Epidemiol Biomark Prev* 17: 1930–1936.

7 Dietary Polyphenols in Diabetes

Yousef Rasmi, Safa Rafique,Quratulain Babar,
Behrokh Daei-Hasani, and Yeganeh Farnamian
E-mail: rasmiy@umsu.ac.ir; safa.sandhu@gmail.com;
ainniebabarrr@gmail.com; hasani.bio@gmail.com;
yfarnamian@gmail.com

CONTENTS

DOI: 10.1201/9781003251538-7

7.1   INTRODUCTION

Polyphenols are considered a large and heterogeneous group of phytochemicals containing multiple rings of phenol obtained from plants (Pandey and Rizvi, 2009). Fruits, coffee, vegetables, tea, whole grains, and nuts are rich sources of polyphenols (Guasch-Ferré et al., 2017). Many fruits like pears, apples, grapes, and berries have a lot of polyphenols, about 200–300 mg per 100 g (Scalbert et al., 2005). Due to the widespread presence of polyphenols in foods, their intake has been associated with a lower incidence of type 2 diabetes (T2D) in people (Guasch-Ferré et al., 2017; Tresserra-Rimbau et al. 2016; Wedick et al., 2012). Diabetes (T2D) and obesity (Obesity Epidemic) are major public health challenges for many countries (Guasch-Ferré et al., 2017). About 90 percent of T2D cases were associated with a combination of several unhealthy lifestyle factors, such as sedentary lifestyle, smoking, excessive alcohol consumption, overweight/obesity, and unhealthy diets (Chen et al., 2012). It has been reported that polyphenols primarily have anti-inflammatory properties and impact diabetes type 2 (T2D) by promoting glucose uptake in tissues and thereby influencing insulin sensitivity (Guasch-Ferré et al., 2017; Scalbert et al., 2005). Therefore, lifestyle changes and promoting healthy diets to prevent diabetes have been considered the cornerstone of policymakers and researchers alike (Guasch-Ferré et al., 2017). With a focus on polyphenol-rich food characteristics, this chapter summarizes the relevant evidence for a link between dietary polyphenols and type 2 diabetes. Furthermore, the present study aims to describe some of the important properties of polyphenols and their anti-oxidant properties to understand polyphenols' role in T2D.

7.2   DIETARY POLYPHENOLS

We get polyphenols from fruits, vegetables, cereal, dry beans, chocolate, tea and coffee, and berry fruits like apples, pear, grapes, cherries, and blackberries contain up to 200–300 mg polyphenol per 100 g fresh weight, and a cup of tea or coffee contains about 100 mg.

7.2.1   STRUCTURE, CLASSIFICATION, AND SOURCES

Polyphenols are chemical compounds containing an aromatic ring and a hydroxyl group (Han et al., 2007), Polyphenols are extremely important for their antioxidant properties. These properties are essential for diabetic management. Pharmacologically, these substances possess anti-inflammatory, anti-microbial, antihypertensive, and anti-microbial effects. Polyphenols show promise in preventing diabetes in peripheral tissues, as shown in Figure 7.6. A rise in insulin secretion, an inhibition of DPP-4 enzyme activity, and an increase in release of glucagon-like peptide-1 (GLP-1) are associated with these effects. Basic and clinical research could lead to new chemical families that would prevent or reduce type 2 diabetes mellitus's effects on human health. To assess polyphenols' anti-diabetic effect, more clinical studies are needed (Cardona et al., 2013).

7.2.2   FLAVONOIDS

In-plant tissues, Polyphenols are found as glucosides which are made by plants themselves. Primarily flavonoids exist as aglycones which are the non-sugar part of a glucoside. But flavonoids may exist

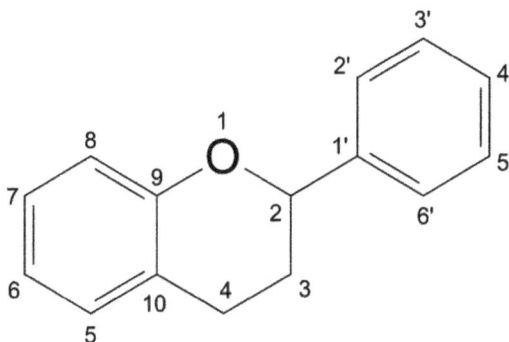

FIGURE 7.1 Common structure of flavonoids. Three rings are labeled as A, B, and C, while C is the heterocyclic ring. The oxygen atom is numbered in the first position, and the remaining carbon atoms are numbered from C2 to C10. Ring B shows six positions from C1` to C6`.

as glucosides or aglycones. Structurally, flavonoids exist as diphenyl propane (C6-C3-C6), with phenolic rings A and B that are connected to each other through a heterocyclic ring (Khoddami et al., 2013). The heterocyclic ring has been labeled as ring C in Figure 7.1. Different types of compounds categorized under polyphenols, that is, anthocyanins, isoflavone, flavonols, flavanones, Flavones, flavanonols, and so forth are formed by changes in oxidation state and hydroxylation pattern of heterocyclic pyran ring (as shown in Figure 7.1). The carbonyl group formed by C4 and the existence or non-existence of a double bond in between C2 and C3 determines the changes in oxidation state and hydroxylation pattern. The major subclasses of polyphenols are Flavanones, flavones, and flavonols therefore most of the flavonoids comprise these compounds (D' Archivio et al., 2007).

7.2.2.1 Flavonols

Flavonols are dihydroflavonols that are derived from flavanones as their 3-hydroxy resultants. The main difference between flavones and flavonols is that flavonols have a hydroxyl group at the C3 position (Oliveira et al., 2014). Kaempferol and quercetin are the most popular flavonols found as aglycones (Figure 7.2). Flavonols are found as over 270 and 340 glucosidic structures (Tsao, 2010). Flavonols are abundantly found in broccoli, onions (1.2 g per kg), blueberries, and leeks; 45 mg per liter of flavonols are found in tea (Manach et al., 2004).

7.2.2.2 Flavanols

The saturated heterocyclic ring is present in Flavanols (flavan-3-ols), a hydroxyl group at carbon number 3, and a lack of double bond at positions 2 and 3 on the carbon atom. Flavanols have existed in food items as aglycones, unlike several classes of flavonoids. They are found in two forms, polymerics and monomeric, which include tannins, catechins, and epicatechins, as mentioned in Figure 7.2 (Lattanzio, 2013). Flavanol contains a characterization of both catechins and epicatechins due to the presence of the hydroxyl (OH) group at carbon number 3 and is considered as diastereoisomers. Epicatechin consists of 2 isoforms which are in cis configuration (–) -epicatechin and (+) -epicatechin, while catechin also shows stereoisomers but in trans configuration (Giada, 2013). These structures are considered tannins because of several unique properties of polymeric phenols like high molecular weight and water solubility. They are of two types condensed tannins and hydrolyzable tannins when they form complex with proteins, polysaccharides, and alkaloids (Crozier et al., 2008).

Condensed tannins are most abundant in woody plants, which are polymers of epicatechin, catechin, and leucoanthocyanidin. Due to oxidative stress, condensed tannins are converted to

Luteonin   Vanillic acid   Erodictyol

Protocatechuic acid

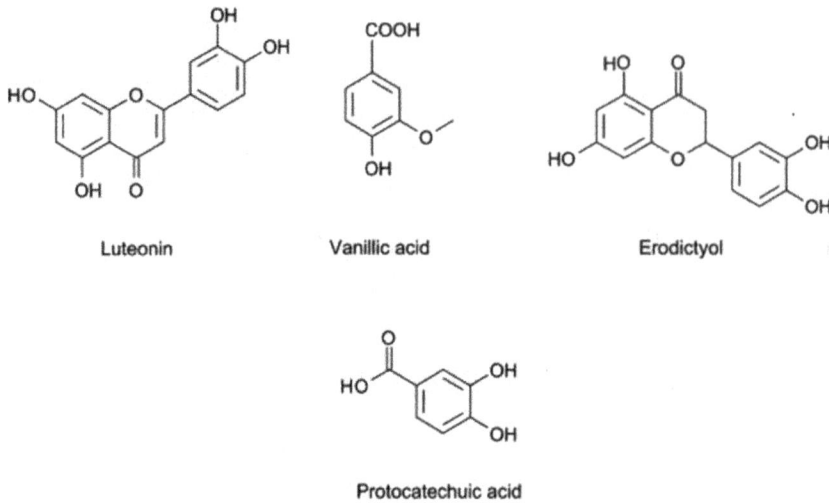

FIGURE 7.2   Structures of erodictyol, luteolin, protocatechuic and vanillic acid.

anthocyanidins (Vladimir-Knežević et al., 2012). Hydrolyzable tannins are esters of ellagic and gallic acid in which the OH group through esterification yields ellagitannins and also tannins (Ozcan et al., 2014).

7.2.2.3   Isoflavone

Soya and its products are the most common source of isoflavones. These components are structurally similar to estrogens. Due to the ability to connect to estrogen receptors, it is classified as phyto-estrogen and mainly contains Genistein, daidzein, and genistein.

7.2.2.4   Flavone

Flavone is a double-bonded molecule that contains a carbonyl group and ring B at C4 and C2 positions, respectively. As mentioned in Figure 7.2 luteolin and apigenin are considered important flavones. On the 4th position of the carbon atom oxygen is placed, and as compared to other classes ring B is attached to carbon at the 3rdposition. They are found in chicory, parsley, celery, chamomile, and dry tea leaves (Hostetler et al., 2017).

7.2.2.5   Flavanone

Flavanones are saturated compounds with three carbon chains along with the carbonyl group at position 4 of the carbon atom in the heterocyclic ring. They have a very unique substitution pattern and several derivatives are found through them which include prenylated flavanones and benzylated flavanones (Giada, 2013). Glycosylated flavanones are also substituted products of flavanone in which disaccharide takes position at C7 after substitution reaction. In fruits, hesperidin along with eriodictyol are in form of aglycone. Flavanone is richly present in eriodictyol, hesperidin, naringin, and citrus fruits (Brett et al., 2008).

7.2.2.6   Anthocyanins and anthocyanidins

Anthocyanins and anthocyanidins are unsaturated compounds due to the presence of a double bond in their heterocyclic ring. On ring B in anthocyanins, methoxylation and hydroxylation patterns are present because they are glycosylated forms of anthocyanidins (Giada, 2013). Anthocyanins are formed due to several variations in the OH group along with several sugar units. Arabinose, glucose, and galactose are present in them as monosaccharide sources (Ozcan et al., 2014). Glycosylated anthocyanins are the most colorful compounds and water-soluble pigments found in fruits and colorful petals of flowers (Crozier et al., 2008). Eggplant, cherry, red fruits, black grape, and strawberry are the main source of anthocyanins (Manach et al., 2004). They impart various bright colors to plants like red, purple, and purple-pink.

7.2.3 NONFLAVONOIDS

Nonflavanoids are different from polyphenols because of their simple structure having a single aromatic ring only. Phenolic acids, lignans, and stilbenes are included in this category.

7.2.3.1 Phenolic acids

Phenolic acids are found in conjugation with glucose, polyphenols, and quinic acid. They have various classes like cinnamic acid and benzoic acid derivatives. It is very rare to find phenolic acid in free form (Chang et al., 2005). Phenolic acids consist of hydroxycinnamic and hydroxybenzoic acids in which the location, number, and position of the hydroxyl group are varied on the aromatic ring.

Hydroxybenzoic acids consist of different types, like vanillic acids, syringic, *p*-hydroxybenzoic, gallic, and protocatechuic as mentioned in Figure 7.2 (Singla et al., 2019). They are found in conjugated forms like esters and glycosides along with free acids as well. The side chain of three carbon (C6–C3) is present in an aromatic ring of hydroxycinnamic acids. They are commonly present in our food in form of sinapic acids, ferulic, caffeic, and p-coumaric (Oliveira et al., 2014) (Bankar et al., 2011). The concentration of hydroxybenzoic acid in plants is usually very small that why they are richly found in red fruits, black radish, and onion (Manach et al., 2004).

7.2.3.2 Stilbenes

In stilbenes, two moieties of phenyl are linked together by a group of 2-carbon methylene 1,2-diarylethenes, which is considered a small group of nonflavanoids (Figure 7.3). It consists of two rings A and B in which hydroxyl and methoxyl groups are substituted at various positions. It consists of various forms like trans and cis configuration. Majorly it is found in glycosylated form, while a very minor amount has occurred in the free state, as mentioned in Figure 7.3 (Chang et al., 2005). Grapes, berries, and peanuts contain resveratrol, having structural formula (3,5,4'- trihydroxystilbene) which is considered well-known stilbenes (Han et al., 2007). Stilbene is very important according to the medical point of view due to its remarkable role in cancer and neurodegenerative disorders, but it is present in a very small amount in our diet.

7.2.3.3 Lignans

Lignans are classified into different subgroups in which eight groups mainly include dibenzylbutane, furan, and aryltetralin. Lignans are a non-flavonoid class having a wide variety of structural forms due to substituted patterns of different moieties at positions C9 and C9'. On the 8th position of carbon of the propane side chains mainly two propylbenzene units are linked together at β-position (Suzuki and Umezawa, 2007). Lignans are mainly present in free form in vegetable oil, seed, and legume; while rarely found in glycosylated form (MacRae and Towers, 1984) (Axelson et al., 1982). Lignan is abundantly present in flaxseed where it is metabolized to enterodiol and enterolactone by anaerobic bacteria present in the small intestine.

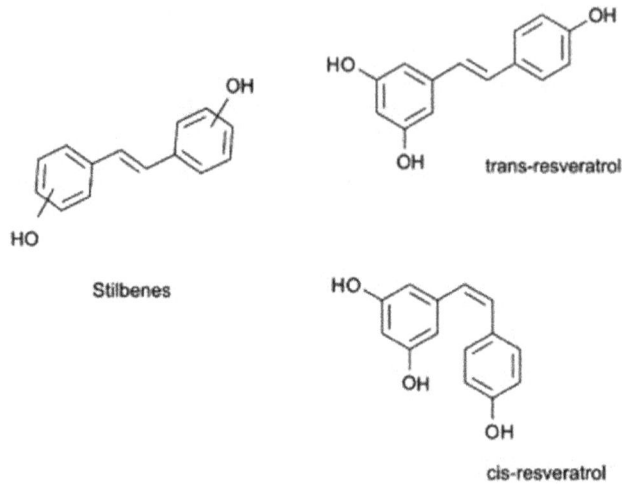

FIGURE 7.3   Structure of *cis*- and *trans*-resveratrol (right) and stilbenes (1,2-diarylethenes; left).

7.3   THE BIOLOGICAL AND BIOCHEMICAL FUNCTION OF POLYPHENOLS

Polyphenolic compounds play an important role in various biological and biochemical activities due to their antioxidant properties (Ferguson, 2001). They play an important role in the prevention of diabetes, cancer, neurodegenerative diseases, cardiovascular diseases, and osteoporosis (Zhou et al., 2016).

7.3.1   ANTI-HYPERTENSIVE EFFECT

Several flavonols are found in cocoa beans, especially catechins, epicatechins, and procyanidins. The flavanol content of cocoa products significantly improves endothelial function, reduces low-density lipoprotein (LDL) oxidation sensitivity, and promotes endothelium-dependent vasoconstriction (Williamson, 2017) (Grassi et al., 2008). According to the European Food Safety Authority (EFSA), cocoa flavanols have been shown to have beneficial effects on endothelium-dependent vasodilation (Williamson, 2017).

7.3.2   PRO-OXIDANT ACTIVITY

Despite the polyphenols' antioxidant properties, high doses of some polyphenols can cause DNA damage, apoptosis, and cell death. The Fenton Reaction causes DNA degradation due to OH radical formation when copper-containing flavonoids and tannic acid are present (Ferguson, 2001). Numerous flavonoids exhibit pro-oxidant activity when in contact with nitric oxide free radicals, and may damage DNA (Ferguson, 2001) (Ohshima et al., 1998).

   The polyphenols in tea have been known to have antioxidant properties, but some studies have shown that they are pro-oxidants as well. Several studies have shown that high doses of tea polyphenols exhibit pro-oxidant activity in hepatocytes, which is toxic (Galati et al., 2006). Polyphenols inhibit carcinogenesis by initiating apoptosis at high doses. There has been evidence that carvacrol and tea polyphenols promote oxidative stress in cancer cells (Mao et al., 2017),which is achieved via a dose-dependent mechanism (Günes-Bayir et al., 2017). This is why it is extremely important to determine the right amount of antioxidants to use in treating a specific disease.

7.3.3 Anti-atherosclerotic effect

Polyphenols prevent oxidation of LDL, increasing high-density lipoprotein (HDL), reducing LDL and triglycerides, which leads to protecting the heart from various cardiovascular disorders. Anti-atherosclerotic properties are shown by extra virgin olive oil, black and green tea, bitter chocolate, forest fruits, citrus fruits, and grapes. Oxidized LDL causes atherosclerotic effects. Several studies have shown that some polyphenols protect LDL from oxidation (Manach et al., 2004). Polyphenols present in olive oil also contribute to the downregulation of gene expression along with the protection of LDL from oxidation (Castaner et al., 2012). Anti-atherosclerotic effects are shown by tannic acid and resveratrol by improving endothelial dysfunction in cell cultures and in animal and human experiments (Wiciński et al., 2017; Xu et al., 2017).

7.3.4 Anti-metabolic effect

As a consequence of the clustering of factors like hypertension, dyslipidemia, and blood-sugar regulation disorders, several metabolic diseases may be initiated (Group, 2005). Polyphenols play an important role in improving such factors to get relief from metabolic syndrome.

7.3.5 Anti-obesity effect

Polyphenols can suppress obesity by inhibiting pre-adipocyte differentiation, stimulating adipocyte apoptosis, suppressing lipogenesis, reducing adipocyte proliferation, and contributing to lipolysis and beta-oxidation of fats *in vitro*. Numerous laboratory and animal studies have shown that green tea has anti-obesity properties (Basu et al., 2010) (Sung et al., 2010). In the context of obesity, green tea has been shown to reduce the destructive nature of ROS molecules as well as suppress inflammation and associated damage (Wang et al., 2014).

7.3.6 Microbial suppression effect

Polyphenols have shown anti-microbial effects towards various species of bacteria and fungi through suppression of microbial virulence and synergism with antibiotics (Daglia, 2012).

7.3.7 Immunomodulatory effect

Inhibition of cytokine and cytokine receptors is an effect of some plant polyphenols, while others work on secretory processes (Ferguson, 2001). Caffeic acid and quinic acid esterification lead to the formation of chlorogenic acid, which has shown anti-inflammatory effects in various animal experiments.

7.3.8 Anti-oxidant activity

Oxygen derived from molecular oxygen is known as reactive oxygen species (ROS) (Hancock et al., 2001). Production of ROS takes place in most aerobic organisms and is triggered by ultraviolet rays, radiations, cigarette smoke, aging, and pathogenic invasion (KOÇYİĞİT and SELEK, 2016) (Mao et al., 2017). Apoptosis, gene expression, and activation of cell signaling cascades occur when ROS are present in low concentrations in the body.

However, excessive ROS causes some harmful effects. After the accumulation of ROS, the function of cells and tissues is affected and leads to causing pathogenesis of various serious disorders such as cancer, atherosclerosis, and neurodegenerative diseases (Aruoma, 1998) (Alagawany et al., 2015). ROS production occurs not only in animal cells but also in plant cells, while plants play an important role to save these species from oxidative stress (Mahalingam and Fedoroff, 2003). Endogenous antioxidants as well as exogenous antioxidants together play an important role in minimizing oxygen-induced oxidation.

In terms of human health, polyphenols play an important role just because of their health benefits due to antioxidant properties (Koçyiğit and Selek, 2016). By acting as antioxidant polyphenols protect cells and tissues from the accumulation of oxidative stress and free radicals. They reduce the number of radicals and protect DNA from oxidative damage along with peroxidation production (Ferguson, 2001).

In the Lamiaceae family, carvacrol is a major antioxidant reagent that plays a critical role in cancer prevention (Günes-Bayir et al., 2017). It has been found that the drug cisplatin (CP) hastoxic effects on the liver, kidneys, and several genes when used in the treatment of cancer. Researchers have found that when rats were treated with naringenin-oxime (NOX), they demonstrated an antioxidant effect, which also reduced ROS production and DNA damage (Koyuncu et al., 2017).

7.3.9 ANTI-CANCER EFFECT

In cancer treatment, polyphenols inhibit tumor growth by imparting cytotoxic effects, inducing apoptosis, and activating enzymes for apoptosis (Nirmala et al., 2018). Pro-oxidant activities of polyphenols are interlinked with cytotoxic effects on cancer cells. Pro-oxidant activity is related to polyphenolic compounds, the presence of free iron and copper, as well as the oxidative state of the environment.

7.3.10 HYPOGLYCEMIC EFFECT

Through the regulation of insulin and glucagon-like peptide-GLP-1 pathways, polyphenols reduce the risk of developing diabetes. Polyphenols increase insulin sensitivity in peripheral tissues (Domínguez Avila et al., 2017). The effects of cinnamon on hyperlipidemia and glucose utilization were described in a study involving rats. In that study, the cinnamon extract reduced visceral fat mass, serum glucose levels, liver weight, and insulin concentrations (Tuzcu et al., 2017). A very common fruit in some communities, the palm is known for its antihyperglycemic, anti-cancer, gastric-protective, and anti-inflammatory properties. Among the important phenolic acids found in the palm are lignans, isoflavones, flavonoids, and tannins. It inhibits the enzymes like α-glucosidase and α-amylase that is responsible for glucose absorption from the intestines. This makes the palm fruit useful in treating and preventing diabetes mellitus (Vayalil, 2012).

7.4 ANTIDIABETIC POLYPHENOLS

Insulin dysfunction or shortage causes diabetes mellitus, a metabolic condition characterized by hyperglycemia. According to the World Health Organization (WHO), 9 percentof people over the age of 18 suffered from Diabetes Mellitus (DM). This scenario is expected to lead to an estimated 552 million diabetics in 2030. According to research, about one-third of diabetics suffer from diabetic nephropathy (DN), a condition that increases the total cost of therapy. Diabetes mellitus type 1 (DM1), also known as insulin-dependent DM, is caused by the autoimmune destruction of the pancreas cells (Reno et al., 2015). Diabetic type 2 (non-insulin-dependent DM) is characterized by insulin resistance. Approximately 90 to 95 percent of diabetics are affected by this kind of disease. Insulin resistance in the peripheral bloodstream, and beta-cell dysfunction, define type 2 diabetes. Human type 2 diabetes may be triggered or exacerbated by a combination of genetic and environmental factors, such as chronic hyperglycemia and obesity. Irregular blood sugar levels, abnormally high amounts of glucose, and altered lipid metabolism are all linked to insulin resistance (IR) (Matias and Sousa, 2017). Additionally, it has been related to increased subclinical inflammation and oxidative stress. These metabolic problems cause long-term pathogenic diseases, including macro and microvascular issues, neuropathy, nephropathy, retinopathy, and nephropathy – therefore, lower quality of life and an increased risk of death. Diabetic patients may benefit from using polyphenols as nutraceuticals and supplements because of their beneficial biological effects. Based on several animal models and human research, polyphenols may have an important function to play in several aspects of metabolism. Attenuate hyperglycemic conditions and insulin resistance; increased adipose tissue metabolism; reduced oxidative stress and stress-sensitive pathways; and inflammatory processes (McIntyre et al., 2019).

7.4.1 Curcumin

Anti-diabetic medications may be used with phytochemicals to treat DM and other hyperglycemia conditions. Diferuloylmethane (diferuloylmethane) is the yellow pigment extracted from the dried rhizomes of *Curcuma longa* L. (turmeric). Curcumin's capacity to reduce hyperglycemia and improve a variety of macro and microvascular consequences, including cardiovascular disease, nephropathy, retinopathy, and endothelial dysfunction, has drawn interest in the field of diabetes. Curcumin, with antidiabetic drugs or other phytochemicals, has shown promising results in the management of glycemia and the reduction of other diabetic problems (Kianbakht et al., 2013).

It is seen that the use of curcumin supplements dramatically slowed the development and damage of kidney lesions. In addition, urinary enzyme levels (ALT, ALP, AST, and AP) were lowered with curcumin supplementation, and these enzymes were shown to be present in lower concentrations in the urine of curcumin supplementation participants. Cutting down on the amount of glucose-6-phosphatase and lactate dehydrogenase activity in the kidneys was achieved by supplementing with curcumin. Curcumin administration raised the polyunsaturated fatty acid/saturated fatty acid (PUFA/SFA) ratio while decreasing blood phospholipid and triglyceride levels. As a whole, the addition of curcumin to diabetic nephropathy was beneficial (Ali et al., 2020) (Figure 7.4).

7.4.2 Resveratrol

As a primary source of glucose intake in our bodies, our muscles are the primary target of resveratrol (Figure 7.5). To improve glucose transport, this chemical increases the number of GLUT4 transporters plasma membrane of myocytes as well as in muscles, through the PI3K-Akt pathway, which increases intracellular glucose uptake and utilization. It also boosts fatty acid oxidation, preventing the lipotoxicity that causes insulin resistance, therefore reducing the risk of diabetes (Gocmez et al., 2019).

Antioxidant defense in pancreatic tissue is improved by resveratrol in part because of an increase in antioxidant enzyme activity in this organ – (superoxide dismutase(SOD), catalase (CAT) and glutathione peroxidase (GPx)), and glutathione S-transferase. It is also capable of reversing beta-cell degradation in diabetic rats and preventing beta-cell death owing to the suppression of caspase-3 by this polyphenol. As a result, insulin levels in the blood rise (Do et al., 2012). When resveratrol is taken by diabetics it improves glucose homeostasis because it reduces inflammation, which is linked to diabetes development. Resveratrol has been shown to reduce the production of pro-inflammatory cytokines (IL-1 and IL-6) and transcription factor NF-B in the liver (Kosuru et al., 2018).

7.4.3 Catechins

Naturally occurring polyphenols, such as catechins found in tea leaves, are found in a variety of foods and medicinal plants. Catechin is made up of two benzene rings (A and B) and a dihydropyran

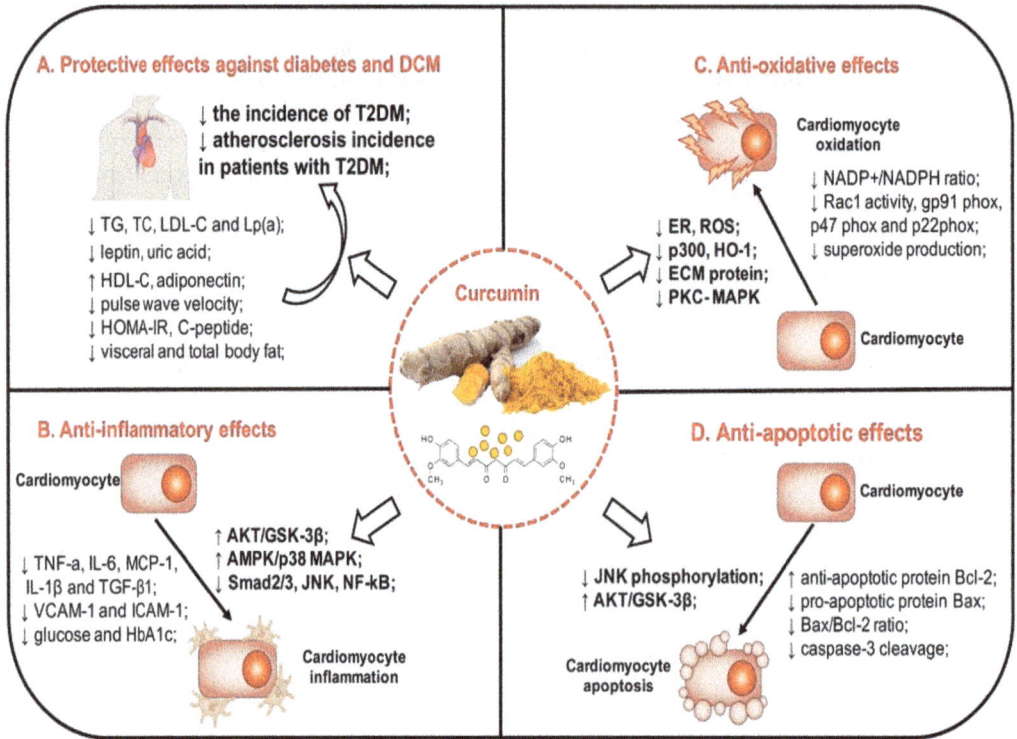

FIGURE 7.4   The effect of curcumin in the diabetic patient (Zheng et al., 2018).

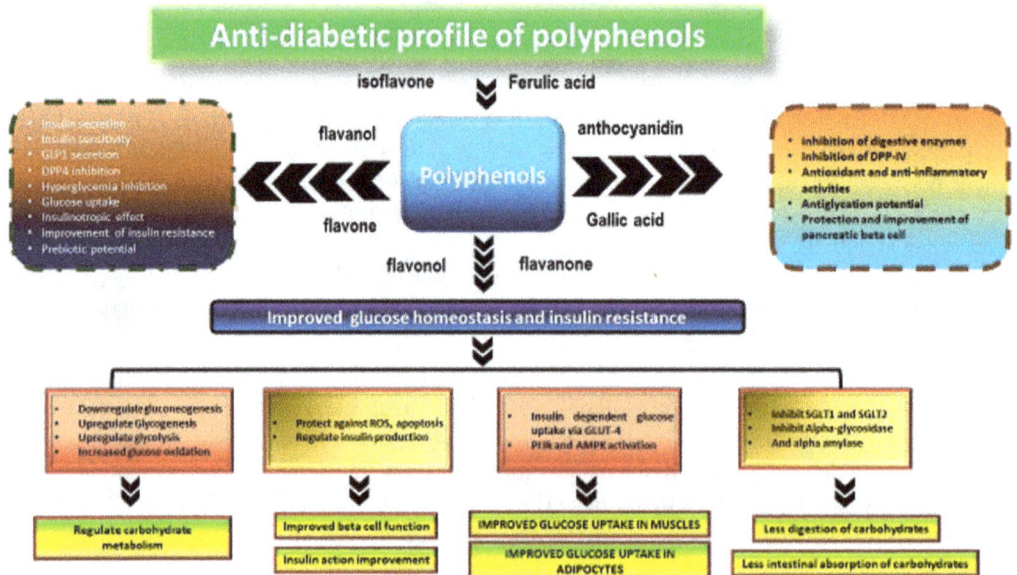

FIGURE   7.5 a-Chemical   structure   of   resveratrol;   b-catechins;   c-procyanidin;   d-anthocyanins;
e-Caffeoylquinic acids; f-isoflavones.

heterocycle (C) containing a hydroxyl group on carbon 3, Figure 7.5. On carbons 2 and 3, there are two chiral centers. Flavan-3-of substances are catechin stereoisomers having carbons 2 and 3 in the cis (-epicatechin) or trans ((+)-catechin) structure (Khan and Mukhtar, 2019).

The bioavailability of catechins is quite low. Plasma catechin concentrations rise fast following a cup of tea, peaking after 2 hours, and then gradually decreasing to their baseline levels within 8 hours. Catechins are taken up and excreted by human cells via p-glycoprotein, however, because of individual differences in human metabolism, absorption of catechins might vary greatly from person to patient (Ullmann et al., 2003). Several studies have demonstrated that green tea may enhance insulin sensitivity in the laboratory. In animal tests, catechins found in green tea extract have been shown to have a hypoglycemic impact. Green tea drinking has been linked to a lower risk of developing type 2 diabetes in epidemiological studies. Glycated hemoglobin is reduced in persons with high blood sugar who take green tea extract supplements, according to a clinical study. EGCG has been shown to improve insulin sensitivity and boost glucose uptake into cells (Asbaghi et al., 2019).

Tea polyphenols have been shown to reduce hyperglycemia in diabetic patients, as well as to be useful in the treatment of diabetes mellitus. Prospective cross-sectional research was conducted, among women 45 years of age and older who were free of cardiovascular disease, cancer, or diabetes. Type 2 diabetes mellitus risk was shown to be reduced by 30 percent when four cups of green tea were consumed daily. Over a 12-week period it is found that supplementing mice with EGCG reduced fasting insulin levels via altering the expression of the GLUT4. Green tea extract supplementation (856 mg EGCG) for 16 weeks reduced fasting insulin, insulin resistance, and glycated hemoglobin in obese adults with type 2 diabetes (Karthikeyan and Pawar, 2021).

7.4.4 PROCYANIDIN

To detect the antidiabetic mechanism of procyanidin, several *in vivo* and in vitro studies have been performed that include a reduction in the digestion of carbohydrate, maintenance of pancreatic cell function, modification in the metabolism of hepatic glucose, and impact of skeletal muscle on glucose absorption, as mentioned in Figure 7.5. There are reports that procyanidin may reduce carbohydrate digestion and hence glucose absorption in the gut by inhibiting the enzyme glycosidase, which metabolizes disaccharides and starch to glucose (Johnson et al., 2013).

It is has been proposed that, as the molecular weight of the polyphenol increases, its activity rate also increases. Some researchers have discovered that amylase is responsible for hydrolyzing the glycosidic linkages at C1-C4 in both amylose and amylopectin. Procyanidin influences hepatic glucose metabolism by increasing glucose absorption and decreasing glucose output. Glucokinase activity in the liver is increased by several substances. Phosphorylation by this enzyme facilitates the storing of glucose in the form of glycogen and the subsequent digestion and elimination of glucose through the glycolytic process (Bang and Choung, 2014). One of the most important steps in glucose production is the irreversible carboxykinase of phosphoenolpyruvate carboxykinase, which is reduced by these polyphenols. Aside from activating the AMPK and PI3K/Akt pathways, procyanidin reduces glucose-6-phosphatase via activating these pathways (Gao et al., 2018).

Procyanidin is regarded as insulin secretagogues when it comes to pancreatic-cell activity. Antioxidative stress is prevented, insulin secretion is increased, and cell survival is improved by the presence of procyanidin in the pancreas. Several elements influence this capacity to induce insulin secretion. Those substances with a larger number of hydroxyl groups in their chemical structure are better at treating the disease (Kurimoto et al., 2013). The dosage is also a significant component. As a result, in long-term therapy, the range of dose 24 mg/kg body weight was found to be less effective than those with a range of 14 mg/kg. Human myotubes can be stimulated by procyanidin with a dose range of 0.1–10 M. (Pinna et al., 2017).

The specific advantages of grape seed procyanidin for diabetes sequelae include reducing diabetic nephropathy avoiding early morphological and functional abnormalities in those nerves which reside in the peripheral area, ameliorating pancreatic cell dysfunction and glycation-associated heart damage, and guarding against early diabetic peripheral neuropathy by moderating glycation-induced neuropathy (Johnson et al., 2013).

7.4.5 ANTHOCYANIDINS

As far as health benefits go, anthocyanidins (Figure 7.5) have been thoroughly examined, including their antioxidant and anti-inflammatory properties. Anti-diabetic effects of anthocyanidins were primarily attributed to their ability to reduce oxidative stress, improve insulin resistance, and increase insulin production (Zhu et al., 2012).

It is a special class of OH donors with momentous scavenging abilities to reduce oxidative stress that is made up of anthocyanidins. SOD, malonic dialdehyde levels, and thiobarbituric acid reactive substances (TBARS) may all be improved by anthocyanidins, which can decrease the level of plasma glucose by strengthening the protective antioxidant capacity of diabetic patients. C3G protects cells against high-glucose stress by stimulating the production of glutathione, according to in vitro studies. Islet cells are very vulnerable to oxidative stress because of the poor expression of CAT, SOD, and GPx in islets that are major antioxidant enzymes. To combat hyperglycemia, AMPK/ACC/mTOR pathway oxidative stress was induced in islet cells using an anthocyanin-rich mulberry extract (Zhu et al., 2012).

Anthocyanidin might reduce insulin resistance by modulating lipid levels, lowering cholesterol, triglycerides, and low-density cholesterol, and raising apolipoprotein and high-density cholesterol. Excess anthocyanidins in extracts of bilberry enhanced resistance of insulin in mice and lowered total triglycerides and cholesterol in the liver and blood. TNF-α and IL-6 were discovered to be linked to insulin resistance in a study (Shi et al., 2017). Additionally, studies have demonstrated that Cyanidin-3-glucoside is responsible for the suppression of adipocytes, activation of insulin signaling pathway following a reduction in TNF-α -induced resistance of insulin, and activation of FoxO1 (Guo et al., 2012).

In a variety of methods, anthocyanins boosted insulin production. Insulin secretion was increased by enhancing the glucose transport-related gene (Glut2) in islet beta cells along with $Ca2^+$ expression with cyanidins (Kato et al., 2015). Delphinidin 3-arabinoside (DPPIV) and its substrate GLP-1 were modulated by anthocyanidins (delphinidin 3-arabinoside) in berry drinks, which elevated secretion of insulin and mRNA expression of some genes, especially of the insulin receptor. A calcium-dependent kinase mechanism might be used to liberate GLP-1 from Delphinidin 3-rutinoside (Johnson and De Mejia, 2016).

7.4.6 CAFFEOYLQUINIC ACIDS

5-caffeoylquinic acids (Figure 7.5) have been shown to block both salivary and pancreatic amylase, indicating that these chemicals may help to limit the absorption of glucose into the bloodstream after a meal in obese or diabetic individuals (Islam et al., 2013). Despite research showing the poor digestion and absorption of caffeoylquinic acids, various experiments have been conducted to examine if caffeoylquinic acids may affect the metabolism of peripheral tissues. In vitro studies have shown that caffeoylquinic acids may block gluconeogenesis enzyme hepatic glucose-6-phosphatase. There is evidence that caffeoylquinic acids may exert their hypoglycemic action by reducing the amount of glucose produced in the liver (Henry-Vitrac et al., 2010). A high concentration of 5-caffeoylquinic acid in instant coffee did not affect rats' glucose adsorption capacity. Caffeoylquinic acid concentrations in liver cells may have been insufficient to block the expression of G-6-Pase (Bassoli et al., 2015).

Resistance of insulin in peripheral tissues like muscle, liver, and adipocytes may be alleviated by increasing insulin sensitivity in PTP1B. Noncompetitive inhibitors of PTP1B have been found to include chlorogenic acid and caffeoylquinic acid derivatives. According to various studies caffeoylquinic acids are considered as most powerful inhibitor having a Ki value of roughly 15 micro mols. These findings imply that this plant might be utilized to increase insulin sensitivity in diabetic patients with obesity. If so, chlorogenic acid therapy has been shown to increase glucose absorption in both adipocytes that are insulin sensitive and those that are resistant to insulin (Reis et al., 2019).

The long-term consumption of coffee or caffeoylquinic acid-rich extracts has been shown to reduce blood glucose, increase insulin response, decrease hepatic insulin resistance, reduce serum lipids, and promote weight loss in humans (Lecoultre et al., 2014). Caffeoylquinic acids have a profound effect on human energy metabolism, and this might explain why traditional medicine recommends the use of caffeoylquinic acid-rich extracts for the treatment of obesity and diabetes (Xie et al., 2018).

7.4.7 Isoflavones

Epidemiological evidence studies show that soy isoflavone (Figure 7.5) consumption is associated with a decreased risk of diabetes. By suppressing carbohydrate digestion and glucose absorption in the small intestine, soy isoflavones have a hypoglycemic impact on Goto–Kakizaki diabetic rats, and they have a protective effect on diabetic nephropathic rats.

Puerarin has been shown to improve glucose levels among all types of soy isoflavones, increasing insulin sensitivity and resistance, inhibiting Maillard reaction, protecting pancreatic cells, exerting anti-inflammatory activity, reducing ROS species along with inhibition of AGE, which is known as advanced glycation end products formation (Konishi et al., 2019). Additionally, puerarin reduces diabetes-related problems such as diabetic nephropathy, retinopathy, neuropathy, and cardiovascular disorders. There are numerous ways in which Genistein benefits type 2 diabetes. First, it significantly lowers blood sugar levels, which in turn alleviates cell proliferation and reduces the apoptosis of various cells (Chen et al., 2018). Second, it reduces inflammation and oxidative stress in the heart, which in turn improves fracture resistance. Third, it improves vascular function, which in turn improves diabetic nephropathy, angina, and limb ischemia (Odle et al., 2017).

According to research in non-obese diabetic mice, daidzein induces an immunomodulatory effect like insulin resistance, improvement of hyperglycemia, dyslipidemic state, obesity, inflammation, and other complications; while no effect on blood glucose or glucose tolerance was observed (Fu et al., 2012). According to a cohort study, it was reported that a 6-month intake of daidzein was unable to affect insulin sensitivity and glycemic control in Chinese women having poor glucose regulation. Daidzein is rich in polyphenols improves glucose homeostasis in type 2 diabetic mice.

Researchers found that streptozotocin-induced diabetic rats treated with Biochanin A had lower blood sugar. Decreased insulin resistance and enhanced inulin sensitivity and lipid profile were seen in diabetic rats treated with Biochanin A. It also reduced neuropathic pain in diabetic rats. SIRT1 expression in pancreatic tissues may be a factor in reducing insulin resistance and reducing hyperglycemia in T2D. In terms of antidiabetic properties, methylated isoflavones seem to be superior to their corresponding non-methylated counterparts. In the meanwhile, greater research is needed on both animals and humans (Hasanein and Mahtaj, 2015).

7.5 ANTIDIABETIC POLYPHENOLS IN CLINICAL TRIALS AND MARKETED PRODUCTS

Although a lot of in vitro and in vivo studies have proved that polyphenols are effective against T2D, not many clinical reports are available in this regard. To see the effectiveness of polyphenols against diabetes in humans, we shall discuss some randomized and concurrent clinical evidence in this article.

A randomized, double-blind clinical trial showed that if a diabetic patient takes 1500 mg of curcumin per day, then there is a reduction in his weight and fasting blood glucose level. Also, a reduction in the complication of the case in type 2 diabetic patients were seen in this study (Adibian et al., 2019). Another clinical study revealed that a 9-months supplementation of curcumin helps in the better functioning of beta cells and prevents the development of T2D in vulnerable pre-diabetic patients (Chuengsamarn et al., 2012). A rise in levels of adiponectin, a reduced level of leptin, and a reduction in inflammation were seen in 118 T2D patients who were given a dose of 1000 mg curcumin and 10 mg of piperine daily for 12 weeks. The efficiency of resveratrol against diabetes due to better control of glycemic index and insulin resistance is evident from some clinical research. It was shown in the very first randomized clinical trial in this regard, that there was a decrease in oxidative stress and insulin sensitivity of patients of T2D who were supplemented with resveratrol (Brasnyó et al., 2011). Improved and better levels of HbA1c, insulin, fasting blood glucose and systolic pressure were seen in T2D patients after administration of 1g resveratrol each day for 45 days in a study (Movahed et al., 2013). Opposite results were seen in some other clinical trials. The glycemic levels were not controlled by the daily dose of 500 mg of resveratrol 2 times a day for 5 weeks given to T2D patients in another randomized clinical study by Thazhath et al. (2016). In 2016, a randomized clinical trial in which placebo was controlled and the study was double blind, it was shown that there was no visible improvement in metabolic indicators of T2D in patients who were supplemented for six months with 500 mg of resveratrol each day (Bo et al., 2016). A positive role of resveratrol in cognitive function, neurovascular coupling, and a better glycemic level was seen in T2D patients (Bhatt et al., 2012; Wong et al., 2016). According to a similar study it was observed that giving resveratrol with a dose of 500 mg 2 times a day for approximately 5 weeks to patients discovered that no significant alteration and betterment in body weight or GLP-1 secretion was observed. It was also observed that glycemic index and energy intake along with gastric emptying were not improved efficiently (Thazhath et al., 2016; Timmers et al., 2016). This study concludes that resveratrol treatment is not effective for hepatic and peripheral insulin sensitive people.

Another randomized controlled clinical study revealed that in 60 patients who were supplemented with powdered green tea extract for two months every day, the diastolic blood pressure decreased along with HbA1C. However, there was no significant effect on the body weight and fasting blood glucose level of patients, also the systolic blood pressure, HOMA level, and lipid profile remained unchanged (Fukino et al., 2008). Another clinical study showed opposite results when T2D patients were given supplementation of 1500 mg green tea, as the blood glucose level, fasting HbA1C level and lipoprotein profile of patients remained unaffected (Hua et al., 2011).

Berberine is found in barberry, which helps to improve the regulation metabolism of carbohydrates and fat in several in vivo and in vitro models; it is an important isoquinoline alkaloid. It was shown by the in-depth analysis of 14 clinical trials that berberine intake gives similar results as some drugs like metformin, glipizide, and rosiglitazone in carbohydrate and fat metabolism (Dong et al., 2012). It was seen that intake of berberine had a positive effect on the antioxidant activity of plasma and in decreasing the blood glucose level, along with bringing some changes to the lipid profile of diabetic patients. According to a clinical trial in which 36 patients who had recently acquired diabetes, were given a dose of 1.5 grams of berberine daily for three months and it was concluded that berberine supplementation gave similar results as of metformin against T2D. 48 other patients in the same research were given 500 mg of berberine three times a day for up to three months and their blood glucose levels were lowered within one week (Yin et al., 2008).

Clinical studies have also been performed to see the effect of quercetin intake on type 2 diabetes patients. In type 2 diabetes patients with high levels of maltose due to the inhibited activity of alpha glycosidase, 400 mg single intake of quercetin lowered the postprandial blood glucose levels (Hussain et al., 2012). In a clinical study, a better blood glucose level and lipid profile along with improved blood pressure and lowered vulnerability of cardiovascular disease in male smokers was seen in 49 patients who were supplemented with 100 mg capsules of quercetin for 10 weeks, however when 47 patients of T2D were given a diet fortified with 250 mg quercetin per day or similar placebo of cellulose capsules for 8 weeks, it did not reduce the blood glucose levels, nor were the lipid profiles improved (Lee et al., 2011; Mazloom et al., 2014).

Anthocyanins have anti-diabetic properties by lowering blood glucose and HbA1c levels, boosting insulin production and decreasing insulin resistance. The stimulation of adiponectin by anthocyanins enhanced endothelium-dependent vasodilation in diabetic patients. Cinnamon's health benefits have been linked to its impact on hyperlipidemia and glucose utilization in laboratory tests. Cinnamon supplementation (1–6 g/day) lowered serum triglyceride, blood glucose along with LDL and total lipid and cholesterol in diabetic patients in a clinical study of 60 patients (Khan et al., 2003). A randomized controlled experiment discovered that a polyphenol-rich diet enhances glucose metabolism in patients that are at high risk of diabetes (Vetrani et al., 2018).

In chili peppers, Capsaicin is found as a source of polyphenols that have been used to treat diabetes, heart disease, and cancer. That is why supplementing T2D patients with a dose of 5 g led to a drop in plasma level of glucose and maintenance of insulin in a clinical experiment (Sun et al., 2020). In double-blind randomized controlled crossover research, high-polyphenol chocolate supplements might protect diabetic patients against dysfunction of endothelial cells and oxidative stress during severe hyperglycemia generated by a 75 g glucose administrated orally (Mellor et al., 2013). In overweight T2D patients, daily use of polyphenol-rich extra-virgin olive oil may enhance metabolic regulation and the profile of circulating inflammatory adipokines (Santangelo et al., 2016).

Grape seed extract that is rich in flavonoids significantly reduces inflammatory and glycemic markers as well as oxidative stress. A double-blind, placebo-controlled, randomized, double-dummy study was conducted on obese type 2 diabetics following obesity (Cao et al., 2019). Those with type 2 diabetes who consumed flavanol-rich cocoa daily experienced improved vascular function. Pycnogenol was effective in managing diabetes, reducing cardiovascular disease risk factors, and reducing the need for antihypertensive medications. In Pycnogenol, an extract from French maritime pine bark rich in procyanidins and bioflavonoids, phytocyanin is a component. Results of randomized, double-blind, placebo-controlled studies of 48 diabetic patients were obtained (Zibadi et al., 2008). Researchers found that Brazilian green propolis dosed at 226.8 grams/day protected diabetic patients from uric acid changes and changes to estimated glomerular filtration rate in a double-blind, randomized, and controlled study. These herbs contained a lot of polyphenols and flavonoids (Fukuda et al., 2015). Efficacy of acacia polyphenol (250 mg/day) in improving glucose homeostasis in non-diabetics was studied by Okawa et al. (2013) in a randomized, double-blind, placebo-controlled trial.

When glucose levels are raised in healthy males, polyphenols present in coffee may enhance peripheral endothelial function (Ochiai et al., 2014). Polyphenols from coffee have no significant effect on glucose or insulin in healthy adults, regardless of their chlorogenic acid content (Rakvaag and Dragsted, 2016). Red wine polyphenols improved insulin resistance and decreased lipoprotein plasma concentrations when administered to 67 men with high cardiovascular risk (Chiva-Blanch et al., 2013). Researchers found that anthocyanins in whortleberries lower blood glucose levels and increase HbA1c levels in randomized, double-blind, placebo-controlled clinical studies (Kianbakht et al., 2013).

Powdered dry leaves of Eugenia punicifolia (Kunth) DC (Myrtaceae) with high flavonol glycoside levels have shown to be effective in the treatment of diabetic patients. In a non-controlled, pilot study and after a three-month therapy with *E. punicifolia* powdered dry leaves was shown to lower the glycosylated hemoglobin and basal insulin levels at the dramatic level (Sales et al., 2014). In diabetic and prediabetic volunteers with a sample size of 15, an aqueous extract of *Bauhinia forficata (L) subsp. pruinose (Fabaceae)* with 0.15 percent infusions rich in rutin was found to dramatically lower the level of HbA1c (Toloza-Zambrano et al., 2015). An oral glucose tolerance test on healthy participants revealed that the onion extract had no hypoglycemic impact. Polyphenols found in coffee, whortleberry, olive oil, guava tea, propolis, chocolate, red wine, grape seed, and cocoa have reportedly improved glucose metabolism and reduced insulin resistance along with enhanced vascular function in T2D patients. In Table 7.1 a summary of polyphenols with their mechanism of action observed by clinically based studies is mentioned.

The antioxidant properties of polyphenols make them extremely important. Their role in diabetes management cannot be overstated. The pharmacological properties of these substances include anti-inflammation, antimicrobial, anti-hypertensive, and anti-microbial functions. It was discovered that polyphenols can have potential anti-diabetic effects in peripheral tissues (Figure 7.6). These effects are associated with an increased production of the hormone glucagon-like peptide-1 (GLP-1), an inhibition of the enzyme dipeptidyl peptidase-4 (DPP-4), and an increase in insulin secretion (Figure 7.6). Researchers at the basic and clinical levels could develop chemical families that

TABLE 7.1
Polyphenols and their mechanism of action found in clinical trials

Polyphenol	Mechanism of action	Trial details	Reference
Curcumin	decrease fasting blood glucose and weight	double-blind	(Adibian et al., 2019)
Resveratrol	Controlled level of glucose and insulin resistance	randomized clinical trial	(Brasnyó et al., 2011)
green tea	reduction in diastolic blood pressure declined HbA1c level	randomized controlled trial	(Fukino et al., 2008)
Berberine	regulates carbohydrate and fat metabolism	meta-analysis	(Yin et al., 2008).
Quercetin	α-glucosidase inhibition reduced cardiometabolic risks	Multiple Clinical trials	(Hussain et al., 2012) (Lee et al., 2011; Mazloom et al., 2014)
Anthocyanins	Controlled level of glucose and insulin resistance	Clinical study	(Khan et al., 2003).
Cinnamon	Improves hyperlipidemia and glucose utilization	clinical study	(Khan et al., 2003).
Capsaicin	Controlled level of glucose and insulin resistance	Randomized clinical trial	(Sun et al., 2020).
polyphenols	improves glucose metabolism	Randomized clinical trial	(Vetrani et al., 2018).
Extracts of grape seed	improved inflammation, declined ROS	placebo-controlled trial	(Cao et al., 2019).
flavanol-containing cocoa	improve the vascular function	randomized controlled trial	(Balzer et al., 2008)
Pycnogenol	reduce risk factors of cardiovascular disorders	placebo-controlled trial	(Zibadi et al., 2008)
Brazilian green propolis	Improved diabetic neuropathies	randomized controlled	(Fukuda et al., 2015)
acacia polyphenol	Improved glucose hemostasis	double-blind trial	(Ogawa et al., 2013).
Red wine	insulin resistance	clinical study	(Chiva-Blanch et al., 2013).
whortleberry	Hypoglycemia	Randomized clinical trial	(Kianbakht et al., 2013).
high-polyphenol chocolate	protect against oxidative stress and hyperglycemia	a randomized controlled crossover study	(Mellor et al., 2013).
Eugenia punicifolia	reduce the level of glycosylated hemoglobin	pilot non-controlled study	(Sales et al., 2014)

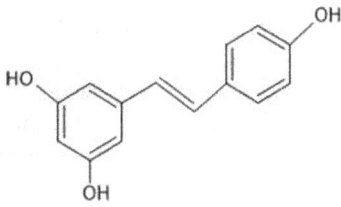

a. Resveratrol

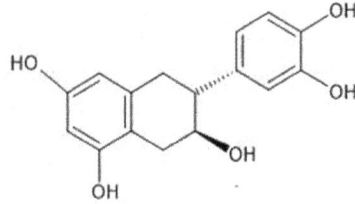

b. Catechin

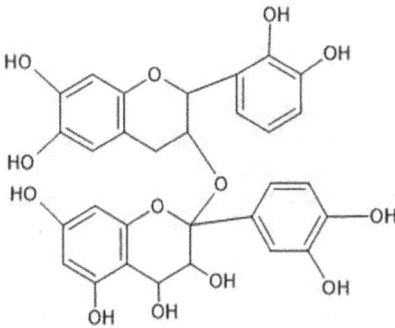

c. Procyanidin

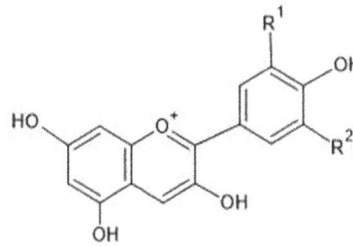

d. Anthocyanidins

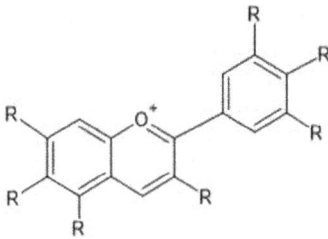

e. Caffeoylquinic acids

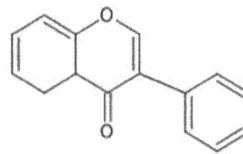

f. Isoflavones

FIGURE 7.6 Summary of antidiabetic activities of polyphenols and their interplay in glucose homeostasis and insulin resistance.

prevent or lessen the effects of type 2 diabetes mellitus on human health. Thus, further clinical research is needed to evaluate polyphenols' anti-diabetic potential.

7.6 CONCLUSION

Polyphenols are natural organic compounds present in most plants and eaten every day. They have attracted attention due to their health benefits. Due to the therapeutic potentials of polyphenols, they have been widely used to treat various diseases, including diabetes. Positive effects of polyphenols on glucose homeostasis have been found in a large number of animal models, in vitro studies, and in several human trials. To confirm the utility of polyphenol consumption in preventing diabetes, more human trials with well-defined dietary patterns, controlled study designs, and investigation of molecular pathways involved in glucose homeostasis are needed. However, the mechanisms involved in the therapeutic effects of polyphenols in DM have not yet been fully elucidated, as they are multifunctional agents that can regulate a number of processes at the cellular and enzymatic levels and may have epigenetic effects. Further evaluation of polyphenols as potential nutrients is needed, and also it is necessary to answer the question of their bioavailability at concentrations that permit pharmacological action. This is the main limiting challenge to the implications of polyphenols as therapeutic agents.

REFERENCES

Adibian, M., Hodaei, H., Nikpayam, O., Sohrab, G., Hekmatdoost, A., and Hedayati, M.J.P.R. 2019. The effects of curcumin supplementation on high-sensitivity C-reactive protein, serum adiponectin, and lipid profile in patients with type 2 diabetes: A randomized, double-blind, placebo-controlled trial. *Phytother Res.* 33: 1374–1383.

Alagawany, M., El-Hack, M., Farag, M.R., Tiwari, R., and Dhama, K. 2015. Biological effects and modes of action of carvacrol in animal and poultry production and health – A review. *Adv Anim Vet Sci.* 3: 73–84.

Ali, T.M., Abo-Salem, O.M., El Esawy, B.H., and El Askary, A. 2020. The potential protective effects of diosmin on streptozotocin-induced diabetic cardiomyopathy in rats. *Am J Med Sci.* 359: 32–41.

Aruoma, O.I. 1998. Free radicals, oxidative stress, and antioxidants in human health and disease. J *Am Oil Chem Soc* 75: 199–212.

Asbaghi, O., Fouladvand, F., Gonzalez, M.J., Aghamohammadi, V., Choghakhori, R., and Abbasnezhad, A. 2019. The effect of green tea on C-reactive protein and biomarkers of oxidative stress in patients with type 2 diabetes mellitus: A systematic review and meta-analysis. *Complement TherMed.* 46: 210–216.

Axelson, M., Sjövall, J., Gustafsson, B., and Setchell, K. 1982. Origin of lignans in mammals and identification of a precursor from plants. *Nature* 298: 659–660.

Balzer, J., Rassaf, T., Heiss, C., Kleinbongard, P., Lauer, T., Merx, M., Heussen, N., Gross, H.B., Keen, C.L., Schroeter, H., and Kelm, M. 2008. Sustained benefits in vascular function through flavanol-containing cocoa in medicated diabetic patients: A double-masked, randomized, controlled trial. *J Am Coll Cardiol.* 51: 2141–2149.

Bang, C.-Y., and Choung, S.-Y. 2014. Enzogenol improves diabetes-related metabolic change in C57BL/KsJ-db/db mice, a model of type 2 diabetes mellitus. *J Pharm Pharmacol.* 66: 875–885.

Bankar, G.R., Nayak, P.G., Bansal, P., Paul, P., Pai, K., Singla, R.K., and Bhat, V.G. 2011. Vasorelaxant and antihypertensive effect of Cocos nucifera Linn. endocarp on isolated rat thoracic aorta and DOCA salt-induced hypertensive rats. *J Ethnopharmacol.* 134: 50–54.

Bassoli, B.K., Cassolla, P., Borba-Murad, G.R., Constantin, J., Salgueiro-Pagadigorria, C.L., Bazotte, R.B., and de Souza, H.M. 2015. Instant coffee extract with high chlorogenic acids content inhibits hepatic G-6-Pase in vitro, but does not reduce the glycaemia. *Cell Biochem Func* 33: 183–187.

Basu, A., Sanchez, K., Leyva, M.J., Wu, M., Betts, N.M., Aston, C.E., and Lyons, T.J. 2010. Green tea supplementation affects body weight, lipids, and lipid peroxidation in obese subjects with metabolic syndrome. *J Am Coll Nutr.* 29: 31–40.

Bhatt, J.K., Thomas, S., and Nanjan, M.J. 2012. Resveratrol supplementation improves glycemic control in type 2 diabetes mellitus. *Nutr Res.* 32: 537–541.

Bo, S., Ponzo, V., Ciccone, G., Evangelista, A., Saba, F., Goitre, I., Procopio, M., Pagano, G.F., Cassader, M., and Gambino, R. 2016. Six months of resveratrol supplementation has no measurable effect in type 2 diabetic patients. A randomized, double blind, placebo-controlled trial. *Pharmacol Res.* 111: 896–905.

Brasnyó, P., Molnár, G.A., Mohás, M., Markó, L., Laczy, B., Cseh, J., Mikolás, E., Szijártó, I.A., Mérei, A., and Halmai, R. Mészáros, L.G., Sümegi, B., Wittmann, I. 2011. Resveratrol improves insulin sensitivity, reduces oxidative stress and activates the Akt pathway in type 2 diabetic patients. Br J Nutr. 106: 383–389.

Brett, G.M., Hollands, W., Needs, P.W., Teucher, B., Dainty, J.R., Davis, B.D., Brodbelt, J.S., and Kroon, P.A. 2008. Absorption, metabolism and excretion of flavanones from single portions of orange fruit and juice and effects of anthropometric variables and contraceptive pill use on flavanone excretion. *Br J Nutr.* 101: 664–675.

Cao, H., Ou, J., Chen, L., Zhang, Y., Szkudelski, T., Delmas, D., Daglia, M., Xiao, J. 2019. Dietary polyphenols and type 2 diabetes: Human Study and Clinical Trial. *Crit Rev Food Sci Nutr.* 59: 3371–3379.

Cardona, F., Andrés-Lacueva, C., Tulipani, S., Tinahones, F.J., and Queipo-Ortuño, M.I. 2013. Benefits of polyphenols on gut microbiota and implications in human health. *J Nutr Biochem.* 24: 1415–1422.

Castaner, O., Covas, M.-I., Khymenets, O., Nyyssonen, K., Konstantinidou, V., Zunft, H.-F., de la Torre, R., Munoz-Aguayo, D., Vila, J., and Fito, M. 2012. Protection of LDL from oxidation by olive oil polyphenols is associated with a downregulation of CD40-ligand expression and its downstream products in vivo in humans. *Am J Clin Nut.* 95: 1238–1244.

Chang, J., Reiner, J., and Xie, J. 2005. Progress on the chemistry of dibenzocyclooctadiene lignans. *Chem Rev.* 105: 4581–4609.

Chen, L., Magliano, D.J., and Zimmet, P.Z. 2012. The worldwide epidemiology of type 2 diabetes mellitus – present and future perspectives. *Nat Rev Endocrinol.* 8: 228–236.

Chen, X., Wang, L., Fan, S., Song, S., Min, H., Wu, Y., He, X., Liang, Q., Wang, Y., and Yi, L. 2018. Puerarin acts on the skeletal muscle to improve insulin sensitivity in diabetic rats involving μ-opioid receptor. *Eur J Pharmacol.* 818: 115–123.

Chiva-Blanch, G., Urpi-Sarda, M., Ros, E., Valderas-Martinez, P., Casas, R., Arranz, S., Guillén, M., Lamuela-Raventós, R.M., Llorach, R., Andres-Lacueva, C., and Struth, R. 2013. Effects of red wine polyphenols and alcohol on glucose metabolism and the lipid profile: a randomized clinical trial. *Clin Nutr.* 32: 200–206.

Chuengsamarn, S., Rattanamongkolgul, S., Luechapudiporn, R., Phisalaphong, C., and Jirawatnotai, S.J. 2012. Curcumin extract for prevention of type 2 diabetes. *Diabetes Care* 35: 2121–2127.

Crozier, A., Clifford, M.N., and Ashihara, H. 2008. *Plant Secondary Metabolites: Occurrence, Structure and Role in the Human Diet.* John Wiley.

Daglia, M. 2012. Polyphenols as antimicrobial agents. *Curr Opin Biotechnol.* 23: 174–181.

D' Archivio, M., Filesi, C., Di Benedetto, R., Gargiulo, R., Giovannini, C., Masella, R. 2007. Polyphenols, dietary sources and bioavailability. *Annali-Istituto Superiore di Sanita* 43: 348.

Do, G.M., Jung, U.J., Park, H.J., Kwon, E.Y., Jeon, S.M., McGregor, R.A., and Choi, M.S. 2012. Resveratrol ameliorates diabetes-related metabolic changes via activation of AMP-activated protein kinase and its downstream targets in db/db mice. *Mol Nutr Food Res.* 56: 1282–1291.

Domínguez Avila, J.A., Rodrigo García, J., González Aguilar, G.A., and De la Rosa, L.A. 2017. The antidiabetic mechanisms of polyphenols related to increased glucagon-like peptide-1 (GLP1) and insulin signaling. *Molecules* 22: 903.

Dong, H., Wang, N., Zhao, L., and Lu, F. 2012. Berberine in the treatment of type 2 diabetes mellitus: A systemic review and meta-analysis. *Evid Based Complement Alternat Med.* 2012: 591654.

Ferguson, L.R. 2001. Role of plant polyphenols in genomic stability. *Mutat. Res.- Fundam. Mol. Mech. Mutagen.* 475: 89–111.

Fu, Z., Gilbert, E.R., Pfeiffer, L., Zhang, Y., Fu, Y., and Liu, D. 2012. Genistein ameliorates hyperglycemia in a mouse model of nongenetic type 2 diabetes. *App Physiol Nutr Metab.* 37: 480–488.

Fukino, Y., Ikeda, A., Maruyama, K., Aoki, N., Okubo, T., and Iso, H. 2008. Randomized controlled trial for an effect of green tea-extract powder supplementation on glucose abnormalities. *Eur J Clin Nutr.* 62: 953–960.

Fukuda, T., Fukui, M., Tanaka, M., Senmaru, T., Iwase, H., Yamazaki, M., Aoi, W., Inui, T., Nakamura, N., and Marunaka, Y. 2015. Effect of Brazilian green propolis in patients with type 2 diabetes: A double-blind randomized placebo-controlled study. *Biomed Rep.* 3: 355–360.

Galati, G., Lin, A., Sultan, A.M., and O'Brien, P.J. 2006. Cellular and in vivo hepatotoxicity caused by green tea phenolic acids and catechins. *Free Radic Bio Med.* 40: 570–580.

Gao, Z., Liu, G., Hu, Z., Shi, W., Chen, B., Zou, P., and Li, X. 2018. Grape seed proanthocyanidins protect against streptozotocin-induced diabetic nephropathy by attenuating endoplasmic reticulum stress-induced apoptosis. *Mol Med Rep.* 18: 1447–1454.

Giada, M. 2013. Food phenolic compounds: Main classes, sources and their antioxidant power. *Oxidative Stress and Chronic Degenerative Diseases – A Role for Antioxidants*: 87–112.

Gocmez, S.S., Şahin, T.D., Yazir, Y., Duruksu, G., Eraldemir, F.C., Polat, S., and Utkan, T. 2019. Resveratrol prevents cognitive deficits by attenuating oxidative damage and inflammation in rat model of streptozotocin diabetes induced vascular dementia. *Physiol behav.* 201: 198–207.

Grassi, D., Desideri, G., Necozione, S., Lippi, C., Casale, R., Properzi, G., Blumberg, J.B., and Ferri, C. 2008. Blood pressure is reduced and insulin sensitivity increased in glucose-intolerant, hypertensive subjects after 15 days of consuming high-polyphenol dark chocolate. *J Nutr.* 138: 1671–1676.

Group, I.E.T.F.C. 2005. *International Diabetes Federation: The IDF consensus worldwide definition of the metabolic syndrome.* www. idf. org/webdata/docs/Metabolic_syndrome_def. pdf.

Guasch-Ferré, M., Merino, J., Sun, Q., Fitó, M., and Salas-Salvadó, J. 2017. Dietary polyphenols, Mediterranean Diet, prediabetes, and type 2 diabetes: A narrative review of the evidence. *Oxid Med Cell Longev.* 2017: 6723931.

Günes-Bayir, A., Kiziltan, H.S., Kocyigit, A., Güler, E.M., Karataş, E., and Toprak, A. 2017. Effects of natural phenolic compound carvacrol on the human gastric adenocarcinoma (AGS) cells in vitro. *Anticancer Drugs* 28, 522–530.

Guo, H., Xia, M., Zou, T., Ling, W., Zhong, R., and Zhang, W. 2012. Cyanidin 3-glucoside attenuates obesity-associated insulin resistance and hepatic steatosis in high-fat diet-fed and db/db mice via the transcription factor FoxO1. *J Nutr Biochem.* 23: 349–360.

Han, X., Shen, T., and Lou, H. 2007. Dietary polyphenols and their biological significance. *Int J Mol Sci.* 8: 950–988.

Hancock, J., Desikan, R., and Neill, S. 2001. Role of reactive oxygen species in cell signalling pathways. *Biochem Soc Trans.* 29: 345–349.

Hasanein, P., and Mahtaj, A.K. 2015. Ameliorative effect of rosmarinic acid on scopolamine-induced memory impairment in rats. *Neurosci lett.* 585: 23–27.

Henry-Vitrac, C., Ibarra, A., Roller, M., Mérillon, J.-M., and Vitrac, X. 2010. Contribution of chlorogenic acids to the inhibition of human hepatic glucose-6-phosphatase activity in vitro by Svetol, a standardized decaffeinated green coffee extract. *J Agric Food Chem.* 58: 4141–4144.

Hostetler, G.L., Ralston, R.A., and Schwartz, S.J. 2017. Flavones: Food sources, bioavailability, metabolism, and bioactivity. *Adv Nutr.* 8: 423–435.

Hua, C., Liao, Y., Lin, S., Tsai, T., Huang, C., and Chou, P. 2011. Does supplementation with green tea extract improve insulin resistance in obese type 2 diabetics? A randomized, double-blind, and placebo-controlled clinical trial. *Altern Med Rev.* 16: 157–163.

Hussain, S., Ahmed, Z., Mahwi, T., and Aziz, T.J.I.J.o.D.R. 2012. Quercetin dampens postprandial hyperglycemia in type 2 diabetic patients challenged with carbohydrates load. *Int J Diabetes Res.* 1, 32–35.

Islam, M.N., Jung, H.A., Sohn, H.S., Kim, H.M., and Choi, J.S. 2013. Potent α-glucosidase and protein tyrosine phosphatase 1B inhibitors from Artemisia capillaris. *Arch Pharmacol Res.* 36: 542–552.

Johnson, M.H., and De Mejia, E.G. 2016. Phenolic compounds from fermented berry beverages modulated gene and protein expression to increase insulin secretion from pancreatic β-cells in vitro. *J Agric Food Chem.* 64: 2569–2581.

Johnson, M.H., De Mejia, E.G., Fan, J., Lila, M.A., and Yousef, G.G. 2013. Anthocyanins and proanthocyanidins from blueberry–blackberry fermented beverages inhibit markers of inflammation in macrophages and carbohydrate-utilizing enzymes in vitro. *Mol Nutr Food res.* 57: 1182–1197.

Karthikeyan, E., and Pawar, M.R.V. 2021. Role of Catechins in Diabetes Mellitus. *European J Mol Clin Med.* 7: 7604–7609.

Kato, M., Tani, T., Terahara, N., and Tsuda, T. 2015. The anthocyanin delphinidin 3-rutinoside stimulates glucagon-like peptide-1 secretion in murine GLUTag cell line via the Ca2+/calmodulin-dependent kinase II pathway. *PLoS One* 10, e0126157.

Khan, A., Safdar, M., Khan, M.M.A., Khattak, K.N., and Anderson, R.A. 2003. Cinnamon improves glucose and lipids of people with type 2 diabetes. *Diabetes Care* 26: 3215–3218.

Khan, N., and Mukhtar, H. 2019. Tea polyphenols in promotion of human health. *Nutrients* 11, 39.

Khoddami, A., Wilkes, M.A., Robertis, T.H. 2013. Techniques for analysis of plant phenolic compounds. *Molecules* 18: 2328–2375.

Kianbakht, S., Abasi, B., and Dabaghian, F.H. 2013. Anti-hyperglycemic effect of Vaccinium arctostaphylos in type 2 diabetic patients: A randomized controlled trial. *Forsch Komplementmed.* 20: 17–22.

Koçyiğit, A., and Selek, Ş. 2016. Eksojen Antioksidanlar İki Yönü Keskin Kılıçlardır. *Bezmialem Sci.* 2: 70–75.

Konishi, K., Wada, K., Yamakawa, M., Goto, Y., Mizuta, F., Koda, S., Uji, T., Tsuji, M., and Nagata, C. 2019. Dietary soy intake is inversely associated with risk of type 2 diabetes in Japanese women but not in men. *J Nutr.* 149: 1208–1214.

Kosuru, R., Kandula, V., Rai, U., Prakash, S., Xia, Z., and Singh, S. 2018. Pterostilbene decreases cardiac oxidative stress and inflammation via activation of AMPK/Nrf2/HO-1 pathway in fructose-fed diabetic rats. *Cardiovasc Drugs Ther.* 32: 147–163.

Koyuncu, I., Kocyigit, A., Gonel, A., Arslan, E., and Durgun, M. 2017. The protective effect of naringenin-oxime on cisplatin-induced toxicity in rats. *Biochem Res Int.* 2017.

Kurimoto, Y., Shibayama, Y., Inoue, S., Soga, M., Takikawa, M., Ito, C., Nanba, F., Yoshida, T., Yamashita, Y., and Ashida, H. 2013. Black soybean seed coat extract ameliorates hyperglycemia and insulin sensitivity via the activation of AMP-activated protein kinase in diabetic mice. *J Agric Food Chem.* 61: 5558–5564.

Lattanzio, V. 2013. Phenolic Compounds: Introduction 50. *Nat. Prod.* 1543–1580.

Lecoultre, V., Carrel, G., Egli, L., Binnert, C., Boss, A., MacMillan, E.L., Kreis, R., Boesch, C., Darimont, C., and Tappy, L. 2014. Coffee consumption attenuates short-term fructose-induced liver insulin resistance in healthy men. *Am J Clin Nutr.* 99: 268–275.

Lee, K.-H., Park, E., Lee, H.-J., Kim, M.-O., Cha, Y.-J., Kim, J.-M., Lee, H., Shin, M.-J. 2011. Effects of daily quercetin-rich supplementation on cardiometabolic risks in male smokers. *Nutr res. Pract.* 5: 28–33.

MacRae, W.D., and Towers, G.N. 1984. Biological activities of lignans. *Phytochemistry* 23, 1207–1220.

Mahalingam, R., and Fedoroff, N. 2003. Stress response, cell death and signalling: The many faces of reactive oxygen species. *Physiol Plant* 119: 56–68.

Manach, C., Scalbert, A., Morand, C., Rémésy, C., and Jiménez, L. 2004. Polyphenols: Food sources and bioavailability. *Am J Clin Nutr.* 79: 727–747.

Mao, X., Gu, C., Chen, D., Yu, B., and He, J. 2017. Oxidative stress-induced diseases and tea polyphenols. *Oncotarget* 8: 81649.

Matias, N.R., and Sousa, M.J. 2017. Mobile health, a key factor enhancing disease prevention campaigns: Looking for evidences in kidney disease prevention. *J Inf Syst Eng Manag.* 2: 3.

Mazloom, Z., Abdollah, Z.S.M., Dabbaghmanesh, M.H., and Rezaianzadeh, A. 2014. The effect of quercetin supplementation on oxidative stress, glycemic control, lipid profile and insulin resistance in type 2 diabetes: a randomized clinical trial. *J Health Sci Surveillance Sys.* 10: 8–14.

McIntyre, H.D., Catalano, P., Zhang, C., Desoye, G., Mathiesen, E.R., and Damm, P. 2019. Gestational diabetes mellitus. *Nat Rev Dis Primers* 5: 1–19.

Mellor, D.D., Madden, L.A., Smith, K.A., Kilpatrick, E.S., and Atkin, S.L. 2013. High-polyphenol chocolate reduces endothelial dysfunction and oxidative stress during acute transient hyperglycaemia in Type 2 diabetes: A pilot randomized controlled trial. *Diebate Med.* 30: 478–483.

Movahed, A., Nabipour, I., Louis, X.L., Thandapilly, S.J., Yu, L., Kalantarhormozi, M., Rekabpour, S.J., and Netticadan, T. 2013. Antihyperglycemic effects of short term resveratrol supplementation in type 2 diabetic patients. *Evid Based Complement Alternat Med.* 2013: 851267.

Nirmala, J. G.,Celsia, S.E., Swaminathan, A., Narendhirakannan, R., and Chatterjee, S. 2018. Cytotoxicity and apoptotic cell death induced by Vitis vinifera peel and seed extracts in A431 skin cancer cells. *Cytotechnology* 70: 537–554.

Ochiai, R., Sugiura, Y., Shioya, Y., Otsuka, K., Katsuragi, Y., and Hashiguchi, T. 2014. Coffee polyphenols improve peripheral endothelial function after glucose loading in healthy male adults. *Nutr Res.* 34: 155–159.

Odle, B., Dennison, N., Al-Nakkash, L., Broderick, T.L., and Plochocki, J.H. 2017. Genistein treatment improves fracture resistance in obese diabetic mice. *BMC Endocr Disord.* 17: 1–8.

Ogawa, S., Matsumae, T., Kataoka, T., Yazaki, Y., Yamaguchi, H.J.E. 2013. Effect of acacia polyphenol on glucose homeostasis in subjects with impaired glucose tolerance: A randomized multicenter feeding trial. *Exp ther Med.* 5: 1566–1572.

Ohshima, H., Yoshie, Y., Auriol, S., and Gilibert, I. 1998. Antioxidant and pro-oxidant actions of flavonoids: Effects on DNA damage induced by nitric oxide, peroxynitrite and nitroxyl anion. *Free RadicBiol Med.* 25: 1057–1065.

Oliveira, L.L., Carvalho, M. V., and Melo, L. 2014. Health promoting and sensory properties of phenolic compounds in food. *Rev. Ceres* 61: 764–779.

Ozcan, T., Akpinar-Bayizit, A., Yilmaz-Ersan, L., and Delikanli, B. 2014. Phenolics in human health. *Int J Chem Eng Appl.* 5: 393.

Pandey, K.B., and Rizvi, S.I. 2009. Plant polyphenols as dietary antioxidants in human health and disease. *Oxid Med Cell Longev.* 2: 270–278.

Pinna, C., Morazzoni, P., and Sala, A. 2017. Proanthocyanidins from Vitis vinifera inhibit oxidative stress-induced vascular impairment in pulmonary arteries from diabetic rats. *Phytomedicine* 25: 39–44.

Rakvaag, E., and Dragsted, L.O. 2016. Acute effects of light and dark roasted coffee on glucose tolerance: a randomized, controlled crossover trial in healthy volunteers. *Eur J Nutr.* 55: 2221–2230.

Reis, C.E., Dórea, J.G., and da Costa, T.H. 2019. Effects of coffee consumption on glucose metabolism: a systematic review of clinical trials. *J Tradit Complement Med.* 9: 184–191.

Reno, F.E., Normand, P., McInally, K., Silo, S., Stotland, P., Triest, M., Carballo, D., and Piché, C. 2015. A novel nasal powder formulation of glucagon: Toxicology studies in animal models. *BMC Pharmacol Toxicol.* 16: 1–16.

Sales, D.S., Carmona, F., de Azevedo, B.C., Taleb-Contini, S.H., Bartolomeu, A.C. D., Honorato, F.B., Martinez, E.Z., and Pereira, A.M.S. 2014. Eugenia punicifolia (Kunth) DC. as an adjuvant treatment for type-2 diabetes mellitus: A non-controlled, pilot study. *Phytother Res.* 28: 1816–1821.

Santangelo, C., Filesi, C., Varì, R., Scazzocchio, B., Filardi, T., Fogliano, V., D'archivio, M., Giovannini, C., Lenzi, A., Morano, S., and Masella, R. 2016. Consumption of extra-virgin olive oil rich in phenolic compounds improves metabolic control in patients with type 2 diabetes mellitus: A possible involvement of reduced levels of circulating visfatin. *J Endocrinolo Invest.* 39: 1295–1301.

Scalbert, A., Manach, C., Morand, C., Rémésy, C., and Jiménez, L. 2005. Dietary polyphenols and the prevention of diseases. *Crit Rev Food Sci Nutr.* 45: 287–306.

Shi, M., Loftus, H., McAinch, A.J., and Su, X.Q. 2017. Blueberry as a source of bioactive compounds for the treatment of obesity, type 2 diabetes and chronic inflammation. *J Func Food.* 30, 16–29.

Singla, R.K., Dubey, A.K., Garg, A., Sharma, R.K., Fiorino, M., Ameen, S.M., Haddad, M.A., and Al-Hiary, M. 2019. Natural polyphenols: Chemical classification, definition of classes, subcategories, and structures. *J AOAC Int.* 102: 1397–1400.

Sun, C., Zhao, C., Guven, E.C., Paoli, P., Simal-Gandara, J., Ramkumar, K.M., Wang, S., Buleu, F., Pah, A., and Turi, V., Damian, G., Dragan, S., Tomas, M., Khan, W., Wang, M., Delmas, D., Puy Portillo, M., Dar, P., Chen, L., and Xiao, J. 2020. Dietary polyphenols as antidiabetic agents: Advances and opportunities. *Food Frunt* 1: 18–44.

Sung, H.-Y., Hong, C.-G., Suh, Y.-S., Cho, H.-C., Park, J.-H., Bae, J.-H., Park, W.-K., Han, J., and Song, D.-K. 2010. Role of (−)-epigallocatechin-3-gallate in cell viability, lipogenesis, and retinol-binding protein 4 expression in adipocytes. *Naunyn-Schmiedeb Arch Pharmacol.* 382: 303–310.

Suzuki, S., and Umezawa, T. 2007. Biosynthesis of lignans and norlignans. *J Wood Sci.* 53: 273–284.

Thazhath, S.S., Wu, T., Bound, M.J., Checklin, H.L., Standfield, S., Jones, K.L., Horowitz, M., and Rayner, C.K. 2016. Administration of resveratrol for 5 wk has no effect on glucagon-like peptide 1 secretion, gastric emptying, or glycemic control in type 2 diabetes: A randomized controlled trial. *Am J Clin Nutr.* 103: 66–70.

Timmers, S., De Ligt, M., Phielix, E., Van De Weijer, T., Hansen, J., Moonen-Kornips, E., Schaart, G., Kunz, I., Hesselink, M.K., Schrauwen-Hinderling, V.B. and Schrauwen P. 2016. Resveratrol as add-on therapy in subjects with well-controlled type 2 diabetes: a randomized controlled trial. *Diabetes Care* 39: 2211–2217.

Toloza-Zambrano, P., Avello, M., and Fernández, P. 2015. Determinación de rutina y trigonelina en extractos de hojas de Bauhinia forficata subsp. pruinosa y evaluación del efecto hipoglicemiante en humanos. *Bol Latinoam CaribePlant Med Aromat.* 14: 21–32.

Tresserra-Rimbau, A., Guasch-Ferre, M., Salas-Salvado, J., Toledo, E., Corella, D., Castaner, O., Guo, X., Gomez-Gracia, E., Lapetra, J., and Aros, F. Investigators Ps. 2016. Intake of total polyphenols and some classes of polyphenols is inversely associated with diabetes in elderly people at high cardiovascular disease risk. *J Nutr.* 146: 767–777.

Tsao, R. 2010. Chemistry and biochemistry of dietary polyphenols. *Nutrients* 2: 1231–1246.

Tuzcu, Z., Orhan, C., Sahin, N., Juturu, V., and Sahin, K. 2017. Cinnamon polyphenol extract inhibits hyperlipidemia and inflammation by modulation of transcription factors in high-fat diet-fed rats. *Oxid Med Cell Longev.* 2017:1583098.

Ullmann, U., Haller, J., Decourt, J., Girault, N., Girault, J., Richard-Caudron, A., Pineau, B., and Weber, P. 2003. A single ascending dose study of epigallocatechin gallate in healthy volunteers. *J Int Med Res.* 31: 88–101.

Vayalil, P.K. 2012. Date fruits (Phoenix dactylifera Linn): an emerging medicinal food. *Crit Rev Food Sci Nutr.* 52: 249–271.

Vetrani, C., Vitale, M., Bozzetto, L., Della Pepa, G., Cocozza, S., Costabile, G., Mangione, A., Cipriano, P., Annuzzi, G., and Rivellese, A.A. 2018. Association between different dietary polyphenol subclasses and the improvement in cardiometabolic risk factors: evidence from a randomized controlled clinical trial. *Acta Diabetol.* 55: 149–153.

Vladimir-Knežević, S., Blažeković, B., Štefan, M.B., and Babac, M. 2012. Plant polyphenols as antioxidants influencing the human health. Intechopen.

Wang, S., Moustaid-Moussa, N., Chen, L., Mo, H., Shastri, A., Su, R., Bapat, P., Kwun, I., and Shen, C.-L. 2014. Novel insights of dietary polyphenols and obesity. *J Nutr Biochem.* 25: 1–18.

Wedick, N.M., Pan, A., Cassidy, A., Rimm, E.B., Sampson, L., Rosner, B., Willett, W., Hu, F.B., Sun, Q., and van Dam, R.M.2012. Dietary flavonoid intakes and risk of type 2 diabetes in US men and women. *Am J Clin Nutr.* 95: 925–933.

Wiciński, M., Malinowski, B., Węclewicz, M.M., Grześk, E., and Grześk, G. 2017. Anti-atherogenic properties of resveratrol: 4-week resveratrol administration associated with serum concentrations of SIRT1, adiponectin, S100A8/A9 and VSMCs contractility in a rat model. *Exp Ther Med.* 13: 2071–2078.

Williamson, G. 2017. The role of polyphenols in modern nutrition. *Nutr Bull.* 42: 226–235.

Wong, R.H., Raederstorff, D., and Howe, P.R.C.2016. Acute resveratrol consumption improves neurovascular coupling capacity in adults with type 2 diabetes mellitus. *Nutrients* 8: 425.

Xie, M., Chen, G., Wan, P., Dai, Z., Zeng, X., and Sun, Y. 2018. Effects of dicaffeoylquinic acids from ilex kudingcha on lipid metabolism and intestinal microbiota in high-fat-diet-fed mice. *J Agric Food Chem.* 67: 171–183.

Xu, Y., Liu, P., Xu, S., Koroleva, M., Zhang, S., Si, S., and Jin, Z.G. 2017. Tannic acid as a plant-derived polyphenol exerts vasoprotection via enhancing KLF2 expression in endothelial cells. *Sci. Rep.* 7: 1–9.

Yin, J., Xing, H., and Ye, J. 2008. Efficacy of berberine in patients with type 2 diabetes mellitus. *Metabolism* 57: 712–717.

Zheng, J., Cheng, J., Zheng, S., Feng, Q., and Xiao, X. 2018. Curcumin, a polyphenolic curcuminoid with its protective effects and molecular mechanisms in diabetes and diabetic cardiomyopathy. *Front Pharmacol.* 9: 472.

Zhou, Y., Zheng, J., Li, Y., Xu, D.-P., Li, S., Chen, Y.-M., and Li, H.-B. 2016. Natural polyphenols for prevention and treatment of cancer. *Nutrients* 8: 515.

Zhu, W., Jia, Q., Wang, Y., Zhang, Y., and Xia, M. 2012. The anthocyanin cyanidin-3-O-β-glucoside, a flavonoid, increases hepatic glutathione synthesis and protects hepatocytes against reactive oxygen species during hyperglycemia: Involvement of a cAMP–PKA-dependent signaling pathway. Free *Radic Biol Med.* 52: 314–327.

Zibadi, S., Rohdewald, P.J., Park, D., and Watson, R.R. 2008. Reduction of cardiovascular risk factors in subjects with type 2 diabetes by Pycnogenol supplementation. *Nutr Res.* 28, 315–320.

8 Dietary Polyphenols in Bacterial and Fungal Infections

Veda Joshi, Swarangi Tambat, Mustansir Bhori,
Jyotirmoi Aich, and Kanchanlata Tungare
E-mail: ved.jos.bt19@dypatil.edu; swarangi1509@gmail.com;
mustansir.bhori@dypatil.edu; jyotirmoi.aich@dypatil.edu;
kanchanlata.singh@dypatil.edu

CONTENTS

8.1 INTRODUCTION

Polyphenols are among the most diverse classes of chemicals that have garnered considerable interest from the scientific community in recent years. Polyphenols are the secondary metabolites produced by plants (Daglia, 2012) that are of immense importance to humans, as they are not produced by animals, making plants the sole producers. However, these natural compounds form a regular integrant of human nutrition. The increase in the focus of research on polyphenols can be attributed due to their structural diversity. This in turn is a result of the variable carbon atom backbone, modifications (e.g., glycosylation) of the core structure (Lewandowska et al., 2013), bioactive properties like antibacterial, antifungal, antitumor, anti-allergic, anti-hypertension, and so forth (Daglia, 2012), and their varied mechanisms of action that qualify them as plant nutraceuticals (Marín et al., 2015). Their unique mechanisms of action and synergism are presently looked upon as a solution to the increasing resistance by microbes to conventional synthetic formulations. Dietary polyphenol can be categorized as a subclass of bioactive phenolic compounds that are produced as secondary metabolites in plants and found in abundance in varied food sources (fruits, green vegetables, cocoa products) and beverages (green tea, coffee, red wine) (Han et al., 2007).

The dietary polyphenols can be classified as phenolic acids, coumarins, flavonoids and non-flavonoids based on their structural complexity as mentioned in Figure 8.1.

The phenolic acids impart antioxidant activity via a radical scavenging mechanism to become a potent antioxidant compound (Kumar and Goel, 2019). This class is further divided into two subclasses: hydroxybenzoic acid (HBA) and hydroxycinnamic acid (HCA) (Kumar and Goel, 2019) as mentioned in Figure 8.2.

Chlorogenic acid, syringic acid, and vanillic acid are prominent examples of HBA. HBA (phenolic acids) are less commonly found than hydroxycinnamic acids (Manach et al., 2004). Apples, cherries, plums, berries, grapes, spinach, potatoes, coffee and tea have a high percentage of HCA like ferulic acid and caffeic acid (Mitra et al., 2022). HBA are found in onions and black radish. Tea is reported to consist of a very high gallic acid content (Manach et al., 2004). Coumarins are the naturally derived simple phenolics majorly obtained from vascular plants (commonly dicots), although some microorganisms do synthesize them (Küpeli Akkol et al., 2020) and are known to be vitamin K antagonists (Schalekamp and De Boer, 2010; Matos et al., 2015).The photochemical properties of coumarins and the ability to subdue the effects of radiotherapy make them a popular treatment for cancers such as leukemia, prostate cancer, and renal cell carcinoma (Küpeli Akkol et al., 2020). Scopoletin, warfarin, coumarin, phenprocoumon are commonly used coumarins. Novobiocin and aflatoxin are some microbial coumaric products (Cooke et al., 1997; Küpeli Akkol et al., 2020). In the recent quest to find naturally available bioactive products, flavonoids have emerged to be of prime importance. Over four thousand flavonoid compounds have been identified and are primarily responsible for vividness in fruit and floral pigmentations (Pandey and Rizvi, 2009). Flavonoids can be classified into different subclasses like flavones, anthocyanidins, chalcones, flavanols, flavonols, isoflavones, flavanones (Singla et al., 2019) as described in Figure 8.3.

Flavonols are eminent components of cocoa (Sokolov et.al,2013) and are the elementary units of proanthocyanidins (Panche et al., 2016). Flavonol glucosides exhibit dual fluorescence due to induced tautomerism resulting in floral pigmentation and shielding the plant from UV rays (Smith and Markham, 1998). Quercetin, kaempferol, rutin, and myricetin belongs to this class. Kaempferol is present in apples, peaches, berries (blackberries and raspberries), grapes, tomatoes, potatoes, broccoli, spinach, lettuce, onions, cucumber, and green tea. Quercetin is present in spices, vegetables like onion, tea and fruits like apple, berries, and pomegranate. Myricetin is present to a considerable extent in red wine, vegetables, berries, tea, and nuts (Mitra et al., 2022). Flavanones are abundantly found in citrus fruits like oranges, lemons, and so forth, and manifest free radical scavenging property to a great extent (Panche et al., 2016). Examples include hesperidin, eriodictyol, naringin, and naringenin. Flavanols include both catechins as well as tannins along with their subclasses. Catechins are monomers exclusively found in tea leaves (*Camellia sinensis*) and beverages like red wine, green tea, and so forth (Isemura, 2019). They exhibit high antioxidant properties, which promote the usage of green tea as a

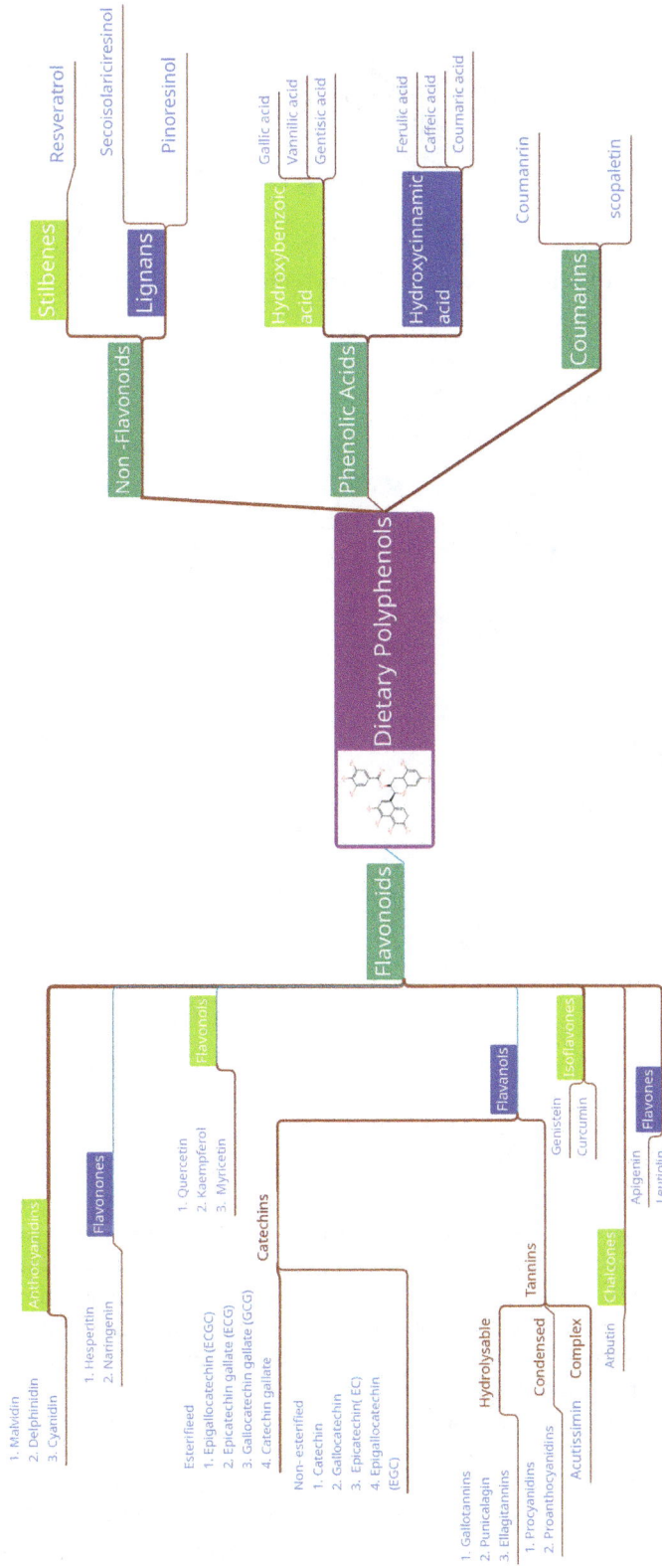

FIGURE 8.1 Classification and examples of polyphenols.

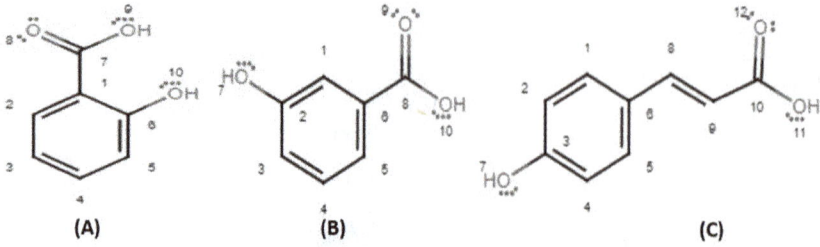

FIGURE 8.2   Structures of (A) Phenolic acids (B) Hydroxybenzoic acid (HBA) (C) Hydroxycinnamic acid.

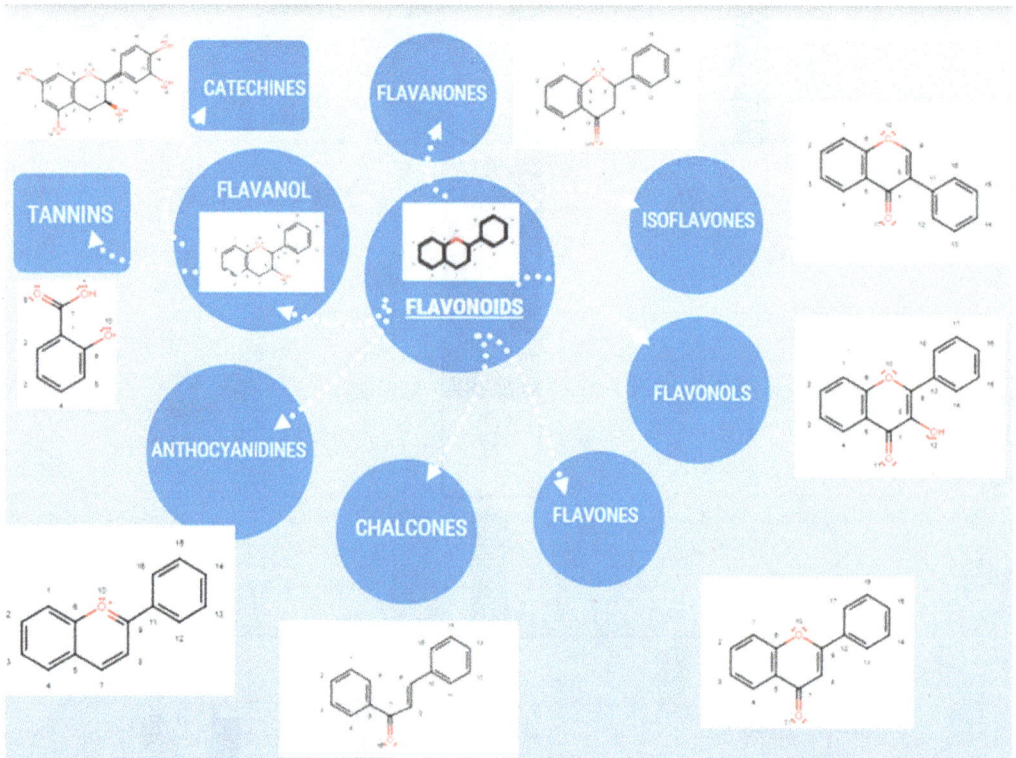

FIGURE 8.3   Classification of flavonoids.

natural health-promoting beverage to be included in one's daily diet. These catechins can be subcategorized as esterified catechins and non-esterified catechins. Epigallocatechin (ECGC), epicatechingallate (ECG), gallocatechingallate (GCG), and catechin gallate belong to the esterified class of catechins, whereas gallocatechin, catechin, epicatechin (EC), and epigallocatechin (EGC) come under non-esterified catechins (Gadkari and Balaraman, 2015; Batista et al., 2021). Tannins are available as natural polymers of phenolic acids and resemble flavonoids to a great extent. They are sub-categorized as condensed, hydrolyzable, and complex tannins (Ree and Khanbabaee, 2001). The condensed ones are proanthocyanidins (in grapes) and procyanidins. When hydrolyzed, the hydrolyzable tannins – gallotannins and ellagitannins – produce phenolic acids such as gallic acid and ellagic acid, respectively (Gadkari and Balaraman, 2015).

Flavones are usually found in plant leaves and floral components. Apigenin and luteolin are examples of this sub-class. Isoflavones are profusely found in leguminous plants and are referred

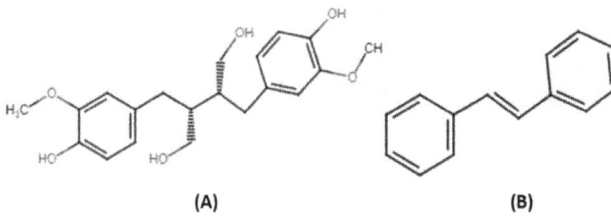

FIGURE 8.4   Structures of non-flavonoids (A) Stilbenes (B) Lignan (secoisolariciresinol).

to as phyto-estrogens owing to their estrogenic activity in animal models (Aoki et al., 2000;Panche et al., 2016). Anthocyanidins present in the exocarp are responsible for the vividness of fruit colors. They impart color based on which hydroxyl groups are acylated or methylated in the aromatic rings and according to various anthocyanidins' pH. They have an auxium ion (o^+) in their heterocyclic chain. Malvidin, peonidin, delphinidin, and pelargonidin are a few examples of this class. Lastly, chalcones can be found in berries and foods containing wheat (Panche et al., 2016). Chalcones defy the structure of the basic flavonoid by eliminating the heterocyclic chain that links the aromatic rings. Hence, they are also called open-chain flavonoids and play a role in a variety of health-related nutritional benefits. Chalco naringenin, arbutin, phloretin, and phlorizin are some of the studied chalcones (Vardhan and Shukla, 2017). In non-flavonoids (Figure 8.4), stilbenes have a broad spectrum of activity against various illnesses. Stilbenes have been found to reduce the occurrence of cancer by interfering with molecular processes at all phases of carcinogenesis, like autophagy or apoptosis (Sirerol et al., 2016). Some of the natural stilbene isolates are resveratrol, pterostilbene, and piceatannol. Numerous studies have demonstrated the antimutagenic, antioxidant, and cancer chemopreventive attributes of resveratrol (Jang et al., 1997). Another non-flavonoid class of dietary polyphenols are lignans (Figure 8.4), which are considered to be of great importance being a bioactive compound. Increased dietary consumption of these polyphenols has been linked to a decrease in the prevalence of estrogen-related malignancies like breast cancer in postmenopausal women (Horn-Ross et al., 2003) . Pinoresinol and secoisolariciresinol serves as an example of this class (Vardhan and Shukla, 2017).

The diverse nature of polyphenols attributed to their configurations and presence in a wide range of plants makes extensive study of them laborious (Pérez-Jiménez et al., 2010). Various studies have confirmed the presence of polyphenols in many food substances, like fruits, vegetables, beans, spices, and seasonings, beverages, and so forth, procured from their respective plant parts. It has been reported that food seasonings like cloves, capers, cinnamon, dried peppermint, dried rosemary, saffron, curcumin, dried basil leaves, and so forth have a very high polyphenolic content followed by seeds, fruits, and vegetables (Pérez-Jiménez et al., 2010). Amongst fruits apple, berries (strawberry, raspberry, cranberry, blueberry, blackberry, mulberry, black chokeberries, and black elderberries), cherry, plum, apricot, grapes, blackcurrant, prune, peach, and citrus fruits (orange, lemon and lime) are known to harbor a rich source of polyphenols. Seeds and nuts also have high polyphenolic content. Vegetables are also reported to serve as a quality source, such as carrots, broccoli, black radish, onion, tomatoes, potatoes, spinach, olives (green and black), shallots, lettuce and so forth. Other cereal grains (wheat, maize, oat, rye, etc.) and cocoa products (chocolates and cocoa powder) have displayed prominent polyphenol contents. Alcoholic beverages (red wine and grape wine), non-alcoholic beverages (tea, coffee, cocoa, green tea) and oils (sesame seed oil, olive oil, etc.) also account for a considerable amount of polyphenols in them (Pérez-Jiménez et al., 2010; Mitra et al., 2022). Studies have confirmed the distribution of polyphenols to be non-uniform across all the plant species (Scalbert and Williamson, 2000). The largest reservoirs of polyphenols being seasonings, spices, fruits, vegetables, and beverages, their presence in the diet should be balanced for a net observable effect. Oregano, broccoli, carrots, green pepper, tea, and oranges are rich in flavones like luteolin and apigenin. Lemons are a rich source of flavanones like eriodictyol and hesperetin. Hesperetin is also present in lime and orange. Further, anthocyanidins are reported to be present in berries (strawberry, cranberry, raspberry and blueberry). Other fruits like grapes and blackcurrant, cocoa, cereals, nuts, wine, and tea are rich sources of delphinidin and cyanidin. Isoflavones such as genistein is reported to be present in legumes, soy, and seeds like soyabean, whereas tannin (flavanols) are present in berries, apple, grapes, plum, pomegranate, peach, walnuts, chocolate, tea, coffee, and wine (Mitra et al., 2022).

Non flavonoid lignans like secoisolariciresinol, pinoresinol, and sesamol are highly present in sesame seeds, flaxseed, soybeans, cereals, broccoli, cabbage, apricots, and berries (Mitra et al., 2022). On the other hand, the most common yet important stilbene, that is, resveratrol, are found in grapes, peanuts, and red wine (Mitra et al., 2022). In order to reap maximum effects from the intake of polyphenols, it is crucial to gain an understanding of the optimum dietary requirements of those polyphenols and their respective food sources. The unavailability of sufficient evaluations does not yield very exact insights into the proper contents of the therapeutically significant polyphenols in food and also their rich sources (Scalbert and Williamson, 2000). Moreover, their rapid elimination and metabolism from the body reduces the bioavailability of polyphenols. It has been reported that intake of 10 to 100 milligrams of any individual polyphenol can scarcely surpass its concentration value of 1 micromolar in the plasma (Scalbert and Williamson, 2000). So, for the ideal utilization of the therapeutic properties of polyphenols, their total intake should be much more than 500 mg/day per person. Also, to avoid any undesirable or toxic effects, the levels should not exceed the permissible intake values that are pre-defined for different types of polyphenols. Different studies have concluded that a daily intake of approximately 1000-1750 mg/day is required by a normal adult human being (Scalbert and Williamson, 2000; Grosso et al., 2014). The intake can increase up to approximately 1755 milligrams/day for a healthy male and the intake in healthy males is slightly higher than in healthy females (1725 mg/day) (Grosso et al., 2014). Young people with high physical exertion and rate of metabolism have shown to consume higher levels of polyphenols than old and weak individuals (Grosso et al., 2014). The human diet is found to be the richest in flavonoids, forming around 2/3rd part, and phenolic acids, forming around 1/3rd part (Scalbert and Williamson, 2000). Of all the flavonoids, flavanols – specifically, proanthocyanidins and catechins – form the majority (Scalbert and Williamson, 2000). On the contrary, lignans and stilbenes have the lowest content in the diet (Grosso et al., 2014). Thus, a balanced diet with the required contents of the different polyphenols is mandatory.

The lifestyle of today's generations has changed arbitrarily and has an overall negative impact on dietary intake, as a consequence of which, the past few years have seen a hike in the demand for supplementary food products as the individual's daily diet is not able to make up for the body's nutritional requirements. These supplements include either synthetic formulations of vitamins, minerals, and so forth, or the food sources rich in them. However, most of the synthetic formulations in the form of tablets or tonics have raised toxicity concerns; therefore, intake of naturally available nutrient-rich food sources is advised by nutritionists and dieticians. Food sources like vegetables, green tea, fruits, flax seeds, cod-liver oil, soy, and so forth are regularly utilized by people as supplementary food (Gollücke et al., 2018). The supplementation of soy and beverages like green tea has come to the top of physicians' recommendation lists, according to a recent study (Gollücke et al., 2018). But, due to green tea's property of being extensively metabolized in the body's systems, investigations are carried out to derive formulations from such dietary polyphenols that would make them stable and increase their bioavailability (Scalbert et al., 2002). A few research studies have reported powder formulation of concentrated pomegranate extract containing high levels of ellagic acid procured by ethanol extraction and other down-streaming processes (Alkayali, 2009; Gollücke et al., 2018). Significant amounts of proanthocyanidins and catechins have been reported to be extracted from the formulation of cold-pressed grape seed oil (Eckert et al., 2007; Gollücke et al., 2018). The dry concentrate obtained from grape juice and wine when taken for approximately 14 days (1.0 mg to 2.0 mg per day) showed improved cardiovascular health in human subjects (Howard et al., 2003; Gollücke et al., 2018). The dosage formulation of many important polyphenols and their safety percentages regarding their toxicity is yet to be determined and has a good scope in future nutrigenetics.

The antioxidant properties of polyphenols dominate with other beneficial properties in particular, antimicrobial, anticarcinogenic, anti-hypertensive, antibacterial, antifungal, antiviral, anti-inflammatory, anti-allergic, and so forth. Various meta-analyses and comprehensive studies of epidemiology have unfolded their favorable manifestations in chronic diseases like asthma, cardiovascular diseases, infections, diabetes, neurodegenerative diseases, hypertension, aging, and so forth (Pandey and Rizvi, 2009). The main contributing property of polyphenols is as an antioxidant, because it ultimately reduces the oxidative damage at the molecular level. Polyphenols exert cardio-protective properties, anti-diabetic properties, for example, anthocyanidins (Cory et al., 2018; Pandey and Rizvi, 2009), anti- carcinogenic effects by lowering the number of free radicals and blocking the progress of cell cycle and, thus, cancer (Cory et al., 2018), antioxidant properties, and so forth (Pandey and Rizvi, 2009). As we know that aging is a result of cellular oxidative stress, the antioxidant properties as well as anti-inflammatory effects of polyphenols can function as anti-aging compounds (Pandey and Rizvi, 2009). Studies have collected evidence of the antioxidant, scavenging, and immunomodulatory mechanisms of resveratrol, EGCG, and curcumin that are reported to suppress any neurotoxic effects on the neurons. Neuroprotective effects of polyphenols are well reported against the agglomeration of β-amyloid protein in Alzheimer's disease and other neurodegenerative diseases (Cory et al., 2018). Curcumin, resveratrol, and catechins also display anti-obesogenic effects by interfering with lipogenesis, causing oxidation of adipocytes and enhancing the cellular expenditure of energy (Cory et al., 2018). As a good diet is related to superior gut function, the different immunobiotic effects of polyphenols lead to a decrease in gut discomfort and also aid in the improvement of gut-associated disorders. Furthermore, a happy stomach maintains a stable mental state. Research on the positive manifestations of polyphenols in the treatment of mental disorders like autism are underway.

The overall disease, or any abnormal condition, results from many factors at the cellular and molecular levels. One major factor is the oxidative stress or damage caused by the disproportion between the number of free radicals or reactive oxygen species (ROS) formed as a result of cellular metabolism and their neutralization by various protective mechanisms. The excess free radicals cause cellular harm by damaging various biomolecules like DNA, lipids, proteins, and so forth. The inflammation triggered by oxidative stress is the cause of many chronic diseases (Hussain et al., 2016). The antioxidant properties of polyphenols are due to their potential to chelate metal ions, the considerable number of -OH (hydroxyl) groups in their structure, and their ability to scavenge a huge array of free radicals (Hussain et al., 2016). Various mechanisms of polyphenols that mediate radical scavenging have been identified (Lv et al., 2021; Perron and Brumaghim, 2009). Some reports suggest that the -OH groups in the polyphenols reduces and thus neutralizes the action of the free radicals (Lv et al., 2021; Perron and Brumaghim, 2009). Lv and colleagues also emphasize inhibition of enzymes involved in ROS generation by polyphenols (Lv et al., 2021). For example, flavonoids (quercetin and luteolin) and stilbenes are observed to restrict the functioning of xanthine oxidase (XO), which is a prime enzyme in free radical production. Similarly, nitric oxide production is hampered by the inhibition of the nitric oxide synthase (NOS) resulting in a decline of free radicals (Hussain et al., 2016). Free metal ions are involved in free radical formation. Chelation of such metal ions by polyphenols restricts this activity (Lv et al., 2021). Moreover, the functions of antioxidant enzymes and proteins present in the body are also enhanced by polyphenols, thus, suppressing the unrequired oxidation reactions in the cell. The synergism between polyphenols and other compounds in the body may also produce desirable antioxidant outcomes (Lv et al., 2021). As flavonoids form the majority of our daily diet, and their antioxidant properties are also remarkable, the fruits and vegetables containing high flavonoids content need to be involved as much as possible in human nutrition.

As anything inadmissible or in excess is a poison, comprehension of the toxic effects or potential risks of polyphenols should be explored in addition to their virtues. Generally, toxicity or any hazards, apart from an individual's susceptibility to food allergies resulting from naturally occurring foods is very low. Normal physiological levels of various parameters like normal fat and cholesterol levels may increase due to the high consumption of chocolates and coffee. Proper ways of targeting the correct population should be kept in mind. Pregnant women, children, the elderly, and people with co-morbidities should not be risked for any experimentation (Mennen et al., 2005). Polyphenols at high doses are reported to develop genotoxic, carcinogenic effects and thyroid toxicity. Flavonoids display the highest thyroid toxicity effects. At high doses, the estrogen-like polyphenols (mainly, isoflavones) may result in antiandrogenic outcomes like infertility and abnormal sexual dysmorphism of a baby in the uterus or post-parturition. They may even exert antinutritional manifestations. For example, high consumption of tea may affect normal iron absorption mechanisms other than by heme. This may further result in iron deficiency and its corollaries (Mennen et al., 2005). So, thorough toxicity testing is mandatory to encourage dietary polyphenol consumption or in the production of its supplements (Table 8.1).

8.2 POLYPHENOLS IN THE PREVENTION OF BACTERIAL AND FUNGAL INFECTIONS

With an increase in the global burden of bacterial and fungal infections, researchers are venturing into novel substitutes to conventional therapeutics. The ongoing research has pioneered ways for polyphenols to be exploited for their antimicrobial properties.

8.2.1 ABSORPTION OF DIETARY POLYPHENOLS

In the surging field of dietetics and nutrigenetics, the importance of inclusiveness of polyphenols in daily diet has emerged to be one of the eccentric ways to heal naturally. Polyphenols as discussed earlier, are available in a diverse range, having a vast structural diversity, and hence differ in their pharmacokinetics properties related to absorption, bioavailability, and bio accessibility to the tissues. Flavonoids, among all the other polyphenols, are the ones that are found in abundance in daily diets and beverages like green tea, which contains catechins like epicatechin, epicatechin-3-gallate, epigallocatechin, and EGCG (Chacko et al., 2010). Also, due to their low molecular weight, they are easily broken down and absorbed. The flavonoid glycosides present in the food matrix are acted upon by the β- glucosidase and are converted to their respective aglycones. The enzyme, β-glucosidase is produced by various microflora residing in the intestinal lumen, but recent studies have given an insight of production of β- glucosidase by the shredded epithelial cell in the oral cavity, resulting in the breakdown of flavonoids to their respective aglycones like quercetin and genistein. The presence of such aglycones in the oral cavity has shown considerable inhibition of oral cancer cell proliferation (Walle et al., 2005).

The acidic environment in the stomach due to low pH affects the stability of polyphenols, but few reports have suggested that proanthocyanidins and flavonoids like resveratrol, catechin, quercetin show unusual stability (Lewandowska et al., 2013). Gastric peristalsis led to the release of more phenolic compounds from the food matrix followed by absorption of ferulic acid and cholinergic acid (Aura, 2008). Malvidin-3-glucoside – an integral flavonoid component in the skin of purple grapes is found in concentrations of $3 \times 10^{-4}M$ in young red wine. This compound is

TABLE 8.1

Sources of dietary polyphenols along with their requirements and therapeutic potential

Polyphenolic Compound	Occurrence/ Sources	Approx. Dietary requirement (mg/day)	Therapeutic Properties	References
Flavones	Oregano, broccoli, celery, peppermint, carrots, green pepper, tea and orange	15.47	Antioxidant, anti-inflammatory,	(Lv et al., 2021), (Hussain et al., 2016), (Mitra et al., 2022), (Grosso et al., 2014)
Flavonols	Apples, peaches, berries, pomegranate, grapes, tomatoes, potatoes, broccoli, spinach, lettuce, onions, cucumber, red wine, nuts and green tea	105.9	Antioxidant, anti-inflammatory, antifungal	(Lv et al., 2021), (Hussain et al., 2016), (Daglia, 2012), (Mitra et al., 2022), (Grosso et al., 2014)
Flavanones	Lemons, lime and orange	103.6	Antioxidant, anti-inflammatory	(Lv et al., 2021), (Hussain et al., 2016), (Mitra et al., 2022), (Grosso et al., 2014)
Flavanols	Berries, apple, grapes, plum, pomegranate, peach, walnuts, chocolate, tea, coffee and wine	636.13	Antioxidant, anti-inflammatory, antiviral, antibacterial, antifungal, anti-cancer	(Lv et al., 2021), (Hussain et al., 2016), (Daglia, 2012), (Mitra et al., 2022), (Grosso et al., 2014)
Anthocyanidins	Berries, blackcurrant, cocoa, cereals, nuts, wine and tea	29.64	Antioxidant, anti-inflammatory	(Lv et al., 2021), (Hussain et al., 2016), (Mitra et al., 2022), (Grosso et al., 2014)
Isoflavones	Legumes, soy and seeds (soyabean)	1.59	Anti-oxidant, anti-inflammatory, anti-cancer, antimicrobial, antibacterial, antifungal	(Lv et al., 2021), (Mitra et al., 2022), (Grosso et al., 2014)
Phenolic acids	Apples, cherries, plums, berries, grape, spinach, potato, coffee, tea, onions and black radish	798.73	Antioxidant, antiviral	(Lv et al., 2021), (Chojnacka et al., 2021), (Manach et al., 2004), (Mitra et al., 2022), (Grosso et al., 2014)
Lignans	Sesame seeds, flaxseed and soybeans, rye, oat and wheat, broccoli, cabbage, apricots and berries	0.59	Antioxidant, anti-inflammatory, anti-cancer	(Lv et al., 2021), (Hussain et al., 2016), (Mitra et al., 2022), (Grosso et al., 2014)
Stilbenes	Grapes, peanuts and red wine	0.19	Antioxidant, anti-inflammatory, anti-cancer	(Lv et al., 2021), (Hussain et al., 2016), (Mitra et al., 2022), (Grosso et al., 2014)

readily absorbed in the stomach by virtue of transporters in the gastric wall a bilitranslocase (TC 2.A.65.1.1) (Passamonti et al., 2005). This is an organic anion transporter that assists the transport of biliverdin and bilirubin and is inclusive of anthocyanidin and flavonoid aglycones to the systemic circulation. They are normally present in the kidney, vascular endothelium, gastric epithelium and liver (Karawajczyk et al., 2007).These phenolic compounds generally are bound to protein, mainly albumin, while circulating through the bloodstream.

In the small intestine, polyphenols mainly undergo glucuronidation reactions catalyzed by two different isoenzymes of Uridine diphosphate glucuronosyltransferase- UGT-1A8 and UGT-1A10. Once inside the enterocytes, the polyphenols are transported to the liver through portal veins, wherein they undergo coupling reactions similar to those of xenobiotics where hydrophobic compounds are made more hydrophilic thereby facilitating their excretion into the bile. This bile, containing all the phenolic metabolites, re-enters the intestinal lumen via enterohepatic circulation, where metabolites like glucuronides and sulfates of polyphenolic compounds that are too hydrophilic to enter the enterocytes by simple diffusion are taken up by ABC transporters. These ATP binding cassettes (ABC) contain one or more than one ATP binding domains where ATP is hydrolyzed to translocate a variety of compounds including polyphenols across the biological membranes (Lewandowska et al., 2013).

On reaching the large intestine, the polyphenols are acted upon by the colonial microflora in the intestinal lumen, which metabolizes many phenolic compounds. To cite an example: *Clostridium orbiscindens* and *Eubacterium ramulus* produce enzymes that bring about the fission of C-ring in quercetin and naringenin. *Enterococcus casseliflavus* – a gram-positive, facultative anaerobe deglycosylates quercetin-3- glucoside (a potent antioxidant flavonoid) in humans (Aura, 2008). Likewise, the colon houses a manifold microbial population of obligate and facultative anaerobes that aid in the degradation of undigested food and associated phenolic compounds that are still attached to the food matrix. A few metabolic conversions are described below

8.2.1.1 Flavonoids

Quercetins are subject to deglycosylation reactions and ring cleavage by *Clostridium* and *Enterococcus* species found in human feces (Schneider et al., 1999; Winter et al., 1991). Flavanols such as luteolin, kaempferol, and naringenin undergo ring cleavage assisted by *Eubacterium ramulus* found in human feces (Schneider and Blaut, 2000). Condensed tannins, which normally have a high degree of polymerization, are not broken down by the gut microbiota, but a low degree of polymerization favors their substantial degradation in the cecum. Proanthocyanidins and flavan-3-ol are converted to phenolic acids (phenylacetic, phenylpropionic, phenylvaleric, benzoic acid, etc.) resulting in the accumulation of such polar metabolites within the body systems (Abia and Fry, 2001; Levrat et al., 1993).

8.2.1.2 Lignans

Pinoresinol, secoisolariciresinol, and lariciresinol are subject to dehydroxylation by *Clostridium scindens* and *Eggerthellalenta* present in human feces (Clavel et al., 2005, 2006a) and demethylation by *Butyribacterium, Eubacteriumcalenderi, Eubacteriumlimosum* residing in the human intestinal lumen (Clavel et al., 2006a, b). Secoisolariciresinoldiglucoside undergo enterolactone and enterodiol formations metabolized by *Lactonifactorlongoviformis* and *Peptostreptococcus* species, respectively, in the colon (Clavel et al., 2007; Wang et al., 2000).

8.2.1.3 Tannins

Ellagtannins are a class of polyphenolic tannins which are high molecular weight compounds generally metabolized by colonial microbiota. In humans, degradation of ellagitannins occurs primarily by its hydrolysis to ellagic acid which is further broken down to lactone subunits-urolithin A and urolithin B (Cerda et al., 2004).

8.2.1.4 Phenolic acids

Phenolic acids are either directly present as a component of our food matrix or formed during flavonoid ring formation. Caffeic acid, chlorogenic acid, caftaric acid are deesterified by human fecal microbiota or decarboxylated and reduced to form 3-(3-hydroxyphenyl) propionic acid and 4-ethylcatechol, respectively(Peppercorn and Goldman 1971). The intestinal tissues and microbial flora both produce intestinal esterases that aid in the uptake of such esterified acids (Aura, 2008). The portal vein absorbs and transports the metabolites produced by the intestinal microbiota to the liver, where they undergo conjugation reactions resulting in mono glucuronides and mono sulfates of varied phenolic acids. These conjugated metabolites are then delivered to the circulation and eventually eliminated in the urine, whereas the unabsorbed metabolites are removed through feces. Glucose transporters like GLUT-3, GLUT-1 and GLUT-2 also play a key role in the absorption of polyphenols (Lewandowska et al., 2013).

Still, the bioavailability and accessibility of polyphenols to the tissues remain quite low as the majority of the phenolic compounds are being eliminated from the body as they become a substrate to many enzymes involved in xenobiotic metabolism. To overcome this, phenolic compounds are combined with other biomolecules in order to stabilize them and escape from their excessive metabolism. This ensures their prolonged presence in systemic circulation and enhances their effectiveness. For example, green tea catechins are complexed with phospholipids or lecithinized curcumin delivery systems to increase the bioavailability of the phenolic compounds. Likewise, silibinin formulated with phosphatidylcholine (silipide) is administered to patients confirmed with colorectal adenocarcinoma to increase the effectiveness of the treatment. Many such modulations are done to maximize the effect of polyphenols.

8.2.2 ANTIMICROBIAL EFFECTS OF DIETARY POLYPHENOLS

Polyphenols, as mentioned earlier, can be broadly classified into flavonoids and non-flavonoids (Table 8.2). Studies have suggested a strong correlation between flavonoids and the antimicrobial property associated with them. All the subclasses of flavonoids like-flavan-3-ol, flavonols and tannins are capable of exhibiting antibacterial, antifungal as well as antiviral effects. These polyphenols have a wide spectrum and disable a number of microbial virulence factors by adopting various strategies like prevention of attachment of viral surface proteins to the host ligands, inhibition of biofilm formation and neutralization of bacterial toxins. Also, administration of polyphenols in a synergistic manner with antibiotics has proved to overcome the antibiotic resistance and multidrug resistance property that is prevalent in many disease-causing microbes, which pose a challenge in defining an appropriate treatment for the same.

TABLE 8.2
Classification of antimicrobial polyphenols

Sl. No.	Target bacteria	Antibacterial polyphenols	References
1.	*E. coli, Streptococcus pneumoniae, B. subtilis, S. aureus, Shigella dysenteriae, Salmonella typhimurium,* etc.	Chlorogenic acid	(Rempe et al., 2017)
2.	*Escherichia coli, Helicobacter pylori,* etc.	Quercetin	(Rempe et al., 2017)
3.	*Helicobacter pylori*	Apigenin	(Rempe et al., 2017)

Sl. No.	Target fungi	Antifungal polyphenols	References
1.	*Candidaalbicans*	Carvacrol, Curcumin	(Ansari et al., 2013)
3.	*Cryptococcus neoformans, Rhizopus* spp.	Thymol	(Ansari et al., 2013)
4.	*Aspergillus* spp.	Eugenol	(Ansari et al., 2013)

A few anti-microbial associated polyphenols are discussed below: prominent catechin, Flavan-3-ol, found in green tea *(Camellia sinensis)* has been known to be a potent antibacterial modulator since the late 1990s. It inhibits the growth of different bacterial species like *Escherichia coli*, *Streptococcus mutans*, *Clostridium perfringes and Clostridium jejuni in vitro* (Diker et al., 1991; Ahn et al., 1991).

One more such tea catechin – epigallocatechin gallate (EGCG) – is well researched to see its effect on growth of *Helicobacter pylori* – an ulcer-forming pathogen found in the gastric lumen of humans. It does so by producing urea enough to disrupt the membrane lining mucus, causing the gastric epithelial cell lining to come in contact with HCl and result in ulcers and gastric cancers. All the researched isolated of *Helicobacter pylori* were found to be highly resistant to antibiotic drugs, like metronidazole and/or clarithromycin, and showed sensitivity to EGCG at MIC (concentration at which 90% of growth of microbe is inhibited) of 100 µg/mL. EGCG is also a well-known anti-virulent and aids in treating viral infections like flu by preventing its attachment to the host cell by virtue of EGCG's binding to the viral hemagglutinin (Daglia 2012).

Flavanols exhibit their antimicrobial effect by penetrating the phospholipid layer of cell membrane. Flavanols such as rhamnetin, myricetin, morin, and quercetin have been shown to have potent antibacterial properties against *Chlamydia pneumonia*e, an obligate intracellular gram-negative bacterium that causes pharyngitis, sinusitis, and pneumonia (Daglia 2012). Tannins, the naturally occurring astringent polyphenols, manifest themselves either in condensed form (e.g., proanthocyanidins) or hydrolyzed form (gallotannins and ellagitannins). Generally, they are dimers, oligomers, or polymers of catechins found in fruits, oak, or bark of trees. Proanthocyanidins that are derived from berries are primarily studied for antibacterial properties and can further be classified as A-type and B-type based on the occurrence of double or single linkage connecting two flavonoid units. They inhibit the growth of uropathogenic and carcinogenic bacteria like *Escherichia coli* and *Streptococcus mutans respectively* (Côté et al., 2010). Type-A proanthocyanidins exhibit their antimicrobial activity by destabilizing plasma membrane, inhibiting extracellularly released microbial enzymes, modulation of microbial metabolism by chelating the important metal cofactors like Fe^{+2} or Zn^{+2} and inhibiting the rate-limiting substrate for microbial growth thus acting as a bacteriostatic compound.

Proanthocyanidins also exhibit antiviral properties against influenza A virus and type 1 herpes simplex virus, in addition to antibacterial activities (HSV). The mechanism of action, in this case, appears to be blocking the virus from entering the host cell, which is the first and most important stage in initial HSV-1 infection (Gescher et al., 2011; Daglia 2012).

Ellagitannin is reported to be an antiviral agent with specific antiviral activity against HIV infections (Daglia 2012) by imposing inhibitory effects on HSV-2 and/or HSV-1 reproduction, as well as on Epstein-Barr virus replication (Ito et al., 2007). Ellagitannin's anti-herpes virus action appears to be owing to a strong inhibitory impact on HSV-1 and HSV-2 proliferation. They also act synergistically with aciclovir (the first specific antiviral drug against herpes simplex virus) resistant strains, thus rendering an effective treatment for herpes (Vilhelmova et al., 2011).

Non-flavonoids also exhibit limited types of microbial growth inhibitions. Lignans (precursors of phytoestrogen) have established themselves to be a formidable antibacterial polyphenol for such pathogenic bacterial strains, where conventional antibiotics and synthetic drug formulations have failed. This can be exemplified in *Aristolochia taliscana* roots. *Aristolochia taliscana*, a plant used in traditional Mexican medicine, were discovered to contain neolignans (C6-C3 linkage lignans formed by shikimic acid pathway), the most powerful of which is Licarin A, which was found with MICs ranging from 3.120 to 12.50 mg/ml against four mono-resistant versions and twelve clinical isolates of Mycobacterium TB strains (León-Díaz et al., 2010). Lignans, therefore, represent themselves as potent bioactive compounds in therapeutics for tuberculosis.

8.2.3 MECHANISM OF ACTION OF POLYPHENOLS ON BACTERIA AND FUNGUS

8.2.3.1 Anti-bacterial mechanisms of polyphenols

The increasing incidences of drug resistance shown by many bacteria have forced researchers to search for novel therapeutics. Natural polyphenols have captured much attention with regard to their several prominent bioactivities and their diverse anti-bacterial mechanisms like cell membrane disruption and non-membrane disruption mechanisms like inhibition of the enzymatic activities of vital cellular enzymes like NADH-cytochrome c reductase, topoisomerase, and ATP synthase. The various mechanisms can be grouped as either membrane disruption mechanisms or non-membrane mechanisms (Rempe et al., 2017) (Figure 8.5).

8.2.3.1.1 Membrane disruption by polyphenolic compounds
Antibacterial activity of polyphenols results from the cell membrane damage of the bacterial cells regardless of their gram staining nature. Ferulic acid, gallic acid, and epigallocatechin gallate are reported to be involved in the membrane disruption of bacteria. A recent study has confirmed epigallocatechin gallate's involvement against some specific bacterial strains like the methicillin-resistant *Staphylococcus aureus* (MRSA) strain and makes the cells defenseless to beta-lactam. To date, the exact mechanisms for the specific polyphenols remain unprobed (Rempe et al., 2017).

Yet another study involving lipid profiles generated using high-resolution gas chromatography (HRGC) was done to evaluate the modifications of a few specific lipids due to the action of polyphenols, namely, carvacrol, eugenol, or thymol in both Gram-positive and Gram-negative bacteria. All three phenolics elevated the levels of the crucial fatty acids (oleic acid, palmitic acid, etc.) in Gram-negative bacteria while the lipid profiles of one of the Gram-positive bacteria (*S. aureus*) became altered for many of the fatty acids. However, treatment of *S. aureus* with thymol resulted in an eminent increase in the saturated fatty acids and a decline in the unsaturated fatty acids, which

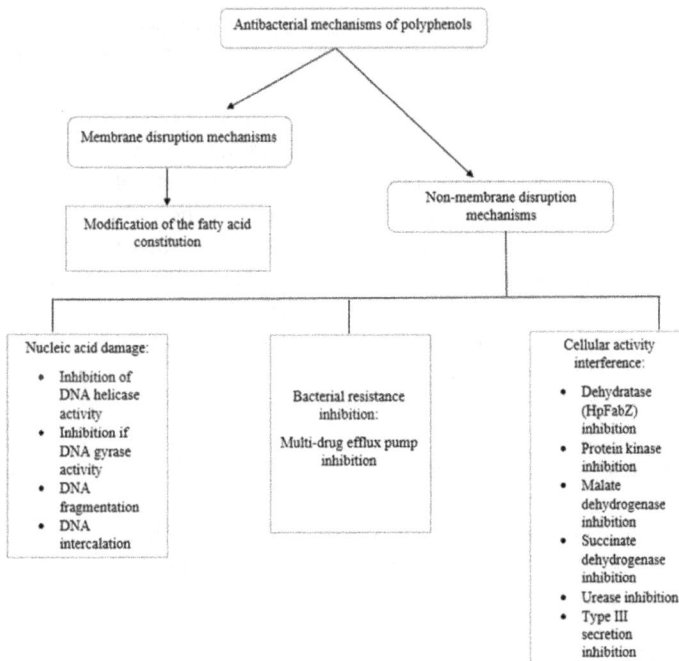

FIGURE 8.5 Different antibacterial mechanisms.

can be accredited to a desaturase response to the leakage of the cellular components(Rempe et al., 2017). Quercetin and chlorogenic acid have been reported to act as anti-bacterial polyphenols by cell disruption mechanisms (Rempe et al., 2017).

8.2.3.1.2   Non-membrane disintegration mechanisms by polyphenolic compounds
A plethora of reports have shed light on the non-membrane disintegration mechanisms of polyphenols. Different polyphenolic compounds have reported various mechanisms such as inhibition of DNA gyrase, Type III secretion, DNA helicase, multi-drug efflux pumps, dehydratase (HpFabZ), protein kinase, urease, succinate dehydrogenase, and malate dehydrogenase. DNA intercalation and induction of DNA fragmentation are the other major non-membrane antibacterial mechanisms of polyphenols. For example, apigenin is involved in dehydratase inhibition (HpFabZ) and protein kinase inhibition. Quercetin has been shown to be involved in dehydratase inhibition (HpFabZ), and so forth (Rempe et al., 2017; Makarewicz et al., 2021).

8.2.3.2   Antifungal mechanisms of polyphenols

The need for novel antifungal therapeutics is firstly due to multi-drug resistance (MDR) acquired by the fungus and, secondly, due to higher incidences of opportunistic or secondary fungal infections in immunocompromised individuals suffering from AIDS, HIV, Cancer, and so forth (Ansari et al., 2013). Polyphenols have been studied for their antifungal properties as well. They target the required fungal population through varied mechanisms like interfering with their ergosterol biosynthesis, phosphorylation of kinases involved in signal transduction, induction of apoptosis, disruption of fluidity of the cell membrane, inhibition of morphogenesis, and reversing the multi-drug resistance shown by the fungi (Ansari et al., 2013; Teodoro et al., 2015).

Polyphenols exert their antifungal properties by undertaking specific mechanisms of action against peculiar fungal species. Discussed below are a few different mechanisms implemented by different polyphenols against the most commonly encountered fungal species:

(a) *Candida albicans*: It is the causative agent of the common skin infection called cutaneous candidiasis characteristic of the skin and nails. Most of the polyphenolic compounds act by disintegrating the fungal cell membrane by interfering with the biosynthesis of ergosterol. Ergosterol, similar to cholesterol in mammalian cell membranes, plays a great role in maintaining the plasma membrane integrity and morphology. Pterostilbene suppresses the Ras/cAMP signaling pathway and also blocks the ergosterol biosynthesis. Induction of apoptosis is another mechanism employed by the polyphenols. The nucleic acid biosynthesis is hindered by the action of epigallocatechin-3-gallate on dihydrofolate reductase, a crucial enzyme in the purine biosynthesis pathway. It has displayed synergism with ergosterol biosynthesis pathway inhibitors as well (Simonetti et al., 2020).

(b) *Trichophyton rubrum*: It belongs to the dermatophytic fungal genus of *Trichophyton*, which includes the fungal species causing different infections of the skin, scalp, nails, and so forth. Ergosterol biosynthesis in *Trichophyton rubrum* is curtailed by gallic acid, as it interferes with the enzyme involved in the ergosterol biosynthesis pathway (squalene epoxidase). Quercetin is also involved to affect the normal cellular levels of ergosterol (Simonetti et al., 2020).

(c) *Aspergillus:* It is known to cause aspergillosis or the respiratory disease affecting primarily the lungs. Polyphenolic compounds in the *Aspergillus* spp. mainly act by either reversing the fungi's MDR or disintegrating the cytoplasmic membrane of the fungal cells. Polyphenols including chalcones, xanthones, stilbenes, phenolic acids, flavonoids, anthocyanins, catechins, and tannins are hydrophobic compounds. They attach to the ABC (ATP-binding cassette) transporters directly and restrict the formation of the tertiary structures of the proteins. Hence, their normal functionality and the function of interest, that is, provision of drug resistance to the administered drug is repressed (Ansari et al., 2013).

(d) *Rhizopus:* Thymol directly interacts with ergosterol in the *Rhizopus* spp., which disintegrates the fungal cytoplasmic membrane. The disintegration of the membrane effectuates the loss of morphology of the membrane and expulsion of all the cytosolic contents that ultimately results in the death of the fungi (Ansari et al., 2013).

8.2.4 SYNERGISM

Synergism observes the amplified effect when two or more drugs are administered together as compared to a lesser effect when a single drug is administered. In most of the therapies targeted to cure cancer, harmful chemical compounds are administered in higher dosages, which are toxic to many healthy tissues and cells and lead to many other complications. Use of such a synergistic approach, which follows the requirement of low dosages of combined drugs, can profoundly establish a novel cancer treatment with much-decreased side effects (Tallarida, 2011).

The recent advancements in the synergistic approach concerning polyphenols have used polyphenols in combination with antibiotics to enhance their efficacy at lower dosages and invalidate the multidrug resistance hallmark of many infectious pathogens (Mitra et al., 2022).

Curcumin, a natural flavonoid complexed with vincristine (an anticancer drug) shows considerable decrease in cancer cell proliferation and induction of apoptosis (Sreenivasan and Krishnakumar, 2015; Mitra et al., 2022). In the same way, curcumin combined with piperine exhibits neuroprotective ability in SH-SY5Y cells (Abdul Manap et al., 2019; Mitra et al., 2022). Cisplatin (a chemotherapeutic drug), when administered with resveratrol (a flavonoid), results in induction of apoptosis in A549 cells (human lung carcinoma cells) by bringing about alterations in autophagy mechanism and confers better results as compared to a lone cisplatin administration (Hu et al., 2016; Mitra et al., 2022). Recent findings have also suggested that two or more polyphenols when combined may result in a synergistic effect. Using the microbial test system (MTS) – an antioxidant activity assay – it was recently discovered that *Potentillafruticosa* leaf extracts (PFE, when coupled with green tea polyphenols (GTP), demonstrated synergistic effects. The results showed that the combination of PFE and GTP in the ratio (3:1) exhibited strong synergism in accordance with H_2O_2 production rate by promoting CAT (common antioxidant enzyme) and SOD (superoxide dismutase) enzyme activity by enhancing their gene expression (Liu et al., 2018). Many such phenolic interactions can be studied and taken further for *in-vivo* analysis to procure satisfactory results for combating infectious diseases.

8.3 POLYPHENOLS IN IMMUNOLOGY

The diverse immunologically important bioactivities of polyphenols can be attributed to the various mechanisms and pathways followed by different types of polyphenols. Studies have confirmed the expression of a number of polyphenolic cell-surface receptors on the surfaces of different immune cells in the host body. The receptors, being specific, can precisely allow the cellular uptake of the respective polyphenol as a consequence of which several signaling pathways become activated in order to produce the required immune response(s). The different groups of polyphenols regulate different gut mucosal immune responses are involved in regulation of anti-allergic immune responses, and their anti-tumorigenic activities are currently being studied and exploited to a huge extent (Ding et al., 2018).

Polyphenols like flavonoids (curcumin, anthocyanidins, quercetin), phenolic acids (ferulic acid, *p*-Hydroxybenzoic acid) and stilbenes (resveratrol) are reportedly studied for their pharmacologically important and immunomodulatory activities. The different absorbed polyphenol forms interact with the intestinal immune system, eliciting the desired immune responses. The different

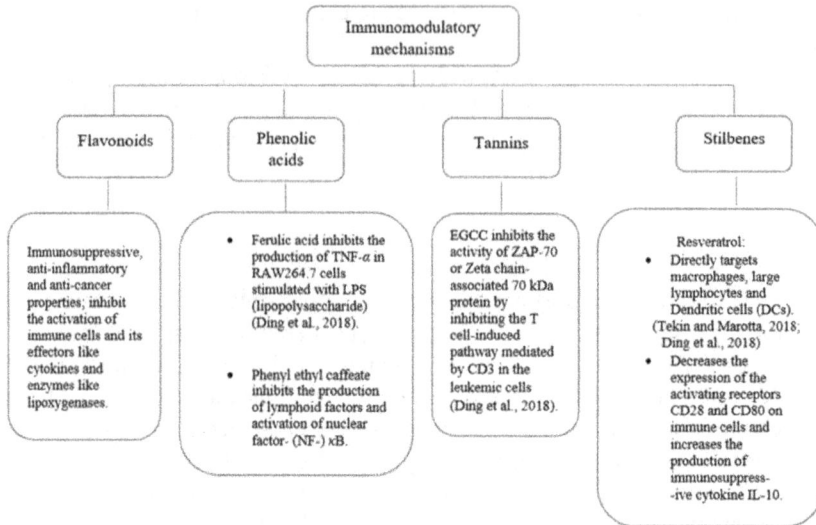

FIGURE 8.6 Different immunological mechanisms of polyphenols.

mechanisms are, in turn, a consequence of binding of polyphenols to a single or multiple receptors on the surface of the immune cells. This binding activates characteristic intracellular signaling pathways that control the host's immune responses and, hence, signifies the ability of polyphenols to use a repertoire of immunomodulatory pathways. The different immunomodulatory mechanisms and regulatory effects of polyphenols on the immune system are exemplified in Figure 8.6.

8.3.1 ROLE OF POLYPHENOLS IN INNATE AND ADAPTIVE IMMUNITY

8.3.1.1 Effect of polyphenols on innate immune system

Various polyphenolic compounds are reported to influence the different inflammatory pathways by interfering with the actions of immune cells and cytokines involved in the innate immune system. Several polyphenols interfere with the differentiation, maturation and activation of the dendritic cells (DCs). Polyphenols interfering with the Toll-Like Receptors (TLRs) induce the activation of the mitogen-activated protein kinase (MAPK), Akt, and nuclear factor- (NF-) κB pathways, leading to DC activation and further elicitation of the immune response (Ding et al., 2018). Phenolic acid like ferulic acid inhibits or reduces the secretion of proinflammatory cytokines like TNF-α, IL-1β, and IL-6 (Ding et al., 2018). Flavonoids such as quercetin interfere in the inflammatory pathways by inhibiting lipoxygenase and the other inflammatory mediators. Quercetin is an inhibitor of mast cell activation as it inhibits Ca2+ influx, histamine, leukotrienes, release of prostaglandins and protein kinase activation (Tekin and Marotta, 2018).

8.3.1.2 Effect of polyphenols on adaptive immune system

Polyphenols are involved in anti-allergic responses by inhibiting the formation of antigen-specific IgE antibody either by interfering in the formation of the allergen-IgE complex or affecting the binding of this complex to its receptors (FceRI) on mast cells and basophils (Ding et al., 2018). Tannins can be used to reduce food allergies as it may be associated with an increase in the proportion of $\gamma\delta$TCR T cells in intestinal intraepithelial lymphocytes (Ding et al., 2018). Polyphenols are reported to alter some signaling pathways mediated by tyrosine and serine-threonine protein kinases, which have been known to be involved in the B-lymphocyte activation and T-cell proliferation (Tekin and Marotta, 2018). Some of the polyphenols are known to regulate the activity of

Regulatory T cells (Tregs) which help in maintaining the immune responses mediated by T-cells (Ding et al., 2018).

8.3.2 Immunobiotics and their interactions with polyphenols

Probiotics is a blanket term for all the beneficial microbiota with diverse functionalities in a human host. There are some essential microbes that can beneficially modulate and elevate the functions of the mucosal immune system. They are termed "immunobiotics." Remarkable evidence was found with respect to interactions between dietary polyphenols and these immunobiotics. These interactions have been shown to be useful in treating and preventing many diseases and also in augmenting normal good health.

Polyphenols exert their beneficial effects via increasing the survival, proliferation, and metabolism of the probiotic families of bacteria such as *Lactobacillaceae* and *Bifidobacteriaceae*, and reducing the growth and survival of bacteria such as *Escherichia coli*, *Clostridium perfringens*, and *Helicobacter pylori*, which are pathogenic to the host (Plamada and Vodnar, 2021).

There are numerous instances of specific polyphenols enhancing the growth of the gut flora. Polyphenols present in red grapes or in their seeds have been observed to support the growth of *Lactobacillus reuteri*, *Lactobacillus acidophilus*, *Clostridiales* and *Ruminococcus* (Pozuelo et al., 2012). Polyphenols in carrots can enhance the growth of *L. rhamnosus* and *Bacteroides* (Ben-Othman et al., 2020). Polyphenolic compounds like hydroxycinnamic acid, ferulic acid, and coumaric acid present in cereals support and enhance the growth of *Lactobacillus* spp. and *Bidifobacteria* spp. (Dias et al., 2021). Polyphenolic compounds in beans in particular, hydroxycinnamic acids, hydroxybenzoic acids and ferulic acids have been found to enhance the growth of *Lactobacillus* spp., namely, *L. acidophilus*, *L. casei* and *L. delbrueckii* (Bertelli et al., 2021. Anthocyanins have been proved to trigger the proliferation of many bacterial species, namely, *Lactobacillus-Enterococcus* spp. and *Bifidobacterium* spp. (Plamada and Vodnar, 2021). Stilbenes found in red grapes, red wine, teaberries, and peanuts support the flourishing of *Lactobacilli* species and *Bifidobacteria* involved in butyrate production – in particular, *F. prasnitzii* (Plamada and Vodnar, 2021. Tannins are defensive in nature, providing protection against harmful pathogens like *S. aureus*, *C. botulinum*, *Penicillium* spp. and HIV. In recent findings, hydrolysable tannins are found to be defensive against *H. pylori* (Plamada and Vodnar, 2021). Polyphenols in berries can enhance the flourishing of many gut microbial species, in particular, *Akkermansia*, *Eubacterium*, *Bacteroides*, *Bifidobacterium*, *Lactobacillus*, and subside the levels of harmful pathogens, including *Staphylococcus*, *Salmonella*, *Pseudomonas*, and *Bacillus* (Plamada and Vodnar 2021).

8.3.3 Polyphenols as immune boosters in viral infections

In addition to their antimicrobial, anti-cancer, anti-inflammatory, and antioxidant activities, polyphenols have been studied to improve the host's immunity by exercising their anti-viral properties and also boosting the immunity of the host. Through their various anti-viral mechanisms (Figure 8.7), polyphenols have promising results in yet another domain, for example, gallic acid, quercetin, and epigallocatechin (EGC) are found to curtail the mechanism of multiplication of viruses. Black elderberry, *Sambucus nigra*, has proven anti-influenza activity, and so forth (Chojnacka et al., 2021).

p-Coumaric acid, caffeic acid, gallic acid and chlorogenic acid from mulberry, *Morus alba*, exhibit antiviral activity at the earliest stage, either by impeding the attachment of viruses through their surface proteins to their specific receptors on the host cells or as a result of the internalization of cell surface receptors that facilitate the adsorption of viruses to the host cells. The abovementioned polyphenols are effective against IAV and IBV varieties (Chojnacka et al., 2021). Theaflavins from black tea inhibit viral attachment to the respective host cells or to their specific

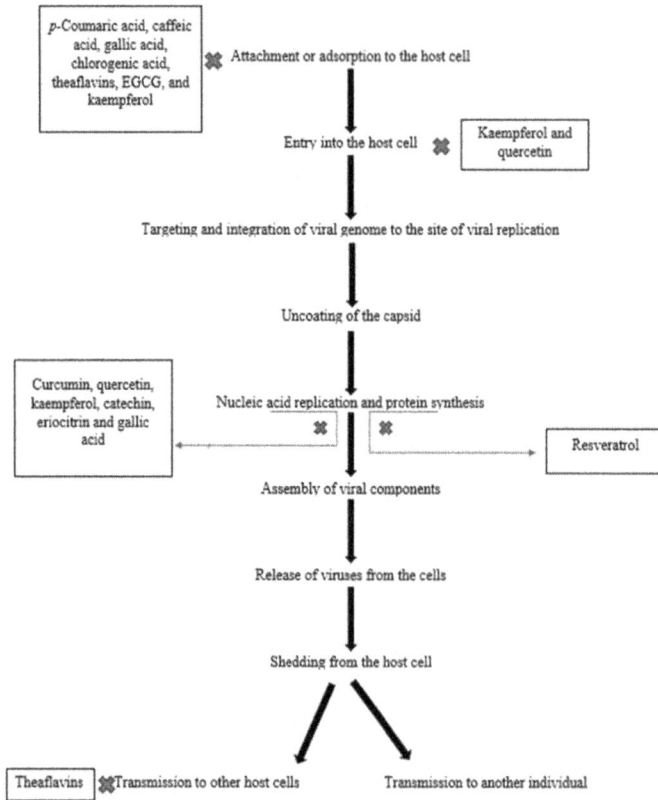

FIGURE 8.7   Inhibitory action of different polyphenols in antiviral mechanism.

receptors thus, ceasing their ability to infect the nearby cells. They are effective against HCV (Chojnacka et al., 2021). Curcumin, quercetin, kaempferol, catechin, and gallic acid from *Morus alba* and *Euphorbia cooperi* inhibit viral replication or viral genome synthesis. These are the most effective against HSV-1. EGCG from green tea and cranberry blocks or alters the antigenic determinants or surface ligands present on the surface of the viruses, thus reducing the extent of adsorption of viruses to the susceptible cell. EGCG is the most effective against viruses like Rotavirus strains and Coliphage virus strains. Quercetin and eriocitrin have anti-viral properties against the Dengue virus. The most common mechanism is by inhibiting the replication of viruses, which is brought about by the breakdown of the protein complex NS2B-NS3 (Chojnacka et al., 2021). Kaempferol and quercetin have anti-viral effects against measles virus. They deactivate the virions, inhibiting the early-onset stages of viral replication, namely adsorption and host-cell penetration (Chojnacka et al., 2021). Resveratrol is effective against the Epstein-Barr virus as it decreases the amount of reactive oxygen species (ROS), blocks the synthesis of proteins required for viral assembly and replication, and inhibits the activation of transcription factors, which is in turn induced by the virus, ultimately obstructing the viral replication process. (Chojnacka et al., 2021).

The most recent and crucial findings that have emerged as a result of the coronavirus pandemic are the antiviral effects of polyphenols against the SARS-CoV-2 virus. Polyphenols (quercetin, curcumin, EGCG, etc.) (Chojnacka et al., 2021) bring about the antiviral effects by either one or a combination of the mechanisms such as arresting the process of replication of the infecting virus, disintegration or breakdown of the viral attachment surface protein or the spike (S) protein and suppression of the effect of the SARS-CoV-2 protease enzyme (Galanakis et al., 2020).

8.4   IMPLICATIONS OF FOOD SAFETY SYSTEMS AND POLICY

Although there are plenty of findings and literature that strongly support the idea of dietary polyphenols being potent bioactive compounds, these studies are limited *to in vitro* assays and lack sufficient clinical trials in human models. Also, their extensive metabolism in human systems and low bioavailability are major disadvantages. Such implications related to their safe utilization is worth discussing.

8.4.1   Food processing

The desire for increased food production to meet the demands of an expanding global population has compelled food manufacturers to investigate techniques to extend the shelf life of consumable goods. Moreover, unlike fruits that are consumed raw, vegetables and beverages undergo cooking processes like heating, stir-frying, boiling, and so forth. These activities have been shown to reduce the antioxidant potency of certain heat-labile compounds, like carotenoids or ascorbic acid due to the release and destruction of antioxidants resulting in the production of redox-active metabolites (Wachtel-Galor et al., 2008; Şengül et al., 2014). It is well documented that food cooked in the microwave contained the least antioxidant content as compared to boiled and steam-cooked food. Furthermore, longer duration of cooking deteriorated the nutritional value drastically irrespective of the cooking style (Wachtel-Galor et al., 2008).

Polyphenols, however have shown unusual stability at higher temperatures (Faller and Fialho, 2009). Hence, the addition of polyphenols to perishable foods like beers and meat deploys antioxidant mechanisms of scavenging free radicals, thereby relinquishing themselves. Thus, they aid in the protection of other compounds from being oxidized and increase the shelf life of food substances (Cory et al., 2018). Reportedly, peeling off the exocarp of fruits or vegetables devoid them of maximum polyphenolic content. Therefore, maceration is performed in the production of red wines, where the skin of grapes (having high phenolic content) is used to leach out tannins, catechins, and other color and flavor-giving compounds. This increases the phenolic content in red wines by ten-fold as compared to white wines (Miglio et al., 2008; Cory et al., 2018). Therefore, the use of polyphenols is increasing in the food industry day by day to meet the demands of the population for food having higher nutritional values.

8.4.2   Marketing and regulation

The recent advancements established in health benefits associated with polyphenols and awareness awakened in people for consumption of foods with nutritional value have defined many strategies to merchandise polyphenols at a global level.

In the global market, polyphenols are currently merchandised as the "foods that have a potentially positive effect on health beyond basic nutrition" (Crowe and Francis, 2013). It has been reported that the market value of dietary, cosmetic, and pharmaceutical products inclusive of polyphenols has exceeded seven hundred million USD in 2015 and would even cross a mark of approximately one billion USD in the near future (Cory et al. 2018). Researchers are working on providing better evaluation systems to validate the use of polyphenolic products for a longer period and have developed new assays to determine their content in different food products.

As such, no regulatory guidelines are made available regarding their daily consumption values specific for each type of dietary polyphenol. The formulation of toxicity range and safety value range for each phenolic compound remains undetermined due to lack of study in animal models, and regular clinical trials. Also, during testing, under-provisioned standardization methods and expensive analytical techniques serve as a stumbling block (Cory et al., 2018).

So far, for many phenolic compounds, only *in vitro* studies have backed their various health-promoting effects. The United States Food and Drug Administration (FDA) has provided a mandate of permitting the antioxidant compounds and nutrients to be merchandised on the claim of health only if they have well-formulated Recommended Daily Intake (RDI) values labeled on the product (Vinson and Motisi, 2015). Polyphenols lack such standardized formulations and, thus, many proposed phenolic supplementations are in the course of experimentation only. Reports have stated that green tea that is exclusively marketed as a naturally derived weight-loss product, results in liver damage (hepatotoxicity) if consumed for a longer period due to oxidative stress imposed by EGCG metabolites on liver cells (Navarro et al., 2014; Mazzanti et al., 2009). Improvisations in standardizing the phenolic contents precisely is a prerequisite in enhancing the usage of polyphenols and paving a natural way for curing ailments.

8.4.3 FORTIFICATION AND SUPPLEMENTATION

In the food industry, the processes of fortification and supplementation have become extensively prominent due to the advancements in technology, and food-processing. "fortification" of food refers to the steps involved in improving food quality in terms of its nutritional value as well as its shelf life. Fortified food needs to have stable interactions and overall structural stability of, and between, the enriched components. Thus, it is a relatively difficult process of ensuring the proper physical and biochemical properties of the constituents. Polyphenol-fortified food is focused upon enhancing the polyphenolic content in the food. The increased polyphenol content is directed towards improved bioactive properties of polyphenols. On the other hand, "supplementation" refers to the provision of specific nutrients to the population with the aim of eradicating their nutritional deficiencies. Polyphenolic supplements are specific with respect to their DRI, and their concentrations as individual supplements is higher than in polyphenol-fortified food. Supplementation of dietary polyphenols is relatively an easier process than fortification, as supplementation does not include considering the stability and interactions of all individual food components and the food as a whole. Rather, supplementation deals with the addition of a specific nutrient. Polyphenols as supplements reach directly to the targeted population, whereas polyphenol-fortified foods supply the nutrient mixture. Generally, polyphenol fortification is carried out in dairy products (cheese, yogurt, milk, etc.), bakery products (cake and biscuits), and cocoa products (chocolate) (Sik et al., 2022). This is so because these food products are consumed by the population on a regular basis.

Although fortification and supplementation by polyphenols has been aimed to deliver maximum beneficial polyphenol concentrations, there are few shortfalls of the same. The natural synergistic effects of polyphenols might become affected in supplements or in fortified food. The various interactions of different polyphenols present in the normally consumed food produce many beneficial results, which are still not fully contemplated. These results might not be replicated by the fortified food or the supplements (Cory et al., 2018). Moreover, the fortified food produced is denser with respect to the energy content rather than the nutritional content. This can affect the efficiency of the polyphenols in the fortified food or supplements to reduce obesity. Proper dosage in humans is yet to be confirmed as higher dose administration of the supplements will cause toxicity in recipients. As human bodies react differently to the different pharmaceuticals, the administration of polyphenolic supplements and the fortified consumption will derive diverse responses from the population. In addition, the interactions of these polyphenols, whose concentrations are more than in the normal diet, with the other prescribed drugs, need to be explored. The diversity of such responses and their possible explanations remain uncertain to date (Cory et al., 2018).

8.5 CONCLUSION AND FUTURE PROSPECTIVE

Recent endeavors in establishing well-defined and targeted therapeutics to conquer life-threatening diseases with minimal adversity has been a constant revolution. In such a triviality, polyphenolics have established the most promising results and are paving the way to novel remedials in treating cancers. Polyphenolics, being produced as the secondary metabolites of plants, have an extensive inclusiveness in supplementary diets – alongside, the antimicrobial properties and immunological interactions briefly studied here. Owing to their immense structural diversity, different classes of phenolics have developed different approaches to combat different microbial and viral infections. Since phenolics are a natural by-product, the occurrence of adverse effects post-therapy as observed in chemically induced therapeutics are subdued. Such distinctive classes of bioactive compounds, which bear immense potential in redefining the current therapeutic approaches, have captivated the attention of many who have the urge to research polyphenols to acquire a deeper insight into them.

Health consciousness has compelled many people to find new alternatives to chemically preserved food. This has unleashed the antioxidant properties of polyphenols. Much literature has supported the idea of polyphenols being used as natural preservatives since chemical preservatives pose a potential threat for development of cancer in the near future. But polyphenols' dosage optimizations and *in vivo* testing are yet to be performed.

Polyphenols combined with antibiotics are capable of potentiating the effectiveness at lower doses of antibiotics as discussed here under the term synergism. But a very limited number of such synergistic combinations have been determined, and still only a low number of *in vivo* tests have been done. Many researchers have limited their studies to *in vitro* testing, and results might differ in animal models and clinical trials. At present, there exist many plant species (few even exotic) or their consumables whose polyphenol profiles are yet to be generated and understood. For the qualitative analysis and quantitation of these polyphenols, there are many detection techniques, like NMR, HPLC, UV-Vis, UPLC, LC-MS and TLC, and methods for isolation like hydrolysis, extraction, and so forth, which can be utilized (Chojnacka et al., 2021).

The literature studies have strongly supported the correlation of gut microbiota and polyphenols in boosting immunity and overall health. However, the interactions taking place are not yet clear and, as a matter of fact, since different sets of gut microbiota present at different ages, generalization has to be worked upon. The effectiveness of specific polyphenols on interacting with specific microbes should possibly modulate the immunity and overall health positively. Advanced *omics* technologies like genomics, metagenomics, transgenomics, metabolomics, or proteomics could provide a better understanding of the action of polyphenols in living organisms and their resulting metabolites (Plamada and Vodnar, 2021).

REFERENCES

Abia, R., and Fry, S.C. 2001. Degradation and metabolism of Of14C-labelled proanthocyanidins from carob (*Ceratonia siliqua*) pods in the gastrointestinal tract of the rat. *J Sci Food Agric* 81: 1156–1165.

Abdul Manap, A.S., Wei Tan, A.C., Leong, W.H., Yin Chia, A.Y., Vijayabalan, S., Arya, A., Wong, E.H., Rizwan, F., Bindal, U., Koshy, S., and Madhavan, P. 2019. Synergistic effects of curcumin and piperine as potent acetylcholine and amyloidogenic inhibitors with significant neuroprotective activity in SH-SY5Y cells via computational molecular modeling and in-vitro assay. *Front Aging Neurosci* 11: 206.

Ahn, Y.J., Kawamura, T., Kim, Mujo, Yamamoto, T., and Mitsuoka, T. 1991. Tea polyphenols: Selective growth inhibitors of Clostridium spp. *Agric Biol Chem* 55: 1425–1426.

Alkayali, A. 2009. Ellagic acid food supplement prepared from pomegranate seed (US Patent US20060280819A1).

Ansari, M., Anurag, A., Fatima, Z., and Hameed, S. 2013. Natural phenolic compounds: A potential antifungal agent. *MicrobPathogStrateg Combat them SciTechnolEduc* 1: 1189–1195.

Aoki, T., Akashi T., and Ayabe, Si. 2000. Flavonoids of leguminous plants: Structure, biological activity, and biosynthesis. *J Plant Res* 113: 475–488.

Aura, A.M. 2008. Microbial metabolism of dietary phenolic compounds in the colon. *Phytochem Rev* 7: 407–429.

Batista, A.G., da Silva-Maia, J.K., and Marostica, M.R. 2021. Generation and alterations of bioactive organosulfur and phenolic compounds. In: Delia, B.R., Jaime, A.F., eds., *Chemical Changes during Processing and Storage of Foods*. Elsevier; p. 537–577.

Ben-Othman, S., Joudu, I., and Bhat, R. 2020. Bioactives from Agri-food wastes: Present insights and future challenges. *Molecules* 25: 510–544.

Bertelli, A., Biagi, M., Corsini, M., Baini, G., Cappellucci, G., and Miraldi, E. 2021. Polyphenols: From theory to practice. *Foods* 10: 2595.

Cerda, B., Espin, J.C., Parra, S., Martinez, P., and Tomas-Barberan, F.A. 2004. The potent in vitro antioxidant ellagitannins from pomegranate juice are metabolised into bioavailable but poor antioxidant hydroxy-6H-Dibenzopyran-6- one derivatives by the colonic microflora of healthy humans. *Eur J Nutr* 43: 205–220.

Chacko, S.M., Thambi, P.T., Kuttan, R., and Nishigaki, I. 2010. Beneficial effects of green tea: A literature review. *Chin Med* 5:13.

Chojnacka, K., Skrzypczak, D., Izydorczyk, G., Mikula, K., Szopa, D., and Witek-Krowiak, A. 2021. Antiviral properties of polyphenols from plants. *Foods* 10: 2277.

Clavel, T., Borrmann, D., Braune, A., Doré, J., and Blaut, M. 2006. Occurrence and activity of human intestinal bacteria involved in the conversion of dietary lignans. *Anaerobe* 12: 140–147.

Clavel, T., Henderson, G., Alpert, C.A., Philippe, C., Rigottier-Gois, L., Dore, J., and Blaut, M. 2005. Intestinal bacterial communities that produce active estrogen-like compounds enterodiol and enterolactone in humans. *Appl Environ Microbiol* 71: 6077–6085.

Clavel, T., Henderson, G., Engst, W., Dore, J., and Blaut, M. 2006. Phylogeny of human intestinal bacteria that activate the dietary lignanssecoisolariciresinoldiglucoside. *FEMS Microbiol* 55: 471–478.

Clavel, T., Lippman, R., Gavini, F., Doré, J., and Blaut, M. (2007). Clostridium saccharogumia sp. nov. and Lactonifactor longoviformis gen. nov., sp. nov., two novel human faecal bacteria involved in the conversion of the dietary phytoestrogen secoisolariciresinol diglucoside. *Syst. Appl. Microbiol.*, 30: 1, 16–26. https://doi.org/10.1016/j.syapm.2006.02.003.

Cooke, D., Fitzpatrick, B., O'Kennedy, R., McCormack, T., and Egan, D. 1997. Coumarin biochemical profile and recent developments. *John W & S* 3: 311–322.

Cory, H., Passarelli, S., Szeto, J., Tamez, M., and Mattei, J. 2018. The role of polyphenols in human health and food systems: A mini-review. *Front Nutr* 87: 1–9.

Cote, J., Caillet, S., Doyon, G., Sylvain, J.F., and Lacroix, M. 2010. Bioactive compounds in cranberries and their biological properties. *Crit Rev Food SciNutr* 50: 666–679.

Crowe, K.M., and Francis, C. 2013. Position of the Academy of Nutrition and Dietetics: Functional foods. *J AcadNutr Diet* 113: 1096–1103.

Daglia, M. 2012. Polyphenols as antimicrobial agents. *CurrOpinBiotechnol* 23: 174–181.

Dias, R., Pereira, C.B., Perez-Gregorio, R., Mateus, N., and Freitas, V. 2021a. Recent advances on dietary polyphenol's potential roles in celiac disease. *Trends Food Sci Technol* 107: 213–225.

Diker, K.S., Akan, M., Hascelik, G., and Yurdakok, M. 1991. The bactericidal activity of tea against *Campylobacter jejuni* and *Campylobacter coli. Lett ApplMicrobiol* 12: 34–35.

Ding, S., Jiang, H., and Fang, J. 2018. Regulation of immune function by polyphenols. *JImmunol Res* 2018: 1264074.

Eckert, P., Heinen, W., and Knaudt, C. 2007. Grapeseed, cold-pressed grape oil, crushed grape and grape flour (US Patent US7226627B1).

Faller, A.L.K., and Fialho, E. 2009. The antioxidant capacity and polyphenol content of organic and conventional retail vegetables after domestic cooking. *Int Food Res J* 42: 210–215.

Gadkari, P.V., and Balaraman, M. 2015. Catechins: Sources, extraction, and encapsulation: A review. *Food Bioprod Process* 93: 122–138.

Galanakis, C.M., Aldawoud, T.M.S., Rizou, M., Rowan, N.J., and Ibrahim, S.A. 2020. Food ingredients and active compounds against the coronavirus disease (COVID-19) pandemic: A comprehensive review. *Foods* 9: 1701.

Gescher, K., Hensel, A., Hafezi, W., Derksen, A., and Kühn, J. 2011. Oligomericproanthocyanidins from *Rumexacetosa L.* inhibit the attachment of herpes simplex virus Type-1. *Antiviral Res* 89: 9–18.

Gharras, H.E.l. 2009. Polyphenols: Food sources, properties and applications – a review. *Int. J. Food Sci. Technol* 12: 2512–2518.

Gollucke, B.P., Correa Peres, R., Ribeiro, D.A., Aguiar, O. 2018. Polyphenols as supplements in foods and beverages: Recent discoveries and health benefits, an update. In: Ronald, R.W., Victor, R.P., Sherma, Z., eds. *Polyphenols: Mechanisms of Action in Human Health and Disease.* Academic Press; pp. 11–18.

Grosso, G., Stepaniak, U., Topor-Mądry, R., Szafraniec, K., and Pająk, A. 2014. Estimated dietary intake and major food sources of polyphenols in the polish arm of the HAPIEE study. *Nutr.* 30: 1398–1403.

Han, X., Shen, T., and Lou, H. 2007. Dietary Polyphenols and their Biological Significance. *Int J Mol Sci* 8: 950–988.

Horn-Ross, P. L., John, E. M., Canchola, A. J., Stewart, S. L., & Lee, M. M. (2003). Phytoestrogen intake and endometrial cancer risk. *J. Natl. Cancer Inst*, 95(15), 1158–1164. https://doi.org/10.1093/jnci/djg015.

Howard, A.N., Nigdikar, Shaja V., Rajput-Williams, J., and Williams, N.R. 2003. Food supplements containing polyphenols (US Patent US6642277B1).

Hussain, T., Tan, B., Yin, Y., Blachier, F., Tossou, M. C., and Rahu, N. 2016. Oxidative stress and inflammation: What polyphenols can do for us? *Oxid Med Cell Longevity2016*: 7432797.

Hu, S., Li, X., Xu, R., Ye, L., Kong, H., Zeng, X., Wang, H., and Xie, W. 2016. The synergistic effect of resveratrol in combination with cisplatin on apoptosis via modulating autophagy in A549 cells. *ActaBiochimBiophys Sin* 48: 528–535.

Isemura, M. 2019. Catechin in human health and disease. *Molecules* 24: 528.

Ito, H., Miyake, M., Nishitani, E., Miyashita, K., Yoshimura, M., Yoshida, T., Takasaki, M., Konoshima, T., Kozuka, M., and Hatano, T. 2007. Cowaniin, a C-glucosidicellagitannin dimer linked through Catechin from Cowaniamexicana. *Chem Pharm Bull* 55: 492–494.

Jang, M., Cai, L., Udeani, G.O., Slowing, K.V., Thomas, C.F., Beecher, C.W.W., Fong, H.H.S., Farnsworth, N.R., Kinghorn, A.D., Mehta, R.G., Moon, R.C., and Pezzuto, J.M. 1997. Cancer chemopreventive activity of resveratrol, a natural product derived from grapes. *Science* 275: 218–220.

Karawajczyk, A., Drgan, V., Medic, N., Oboh, G., Passamonti, S., and Novic, M. 2007. Properties of flavonoids influencing the binding to bilitranslocase investigated by neural network modelling. *BiochemPharmacol* 73:308–320.

Khanbabaee, K., and Van Ree, T. 2001. Tannins: Classification and definition. *Nat Prod Rep* 18: 641–649.

Kumar, N., and Goel, N. 2019. Phenolic acids: Natural versatile molecules with promising therapeutic applications. *Biotechnology Reports* 24: e00370.

KupeliAkkol, E., Genç, Y., Karpuz, B., Sobarzo-Sánchez, E., and Capasso, R. 2020. Coumarins and coumarin-related compounds in pharmacotherapy of cancer. *Cancers* 12: 1959.

Leon-Díaz, R., Meckes, M., Said-Fernández, S., Molina-Salinas, G.M., Vargas-Villarreal, J., Torres, J., Luna-Herrera, J., and Jimenez-Arellanes, A. 2010. Antimycobacterial neolignans isolated from Aristolochia Taliscana. *Mem Inst Oswaldo Cruz* 105: 1, 45–51.

Levrat, M.-A., Texier, O., Régerat, F., Demigné, C., and Rémésy, C. 1993. Comparison of the effects of condensed tannin and pectin on cecal fermentations and lipid metabolism in the rat. *Nutr Res* 13: 427–433.

Lewandowska, U., Szewczyk, K., Hrabec, E., Janecka, A., and Gorlach, S. 2013. Overview of metabolism and bioavailability enhancement of polyphenols. *Agri Food Chem* 61: 12183–12199.

Liu, Z.H., Luo, Z.W., Li, D.W., Wang, D.M., and Ji, X. 2018. Synergistic effects and related bioactive mechanisms of *Potentillafruticosa* Linn. leaves combined with green tea polyphenols studied with microbial test system (MTS). *Natur Prod Res* 32: 1287–1290.

Lv, Q., Long, J., Gong, Z., Nong, K., Liang, X., Qin, T., Huang, W., and Yang, L. 2021. Current state of knowledge on the antioxidant effects and mechanisms of action of polyphenolic compounds. *Natur Prod Comm* 16: 1–13.

Makarewicz, M., Drożdż, I., Tarko, T., and Duda-Chodak, A. 2021a. The interactions between polyphenols and microorganisms, especially gut microbiota. *Antioxidants* 10: 188.

Manach, C., Scalbert, A., Morand, C., Remesy, C., and Jimenez, L. 2004. Polyphenols: Food sources and bio-availability. *American ClinNutr* 79: 727–747.

Marin, L., Miguelez, E.M., Villar, C.J., and Lombo, F. 2015. Bioavailability of dietary polyphenols and gut microbiota metabolism: Antimicrobial properties. *BioMed Res Intern* 2015: 905215.

Matos, M.J., Santana, L., Uriarte, E., Abreu, O.A., Molina, E., and Yordi, E.G. 2015. Coumarins: An important class of phytochemicals. In: Rao, A.V., Rao, L.G., (eds.) Phytochemicals – Isolation, characterization and role in human health. InTech; p. 113–140.

Mazzanti, G., Menniti-Ippolito, F., Moro, P.A., Cassetti, F., Raschetti, R., Santuccio, C., and Mastrangelo, S. 2009. Hepatotoxicity from green tea: A review of the literature and two unpublished cases. *EurClinPharmacol* 65: 331–341.

Mennen, L.I., Walker, R., Bennetau-Pelissero, C., and Scalbert, A. 2005. Risks and safety of polyphenol consumption. *AmerClinNutr* 81: 326S–329S.

Miglio, C., Chiavaro, E., Visconti, A., Fogliano, V., and Pellegrini, N. 2008. Effects of different cooking methods on nutritional and physicochemical characteristics of selected vegetables. *Agri Food Chem* 56: 139–147.

Mitra, S., Tareq, A.M., Das, R., Emran, T.B., Nainu, F., Chakraborty, A.J., Ahmad, I., Tallei, T.E., Idris, A.M., and Simal-Gandara, J. 2022. Polyphenols: A first evidence in the synergism and bioactivities. *Food Reviews Intern* 1: 1–23.

Navarro, V.J., Barnhart, H., Bonkovsky, H.L., Davern, T., Fontana, R.J., Grant, L., Reddy, K.R., Seeff, L.B., Serrano, J., Sherker, A.H., Stolz, A., Talwalkar, J., Vega, M., and Vuppalanchi, R. 2014. Liver injury from herbals and dietary supplements in the US Drug-Induced Liver Injury Network. *Hepatol* 60: 1399–1408.

Panche, A.N., Diwan, A.D., and Chandra, S.R. 2016. Flavonoids: An overview. *NutrSci* 5: e4.

Pandey, K.B., and Rizvi, S.I. 2009. Plant polyphenols as dietary antioxidants in human health and disease. *OxiMed Cell Longevity* 2: 270–278.

Passamonti, S., Vanzo, A., Vrhovsek, U., Terdoslavich, M., Cocolo, A., Decorti, G., and Mattivi, F. 2005a. Hepatic uptake of grape anthocyanins and the role of bilitranslocase. *Food Res Intern* 38: 953–960.

Peppercorn, M.A., and Goldman, P. 1971. Caffeic acid metabolism by bacteria of the human gastrointestinal tract. *Bacteriol* 108: 996–1000.

Perez-Jimenez, J., Neveu, V., Vos, F., and Scalbert, A. 2010. Identification of the 100 richest dietary sources of polyphenols: An application of the phenol-explorer database. *EurClinNutr* 64: 1–23

Perron, N.R., and Brumaghim, J.L. 2009. A review of the antioxidant mechanisms of polyphenol compounds related to iron binding. *Cell BiochemBiophy* 53: 75–100.

Plamada, D., and Vodnar, D.C. 2021. Polyphenols – Gut microbiota interrelationship: A transition to a new generation of prebiotics. *Nutrients* 14: 137.

Pozuelo, M.J., Agis-Torres, A., Hervert-Hernández, D., Elvira Lopez-Oliva, M., Munoz-Martínez, E., Rotger, R., and Goni, I. 2012. Grape antioxidant dietary fiber stimulates Lactobacillus growth in rat cecum. *Food Sci* 77: 59–62.

Rempe, C.S., Burris, K.P., Lenaghan, S.C., and Stewart, C.N. 2017. The potential of systems biology to discover antibacterial mechanisms of plant phenolics. *Fronts Microbiol* 8: 422.

Scalbert, A., and Williamson, G. 2000. Dietary intake and bioavailability of polyphenols. *Nutrition* 130: 2073S–2085S.

Scalbert, A., Morand, C., Manach, C., and Remesy, C. 2002. Absorption and metabolism of polyphenols in the gut and impact on health. *Biomed Pharmacother* 56: 276–282.

Schalekamp, T., and de Boer, A. 2010. Pharmacogenetics of oral anticoagulant therapy. *CurrPharmaceut Design* 16: 187–203.

Schneider, H., and Blaut, M. 2000. Anaerobic degradation of flavonoids by *Eubacterium ramulus*. *Arch Microbiol* 173: 71–75.

Schneider, H., Schwiertz, A., Collins, M.D., and Blaut, M. 1999. Anaerobic transformation of quercetin-3-glucoside by bacteria from the human intestinal tract. *Arch Microbiol* 171: 81–91.

Şengul, M., Yildiz, H., and Kavaz, A. 2014. The effect of cooking on total polyphenolic content and antioxidant activity of selected vegetables. *Food Prop Intern* 17: 481–490.

Sik, B., Szekelyhidi, R., Lakatos, E., Kapcsandi, V., and Ajtony, Z. 2022. Analytical procedures for determination of phenolics active herbal ingredients in fortified functional foods: An overview. *Eur Food Res Technol* 248: 329–344.

Simonetti, G., Brasili, E., and Pasqua, G. 2020. Antifungal activity of phenolic and polyphenolic compounds from different matrices of *Vitis vinifera L.* against human pathogens. *Molecules* 25: 16–24.

Singla, R.K., Dubey, A.K., Garg, A., Sharma, R.K., Fiorino, M., Ameen, S.M., Haddad, M.A., and Al-Hiary, M. 2019. Natural Polyphenols: Chemical, classification, definition of classes, subcategories, and structures. *AOAC Internat* 102: 1397–1400.

Sirerol, J.A., Rodríguez, M.L., Mena, S., Asensi, M.A., Estrela, J.M., and Ortega, A.L. 2016. Role of natural stilbenes in the prevention of cancer. *Oxid Med Cell Longevity2016*: 3128951.

Smith, G.J., and Markham, K.R. 1998. Tautomerism of flavonol glucosides: Relevance to plant UV protection and flower color. *PhotochemPhotobiol A* 118: 99–105.

Sokolov, A.N., Pavlova, M.A., Klosterhalfen, S., and Enck, P. 2013. Chocolate and the brain: Neurobiological impact of cocoa flavanols on cognition and behavior. *NeurosciBiobehav Rev* 37: 2445–2453.

Sreenivasan, S., and Krishnakumar, S. 2015. Synergistic efect of curcumin in combination with anticancer agents in human retinoblastoma cancer cell Lines. *Curr Eye Res* 40: 1153–1165.

Tallarida, R.J. 2011. Quantitative methods for assessing drug synergism. *Gen Canc* 2: 1003–1008.

Tekin, I.O., and Marotta, F. 2018. Polyphenols and immune system. *Preven treat human dis 2:* 263–276.

Teodoro, G.R., Ellepola, K., Seneviratne, C.J., and Koga-Ito, C.Y. 2015. Potential use of phenolic acids as anti-candida agents: A review. *Front Microbiol* 6:1420.

Vardhan, P.V., and Shukla, L.I. 2017. Gamma irradiation of medicinally important plants and the enhancement of secondary metabolite production. *RadiatBiol Intern* 93: 967–979.

Vilhelmova, N., Jacquet, R., Quideau, S., Stoyanova, A., and Galabov, A.S. 2011. Three-dimensional analysis of combination effect of ellagitannins and acyclovir on herpes simplex virus Types 1 and 2. *AntivirRes* 89: 174–181.

Vinson, J.A., and Motisi, M.J. 2015. Polyphenol antioxidants in commercial chocolate bars: Is the label accurate? *Functional Foods* 12: 526–529.

Wang, L.Q., Meselhy, M.R., Li, Y., Qin, G.W., and Hattori, M. 2000. Human intestinal bacteria capable of transforming secoisolariciresinoldiglucoside to mammalian lignans, enterodiol and enterolactone. *ChemPharmaceutBulletein* 48: 1606–1610.

Wachtel-Galor, S., Wong, K.W., and Benzie, I.F.F. 2008. The effect of cooking on Brassica vegetables. *Food Chem* 110: 706–710.

Walle, T., Browning, A.M., Steed, L.L., Reed, S.G., and Walle, U.K. 2005. Flavonoid glucosides are hydrolyzed and thus activated in the oral cavity in humans. *Nutrition* 135: 48–52.

Winter, J., Popoff, M.R., Grimont, P., and Bokkenheuser, V.D. 1991. *Clostridium orbiscindens* Sp. Nov., a human intestinal bacterium capable of cleaving the flavonoid C-Ring. *Systematic Bacteriol. Intern* 41: 355–357.

9 Dietary Polyphenols in Viral Infections

Vishal K. Singh, Jayati Dwivedi, Aditya K. Yadav, and Ramendra K. Singh
E-mail: vishalkumarsingh922@gmail.com; jayati.dwivedi05@gmail.com; adityakumarantu@gmail.com; singhramendrak@gmail.com

CONTENTS

9.1   INTRODUCTION

Viruses with RNA as their genetic material are capable of quickly adapting and exploiting various organisms because of their high mutation rates. In addition to viral genetic variation, which includes mutation, recombination, and reassortment, several environmental factors, including social, ecological, behavioral influences, and health care also play significant roles. We live in an era of swiftly changing local environments and global landscapes, therefore, it comes as no surprise that RNA viruses are the main causative agent for some prominent emerging or re-emerging diseases (Domingo et al., 1997). Abiotic factors, coupled with the significant increase in the human population and urbanization in several developing countries, have notably broadened the number of sampling events that lead to the fitness of RNA virus variants in varied human cell backgrounds (Mandary et al., 2019).

DOI: 10.1201/9781003251538-9

Dietary items derived from plants are gaining ample attention because of their safety and therapeutic potentials. Several traditional folk medicines derived from plants are rich in polyphenols and have been used from the earliest times to cure viral effects, disorders of the bloodstream (blood pressure), stomach disorders, dressing of burns, antiseptic action, and inflammation. A plenteous number of polyphenols are present as antioxidants in the diet and are extensively found in fruits, cereals, dry legumes, vegetables, and chocolate, and in beverages such as tea, coffee, or wine. Polyphenols are classified into two general classes; the first being flavonoids and the other is phenolic acid. Later, these flavonoids were further divided into flavones, flavanols, isoflavones, flavononse, flavonols, and classification of phenolic acids is divided into hydroxybenzoic and hydroxycinnamic acids(Abbas et al., 2017).

Modern medicine is also increasingly turning its attention to folk medicines as a source of novel drug research and to investigate the basis of old remedies, both chemically and biologically. Increasingly, polyphenols are becoming the topic of medical exploration. These compounds have been reported to be active against various viruses like influenza, rabies, herpes, hepatitis A and B, polio, rotavirus, dengue virus, and so forth. Many natural plant-derived polyphenols whose intake can be administered in the human diet and demonstrate their potential antiviral activity or other beneficial properties that may counteract various severe viruses like HIV (Human Immunodeficiency Virus) or SARS-CoV.

With the onset of COVID-19 in the year 2019, the antiviral activity and immunity booster properties of several dietary polyphenols came into the spotlight. One of the most common dietary polyphenols is curcumin – a phenolic compound that is present in Curcuma longa roots – of interest because it is largely extracted and used as a food dye, and can be easily included in everyday diet. Flavanol quercetin and flavanones naringenin have been reported as potential anti-viral candidates. Another important polyphenolic compound is phloretin and Epigallocatechin gallate (present in green tea, apple skin, onion, plum) which is taken into account as an anti-Covid, mainly because it is ubiquitous in vegetables and fruits and besides being endowed with antiviral properties, it has been reported to display potential for the mitigation of ailments characterized by chronic inflammation (Ghidoli et al., 2021).

9.2 STRUCTURE AND CLASSES OF POLYPHENOLS

Polyphenols are a large family of a naturally occurring compound found in plants. More than 8,000 phenolic structures are currently known. They are secondary metabolites derived from a common intermediate, phenylalanine or a close precursor, shikimic acid (De et al., 2000). Polyphenols contain repeating phenolic moieties connected by ester or more stable C-C bonds. The majority of polyphenols in plants exist as glycosides with different sugar units linked to hydroxyl groups. The diversity and wide distribution have led to different ways of categorizing polyphenols. Depending on the number of phenol units within the molecular structure, substituent groups and linkage type between phenol units, they are broadly divided into four classes: Phenolic acids, flavonoids, stilbenes and lignans (Pandey et al., 2009).

9.2.1 PHENOLIC ACIDS

Phenolic acids are derivatives of cinnamic acid and benzoic acid. They are found in fruits, vegetables and cereal grains, mostly attached to the plant cell walls. There are two classes of Phenolic acids: hydroxybenzoic acid and hydroxycinnamic acid (Kumar et al., 2019). Hydroxybenzoic acid inhibits enzymes that are responsible for breaking down the complex carbohydrates, thus keeping blood sugar levels low. The hydroxycinnamic acid acts as a powerful antioxidant and consists chiefly of p-coumaric, caffeic, ferulic and sinapic acids (Razzaghi et al., 2013).

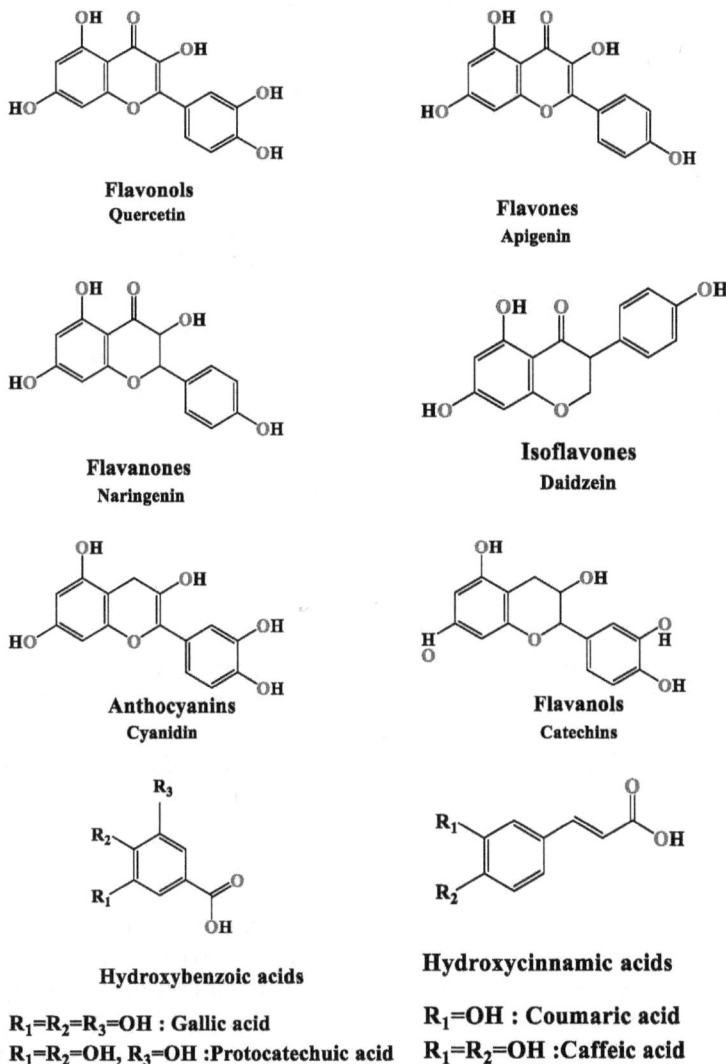

FIGURE 9.1   General structures of flavonoids and phenolic acids.

9.2.2   FLAVONOIDS

Flavonoids are most important class of polyphenols. More than 4,000 varieties of flavonoids have been identified until today. Chemically flavonoids consist of two aromatic rings linked via an oxygenated heterocycle. They are major coloring component of flowers, fruits and leaves. They are divided on basis of variation in the heterocycle involved. Six subgroups of flavonoids are: flavonols, flavones, flavanones, flavanols, anthocyanins, and isoflavones (Figure 9.1). Chalcones are flavonoids with open structure. Variation in number and arrangement of hydroxyl groups and their extent of glycoslylation cause divergence in each group (Panche et al., 2016).

9.2.3   TANNINS

Tannins are the water-soluble plant products found in the bark of trees, in woods, leaves, stems and fruits of the tree. These are the large molecules that easily bind with the protein, starch, cellulose,

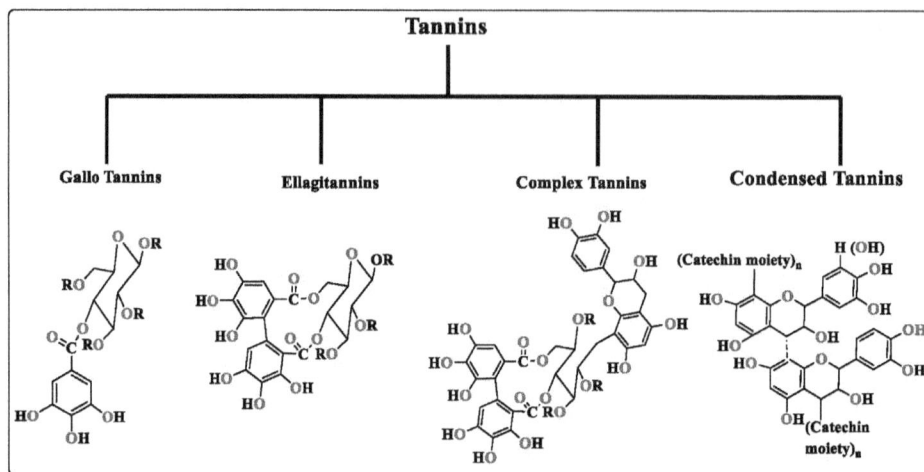

FIGURE 9.2 Some structures of tannins.

and minerals. Tannins are responsible for the minimization of growth rate, feed efficiency, net metabolize energy, and protein digestibility in experimental animals.

Tannins are subdivided into hydrolysable, condensed, and complex. Gallotannins, or ellagitannins, are the hydrolyzable tannins. Gallotannins are basically polyols, substituted with gallic acid units. In gallotannins galloyl units are joined with each other through the ester linkage. Generally, the polyol consists of D-glucose, substituted with gallic acid. Tannic acid is an example of gallotannins (Sharma et al., 2021). Some examples of tannins are presented in Figure 9.2.

9.2.4 STILBENES

Stilbenes are needed in very low quantity in the human diet but it is very much essential for the human body because it is needed to maintain human health (Riviere et al., 2012). Many studies have presented that resveratrol showed many biological activities such as anti-inflammatory, anti-bacterial, antioxidant, and anticancer activities. Resveratrol has been found in red wine and grape juice. Regarding the bioavailability of resveratrol, when it is taken orally, resveratrol is metabolized to its glycosylated form, which enhances its solubility and stability; therefore, it is readily absorbed. Hence, it was concluded that the bioavailability of resveratrol is high, showing great permeability within the human body. So that accumulation of potentially active resveratrol metabolites may produce healthy effects within the human body (Jang et al., 1997). Some examples of stilbene are presented in Figure 9.3.

9.2.5 LIGNANS

Lignans are the low molecular weight polyphenol found in plants, mainly in vegetables, whole grains and seeds. According to their origin, lignans are classified into plant lignans like isolariciresinol, secoisolariciresinoldiglucoside, lariciresinol, and matairesinol, and mammalian lignans such as enterodiol and enterolactone. In lignans, polyphenols are linked with each other through the carbon chain (Al Mamari et al., 2021).

Lignans are also categorized into several groups, like dibenzylfuran, dibenzylbutryolactol, dihydroxybenzylbutane, arylnaphtalene, and aryltetraline lactone derivatives. According to structural type and concentration, several dietary lignans have been shown to possess biological activities

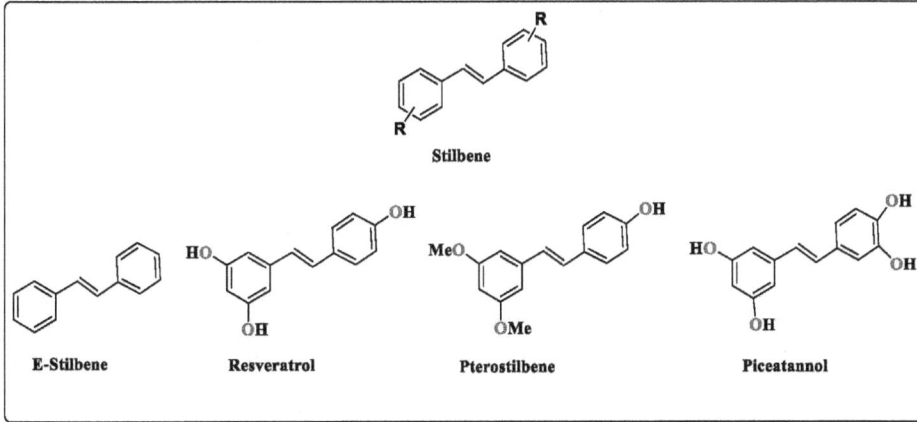

FIGURE 9.3 General structure and some examples of stilbenes.

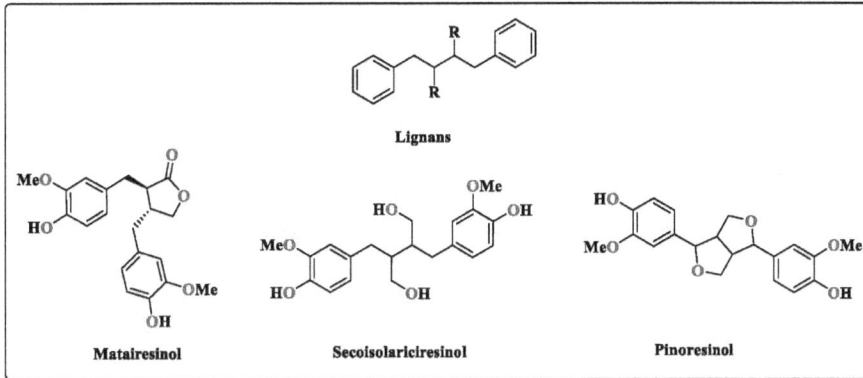

FIGURE 9.4 General structure and some examples of lignans.

such as decreasing the risk of heart attack, breast cancer, and osteoporosis. Also, they shown anti-oxidant properties in human organs, including the liver and the brain (Soleymani et al., 2020). Some examples of lignans are presented in Figure 9.4.

9.3 SIGNIFICANT DIETARY SOURCES OF POLYPHENOLIC COMPOUNDS

Polyphenolic compounds are extensively distributed in plants, like fruits, vegetables, leaves, oils, and so forth. The plant kingdom is an attractive source of natural antioxidants (Moure et al., 2001). Beverages like fruit juices, coffee, tea, wines, and others parts like tobacco, olive oil, and so forth, are a major source of polyphenolics in the human diet. Synthetic antioxidants have shown potential toxicity, therefore, a wide expansion is observed in the research endeavor to discover and develop antioxidants from natural dietary sources.

9.4 BIOAVAILABILITY AND PHARMACOKINETICS OF POLYPHENOLS

Phenols have varying ranges of bioavailability, as no relation was found between quantity of phenols in food and their bioavailability in the consumer's body. The term "bioavailability" is the portion of the nutrients metabolized, digested, and absorbed via normal metabolic processes. Most of the polyphenols usually originate in food items are found as esters, polymers, or glycosides, unable to be absorbed in native form, and only aglycones and few glucosides can easily be absorbed by the small intestine (Zhang et al., 2006).Whereas, some of flavonoids, such as quercetin and few of the anthocyanins also can be metabolized in the intestine and absorbed by stomach. In course of absorption, polyphenols endure up to an extensive alteration along with hydrolyzation by several intestinal enzymes or colonic microflora and get conjugated to intestinal cells and subsequently to the liver by the process of methylation, sulfation, and glucuronidation (Purushotham et al., 2009). Thus, identification and evaluation of resultant circulating metabolites is a very tedious task. But it is imperative to evaluate their nature, biological activity, and position of presence of conjugating groups in structure of polyphenols, as they are responsible for their biological properties. Hence, the rate and extent of its absorption and nature of the metabolites circulating in the plasma depend on the chemical structure of polyphenols and not on their concentration. Absorption of polyphenols is site-dependent, as some of them are well absorbed in the gastro-intestinal tract while others in the intestine or another part of the gut. Blood protein albumin on which metabolites of polyphenols are bound, plays a vital role in bioavailability of polyphenols, and the affinity of binding depends on the chemical structure of polyphenols and their metabolites (Keevil et al., 1998). But it is still unclear as to which form of polyphenols exerts some biological activity, the free one or the albumin-bound phenols (Zhang et al., 2006; Mathur et al., 2002). Polyphenols that accumulate in the tissues, particularly those in which they are metabolized such as intestine and liver, are mainly responsible for exerting the biological activity and their excretion with their derivatives occurs through urine and bile. Mostly, the extensively conjugated metabolites are easily eliminated in bile, whereas small conjugates metabolites like monosulfates are excreted from urine. Amount of metabolites excreted in urine is roughly correlated with maximum plasma concentrations. Usually, flavanones from citrus fruit have quite higher percentage of urinary excretion, and it reduces from isoflavones to flavonols. Therefore, the intake and bioavailability of polyphenols both have their own beneficial effects on human health (Zhang et al., 2006).

9.5 BIOLOGICAL ACTIVITY OF POLYPHENOLIC COMPOUNDS

9.5.1 ANTIMICROBIAL ACTIVITY

Antimicrobial agents kill or slow down the action of viruses and bacteria without inflicting any damage to the surrounding cells and tissues. Up until now, many compounds with the characteristics mentioned above have been found. In this connection, phenolics have also been shown to be potent antibacterial and anti-viral agents. For instance, phenolics constrained the growth and proliferation of hepatitis C virus (HCV); this virus is a primary blood-borne pathogen causing liver cirrhosis and hepatocellular carcinoma (HCC), thus inhibiting infection in primary human hepatocytes (Alter et al., 1997). Kang et al. reported that gallic acid and its derivatives attenuate the growth of cariogenic and periodontopathic bacteria (Kang et al., 2008). Gallic acid and methyl gallate also exhibited strong antibacterial and antiviral potential against Salmonella (Choi et al., 2008). Kratz et al. stated inhibitory activity of gallic acid and methyl gallate against herpes viruses (Kratz et al., 2008). They argued that the attachment of these phenolics to the virus proteins may interfere with their invasion of cells. Kaihatsu et al. reported the suppression of virus type-1 infection of (+)-epigallocatechin 3-O-gallate (Kaihatsu et al., 2018). Moreover, phenolics such as stilbenes, tannins, and isoflavones inhibited the growth of fungi, yeasts, and viruses as well as bacteria such as *Salmonella, Clostridium, Bacillus*, and *E. coli* (Shahidi et al., 2018).

9.5.2 ANTI-SARS-CoV-2

The current pandemic, caused by SARS-CoV-2 virus, is a severe challenge for human health and the world economy. There is an urgent need for development of drugs that can manage this pandemic, as it has already infected more than 250 million people and led to the death of around 5 million people worldwide. *In silico* studies of the three polyphenolic compounds, glucogallin, mangiferin and phlorizin. These compounds are natural compounds so that they are directly isolated from natural sources. These compounds are structurally similar and are known for their antiviral activity. *In silico* studies revealed that these compounds showed the anti-protease activity on SARS-CoV-2 main protease (Mpro) and TMPRSS2 protein. Both the viral protein and the host protein play an important role in the viral life cycle, such as post-translational modification and viral spike protein priming (Chaurasia et al., 2021). From the molecular docking studies, it was found that these compounds displayed significant binding affinity, which was further proved by the molecular dynamics study (Mishra et al., 2021). The molecular dynamics simulation study has predicted that these natural compounds will have a great impact on the stabilization of the binding cavity of the Mpro of SARS-CoV-2. The other physiochemical parameters also indicate that these compounds are expected to have good absorption properties and solubility. Further analysis for these compounds also displayed no involvement in drug-to-drug interaction as well as no toxicity (Singh et al., 2021).

9.5.3 ANTI-HIV ACTIVITY

Phenolic compounds and their derivatives played an important role in inhibition of several steps of HIV-1 life cycle. There are two types of HIV strain, HIV-1 and HIV-2. Of these two, strains, HIV-1 is more common as compared to HIV-2. Therefore, HIV-1 is the main focus of attention for the development of novel molecules against AIDS (Huang et al., 2004). A deep study of the phenolic compounds extracted from natural sources, revealed that these compounds inhibit the different steps of the HIV-1 life cycle, including virus–cell fusion and virus absorption, reverse transcription, integration (IN) and proteolytic cleavage. The various steps of HIV life cycle are inhibited by the phenolic compounds.

Plants containing the excess number of phenolic compounds may be recognized as strong sources of molecules for the treatment of patients suffering from AIDS. Despite the continuous advances made in antiretroviral combination therapy, AIDS is a continuous dreaded killer, the leading cause of death in Africa and the fourth cause worldwide. Many research groups are discovering the bio and chemo diversity of the plant kingdom to find new and better drugs for the treatment of patients suffering from HIV infections, with new targets and new mechanisms of action.

The major classes of polyphenols and bioflavonoids with antiviral activity are active against various virus families such as retroviridae, hepadnaviridae, and hespervirides,

9.5.3.1 Anti HSV activity

Reports have shown that flavonols are comparatively more active than flavones against HSV (herpes simplex virus) type-1. A synergistic effect was also reported between flavonoids and various other antiviral agents. For example, quercetin enhances the effects of 5- ethyl-2'-dioxyuridine against HSV (Amoros et al., 1992).

9.5.3.2 Activity against influenza virus

Quercetin and rutin have been reported to have antiviral activity against influenza virus. Some studies have also demonstrated that these compounds are active against parainfluenza virus, too (Ninfali et al., 2020).

9.5.3.3 Other antiviral activity of polyphenolic compounds

Polyphenolic compounds have been reported to be active against a wide range of viruses like herpes, hepatitis A and B, influenza, rabies, polio, rotavirus, dengue virus, and so forth. Some polyphenolic compounds and their respective antiviral activities have been listed in Table 9.1.

TABLE 9.1
Polyphenolic Compounds with Antiviral Activity

Sl. No.	Types of viruses	Antiviral phenolics	References
1	Rabies virus	Quercetin, Quercetrin, Rutin	Kane et al., 1998
2	Para influenza virus	Quercetin, Rutin	Deca et al., 1987
3	Herpes virus type-1	Galangin, Quercetin, Kempferol, Apigenin	Singh et al., 1988
4	Potato virus	Quercetin, Rutin, Morin	Tahara et al., 1987
5	Influenza virus	Quercetin, Rutin	Tripathi et al., 1981
6	Herpes virus type-2	Quercetin, Chrysin	Bakay et al., 1968
7	Respiratory syncytial virus	Quercetin, Naringin	Feesen et al., 1994
8	Immuno-deficiency virus	Apigenin, Caffeic acid	Tsuchiva et al., 1985; Hayashi et al., 1993
9	Azesky virus	Quercetin, Quercetrin, Morin, Apigenin, Luteolin	Nagai et al., 1992
10	Polio virus	Quercetin	Yu et al., 2010
11	Mango virus	Quercetin	Barros et al., 2010
12	Pseudorabies virus	Quercetin	Morais et al., 2011
13	Dengue virus	Ellagic acid, Rutin, Quercetin	Balde et al., 2011
14	Human cytomegalovirus	Kaempferol, Chrysin	Edziri et al., 2011; Yamada et al., 1991
15	Coxsickie B virus	Galangin, Kaempferol, Chrysin	Edziri et al., 2011; Yamada et al., 1991
16	Hepatitis A and B	Ellagic acid	Kang et al., 2006
17	Porcine epidemic diarrhea virus	Quercetin-7-rhamnoside	Choi et al., 2009
18	Rotavirus	Kaempferol, Chrysin	Cheng et al., 1996
19	Coronavirus	Kaempferol, Chrysin	Cheng et al., 1996
20	Sindabis virus	Catechin	Selway et al., 1986

9.6   CONCLUSION AND FUTURE PERSPECTIVES

The present studies outlined in this chapter offer a vibrant background on the biological effects of polyphenols, mode of metabolism, and their consequences to human health. Foodstuff rich in polyphenols act protectively against the development and progression of several chronic pathological conditions, such as HIV, COVID, and other viral diseases. Although, a number of biological impacts based on epidemiological studies can be well understood scientifically, the mechanism of action of some effects of polyphenols is still unknown. Therefore, a better understanding about the variables of polyphenol bioavailability – such as the kinetics of absorption, accumulation, and elimination – will facilitate the design of their new derivatives, as the role of polyphenols in human health is still a prolific arena of research. Based on these current scientific considerations, polyphenols offer a great hope for protection from and prevention of several chronic human diseases.

ACKNOWLEDGEMENT

Financial assistance to Vishal K. Singh in the form of Senior Research Fellowship (Ref No: 349/CSIR-UGC NET DEC. 2017) by University Grants Commission is sincerely acknowledged.

REFERENCES

Abbas, M., Saeed, F., Anjum, F.M., Afzaal, M., Tufail, T., Bashir, M.S., Ishtiaq, A., Hussain, S. and Suleria, H.A.R. 2017. Natural polyphenols: An overview. *Int J of Food Prope* 20(8): 1689–1699.

Al Mamari, H.H. 2021. Phenolic compounds: Classification, chemistry, and updated techniques of analysis and synthesis. *In Phenolic Compounds. IntechOpen.* DOI: 10.5772/intechopen.98958

Alter, M.J. 1997. The epidemiology of acute and chronic hepatitis C. *Clin Liver Dis* 1(3): 559–568.

Amoros, M., Simõs, C.M.O., Girre, L., Sauvager, F. and Cormier, M. 1992. Synergistic effect of flavones and flavonols against herpes simplex virus type 1 in cell culture. Comparison with the antiviral activity of propolis. *J Nat Prod* 55(12): 1732–1740.

Bakay, M., Mucsi, I., Beladi, I., Gabor, M.M. 1968. Antiviral flavonoids from Alkenaorientalis. *Acta Microb* 15: 223–232.

Balde, A.M., Van Hoof, L., Pieters, L.A., Berghe, D.V. and Vlietinck, A.J. 1990. Plant antiviral agents. VII. Antiviral and antibacterial proanthocyanidins from the bark of Pavetta owariensis. *Phytother Res* 4(5): 182–188.

Barros, A.V., da Conceição, A.O., Simoni, I.C., Arns, C.W. and Fernandes, M.J.B. 2010. Mechanisms of anti-viral action of seeds from Guettarda Angelica Mart against bovine and swine herpes viruses in vitro. *Virus Rev Res* 15(2): 47–52.

Chaurasia, H., Singh, V.K., Mishra, R., Yadav, A.K., Ram, N.K., Singh, P. and Singh, R.K. 2021. Molecular modelling, synthesis and antimicrobial evaluation of benzimidazole nucleoside mimetics. *Bioorg Chem* 115: 105227.

Cheng, P.C., Wong, G. 1996. Honey bee propolis: Prospects in medicine. *Bee World* 77: 8–15.

Choi, H.J., Kim, J.H., Lee, C.H., Ahn, Y.J., Song, J.H., Baek, S.H. and Kwon, D.H. 2009. Antiviral activity of quercetin 7-rhamnoside against porcine epidemic diarrhea virus. *Antiviral Res* 81(1): 77–81.

Choi, J.G., Kang, O.H., Lee, Y.S., Oh, Y.C., Chae, H.S., Jang, H.J., Kim, J.H., Sohn, D.H., Shin, D.W., Park, H. and Kwon, D.Y. 2008. In vitro activity of methyl gallate isolated from galla rhois alone and in combin-ation with ciprofloxacin against clinical isolates of salmonella. *J Microb and Biotech* 18(11): 1848–1852.

Deca, R.T., Gouzalez, I.J., Mactinez, T.M.V., et al. 1987. Soil bio. In vitro antifungal activity of some flavonoids and their metabolites, *Biochem* 223–231.

DeFilippis, V.R., Villareal, L.P. 2000. An introduction to the evolutionary ecology of viruses. *Viral ecology* 125.

Domingo, E.J.J.H., Holland, J.J. 1997. RNA virus mutations and fitness for survival. *Annu Rev Microb* 51(1): 151–178.

Edziri, H., Mastouri, M., Mahjoub, M.A., Ammar, S., Mighri, Z., Gutmann, L. and Aouni, M. 2011. Antiviral activity of leaves extracts of Marrubium alysson L. *J Med Plants Res* 5(3): 360–363.

Fesen, M.R., Pommier, Y., Leteurtre, F., Hiroguchi, S., Yung, J. and Kohn, K.W. 1994. Inhibition of HIV-1 integrase by flavones, caffeic acid phenethyl ester (CAPE) and related compounds. *Biochem Pharm* 48(3): 595–608.

Ghidoli, M., Colombo, F., Sangiorgio, S., Landoni, M., Giupponi, L., Nielsen, E. and Pilu, R. 2021. Food containing bioactive flavonoids and other phenolic or sulfur phytochemicals with antiviral effect: Can we design a promising diet against COVID-19? *Front Nutri* 8: 303.

Hayashi, K., Hayashi, T., Arisawa, M. and Morita, N. 1993. *In vitro* inhibition of viral disease by flavonoids. *Antiviral Chem Chemother* 4(1): 49–53.

Huang, L., Yuan, X., Aiken, C. and Chen, C.H. 2004. Bifunctional anti-human immunodeficiency virus type 1 small molecules with two novel mechanisms of action. *Antimicrob Agents Chemother* 48(2): 663–665.

Jang, M., Cai, L.,Udeani, G. O. 1997. Cancer chemopreventive activity of resveratrol, a natural product derived from grapes. *Science* 275(5297): 218–220.

Kaihatsu, K., Yamabe, M., and Ebara, Y. 2018. Antiviral mechanism of action of epigallocatechin-3-O-gallate and its fatty acid esters. *Molecules* 23(10): 2475.

Kane, C.J., Menna, J.H., Sung, C.C. and Yeh, Y.C. 1988. Methyl gallate, methyl-3, 4, 5-trihydroxybenzoate, is a potent and highly specific inhibitor of herpes simplex virusin vitro. II. Antiviral activity of methyl gallate and its derivatives. *Biosci Rep* 8(1): 95–102.

Kang, E.H., Kown, T.Y., Oh, G.T., Park, W.F., Park, S.I., Park, S.K. and Lee, Y.I. 2006. The flavonoid ellagic acid from a medicinal herb inhibits host immune tolerance induced by the hepatitis B virus-e antigen. *Antiviral Res* 72(2): 100–106.

Kang, M.S., Oh, J.S., Kang, I.C., Hong, S.J. and Choi, C.H. 2008. Inhibitory effect of methyl gallate and gallic acid on oral bacteria. *J Microbiol* 46(6): 744–750.

Keevil, J.G., Osman, H., Maalej, N. and Folts, J.D. 1998. Grape juice inhibits human ex vivo platelet aggrega-tion while orange and grapefruit juices do not. *J Am College Cardiol* 31(2SA): 172A–172A.

Kirchhoff, F. 2013. HIV life cycle: Overview. *Encyclo AIDS* 1–9.

Kratz, J.M., Andrighetti-Fröhner, C.R., Kolling, D.J., Leal, P.C., Cirne-Santos, C.C., Yunes, R.A., Nunes, R.J., Trybala, E., Bergström, T., Frugulhetti, I.C. and Barardi, C.R.M. 2008. Anti-HSV-1 and anti-HIV-1 activity of gallic acid and pentyl gallate. *Memórias do Instituto Oswaldo Cruz* 103(5): 437–442.

Kumar, N. and Goel, N. 2019. Phenolic acids: Natural versatile molecules with promising therapeutic applications. *Biotech Rep* 24: 00370.

Mandary, M.B., Masomian, M. and Poh, C.L. 2019. Impact of RNA virus evolution on quasispecies formation and virulence. *Int J Mol Sci* 20(18): 4657.

Mathur, S., Devaraj, S., Grundy, S.M. and Jialal, I. 2002. Cocoa products decrease low density lipoprotein oxidative susceptibility but do not affect biomarkers of inflammation in humans. *J Nutr* 132(12): 3663–3667.

Mishra, R., Chaurasia, H., Singh, V.K., Naaz, F. and Singh, R.K. 2021. Molecular modeling, QSAR analysis and antimicrobial properties of Schiff base derivatives of isatin. *J Mol Struct* 1243: 130763.

Morais, A.S.M., Marques, M.M.M., Lima, D.M., Santos, S.C.C., Almeida, R.R., Vieira, I.G.P. and Guedes, M.I.F. 2011. Antiviral activities of extracts and phenolic components of two Spondias species against dengue virus, J. Venom. Anim. *Toxins Incl Trop Dis* 17: 4.

Moure, A., Cruz, J.M., Franco, D., Domínguez, J.M., Sineiro, J., Domínguez, H., Núñez, M.J. and Parajó, J.C. 2001. Natural antioxidants from residual sources. *Food Chem* 72(2): 145–171.

Naczk, M., Shahidi, F. 2006. Phenolics in cereals, fruits and vegetables: Occurrence, extraction and analysis. *J Pharm Biomed Anal* 41: 1523–1542.

Nagai, T., Miyaichi, Y., Tomimori, T., Suzuki, Y. and Yamaha, H. 1992. Antiviral activity of two flavonoids from Tanacetum microphyllum. *Antiviral Res* 19: 207–216.

Ninfali, P., Antonelli, A., Magnani, M. and Scarpa, E.S. 2020. Antiviral properties of flavonoids and delivery strategies. *Nutrients* 12(9): 2534.

Panche, A.N., Diwan, A.D., Chandra, S.R. 2016. Flavonoids: An overview. *J Ntri Sci 5*.

Pandey, K.B., and Rizvi, S.I. 2009. Plant polyphenols as dietary antioxidants in human health and disease. *Oxid Med Cell Logev* 2(5): 270–278.

Purushotham, A., Tian, M., Belury, M.A. 2009. The citrus fruit Flavonoid Naringenin suppresses hepatic glucose production from Fao Hepatoma cells. *Mol Nutr Food Res* 53: 300–307.

Razzaghi-Asl, N., Garrido, J., Khazraei, H., Borges, F. and Firuzi, O. 2013. Antioxidant properties of hydroxycinnamic acids: A review of structure-activity relationships. *Curre Med Chem* 20(36): 4436–4450.

Rivière, C., Pawlus, A.D., Mérillon, J.-M. 2012. Natural stilbenoids: Distribution in the plant kingdom and chemotaxonomic interest in Vitaceae. *Nat Prod Rep* 29(11): 1317–1333.

Selway, J.T. 1986. Antiviral activity of flavones and flavans, In: Cody V, Middleton E, Harborne JB, editors. *Plant Flavonoids in Biology and Medicine: Biochemical, Pharmacological, and Structure–Activity Relationships, New York: Alan R. Liss.*

Shahidi, F., Yeo, J. 2018. Bioactivities of phenolics by focusing on suppression of chronic diseases: A review. *Int J Mol Sci* 19(6): 1573.

Sharma, K., Kumar, V., Kaur, J., Tanwar, B., Goyal, A., Sharma, R., Gat, Y. and Kumar, A. 2021. Health effects, sources, utilization and safety of tannins: A critical review. *Toxin Reviews* 40(4): 432–444.

Singh, U.P., Pandey, V.B., Singh, K.N. and Singh, R.O.N. 1988. Structural and biogenic relationships of isoflavonoids. *Canadian J Bot* 166:1901–1910.

Singh, V.K., Srivastava, R., Gupta, P.S.S., Naaz, F., Chaurasia, H., Mishra, R., Rana, M.K. and Singh, R.K. 2021. Anti-HIV potential of diarylpyrimidine derivatives as non-nucleoside reverse transcriptase inhibitors: Design, synthesis, docking, TOPKAT analysis and molecular dynamics simulations. *J Biomol Strut Dyn* 39(7): 2430–2446.

Soleymani, S., Habtemariam, S., Rahimi, R. and Nabavi, S.M. 2020. The what and who of dietary lignans in human health: Special focus on prooxidant and antioxidant effects. *Trends Food Sci Tech* 106: 382–390.

Tahara, S., Hashidoka, Y., Mizutani, J. 1987. Flavonoids as medicines. *Agri Biol Chem* 51: 1039–1045.

Tripathi, V.D., Rastogi, R.P. 1981. In vitro anti-HIV activity of flavonoids isolated from *Garcinia multifolia*. *J Sci Indian Res* 40: 116-121.

Tsuchiya, Y., Shimizu, M., Hiyama, Y., Itoh, K., Hashimoto, Y., Nakayama, M., Horie, T. and Morita, N. 1985. Inhibitory effect of flavonoids on fungal diseases. *Chem Pharm Bull* 33: 3881–3890.

Yamada, H. 1991. Natural products of commercial potential as medicines. *Curr Opin Biotechnol* 2: 203–210.

Yu, C., Yan, Y., Wu, X., Zhang, B., Wang, W. and Wu, Q. 2010. Anti-influenza virus effects of the aqueous extract from *Mosla scabra*. *J Ethnopharmacol* 127(2): 280–285.

Zhang, H., Wang, L., Deroles, S., Bennett, R. and Davies, K. 2006. New insight into the structures and formation of anthocyanic vacuolar inclusions in flower petals. *BMC Plant Biol* 6(1): 1–14.

10 Dietary Polyphenols in Parasitic Diseases and Neglected Tropical Diseases (NTDs)

James H. Zothantluanga, Arpita Paul, Ngurzampuii Sailo, Zonunmawii, H. Lalthanzara, and Dipak Chetia

*E-mail: jameshztta@gmail.com; arpitatua.666@gmail.com; sailongurzampuii11@gmail.com; anunitlau28@gmail.com; hzara.puc@gmail.com; dchetia@dibru.ac.in

CONTENTS

10.1 INTRODUCTION

Malaria, Zika virus, dengue, chikungunya, soil-transmitted helminths, and leishmaniasis are examples of parasitic diseases (PDs). Insect vectors and contaminated food or water carry protozoans, viruses, helminths, and ectoparasites that infect a human host, resulting in millions of infections with a high number of deaths (CDC, 2021a; Karakavuk et al., 2018; WHO, 2021). Chagas disease, trypanosomiasis, Buruli ulcer, leprosy, trachoma, lymphatic filariasis, cysticercosis, dracunculiasis, fascioliasis, mycetoma, schistosomiasis, and onchocerciasis are examples of neglected tropical diseases (NTDs). NTDs are caused by viruses, bacteria, and parasites and are transmitted by insect vectors, along with contaminated food and water (CDC, 2021b; Engels and Zhou, 2020; Molyneux et al., 2017). Among parasitic diseases and NTDs, protozoan infections include malaria, leishmaniasis, Chagas disease, and trypanosomiasis; helminth infections include soil-transmitted helminths, lymphatic filariasis, cysticercosis, dracunculiasis, fascioliasis, schistosomiasis, and onchocerciasis; viral infections include Zika virus, dengue, and chikungunya; bacterial infections include Buruli ulcer, leprosy, and trachoma; while fungal infections include mycetoma (PLOS, 2021).

Polyphenols are chemical compounds whose basic structure consists of more than one phenol ring. Examples of polyphenols include secondary plant metabolites such as flavonoids, phenolic acids, stilbenes, lignans, and tannins (Abotaleb et al., 2020; Cui et al., 2020; Luca et al., 2020; Pandey and Rizvi, 2009; Pietta, 2000; Reinisalo et al., 2015; Serrano et al., 2009). Polyphenols that are present in our daily diets are considered dietary polyphenols (DPs) (Luca et al., 2020; Serrano et al., 2009). Examples of DPs include phenolic acids (protocatechuic acid, vanillic acid, p-coumaric acid, gallic acid, syringic acid, ferulic acid, caffeic acid, chlorogenic acid, sinapic acid); flavonoids (daidzein, formononetin, glycitein, genistein, biochanin, dalbergin, phloretin, xanthohumol, apigenin, luteolin, tangeretin, nobiletin, naringenin, hesperitin, taxifolin, kaempferol, quercetin, myricetin, isorhamnetin, theaflavin, procyanidin, malvidin); tannins (catechin, catechin gallate, gallocatechin, gallocatechin gallate, epicatechin, epicatechin gallate, epigallocatechin, epigallocatechin gallate); and other polyphenols (resveratrol, curcumin, gingerol, secoisolariciresinol, matairesinol, rosmarinic acid, ellagic acid) (Han et al., 2007; Tsao, 2010).

DPs are abundantly found in fruits (blackberries, blueberries, black currant, black grape, elderberries, strawberries, cherries, plums, cranberry, pomegranate juice, raspberry, apples, apricots, grapes, plums, bilberries, gingko biloba, lemon, orange, grapefruit, tangerine juice, celery, olives, pear, peach, peanuts, longan, walnuts, capers); vegetables (celery, capers, chives, onions, dock leaves, fennels, hot peppers, tomatoes, spinach, sweet potato leaves, lettuce, broccoli, Hartwort leaves, kale, fresh parsley, potato, chick pea, lentils, black-eyed peas); cereals (buckwheat, green beans, yellow beans, haricot beans); spices and herbs (dill weed, peppermint, oregano, rosemary, dry parsley, thyme); and are also found in red wine, green tea, black tea, cocoa powder, turnip, endive, leek, chocolate, white wine, soybean, tofu, miso, cider, and turmeric (Han et al., 2007).

DPs scavenge free radicals, inhibit oxidant enzymes, induce endogenous antioxidant enzymes, and modulate signal transduction. DPs showed important pharmacological activities, such as anti-inflammatory, antidiabetic, anti-allergy, anticancer, neuroprotective, cardioprotective, immune stimulant, and many other activities (Han et al., 2007). Moreover, DPs have been reported to show activity against PDs and NTDs (Kedzierski et al., 2007; Mohd et al., 2019; Mounce et al., 2017). Various pathogens have developed resistance against drugs used for the treatment of PDs and NTDs (Christaki et al., 2020; Lampejo, 2020; Lopes, 2019). The pharmacological potential of DPs with their abundance in our daily diet suggests DPs as a promising alternative to treat PDs and NTDs. In the present chapter, DPs that showed activity (*in vitro/in vivo*) against the causative agents of PDs and NTDs are highlighted. Their mechanisms of action and future challenges of DPs are also highlighted.

10.2   DIETARY POLYPHENOLS AGAINST PARASITIC DISEASES

10.2.1   MALARIA

Kaempferol showed in vivo antimalarial activity against chloroquine-sensitive *Plasmodium berghei* ANKA strain. At a dose of 1, 10, and 20 mg/kg, kaempferol showed a percentage inhibition of 16.79 percent, 31.87 percent, and 52.89 percent, respectively. In terms of chemoprophylactic activity, 10 and 20 mg/kg of kaempferol administered to mice before parasite infection reduced parasitemia levels by 24.47 percent and 40.80 percent. Kaempferol also significantly increased the mean survival time of mice. The antimalarial activity of kaempferol is believed to be brought about by its antioxidant activity (Somsak et al., 2018). The dietary sources of kaempferol are apples, asparagus, broccoli, chili pepper, cabbage, lettuce, onions, spinach, oregano, blueberry, cherry, and cranberry (Dabeek and Marra, 2019).

Ellagic acid showed in vitro activity against different strains of *Plasmodium falciparum*, such as W2-Indochina (IC_{50} = 330±24 nM), FcM29-Cameroon (IC_{50} = 180±20 nM), FcB1-Colombia (IC_{50} = 300±17 nM), F32-Tanzania (IC_{50} = 330±27 nM), and Dd2 (IC_{50} = 105±27 nM). Ellagic acid showed synergism with chloroquine, mefloquine, artesunate, and atovaquone. Ellagic acid mainly acts during the erythrocytic stage between the 24th and 40th hours, suggesting that it acted upon either trophozoites or early schizonts stage. Ellagic acid showed in vivo antimalarial activity against *Plasmodium vinckei petteri* infected mice. When the drug was administered through the intraperitoneal route at a dose of 50 and 100 mg/kg/day, 100 percent inhibition of parasites was observed (Soh et al., 2009). Ellagic acid inhibited the formation of β-hematin and reduced glutathione content inside the *Plasmodium* parasite. β-hematin detoxifies the parasite from free heme while glutathione protects the parasite from oxidative stress (Herraiz et al., 2019; Njomnang Soh et al., 2012; Soh et al., 2009). Ellagic acid is found to be present in strawberry, blackberry, pomegranate, red guava, white guava, Surinam cherry, and cambuci (Abe et al., 2012).

Luteolin showed in vitro antimalarial activity against *P. falciparum*. Luteolin prevented the progression of the parasite growth beyond the young trophozoites stage as it arrests the normal cell/life cycle of the parasite in the blood stage. However, it does not inhibit the ring stage of *P. falciparum*. The IC_{50} value of luteolin against chloroquine-sensitive and resistant strains were 11±1 µM and 12±1 µM, respectively (Lehane and Saliba, 2008). Celery, broccoli, onion leaves, peppers, carrots, apple skins, and cabbages are some examples of dietary fruits and vegetables that are rich in luteolin (Lin et al., 2008).

10.2.2 Dengue

Quercetin showed antiviral activity against dengue virus (DENV) type 2. It was observed that after pre-treatment of cells with a dose of 50 µg/ml of quercetin, a prophylactic type effect was produced as the pre-treated cells showed a 14 percent ±1.5 reduction in DENV-2 foci when compared to untreated cells. At the same dose after pre-treatment, quercetin reduced the production of DENV-2 related RNA by more than 67 percent±1. Also, the IC_{50} of quercetin against DENV-2 in the post-adsorption assay was reported to be 35.7 µg/ml. On the other hand, the IC_{50} of quercetin was improved to 28.9 µg/ml when the cells were treated 5 h before infection and 4 days post-infection. Also, the pre- and post-treatment of cells before and after infection with quercetin reduced the production of DENV-2 related RNA by more than 75.7 percent±1.57. Quercetin is believed to exert its antiviral activities by interfering with the viral replication of DENV-2 (Zandi et al., 2011a). The dietary sources of kaempferol are apples, asparagus, broccoli, chili pepper, lettuce, onions, spinach, oregano, blueberry, cherry, and cranberry (Dabeek and Marra, 2019).

Naringenin was reported to show anti-dengue activity against four DENV serotypes. Flow cytometry assay revealed that Huh7.5 cells showed a reduction in DENV infection when the cells were treated with naringenin during and after infection. The IC_{50} of naringenin against DENV-1, DENV-2, DENV-3, and DENV-4 are 35.81, 17.97, 117.1, and 177.5 µM, respectively. It was observed that naringenin was able to inhibit the viral replication of DENV by possibly forming a complex with the non-structural viral proteins of DENV (Frabasile et al., 2017). Naringenin is found in soybean, citrus fruits, tea, apple, cocoa, celery, onions, and berries (Patel et al., 2018).

Fisetin showed antiviral activity against DENV-2 during in vitro studies. After the virus was adsorbed to the Vero cells, fisetin inhibited DENV-2 replication with an IC_{50} value of 55 µg/ml.

When the Vero cells were treated 5 h before infection and 4 days after infection, the IC_{50} value of fisetin was reduced to 43.12 µg/ml. However, fisetin did not show prophylactic activity or direct virucidal activity against DENV-2. It can be hypothesized that fisetin exerts its anti-dengue activity by interfering with certain viral proteins to inhibit viral replication (Zandi et al., 2011b). The dietary sources of fisetin are strawberry, apple, persimmon, lotus root, onion, grape, kiwi, peach, and cucumber (Khan et al., 2013).

10.2.3 CHIKUNGUNYA

Curcumin showed in vitro antiviral activity against the chikungunya virus (CHIKV). The IC_{50} value of curcumin against CHIKV was reported to be 3.89 µM. The possible mechanism of action of curcumin against CHIKV is that it directly interferes with the virus's ability to bind to the surface of cells (Mounce et al., 2017). Curcumin is a dietary polyphenolic compound that is present in turmeric (*Curcuma longa*) (Stanić, 2017).

Apigenin is a dietary polyphenol reported to show in vitro antiviral activity against CHIKV. The IC_{50} value of apigenin against CHIKV was reported to be 22.5 µM. The possible mechanism of action of apigenin against CHIKV is that it either inhibits the cellular entry of the virus or inhibits the replication of CHIKV by interacting with viral proteins (Pohjala et al., 2011). Apigenin is abundantly found in many dietary sources, such as parsley, chamomile, celery, vine-spinach, artichokes, oregano, and garlic (Shankar et al., 2017).

Naringenin is a dietary polyphenol that was reported to show in vitro antiviral activity against CHIKV. The IC_{50} value of naringenin against CHIKV was reported to be 25.8 µM. The possible mechanism of action of apigenin against CHIKV is that it either inhibits the cellular entry of the virus or, inhibits the replication of CHIKV by interacting with viral proteins (Pohjala et al., 2011). Naringenin is found in soybean, citrus fruits, tea, apple, cocoa, celery, onions, and berries (Patel et al., 2018).

10.2.4 LEISHMANIASIS

Gallic acid showed in vitro and in vivo anti-leishmania activity against *Leishmania major*. In vitro studies revealed that the EC_{50} value of gallic acid against *L. major* was 5 µg/ml. In vivo studies showed that gallic acid reduced the size of lesion and parasite burden in female Bagg Albino c mice. The possible mechanism of action of gallic acid against *L. major* is induction of high immunomodulatory activity as evidenced by an increase of phagocytic capability, lysosomal volume, nitrite release, and intracellular calcium $[Ca^{2+}]$ in macrophages (Alves et al., 2017). The dietary sources of gallic acid are blueberry, blackberry, strawberry, plums, grapes, mango, cashew nut, hazelnut, walnut, tea, and wine. Naturally, gallic acid and its derivatives are present in almost every part of a plant, such as the leaf, seed, root, fruit, and bark (Daglia et al., 2014).

Quercetin and its diglycoside rutin showed anti-leishmania activity against *L. tropica*. In vitro studies revealed that the IC_{50} value of quercetin and rutin against *L. tropica* was 182.3 µg/ml and 91.2 µg/ml, respectively. The possible mechanism of action of quercetin and rutin against *L. tropica* is that it damages the DNA by inhibiting Try-R and Try-S (Mehwish et al., 2019). The dietary sources of kaempferol are apples, asparagus, broccoli, chili pepper, lettuce, onions, spinach, oregano, blueberry, cherry, oregano, and cranberry (Dabeek and Marra, 2019). The dietary sources of rutin are tea, green asparagus, onions, buckwheat, wine, eucalyptus, apples, and berries (Riva et al., 2020).

Epigallocatechin-3-gallate (EGCG) and epicatechin gallate (ECG) showed in vitro anti-leishmania activity against *L. infantum*. The IC_{50} value of EGCG and ECG were reported to be 27.7 µM and 75 µM, respectively. *In-silico* studies revealed that both compounds are potent inhibitors

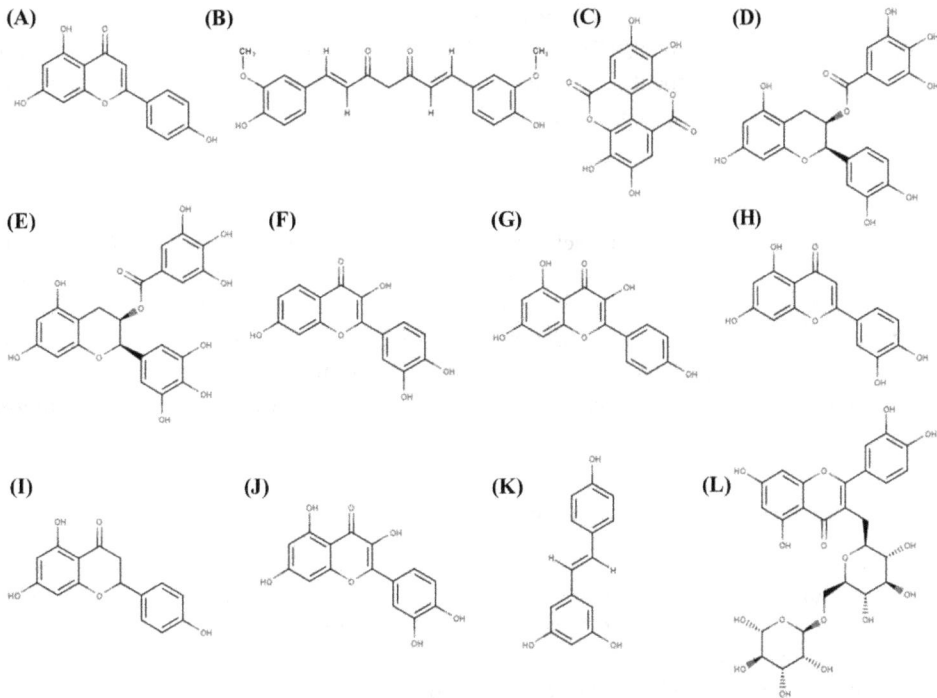

FIGURE 10.1 Chemical structures of (A) Apigenin, (B) Curcumin, (C) Ellagic acid, (D) Epicatechin gallate, (E) Epigallocatechin-3-gallate, (F) Fisetin, (G) Kaempferol, (H) Luteolin, (I) Naringenin, (J) Quercetin, (K) Resveratrol, and (L) Rutin with biological activity against parasitic diseases.

of TS and ARG-L (Khademvatan et al., 2019). The dietary sources of EGCG and ECG are black tea, green tea, black chocolate, rhubarb, broad beans, marrowfat peas, black grapes, blackberries, cherries, apples, and apricots (Arts et al., 2000).

10.2.5 ZIKA VIRUS

Resveratrol showed in vitro antiviral activity against Flavivirus (the causative agent for Zika virus). It was observed that resveratrol treatment resulted in a reduction of Zika virus (ZIKV) titer. In addition, resveratrol also reduces the number of ZIKV mRNA copies when the infected cells are treated with 80 μM of resveratrol post-infection. The possible mechanism of action of resveratrol against ZIKV is through the virucidal activity against ZIKV as well as through inhibition of ZIKV replication (Mohd et al., 2019). The dietary sources of resveratrol are grape, wine, peanut, Itadori tea, and soy (Burns et al., 2002). The chemical structures of DPs with biological activity against parasitic diseases are given in Figure 10.1.

10.3 DIETARY POLYPHENOLS AGAINST NTDS

10.3.1 CHAGAS DISEASE

The potential of epigallocatechin-3-gallate for the treatment of Chagas disease is supported by its in vitro and in vivo activity against *Trypanosoma cruzi*. Trypanocide in vitro studies gave an MIC_{50} value of 311 μM after 120h. In vivo trypanocide studies revealed that parasitemia was significantly diminished in the infected group when treated with 0.8 mg/kg/day of EGCG. The possible mechanism of action of EGCG is it inhibits epimastigotes growth of *T. cruzi* and increases the survival rate in EGCG-treated animals (Güida et al., 2007). The dietary sources of EGCG and ECG are black tea, green tea, black chocolate, rhubarb, broad beans, marrowfat peas, black grapes, blackberries, cherries, apples, and apricots (Arts et al., 2000).

10.3.2 FASCIOLIASIS

Curcumin showed in vitro inhibitory activity against *Fasciola gigantica*, a pathogen that causes Fascioliasis. 60µM of curcumin produced significant (p<0.05) inhibition of worm motility. 40 µM and 60µM of curcumin showed loss of tegumental transport and motility function of treated worms. The possible mechanism of action of curcumin for its inhibitory action against *F. gigantica* is its inhibition of the motility of the worm with severe disruption of the tegumental surface (Ullah et al., 2017). Curcumin is a dietary polyphenolic compound that is present in turmeric (*Curcuma longa*) (Stanić, 2017).

10.3.3 HUMAN AFRICAN TRYPANOSOMIASIS

Quercetin showed in vitro activity against *Trypanosoma brucei* suggesting its potential for the treatment of Human African Trypanosomiasis. The Trypanolysis assay gives dose-dependent destruction of parasites giving an IC_{50} value of 10 µM. Apoptosis assay showed that 20 µM of quercetin promoted time-dependent apoptosis of *T. brucei*. A combination of apoptosis data along with cell viability data showed parasite sensitivity towards quercetin-mediated toxicity giving an IC_{50} < 2 µM. Quercetin significantly inhibited *T. brucei* mediated activation of TNF-α production. Quercetin also decreased the levels of intracellular NO in activated macrophages (Mamani-Matsuda et al., 2004). The dietary sources of kaempferol are apples, asparagus, broccoli, chili pepper, lettuce, onions, spinach, oregano, blueberry, cherry, oregano, and cranberry (Dabeek and Marra, 2019).

Gallic acid showed in vitro activity against *T. brucei*. The in vitro trypanocidal activity of gallic acid against *T. brucei* revealed that the polyphenolic compound has an IC_{50} value of 14.2 µM±1.5 wherein a complete cell death was observed. Gallic acid inhibits parasite growth, which could be due to the presence of extra hydroxyl group and its ability to form reactive oxygen intermediates in the parasite (Amisigo et al., 2019). The dietary sources of gallic acid are blueberry, blackberry, strawberry, plums, grapes, mango, cashew nut, hazelnut, walnut, tea, and wine. Naturally, gallic acid and its derivatives are present in almost every part of a plant such as a leaf, seed, root, fruit, and bark (Daglia et al., 2014).

10.3.4 LEPROSY

EGCG (1%) showed in vivo (topical) activity against *Mycobacterium leprae*. EGCG inhibits reactive oxygen species formation, activates superoxide dismutase, inhibits cytokines, promotes wound healing, and induces collagen deposition (Oktaviyanti et al., 2020). The dietary sources of EGCG are black tea, green tea, black chocolate, rhubarb, broad beans, marrowfat peas, black grapes, blackberries, cherries, apples, and apricots (Arts et al., 2000).

10.3.5 LYMPHATIC FILARIASIS

Naringenin, hesperetin, rutin, naringin, and chrysin are examples of DPs that showed in vitro activity against *Brugia malayi* highlighting their potential for the treatment of lymphatic filariasis. Although the exact mechanism of action of the DPs against *B. malayi* is not completely understood, it was observed that the polyphenolics were exerting their effects against adult worms and microfilariae during in vitro studies. Naringenin exhibited in vivo activity against *B. malayi* and *Mastomys coucha* model (Lakshmi et al., 2010). Naringenin and naringin are predominantly found in fruits of citrus species and tomatoes (Salehi et al., 2019). Hesperetin is mainly present in citrus fruits (Erlund et al., 2002). Chrysin is found in walnuts, leaves, and fruits of doum plants, and in the peel of passion fruits (Stompor-Gorący et al., 2021). The dietary sources of rutin are tea, green asparagus, onions, buckwheat, wine, eucalyptus, apples, and berries (Riva et al., 2020).

10.3.6 TRACHOMA

Tea polyphenols such as epigallocatechin, epicatechin, epigallocatechin gallate, epicatechin gallate, gallocatechin gallate were reported to show in vitro inhibitory activity against *Chlamydia trachomatis*. The DPs were reported to disrupt the lipid bilayer of *C. trachomatis* (Yamazaki et al., 2003).

Baicalein was reported to inhibit *C. trachomatis* during in vitro studies. The possible mechanism of action of baicalein is that it up-regulates the expression of RFX5 gene while it down-regulates the expression of CPAF gene (Hao et al., 2009). The dietary sources of baicalein are wine, tea, citrus fruit, dark chocolate, and herbs (Donald et al., 2012)

10.3.7 SCHISTOSOMIASIS

Resveratrol was reported to disrupt the cell membrane of *Schistosoma mansoni* in *Biomphalaria glabrata* in vivo model (Gouveia et al., 2019). Kaempferol showed in vitro activity against *B. glabrata* by inducing the death of adult worms (Braguine et al., 2012). Quercetin inhibits NAD^+ catabolizing enzyme located in the outer surface of the *B. glabrata* and leads to the death of adult worms in in vitro studies (Braguine et al., 2012). Botulin reduced the locomotor activity of *B. glabrata* in in vitro studies (Cunha et al., 2012). In vivo studies on *S. japonicum* infected mice revealed that limonin causes tubercular disruption, edema, bleeding, and ulcerations leading to the death of the parasite (Eraky et al., 2016). In *S. japonicum* infected mice, hesperidin reduces the number of male and female worms (Allam and Abuelsaad, 2014) while curcumin reduces the egg count (El-Ansary et al., 2007). The chemical structures of DPs with biological activity against NTDs are given in Figure 10.2.

The possible mechanism of action of DPs against parasitic diseases and NTDs are summarized in Figure 10.3. The antioxidant activity of DPs counteracts the potential oxidative stress induced in a host cell by parasite infection. DPs control the gene expression that can get out of hand upon parasite infection. DPs form complexes with essential enzymes of parasites and inhibit replication. DPs can induce oxidative stress within a parasite. DPs can damage the DNA of a parasite, thereby inducing apoptotic cell death. DPs can disrupt the lipid bilayer of the parasite and cause cellular leakage due to compromised cell-wall integrity of the parasite. DPs can also prevent the binding of parasites to cells, thereby preventing cellular entry.

10.4 CURRENT CHALLENGES AND FUTURE SCOPES

Several DPs have been investigated for parasitic diseases. On the other hand, the number of DPs investigated for the potential treatment of NTDs was lesser. It is recommended that more attention needs to be given to the screening of DPs for the treatment of NTDs. The majority of the studies carried out against parasitic diseases and NTDs were limited to in vitro, in vivo, and in silico studies only. Therefore, it is recommended that studies involving humans may be carried out for the clinical applicability of the aforementioned DPs. Many polyphenols, including those that are present in our daily diets, have been found to have low bioavailability. Therefore, it is rational to improve the

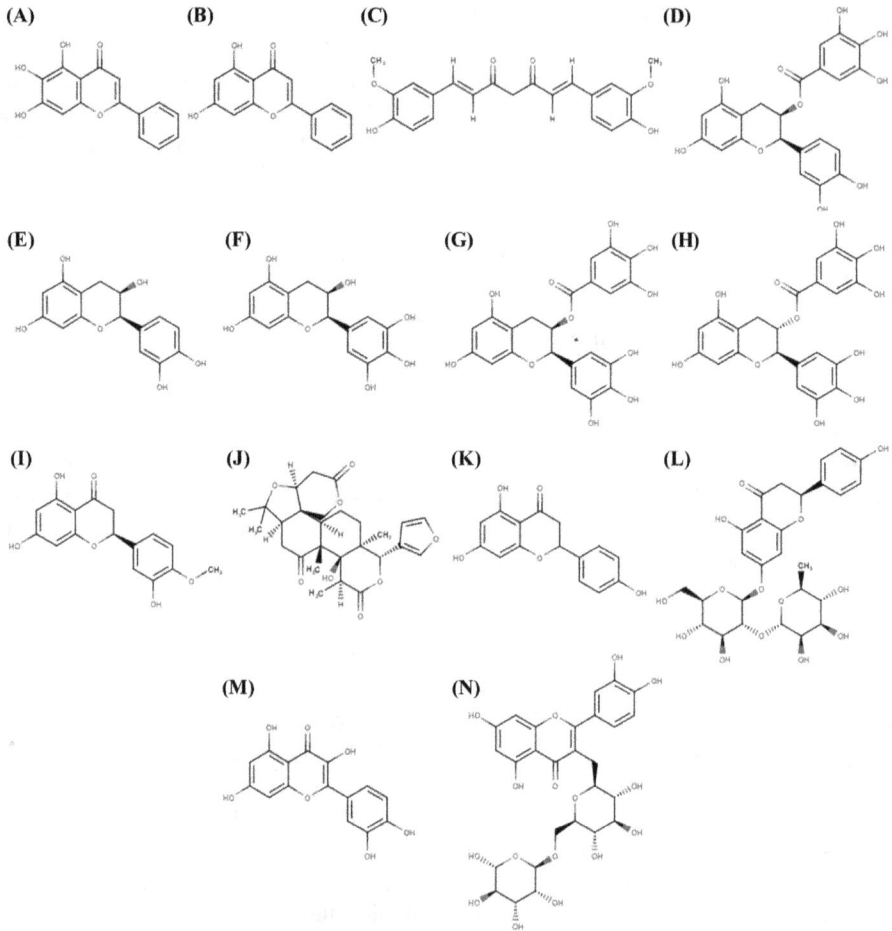

FIGURE 10.2 Chemical structures of (A) Baicalein, (B) Chrysin, (C) Curcumin, (D) Epicatechin gallate, (E) Epicatechin, (F) Epigallocatechin, (G) Epigallocatechin-3-gallate, (H) Gallocatechin gallate, (I) Hesperetin, (J) Limonin, (K) Naringenin, (L) Naringin, (M) Quercetin, and (N) Rutin having biological activity against NTDs.

bioavailability of dietary polyphenols using nanotechnology (Di Lorenzo et al., 2021). Auspiciously, many studies have reported the improvement of the bioavailability of DPs using nanoformulations (Davatgaran-Taghipour et al., 2017; Yang et al., 2020).

10.5 CONCLUSIONS

Several dietary polyphenols have shown promising activity against different pathogens that cause parasitic diseases and NTDs. Dietary polyphenols offer a good alternative to overcome many problems of drug resistance. As compared to NTDs, more dietary polyphenols have been evaluated for their activity against parasitic diseases. Dietary polyphenols achieve their pharmacological activity against parasitic diseases and NTDs by preventing oxidative stress, regulating gene expression, or by inhibiting viral proteins. For dietary polyphenols, more clinical studies and production of clinically consumable products are the need of the hour.

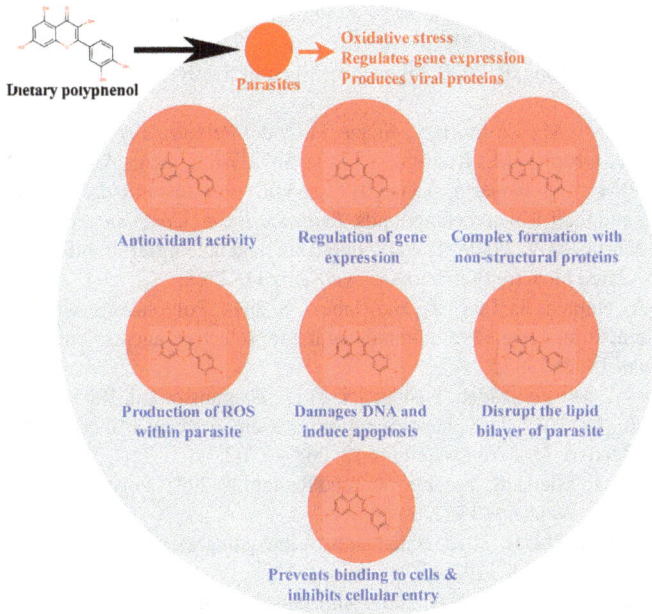

FIGURE 10.3 Possible mechanism of actions of dietary polyphenols against pathogens causing parasitic diseases and NTDs.

REFERENCES

Abe, L.T., Lajolo, F.M., and Genovese, M.I. 2012. Potential dietary sources of ellagic acid and other antioxidants among fruits consumed in Brazil: Jabuticaba (Myrciaria jaboticaba (Vell.) Berg). *J Sci Food Agric* 92: 1679–1687.

Abotaleb, M., Liskova, A., Kubatka, P., and Büsselberg, D. 2020. Therapeutic potential of plant phenolic acids in the treatment of cancer. *Biomolecules* 10: 221.

Allam, G., and Abuelsaad, A.S.A. 2014. In vitro and in vivo effects of hesperidin treatment on adult worms of Schistosoma mansoni. *J Helminthol.* 88: 362–70.

Alves, M.M. de M., Brito, L.M., Souza, A.C., Queiroz, B.C.S.H., de Carvalho, T.P., Batista, J.F., Oliveira, J.S. de S.M., de Mendonça, I.L., Lira, S.R. de S., Chaves, M.H., Gonçalves, J.C.R., Carneiro, S.M.P., Arcanjo, D.D.R., and Carvalho, F.A. de A. 2017. Gallic and ellagic acids: Two natural immunomodulator compounds solve infection of macrophages by Leishmania major. *Naunyn Schmiedebergs Arch Pharmacol* 390: 893–903.

Amisigo, C.M., Antwi, C.A., Adjimani, J.P., and Gwira, T.M. 2019. In vitro anti-trypanosomal effects of selected phenolic acids on Trypanosoma brucei. *PLoS One* 14: e0216078.

Arts, I.C.W., van de Putte, B., and Hollman, P.C.H. 2000. Catechin contents of foods commonly consumed in the Netherlands. 1. Fruits, Vegetables, Staple Foods, and Processed Foods. *J Agric Food Chem* 48: 1746–1751.

Braguine, C.G., Bertanha, C.S., Gonçalves, U.O., Magalhães, L.G., Rodrigues, V., Melleiro Gimenez, V.M., Groppo, M., Silva, M.L.A. e, Cunha, W.R., Januário, A.H., and Pauletti, P.M. 2012. Schistosomicidal evaluation of flavonoids from two species of Styrax against Schistosoma mansoni adult worms. *Pharm Biol.* 50: 925–929.

Burns, J., Yokota, T., Ashihara, H., Lean, M.E.J., and Crozier, A. 2002. Plant Foods and Herbal Sources of Resveratrol. *J Agric Food Chem* 50: 3337–3340.

CDC, 2021a. Parasites. www.cdc.gov/parasites/index.html (accessed December 9, 2021).

CDC, 2021b. Neglected tropical diseases (NTDs) www.cdc.gov/globalhealth/ntd/diseases/index.html (accessed December 10, 2021).

Christaki, E., Marcou, M., and Tofarides, A. 2020. Antimicrobial resistance in bacteria: Mechanisms, evolution, and persistence. *J Mol Evol* 88: 26–40.

Cui, Q., Du, R., Liu, M., and Rong, L. 2020. Lignans and their derivatives from plants as antivirals. *Molecules* 25: 183.

Cunha, N.L., Uchôa, C.J. de M., Cintra, L.S., Souza, H.C. de, Peixoto, J.A., Silva, C.P., Magalhães, L.G., Gimenez, V.M.M., Groppo, M., Rodrigues, V., da Silva Filho, A.A., Andrade e Silva, M.L., Cunha, W.R., Pauletti, P.M., and Januário, A.H. 2012. In vitro schistosomicidal activity of some Brazilian Cerrado species and their isolated compounds. *Evidence-Based Complement. Altern Med.* 1–8.

Dabeek, W.M., and Marra, M.V. 2019. Dietary quercetin and kaempferol: bioavailability and potential cardiovascular-related bioactivity in humans. *Nutrients* 11: 2288.

Daglia, M., Lorenzo, A., Nabavi, S., Talas, Z., and Nabavi, S. 2014. Polyphenols: well beyond the antioxidant capacity: Gallic acid and related compounds as neuroprotective agents: You are what you eat! *Curr Pharm Biotechnol* 15: 362–372.

Davatgaran-Taghipour, Y., Masoomzadeh, S., Farzaei, M.H., Bahramsoltani, R., Karimi-Soureh, Z., Rahimi, R., and Abdollahi, M. 2017. Polyphenol nanoformulations for cancer therapy: experimental evidence and clinical perspective. *Int J Nanomedicine* 12: 2689–2702.

Di Lorenzo, C., Colombo, F., Biella, S., Stockley, C., and Restani, P. 2021. Polyphenols and human health: The role of bioavailability. *Nutrients* 13: 273.

Donald, G., Hertzer, K., and Eibl, G. 2012. Baicalein: An intriguing therapeutic phytochemical in pancreatic cancer. *Curr Drug Targets* 13: 1772–1776.

El-Ansary, A.K., Ahmed, S.A., and Aly, S.A. 2007. Antischistosomal and liver protective effects of Curcuma longa extract in Schistosoma mansoni infected mice. *Indian J Exp Biol* 45: 791–801.

Engels, D., and Zhou, X.N. 2020. Neglected tropical diseases: an effective global response to local poverty-related disease priorities. *Infect Dis Poverty* 9: 10.

Eraky, M.A., El-Kholy, A.A.E.-M., Rashed, G.A.E.-R., Hammam, O.A., Moharam, A.F., Abou-Ouf, E.A.-R., Aly, N.S.M., Kishik, S.M., Abdallah, K.F., and Hamdan, D.I. 2016. Dose-response relationship in Schistosoma mansoni juvenile and adult stages following limonin treatment in experimentally infected mice. *Parasitol Res.* 115: 4045–4054.

Erlund, I., Silaste, M., Alfthan, G., Rantala, M., Kesäniemi, Y., and Aro, A. 2002. Plasma concentrations of the flavonoids hesperetin, naringenin and quercetin in human subjects following their habitual diets, and diets high or low in fruit and vegetables. *Eur J Clin Nutr* 56: 891–8.

Frabasile, S., Koishi, A.C., Kuczera, D., Silveira, G.F., Verri, W.A., Duarte dos Santos, C.N., and Bordignon, J. 2017. The citrus flavanone naringenin impairs dengue virus replication in human cells. *Sci Rep.* 7: 41864.

Gouveia, M.J., Brindley, P.J., Azevedo, C., Gärtner, F., da Costa, J.M.C., and Vale, N. 2019. The antioxidants resveratrol and N-acetylcysteine enhance anthelmintic activity of praziquantel and artesunate against Schistosoma mansoni. *Parasit Vectors* 12: 309.

Güida, M.C., Esteva, M.I., Camino, A., Flawiá, M.M., Torres, H.N., and Paveto, C. 2007. Trypanosoma cruzi: In vitro and in vivo antiproliferative effects of epigallocatechin gallate (EGCg). *Exp Parasitol.* 117: 188–194.

Han, X., Shen, T., and Lou, H. 2007. Dietary polyphenols and their biological significance. *Int J Mol Sci* 8: 950–988.

Hao, H., Aixia, Y., Dan, L., Lei, F., Nancai, Y., and Wen, S. 2009. Baicalin suppresses expression of Chlamydia protease-like activity factor in Hep-2 cells infected by Chlamydia trachomatis. *Fitoterapia* 80: 448–452.

Herraiz, T., Guillén, H., González-Peña, D., and Arán, V.J. 2019. Antimalarial quinoline drugs inhibit β-hematin and increase free hemin catalyzing peroxidative reactions and inhibition of cysteine proteases. *Sci Rep.* 9: 15398.

Karakavuk, M., Aykur, M., Unver, A., and Doskaya, M. 2018. Parasitic diseases that can infect travelers to Africa. *Turkish J Parasitol.* 42.

Kedzierski, L., Curtis, J.M., Kaminska, M., Jodynis-Liebert, J., and Murias, M., 2007. In vitro antileishmanial activity of resveratrol and its hydroxylated analogues against Leishmania major promastigotes and amastigotes. *Parasitol Res.* 102: 91–97.

Khademvatan, S., Eskandari, K., Hazrati-Tappeh, K., Rahim, F., Foroutan, M., Yousefi, E. and Asadi, N., 2019. In silico and in vitro comparative activity of green tea components against Leishmania infantum. *J Glob Antimicrob Resist.* 18: 187–194.

Khan, N., Syed, D.N., Ahmad, N., and Mukhtar, H. 2013. Fisetin: A dietary antioxidant for health promotion. *Antioxid Redox Signal.* 19: 151–162.

Lakshmi, V., Joseph, S.K., Srivastava, S., Verma, S.K., Sahoo, M.K., Dube, V., Mishra, S.K., and Murthy, P.K. 2010. Antifilarial activity in vitro and in vivo of some flavonoids tested against Brugia malayi. *Acta Trop* 116: 127–133.

Lampejo, T. 2020. Influenza and antiviral resistance: an overview. *Eur J Clin Microbiol Infect Dis* 39: 1201–1208.

Lehane, A.M., and Saliba, K.J. 2008. Common dietary flavonoids inhibit the growth of the intraerythrocytic malaria parasite. *BMC Res Notes* 1: 26.

Lin, Y., Shi, R., Wang, X., and Shen, H.M., 2008. Luteolin, a flavonoid with potential for cancer prevention and therapy. *Curr Cancer Drug Targets* 8: 634–646.

Lopes, C.M. 2019. An overview of drug resistance in protozoal diseases. *Int J Mol Sci* 20: 5748.

Luca, S.V., Macovei, I., Bujor, A., Miron, A., Skalicka-Woźniak, K., Aprotosoaie, A.C., and Trifan, A. 2020. Bioactivity of dietary polyphenols: The role of metabolites. *Crit Rev Food Sci Nutr* 60: 626–659.

Mamani-Matsuda, M., Rambert, J., Malvy, D., Lejoly-Boisseau, H., Daulouède, S., Thiolat, D., Coves, S., Courtois, P., Vincendeau, P., and Mossalayi, M.D. 2004. Quercetin induces apoptosis of Trypanosoma Brucei Gambiense and decreases the proinflammatory response of human macrophages. *Antimicrob Agents Chemother* 48: 924–929.

Mehwish, S., Khan, H., Rehman, A.U., Khan, A.U., Khan, M.A., Hayat, O., Ahmad, M., Wadood, A., and Ullah, N. 2019. Natural compounds from plants controlling leishmanial growth via DNA damage and inhibiting trypanothione reductase and trypanothione synthetase: An in vitro and in silico approach. *3 Biotech* 9: 303.

Mohd, A., Zainal, N., Tan, K.K., and AbuBakar, S. 2019. Resveratrol affects Zika virus replication in vitro. *Sci Rep* 9: 14336.

Molyneux, D.H., Savioli, L., and Engels, D. 2017. Neglected tropical diseases: progress towards addressing the chronic pandemic. *Lancet* 389: 312–325.

Mounce, B.C., Cesaro, T., Carrau, L., Vallet, T., and Vignuzzi, M. 2017. Curcumin inhibits Zika and chikungunya virus infection by inhibiting cell binding. *Antiviral Res.* 142: 148–157.

Njomnang Soh, P., Witkowski, B., Gales, A., Huyghe, E., Berry, A., Pipy, B., and Benoit-Vical, F. 2012. Implication of glutathione in the in vitro antiplasmodial mechanism of action of ellagic acid. *PLoS One* 7: e45906.

Oktaviyanti, R.N., Prakoeswa, C.R.S., Indramaya, D.M., Hendradi, E., Sawitri, S., Astari, L., Damayanti, D., and Listiawan, M.Y., 2020. Topical Epigallocatechin Gallate (EGCG) 1% for chronic plantar ulcers in leprosy. *Berk Ilmu Kesehat Kulit dan Kelamin* 32: 134.

Pandey, K.B., and Rizvi, S.I. 2009. Plant polyphenols as dietary antioxidants in human health and disease. *Oxid Med Cell Longev* 2: 270–278.

Patel, K., Singh, G.K., and Patel, D.K. 2018. A review on pharmacological and analytical aspects of Naringenin. *Chin J Integr Med* 24: 551–560.

Pietta, P.G. 2000. Flavonoids as antioxidants. *J Nat Prod* 63: 1035–1042.

PLOS, 2021. PLOS neglected tropical diseases https://journals.plos.org/plosntds/s/journal-information (accessed December 12, 2021).

Pohjala, L., Utt, A., Varjak, M., Lulla, A., Merits, A., Ahola, T., and Tammela, P. 2011. Inhibitors of alphavirus entry and replication identified with a stable Chikungunya Replicon cell line and virus-based assays. *PLoS One* 6: e28923.

Reinisalo, M., Kårlund, A., Koskela, A., Kaarniranta, K., and Karjalainen, R.O. 2015. Polyphenol stilbenes: Molecular mechanisms of defence against oxidative stress and aging-related diseases. *Oxid Med Cell Longev.* 2015: 1–24.

Riva, A., Kolimár, D., Spittler, A., Wisgrill, L., Herbold, C.W., Abrankó, L., and Berry, D. 2020. Conversion of rutin, a prevalent dietary flavonol, by the human gut microbiota. *Front Microbiol* 11.

Salehi, B., Fokou, P., Sharifi-Rad, M., Zucca, P., Pezzani, R., Martins, N., and Sharifi-Rad, J. 2019. The therapeutic potential of Naringenin: A review of clinical trials. *Pharmaceuticals* 12: 11.

Serrano, J., Puupponen-Pimiä, R., Dauer, A., Aura, A.-M., and Saura-Calixto, F. 2009. Tannins: Current knowledge of food sources, intake, bioavailability and biological effects. *Mol Nutr Food Res* 53: S310–S29.

Shankar, E., Goel, A., Gupta, K., and Gupta, S. 2017. Plant flavone apigenin: An emerging anticancer agent. *Curr Pharmacol Reports* 3: 423–446.

Soh, P.N., Witkowski, B., Olagnier, D., Nicolau, M.L., Garcia-Alvarez, M.C., Berry, A., and Benoit-Vical, F. 2009. In vitro and in vivo properties of ellagic acid in malaria treatment. *Antimicrob Agents Chemother* 53: 1100–1106.

Somsak, V., Damkaew, A., and Onrak, P. 2018. Antimalarial activity of kaempferol and its combination with chloroquine in Plasmodium Berghei infection in mice. *J Pathog* 1–7.

Stanić, Z. 2017. Curcumin, a compound from natural sources, a true scientific challenge – A review. *Plant Foods Hum Nutr* 72: 1–12.

Stompor-Gorący, M., Bajek-Bil, A., and Machaczka, M. 2021. Chrysin: Perspectives on contemporary status and future possibilities as pro-health agent. *Nutrients* 13: 2038.

Tsao, R. 2010. Chemistry and biochemistry of dietary polyphenols. *Nutrients* 2: 1231–1246.

Ullah, R., Rehman, A., Zafeer, M.F., Rehman, L., Khan, Y.A., Khan, M.A.H., Khan, S.N., Khan, A.U., and Abidi, S.M.A. 2017. Anthelmintic potential of thymoquinone and curcumin on Fasciola Gigantica. *PLoS One* 12: e0171267.

WHO, 2021. Vector-borne and parasitic diseases www.euro.who.int/en/health-topics/communicable-diseases/vector-borne-and-parasitic-diseases (accessed December 10, 2021).

Yamazaki, T., Inoue, M., Sasaki, N., Hagiwara, T., Kishimoto, T., Shiga, S., Ogawa, M., Hara, Y., and Matsumoto, T. 2003. In vitro inhibitory effects of tea polyphenols on the proliferation of Chlamydia trachomatis and Chlamydia pneumoniae. *Jpn J Infect Dis* 56: 143–145.

Yang, B., Dong, Y., Wang, F., and Zhang, Y. 2020. Nanoformulations to enhance the bioavailability and physiological functions of polyphenols. *Molecules* 25: 4613.

Zandi, K., Teoh, B.T., Sam, S.S., Wong, P.F., Mustafa, M., and Abubakar, S. 2011a. Antiviral activity of four types of bioflavonoid against dengue virus type-2. *Virol J* 8: 560.

Zandi, K., Teoh, B.T., Sam, S.S., Wong, P.F., Mustafa, M.R., and Abubakar, S. 2011b. In vitro antiviral activity of Fisetin, Rutin and Naringenin against Dengue virus type-2. *J Med Plant Res* 5: 5534–5539.

11 Dietary Polyphenols for the Management of Skin Diseases and Wound Healing

*Shiv Kumar Prajapati, Dolly Jain, Sapna Joshi,
Payal Kesharwani, and Siddhartha Maji
*E-mail: shivprajapati1992@gmail.com; dollyjain.btpc@gmail.com;
joshisapna2693@gmail.com; payal.kesharwani705@gmail.com;
siddharthamaji519@gmail.com

CONTENTS

11.1 INTRODUCTION

Apart from their favorable effects on the plant host, phenolic metabolites (polyphenols) have a number of biological features that have a positive impact on human health. Polyphenols are one of the major sources of antioxidants in humans. They are natural compounds that may be found in a variety of fruits (apple, berries, cherries, pear, pomegranate, and grapes) and plants (cocoa, coffee, tea, etc.) (Shahidi and Ambigaipalan, 2015). There are more than 8,000 polyphenolic derivatives recognized. Polyphenols can range from 200 to 300 mg per 100 gm of fresh weight in fruits. Polyphenols are present in substantial concentrations in the products made from fruits. A glass of red wine, a cup of tea, or a cup of coffee typically contains around 100 mg of polyphenols.

Cereals, dried beans, and chocolate are all good sources of polyphenols (Spencer et al., 2008, Scalbert et al., 2005). Collectively, these chemicals are known as phytochemicals. For a variety of reasons, documenting the benefits of polyphenols has proven to be quite challenging; some of the reasons are: a wide range of dietary polyphenols in food, lack of data about polyphenolic content of widely used food compositions, and habitual food consumption is difficult to quantify and characterize. The degree of absorption and metabolic outcome of various polyphenols derived from certain foodstuffs is poorly understood (Spencer et al., 2008). Astringency, bitterness, color, odor and taste are all characteristics of dietary polyphenols. Oxidative stability against oxidative stress is a major trait that plays a prominent role in diseased conditions (Beckman, 2000). Plants are naturally protected by polyphenols from the harmful UV radiation of the sun, so topically applied polyphenols may also favorably supplement sunscreen protection due to their recognized anti-inflammatory, antioxidant, and DNA repair properties. Apart from their favorable effects on the plant host, phenolic metabolites (polyphenols) have a number of biological features that have a positive impact on human health. Regular consumption of polyphenol-rich foods protects the internal cells from free radical damage, which can help to fend off ailments such as Alzheimer's, diabetes, cancer, neurodegenerative and heart-related disease, and so forth (Arts and Hollman, 2005, Graf et al., 2005). Polyphenols, on the other hand, are impressive skin-care agents. These agents, when applied topically, can help heal and rejuvenate skin. Polyphenols have been found in studies to protect skin from cancer and other kinds of UV damage. Plant phenolics, which may be taken through the diet or by topical application, have been shown to reduce symptoms and prevent the occurrence of many skin conditions. Polyphenols also have a preventive effect or can delay the onset of aging of the skin, skin illnesses, such as wrinkles and acne, as well as serious, potentially life-threatening diseases like skin cancer (Działo et al., 2016). Polyphenol's health-promoting functions, such as neutraceuticals, are dependent on not only dietary intake but also on their bioavailability and molecular interactions. However, the polyphenols that are most plentiful in the human diet are not the ones with the highest bioavailability profiles. The bioavailability of polyphenols and polyphenol-comprising foods primarily depends on (1) their chemical structure (e.g., glycosylation, esterification, and polymerization); (2) their intestinal absorption, during which the microflora plays a vital role in polyphenols catabolism; (3) their incorporation inside the food matrix; and (4) their elimination again into the lumen of the intestine (D'Archivio et al., 2007, Manach et al., 2004). The polyphenols that are most plentiful in our diet are often not the ones with the highest bioavailability. Polyphenol bioavailability is, in fact, rather too varied (Manach et al., 2005). Polyphenols administered topically may be a good way to enhance sunscreen protection and other modalities because of their antioxidant, anti-inflammatory, and DNA repair characteristics. Generally, controlled transdermal administration of polyphenols does have benefits over oral or intravenous delivery in terms of improving localized exposure while markedly minimizing toxicity (Morganti et al., 2002, Pinnell, 2003). Over the previous few years, various nanocarrier-based drug delivery systems using SLNs, liposomes, metallic NPs, biodegradable polymers, dendrimers, magnetic NPs, nanoemulsions and magnetic nanoparticles, have attracted a lot of interest (Prajapati et al., 2019, Prajapati and Jain, 2020, Prajapati et al., 2020, Jain et al., 2018, Jain et al., 2020). Although topically applied delivery of nano-polyphenols (e.g., as dermo-therapeutics or cosmeceuticals) may be one way to overcome the problems of bioavailability (e.g., low solubility, low gut absorption) and toxicity (e.g., side effects of a high or accumulated dose of polyphenols) associated with several dietary polyphenols to prevent or treat skin conditions, however, it is necessary to perform bioavailability studies of these formulations.

The skin provides an excellent and accessible region for administering drugs. With a surface area of 1.7 m², the skin is the major organ (16% of body mass). The skin defends our bodies from different harmful agents like allergens, infective agents, and other chemical agents and also prevents harmful contaminations like UV radiation (Liu et al., 2014, Williams and Barry, 2012, Benson and Watkinson, 2012, Torres-Contreras et al., 2022, Rahman et al., 2022). The epidermis, which comprises the stratum corneum, is the outermost layer of skin; the dermis is the intermediate layer; and the hypodermis, the innermost layer (Vestita et al., 2022, Morais et al., 2022, Schoellhammer et al., 2014). Despite being just 15 to 20 µm thick, the stratum corneum (SC) is responsible for the vast majority of the skin's barrier function. It is comprised of dead corneocytes rich with keratin that is densely bound and surrounded by a lipid matrix. This lipid matrix, which is made up mostly of ceramides (50%) and cholesterol (25%) as well as many other free fatty acids, is around 100 nm wide, restricting passive diffusion to small, lipophilic molecules (Madison, 2003, Deli, 2009).

Topical and transdermal therapeutics technologies are appealing techniques for the prevention and treatment of dermatological disorders, where oral and parenteral treatment of drugs does not deliver enough pharmaceuticals to the disease site. This is due to the numerous anatomical and physiological barriers that a drug must cross before reaching a specific layer of the skin. Thus, treating skin disorders like melanoma would be significantly more hopeful and relevant if the drug could be delivered directly to the diseased site (Saleem et al., 2008). The keratinocyte-rich epidermis underneath the SC is around 100 µm thick and formed of keratinocytes. The dermis, the skin's thickest layer, is made up of connective tissue and measures between 1000 and 2000 µm in thickness. Below the SC, the epidermis is made up of keratinocytes, which can release cytokines and chemokines to boost immunological activity at the infection site. Transdermal administration is appropriate for drugs with low molecular weight and high log P values. Large molecular drugs like growth hormones and insulin, on the other hand, are difficult to distribute over the skin (Raza et al., 2013, Barry, 2001). As a result, a variety of penetration-enhancing strategies have been explored and developed to improve skin permeation and thermodynamic activity. Nanocarrier-assisted drug delivery is a passive strategy for improving medication penetration via the transdermal route. Therapeutic macromolecules (>500 Da, polar, hydrophilic) such as peptides and proteins are delivered via active techniques (Barry, 2001, Brown et al., 2006).

11.2   CLASSIFICATION AND SOURCES OF DIETARY POLYPHENOLS

Numerous species of plants have been shown to have over 8,000 polyphenolic components. Phenylalanine is the source of almost all of the phenols obtained from plants. They are most commonly seen in conjugated forms, with one or more sugar residues attached to hydroxyl groups; however direct interactions between the sugar (polysaccharide or monosaccharide) and an aromatic carbon are indeed possible. The classification of polyphenols is done on the basis of their structural elements, a number of the phenyl ring, and the type of linkage between these rings. Polyphenols are mainly categorized into four categories, that is: (a) Phenolic acid, (b) flavonoids, (c) stilbenes and (d) lignans. Plants generate polyphenols, which are highly distributed throughout plant tissues and mostly exist as glycosides. Flavonoids can be found as glycosides or aglycones, despite their primary structure being aglycones. The fundamental structure of all flavonoids is diphenyl propane (C6-C3-C6), wherein phenolic rings are frequently connected by a heterocyclic ring (Khoddami et al., 2013). Other polyphenols, such as anthocyanidins, anthocyanins, and so forth, are synthesized by altering the hydroxylation pattern and oxidation state of the core pyran ring. The existence or absence of a double bond between C2 and C3, as well as the production of a carbonyl group by C4, determine the hydroxylation pattern and oxidation state. Flavones, flavonols, flavanones, and flavanonols form a significant subgroup of all polyphenols and, as a result, they make up the bulk of flavonoid compounds (D'Archivio et al., 2007, Singla et al., 2019). Dietary polyphenols and

TABLE 11.1
Dietary Polyphenols and their Source

Dietary Polyphenol	Class	Source
Cyanidin, Malvidin, Delphinidin, Pelargonidin	Anthocyanidins	Black currant, blueberries, cherries, cranberry, grape, plums, pomegranate, raspberry, strawberries
Naringenin, Hesperetin	Flavanones	Grapes, lemon, orange, orange, peppermint
Genistein, Daidzein	Isoflavones	Grape, soy flour, soy nuts, soybean, tofu, yogurt,
Rutin, Hesperidin, Naringin	Flavonoid glycoside	Grapefruit, lemon, orange juice, orange
Catechin, Epicatechin, Epigallocatechin, Gallocatechin, Procyanidins, Prodelphinidins	Flavanols	Apples, apricots, blackberries, blueberries, cherries, cocoa, grapes, peaches, pears, raisins, tea
Myricetin, Fisetin, Quercetin Kaempferol,	Anthoxanthins	Apples, apricots, beans, blackberries, blueberries, cranberries, broccoli, cherry tomatoes, grapes, hot peppers, lettuce, onions, fennel, spinach, sweet potato leaves,
Apigenin, Luteolin	Flavones	Dry parsley, fresh parsley, olives, oregano, peppers, rosemary
Resveratrol	Stilbenes	Peanuts, grapes
Catechin polymers, Proanthocyanidins, Tannic acids	Tannins	Apple, blackberry, coffee, Grape, pea, peach, plum, pomegranate, raspberries, strawberries, tea, walnuts
Caffeic acid, P-coumaric acid Ferulic acid, Chlorogenic acid,	Hydroxycinnamic acids	Apple, apple, blueberry, cherry, coffee, cranberry, grapefruit, lemon, orange, peach, potato, spinach, tea,
Gallic acid, Ellagic acid	Hydroxybenzoic acid	Grapes, strawberries, raspberries, pomegranate juice

their source are described in Table 11.1. The classification of dietary polyphenols is illustrated in Figure 11.1.

11.2.1 PHENOLIC ACIDS

Phenolic acids cover one-third of all dietary polyphenols and are found in all plants; however, acidic fruits have the highest concentration (Corona et al., 2006). In general, "phenolic acids" refers to phenols that have one carboxylic acid activity. Plant metabolites, on the other hand, relate to a specific group of organic acids (Stalikas, 2007, Herrmann and Nagel, 1989). Cinnamic acids and benzoic acid hydroxyl derivatives are two types of phenolic acids (Incocciati et al., 2022, Morais et al., 2022). The polyphenols containing cinnamic acid derivatives are more common than the benzoic acid derivates (Pandey and Rizvi, 2009). The hydroxycinnamic acids, which are extensively dispersed in the plant kingdom, are a prominent class of phenolic chemicals. Caffeic acid is the very common hydroxycinnamic acid and, in food, it is generally found as chlorogenic acid ester conjugated with quinic acid (5-caffeoylquinic acid) (Han et al., 2007). Various phenolic structures are illustrated in Figure 11.2.

11.2.2 FLAVONOIDS

Flavonoids are a group of polyphenolic chemicals that are found in abundance in nature (vegetables) (González et al., 2011). The basic structure of this group is an oxygenated heterocycle formed by two aromatic rings joined together by three carbon atoms. There are about 4,000 distinct kinds of flavonoids, many of which are responsible for the beautiful bright colors of flowers, fruits, and foliage (Pandey and Rizvi, 2009, de Groot and Rauen, 1998). Flavonoids are further classified into six categories on the basis of heterocycles present in their structure, that is, anthocyanins,

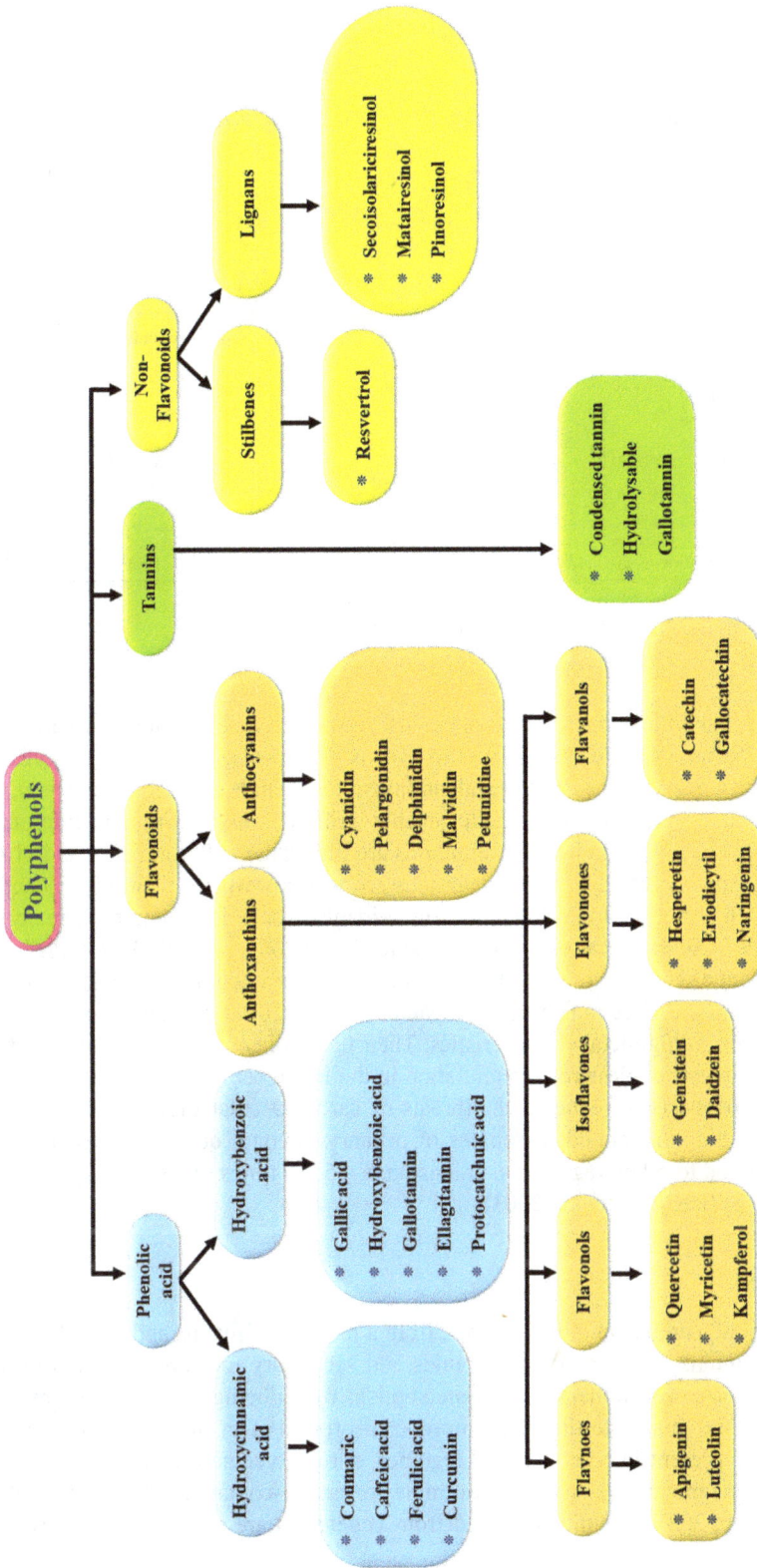

FIGURE 11.1 Classification of polyphenols.

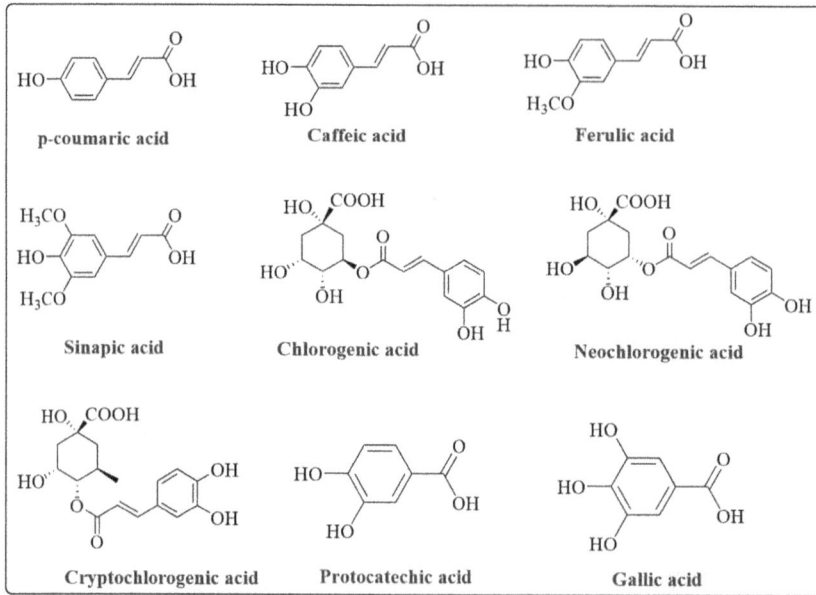

FIGURE 11.2 Chemical structures of some phenolic compounds.

flavanols, flavones, and flavanones, isoflavones. Individual variances within each category are due to differences in the number and arrangement of hydroxyl groups, as well as their alkylation and/or glycosylation levels. Some of the most prevalent flavonoids include quercetin, myricetin, catechins, and others. catechin, kaempferol, epigallocatechin (EPGC), myricetin, apigenin, naringenin, taxifolin, and so forth, and their chemical structures are illustrated in Figures 11.3 and 11.4 (Pandey and Rizvi, 2009, Spencer et al., 2008). Flavonoids are well-known for their antioxidant and anti-radical effects (Amic et al., 2007). While the antioxidant defence is important in many flavonoid effects (Loke et al., 2008). Because of their considerable redox potential, flavonoids have a wide range of antioxidant effects. As a result, the phenolic compounds could function as reducing agents. They act as scavengers of free radicals (Sellappan and Akoh, 2002) and singlet oxygen quenchers, as well as having chelating metal characteristics. Their flavonol concentration significantly reduces atherosclerosis, prevents cholesterol accumulation in the blood serum, and improves vascular wall resistance and also assists in reducing the threats of cardiovascular diseases. In the human diet, onions are one of the most abundant sources of primary flavonol quercetin (Sellappan andAkoh 2002). In comparison to other vegetables, onions have a 5–10 times higher total quercetin concentration (347 mg/kg) (Lachman et al., 2003).

11.2.3 TANNINS

Tannins (tannic acid) can be naturally obtained from a number of plants, and they show good solubility in water. Tannins are phenolic compounds and secondary metabolites found in plants that have favorable effects on ruminant protein metabolism by reducing rumen breakdown of dietary protein and also in the small intestine; they improve absorption of an amino acid (Hassanpour et al., 2011). Tannins are present in practically all legumes, shrubs, vegetables, and fruits on the planet. Tea, wine, and pomegranate are just a few examples of tannin-containing products (Scalbert et al., 2002, Sharma et al., 2021). Tannins obtained from vegetables are easily soluble in water bearing

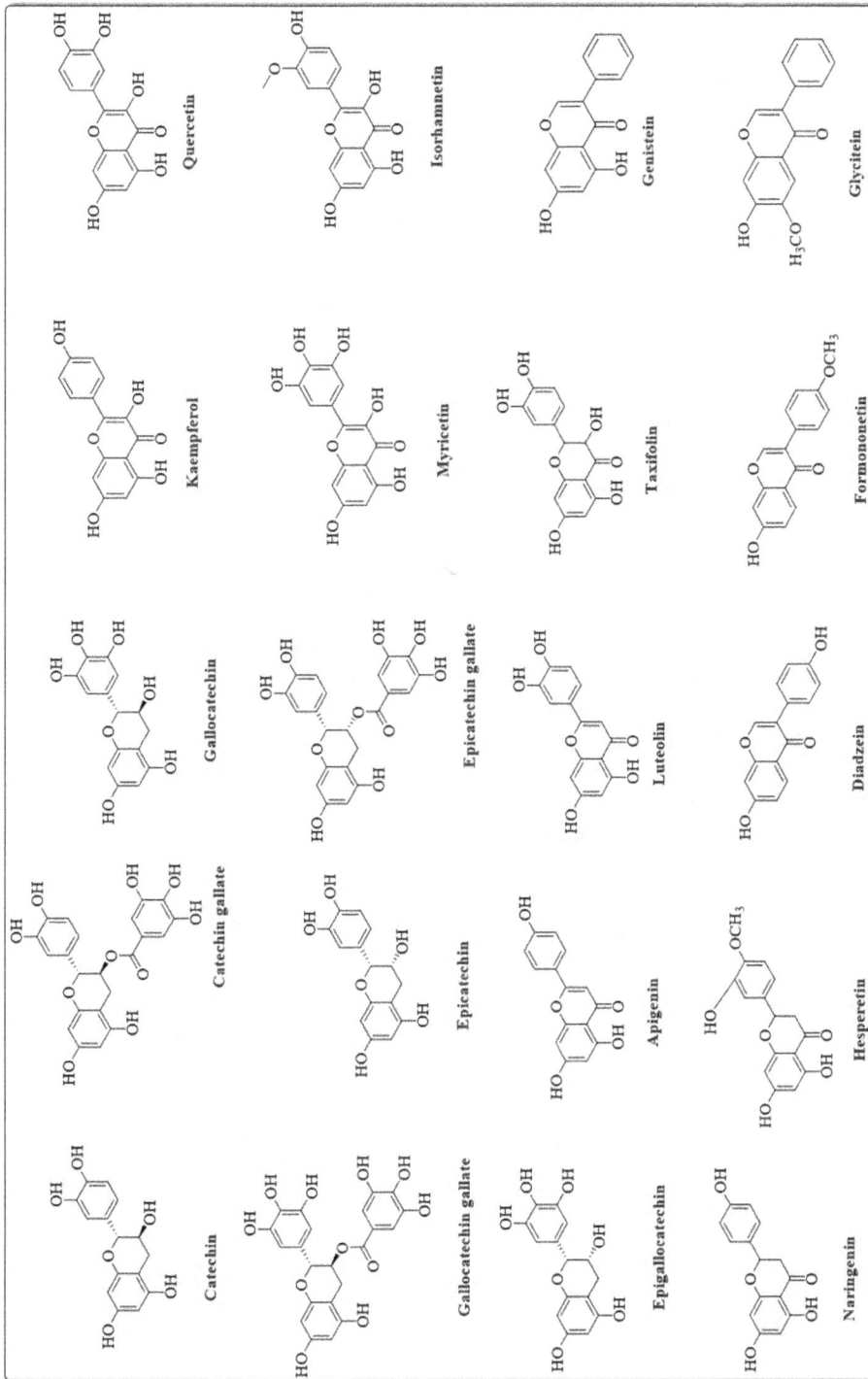

FIGURE 11.3   Chemical structures of different anthoxanthin flavonoids.

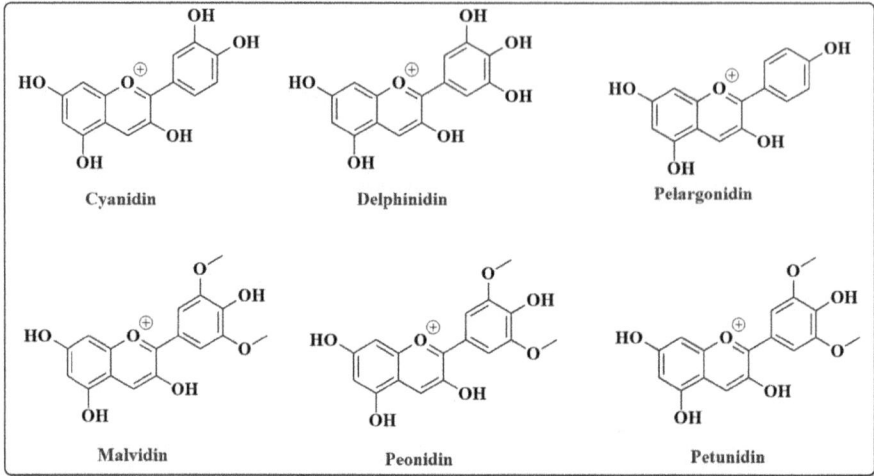

FIGURE 11.4 Chemical structures of different anthocyanin flavonoids.

molecular weight 500–3000D. Because these polyphenols have a high concentration of hydroxyl or other functional groups (1 to 2 per 100 D), they can create cross-links with proteins and other macromolecules. Low-molecular-weight phenolic compounds (molecular weight less than 500) and high-molecular-weight phenolic compounds (molecular weight greater than 3000) are ineffective tanning agents. On reaction with gelatin, alkaloid and protein tannins become precipitated. Tannic acids are categorized into condensed tannins, that is, hydrolysable tannins and nonhydrolyzable tannins. Gallotannins or hexahydroxydiphenic acid esterify the basic structure of polyphenol, such as glucose, and hydroxyl groups into the ellagitannins, that is, hydrolyzable tannins, either partially or entirely. Gallotannins are hydrolyzed by acids, bases, or enzymes to produce glucose and gallic acids. Lactonization transforms ellagitannins hexahydroxydiphenic acid into ellagic acid (Das et al., 2020, Shirmohammadli et al., 2018). Daré et al. (2020) reported the potential of tannic acid. Compared to gallic acid, tannic acid displayed more antioxidant activity. superoxide anion, peroxyl, nitric oxide and peroxynitrite, and reduced ferric ions. Treatment with tannic acid showed the potential against photoaging by depolarizing mitochondrial membrane potential, by inhibiting lipid peroxidation and MMP-1 expression, which is involved in dermis structural degeneration (Daré et al., 2020).

11.2.4 STILBENES

In stilbenes, a two-carbon methylene bridge joins two phenyl moieties. Stilbenes are quite uncommon in the human diet (Pandey and Rizvi, 2009, Shamsi et al., 2021, Dhalaria et al., 2020). Stilbenes are a class of secondary metabolites originating from the phenylpropanoid pathway that is formed in a range of plant species (Orduña et al., 2022, Aleynova et al., 2022). Plant disease resistance and human health are both affected by these chemicals. Plant stilbenes come from the phenylpropanoid pathway in general. Malonyl-CoA and CoA-esters of cinnamic acid derivatives appear to be synthesized by all higher plants; however, stilbenes appear to be produced by only a few plant species. Grape (Vitaceae), pine (Pinaceae), peanut (Fabaceae), and sorghum (Fabaceae) are examples of common plant stilbenes isolated from several plant families (Poaceae) (Chong et al., 2009). The majority of stilbenes in plants are antifungal phytoalexins, which are chemicals produced solely in reaction to infection or injury. One of the most well-studied natural polyphenol stilbenes is resveratrol, also termed as 3,4',5-trihydroxystilbene, which is prevalent largely in grapes (Yan et al., 2022, Li et al., 2022). Red wine is made from grapes and includes a large quantity of resveratrol. Secondary metabolites are important in protecting plants from environmental challenges, and pathogen resistance comprises both inherent and induced defense responses. Both constitutive and inducible defensive responses comprise high quantities of stilbenes, which pre-exist in certain plants or are produced in consequence of microbial invasion (Jeandet et al., 2010). In addition to having a recognized function as phytoalexins, stilbenes may be exploited as chemical signals in allelopathy or response to oxidative stress generated by UV irradiation (Chong et al., 2009, Seigler, 2006, He et al., 2008).

11.3 DIETARY POLYPHENOLS IN SKIN DISEASES AND WOUND INFECTIONS

11.3.1 ANTI-AGING

Aging is thought to be a fundamental physiological phenomenon that is characterized by systemic changes in cell structural integrity induced by changes in metabolic and signal transduction pathways (Li and Zhang, 2017). Natural compounds have sparked a lot of interest in terms of improving health, reducing chronic health diseases, and prolonging the lifespan. Polyphenol extracts have a lot of promise in the investigation of aging and associated disorders in the elderly. In model organisms, plant polyphenols have been shown to moderate the aging process. The ability to scavenge free radicals and antioxidant properties is the key factor to show anti-aging benefits. Resveratrol, a polyphenol molecule found in red wine, has been shown to prevent aging in Caenorhabditis worms via lowering mitochondrial respiration (Wood et al., 2004). In a number of animal models, polyphenols including EGCG, curcumin, and quercetin have been shown to increase lifespan (Prajapati et al., 2021, Shahidi and Ambigaipalan, 2015). Because polyphenols have phenolic hydroxyl groups on their structures, they may directly scavenge reactive oxygen species (ROS). Polyphenol's ability to scavenge ROS is influenced by the amount and location of hydroxyl groups, substituent patterns, and glycosylation of phytochemical compounds (Nenadis et al., 2004, Lu et al., 2002). Polyphenols can serve as antioxidants by influencing the synthesis and activity of endogenous antioxidant and oxidase enzymes. As the main step to neutralize oxidants, two intrinsic enzymes, sodium oxide dismutase 1 (SOD1) and SOD2, present in the cytosol, and matrix of mitochondria, respectively, convert superoxide to hydrogen peroxide (H_2O_2). H_2O_2 is broken down into water and oxygen by catalase and glutathione peroxidases (GSH-Px); 0.027 percent quercetin in diet, 100 mg/kg EGCG, and 8 mg/kg curcumin reversed the oxidative stress-triggered decrease in the level of GSH and SOD levels in mice or rats (Singh et al., 2015, Roy et al., 2012, Uygur et al., 2014). Polyphenols potentially stimulate antioxidant activity in cells through modulating the Nrf2-mediated pathway. Nrf2 "a transcription factor" regulates the formation of detoxifying enzymes such as GPx1, GSH, GST, HO-1, NADP(H) quinone oxidoreductase 1 (NQO1), and SOD by interacting to antioxidant-response elements in the promoter regions of genes (Kobayashi and Yamamoto, 2005, Choi et al., 2021, Venza et al., 2021). Curcumin, EGCG, epicatechin, and luteolin are among the polyphenols that can promote Nrf2 DNA-binding ability or protein expression, and hence increases NQO1, HO-1, and SOD expression (Shah et al., 2010, Rushworth et al., 2006, Wu et al., 2006, Wruck et al., 2007). Curcumin has been found to directly scavenge ROS, enhance endogenous antioxidant expression, activate the Nrf2 pathway, and regulate miRNA (Figure 11.5, A) (Luo et al., 2021).

11.3.2 TOPICAL ALLERGIES

Skin, food, and respiratory allergies are all examples of allergic diseases. When the immune system is sensitized to a typically innocuous allergen, the immune system is skewed toward a major T-helper type 2 response. Polyphenols are a naturally occurring chemical present in foods and plants that have been explored for their anti-allergic capabilities in a variety of skin disease models. Their anti-inflammatory nature is known to influence the migration of immune cells to the skin and the prevention of subsequent infections if the skin barrier is disrupted (Singla et al., 2019). Atopic eczema and dermatitis are allergic skin problems that typically develop in early childhood and persist until the age of 2. They cause the skin to become dry, itchy, inflamed, and erythematous (redness) (Spergel, 2010). Green tea extracts (catechins, epicatechin, epigallocatechin gallate, and derivatives) are anti-inflammatory. In vitro, EGCG inhibits the release of the pro-inflammatory cytokine IL-2, which is a key mediator in allergic skin diseases (Ichikawa et al., 2004, Katiyar et al., 1995). In a study, Hu and Zhou, (2021) reported the potential effect of gallic acid (GA) on atopic dermatitis (AD) like skin inflammation. When GA-treated model mice were compared with the control group, serum IgE and TNF- levels were found to be reduced. In the ear, inflammatory factors (IL-4, IL-5, IL-17, or IL-23) were found to be significantly lower in GA-treated model mice compared to the control group mice, whereas IL-10 and TGF-expression levels were considerably higher in GA-treated model mice. In the GA-treated model group, ROR-t decreased, but SOCS3 expression increased. These findings suggest that GA may help to reduce AD-like skin problems (Hu and Zhou, 2021). In a pilot study, Kojimo et al. (2000) evaluated apple condensed tannins (ACT) for their antiallergic effect in AD patients. Cracking, inflammation, itching, lichenification, sleep disruption, and peripheral blood eosinophil levels were all found to be decreased by an ACT supplement given to the patients at oral dosages of 10 mg/kg every day up to 8 weeks. When compared to the control group, itching and sleep-disturbances ratings were found to be much lower after taking the ACT supplement for just two weeks. The research showed that ACT had an anti-allergic impact and it that reduced the symptoms of AD (Kojima et al., 2000). Polyphenols have also been explained as alleviating the itching and pruritus that are common in these illnesses. Avenanthramide, a polyphenol found in oats, has been observed to relieve the signs of pruritus and itching (Meydani, 2009).

11.3.3 Skin cancer

Skin cancer (SC) is among the most common kind of cancer worldwide; it is caused by mutations in cancer-related genes, such as proto-oncogenes and in skin cells, tumour suppressors lead to an imbalance in abnormal skin proliferation and cell homeostasis. The development of skin tumours can be associated with several causes, while high UV radiation exposure is regarded as the primary risk factor for skin cancer (HeenatigalaPalliyage et al., 2019). Phenolic compounds influence carcinogenesis by inducing cell-defense mechanisms, such as detoxifying and antioxidant enzyme systems, and inhibiting anti-inflammatory and anti-cellular growth signaling pathways as well, which result in cell-cycle arrest and/or apoptosis. These observations indicated that polyphenols have anticancer properties owing to their potential to modify the epigenome of cancer cells. Natural dietary polyphenolic phytochemicals can exhibit anticancer effects via several pathways, including cell-cycle signaling regulation, anticancer drug elimination, antioxidant enzyme activity, apoptosis, and cell-cycle arrest (Sharma et al., 2018, Han et al., 2019, Nasir et al., 2020). These dietary polyphenols chemotherapeutic activity has been confirmed by their excellent prevention of invasion, metastasis, and angiogenesis. Additionally, dietary polyphenols alleviate the risk and progression of malignancies associated with inflammation. Curcumin, resveratrol, quercetin polyphenolic dietary phytochemicals have been broadly investigated as powerful antioxidants that have been shown to induce apoptosis and cell-cycle arrest in a variety of cancer cell lines (Figure 11.5, B).

11.3.4 Wound healing

In living animals, the healing of open wounds is a difficult task that requires a range of physiological and biochemical processes. The physiological reactions of many organs and tissues of the entire organism, as well as a group of cells, enzymes, hormones, and other factors, are commonly involved

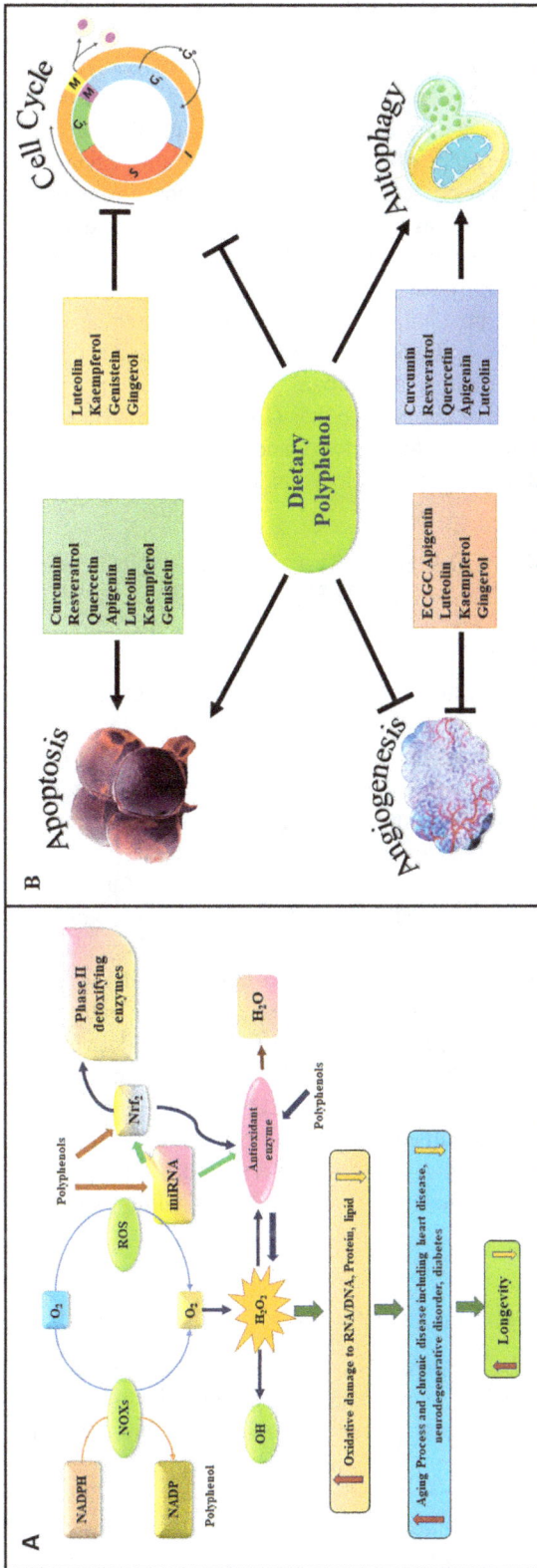

FIGURE 11.5 (A) Polyphenol's antioxidant processes. Nrf2: nuclear factor erythroid 2-related factor 2; ROS: reactive oxygen species. (B) Anticancer mechanism of different dietary polyphenols.

in wound healing. Currently, the mode of action of drugs used to improve wound healing mostly consists of two aspects: Certain drugs will directly activate the body's immune response to release additional immune cells, consequently speeding up the inflammatory response. Moreover, drugs can assist the growth of new tissue by fostering the development and transfer of granulation tissue in the wound, as well as improving the vitality and differentiating potential of granulation cells, thereby significantly improving the emergence of new tissue in the wound (Zhao et al., 2021). The skin is the human body's outermost covering, and it is constantly exposed to external stressors, including UV radiation and harmful substances, as well as mechanical wounds and injury. The skin's ability to repair the damage is critical for survival, and it is affected by a variety of illnesses that result in skin pathologies (Yang et al., 2016). Polyphenols have a profound interest in wound healing because of their antibacterial, regenerative, and antioxidant characteristics (Ghuman et al., 2019). Polyphenols are considered to have significant antioxidant activity, protecting against ROS by donating an electron or a hydrogen atom to free radicals. Furthermore, certain polyphenols possess antibacterial properties against bacteria found in chronic wounds (Liakos et al., 2015). The phenolic substances such as chlorogenic acid, ferulic acid, kaempferol, and resveratrol have strong antioxidant effects and can help improve wound healing in chronic wounds (Guimarães et al., 2021).

11.4   NANOCARRIER-BASED DELIVERY OF DIETARY POLYPHENOLS IN THE MANAGEMENT OF SKIN DISEASES/WOUND HEALING

11.4.1   LIPID-BASED NANOCARRIERS

Considering their propensity to promote drug penetration in the skin layers, lipid-based carriers, particularly liposomes, have been intensively explored for topical applications. They are meant to adhere to the layers of the skin and then destabilize, fuse, or blend with the lipid matrix phase, but they could also function as penetration enhancers by dissolving the stratum corneum's lipid structure and encouraging poor barrier function, resulting in increased partitioning of drug into stratum corneum (El Maghraby et al., 2008, Prasanthi and Lakshmi, 2012). Caddeo et al. (2016) reported the potential effect of liposomes incorporating quercetin and resveratrol. The cellular uptake was significantly increased when incorporated with co-delivered via liposomes, and this resulted in the improved ability to scavenge ROS in fibroblasts. In vivo skin injury was evaluated in the mouse model. The quercetin and resveratrol, when dispersed in oleic acid and buffer, did not show efficacy toward TPA injury (while, when both the polyphenol individually delivered via liposomes, this showed moderate change in the injury (Figure 11.6, D-F). Co-delivery of both polyphenol via liposomes significantly attenuated the lesions and the wound was completely healed (Figure 11.6, G) (Caddeo et al., 2016). Liakopouiou et al. (2021) developed different lipidic nanocarriers, including NLCs, SLNs and nanoemulsions to deliver curcumin. The outcomes revealed that the curcumin-loaded lipidic nanocarrier showed considerable protection against photodegradation. Among all other formulations, NLCs showed rapid wound-healing potential at minimum dose of CUR (Liakopoulou et al., 2021).

11.4.2   POLYMERIC NANOPARTICLES

Biodegradable polymers have been extensively studied for topical applications due to their favorable properties such as biocompatibility, biodegradability, possibilities to reform by functionalization, and so forth,; polymers such as chitosan, hyaluronic acid, guar gum, polycaprolactone, polylactic-co-glycolic acid and so forth are generally used for developing nanocarriers for topical applications. Curcumin-loaded chitosan and polycaprolactone NPs were developed by Zahiri et al. (2020) for wound healing and dermal reconstitution. The developed scaffold showed better biocompatibility for skin and wound healing. The contact angle, pore size, and structural uniformity with good mechanical strength were observed. The cellular attachment and proliferation of PCL/Gela/NCs/ Cur were greater than on the PCL and PCL/Gela scaffolds, indicating that the composite had the ability to repair wounds (Zahiri et al., 2020). Quercetin flavonoid is well known for protection from UV radiation. For this, Nan et al. (2018) developed quercetin-loaded chitosan NPs for topical delivery. The outcomes exhibited that phosphorylation of IkB-α was stimulated due to UV radiation, and this resulted in the deprivation of IkB-α-NF-κB complex. COX-2 expression is then activated by the nucleus NF-B. Abundant prostaglandins induce edema and additional injury by increasing vascular permeability. The results revealed that the developed formulation permeated through the epidermal membrane and showed low toxicity and improved stability quercetin loaded NPs significantly improved the efficacy of quercetin to inhibit NF-kB/COX-2 signaling pathway and skin edema due to UV radiation (Nan et al., 2018).

11.4.3 NANOGEL

The various metabolic deficiencies lead to failure of wound repair. The polyphenols showed excellent repair of the wound by reducing oxidative damage. However, oral intake of polyphenols such as curcumin leads to lower bioavailability and rapid metabolism in the intestine to an inactive product. Therefore, topical use of polyphenols results in a faster wound-healing process. For wound healing, Pathan et al. (2019) devised a curcumin-nanogel composed of fish scale collagen-HPMC. In comparison to those other formulations, the nanogel reflected a sustained drug release and a significant proportion of wound contraction value. The skin irritation study demonstrated a score of less than 2 when compared to the control formulation (Pathan et al., 2019). Xu et al. (2020) created a lignin-based nanogel and employed it as an antioxidant biomaterial for wound healing. Various amount of lignin was incorporated in the more responsive nanogels. It reduced the active oxygen level by protecting the LO_2 cells from the process of apoptosis, which resulted in oxidative stress. The study showed that the nanogel accelerated the healing of wounds produced from burns when studied in mice, as it increased the expression of Ki67, one marker of cell proliferation (Xu et al., 2020). A study conducted by Abbas et al, (2019) studied the effect of curcumin-incorporated nanogel for antibacterial properties and wound healing. The spectrophotometric method confirmed the presence of free radicals, total phenolic and total flavonoids contents. The rabbits were preferred to analyze the wound-healing potential for which different groups; (1) untreated, (2) treated with curcumin and (3) its product made with chitosan-PVA80 were employed. the second-degree burn was treated on rabbits. The chitosan PVA showed excellent antibacterial properties (Abbas et al., 2019). Nain et al, (2021) formulated quercetin-based carbonized nanogels to accelerate chronic wound healing. It possesses NIR-II active copper sulfide (CuS) nanoclusters embedded in a nanogel showing antioxidant nature and endow enhanced physical interaction with bacterial membrane. The delay in the healing of the chronic wound was caused due to microbial pathogenesis leading to inflammation systemic infection, and even to sepsis. The antimicrobial/oxidative/inflammatory property of the quercetin nanogel helps to rapidly accelerate wound healing. Under NIR-II light irradiation, bacteria's minimum inhibitory concentration was ~125 times lower than monomeric Qu or Qu–CNGs. Excellent penetration into the extracellular biofilm matrix was found to be very effective, resulting in the eradication of methicillin-resistant Staphylococcus aureus (MRSA) associated biofilm on diabetic mice wounds. The proinflammatory cytokines (IL-1β) in the infectious wound sites were found to be suppressed (Nain et al., 2021).

11.4.4 HYDROGEL

Hydrogels are one of the hopeful systems for wound dressings due to their exceptional properties, such as the potential to retain a moist microclimate at the wound site, which helps to accelerate healing (Tao et al., 2019). Bhubhanil et al. (2021) developed silver nanoparticle (AgNPs) hydrogel by stabilizing them with guar gum for the topical delivery of curcumin. The developed composite significantly improved the wound healing potential and reduced the bacterial count when compared with antimicrobial marketed gel. The histopathological examination revealed that the composite was able to improve to induce the fibroblast new blood vessel formation, and resulted in wound healing (Bhubhanil et al., 2021). The thermosensitive hydrogel of chitosan conjugating gallic acid was developed by Park et al. (2021) to evaluate their wound-healing potential. The hydrogel formulation of gallic acid showed better self-healing and improved strength of tissue adhesiveness. Besides, the biocompatibility and wound healing were appreciable and significantly encouraged wound closure

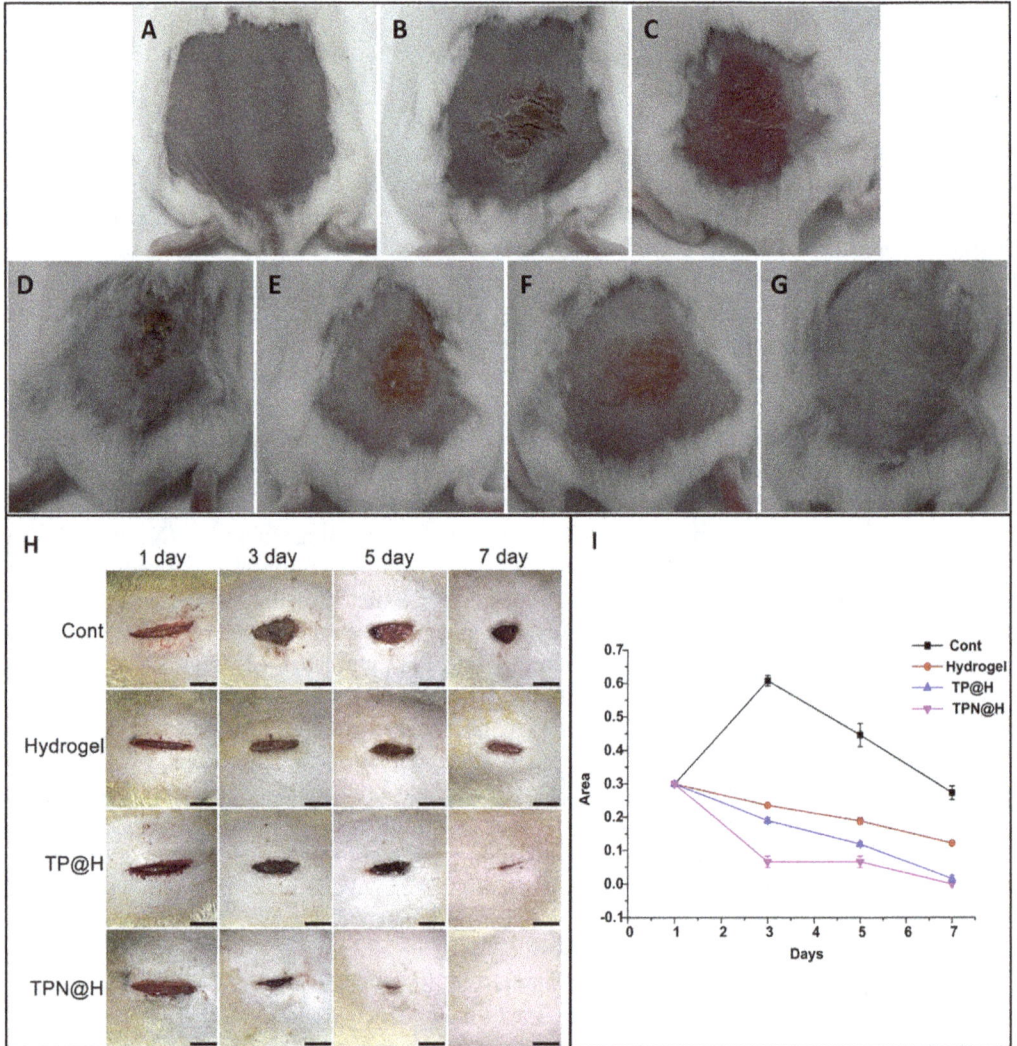

FIGURE 11.6   Illustration of dorsal skin of mice (A) untreated, (B) treated with TPA-only, (C) TPA treated followed by the quercetin and resveratrol in oleic acid/buffer dispersion, (D) empty liposomes, (E) quercetin-loaded liposomes, (F) resveratrol -loaded liposomes (G) quercetin and resveratrol co-loaded liposomes [Reproduced with the permission license number: 5231740350342 (Caddeo et al., 2016)]. (H) The wound healing potential of TPN@H (up to 07 days) (I) The rate of wound site in wounds undergoing different treatments on different days [Reproduced with the permission request number: 5231740689566 (Chen et al., 2020)].

by recruiting fibroblast and upregulating growth factors (Park et al., 2021). Green tea polyphenols have been proven to have significant wound-healing properties. Due to its anti-inflammatory and hypoglycaemic properties, green tea polyphenols are widely used in wound healing. Chen et al. (2020) encapsulated nanospheres of tea polyphenol (TPN) in the PVA/alginate hydrogel matrix (TPN@H) for the treatment of wounds in diabetic conditions. The animal studies and molecular mechanism investigations revealed that TPN@H was able to encourage wound healing by regulating PI3K/AKT signaling pathway in diabetic rats. The wound-closure potential of TPN@H in rats was observed to be fastest when compared with the control group (Figure 11.6, H and I) (Chen et al., 2020).

11.4.5   Miscellaneous NPs

Green synthesis of AgNPs was performed using Curcuma longa extract by Maghimaa and Alharbi, (2020), and their wound-healing and antimicrobial and wound properties were evaluated by coating them on cotton fabric. The fabric was then evaluated against microorganisms (*S. aureus, P. aeruginosa, S. pyogenes*, and *C. albicans)* and showed potent inhibitory effects. The cotton fabric showed remarkable wound healing in the fibroblast (L929) cells (Maghimaa and Alharbi, 2020). Methicillin-resistant *S. aureus* (MRSA)-infected deep-burn wounds (MIDBW) in diabetic individuals are life-threatening to public health. MRSA, which leads to wound infections and has become a serious obstacle to diabetic wound healing. It may potentially result in the patient's limb being

TABLE 11.2
Application Potential of Some Dietary Polyphenol in Different Skin Disease Management

Delivery system	Polyphenol	Skin condition	Remark	Reference
Transferosome	EGCG	Anti-aging, skin damage	The cell viability was found to be improved, reduction in the lipid peroxidation, and showed anti-aging potential	(Avadhani et al., 2017)
Self-emulsifying gel	Curcumin	Wound Closure	Skin permeation improved, fasten the wound closure potential	(Guo et al., 2020)
Hydrogel enclosing AGNPs	Curcumin	Wound healing	Improved wound healing potential, antioxidant property, and was effective against different wound infecting microorganism	(Gupta et al., 2020)
Dendrimer	EGCG and Silibinin		Improved skin permeation and antioxidant properties	(Shetty et al., 2017)
Nanogel	Catechin	UV radiation-induced photo-damage	Nano-gel showed sustained release of catechin, improved photoprotection ability and a decrease in UV-induced oxidative damage was observed.	(Harwansh et al., 2016)
Chitosan NPs	Gallic acid	Wound healing	Improvement in the fibroblast cell migration wound healing wound contraction was observed.	(Kaparekar et al., 2020)
Gel	Gallic acid	skin oxidative stress	Improvement in antioxidant potential and reduction in lipid peroxide was observed.	(Monteiro e Silva et al., 2017)
SLNs	Green tea polyphenol extract	-	The permeation through skin found to be improved	(Dzulhi et al., 2018)
Cold cream	Anthocyanins (Pomegranate)	Antiaging	The cream was non-irritant, and remarkably reduced the skin aging	(Abdellatif et al., 2021)

TABLE 11.3
Clinical Trials on Polyphenols for Management of Skin Diseases

Clinical Trial ID	Disease	Drug	Delivery system	Delivery route	Phase	Stage	Company
NCT03487965	-	Aronia Extract, liqorice root	Extract	Skin	Not Applicable	Completed	Access Business Group
NCT01032031	Skin Cancer	Green tea + vitamin C	Tablet and capsule	Oral	Not Applicable	Completed	University of Manchester
NCT03563365	Acne Vulgaris	Adapalene and Benzoyl peroxide	Cream	Skin	Phase 4	Terminated (Lack of Funding)	Yardley Dermatology Associates
NCT02029352	Basal cell carcinoma	Green tea	Ointment	Skin	Phase 2	Completed	Maastricht University Medical Centre
NCT03902392	Allergic Contact Dermatitis	Red Grape Polyphenol	Capsule	Oral	Not Applicable	Completed	University of Bari
NCT02442453	Periodontitis	Curcuma Longa	Gel	Topical Application	Phase 4	Completed	Tatyasaheb Kore Dental College

amputated (Dang et al., 2003, Katsoulis et al., 2006). For the therapy of such wounds, Xu et al. (2021) fabricated quantum dots (QD) of black phosphorous (BP) (BPQD) and developed EGCG-BPQD nanocomposite. This study conceptualized a strategy of using EGCG customized black phosphorus quantum dots (BPQDs) as nanoplatforms for MIDBW to attain the synergistic functions of NIR (-response, ROS-generation, sterilization, and promoting wound healing. Xu et al. (2021)reported the fabrication of EGCG-BPQDs@Hcomposite, and the sterilization and activation of cell behaviors that can stimulate regenerative activities of skin cells and actively engage in the regeneration of epidermal, and expedite wound healing in diabetic patients (Xu et al., 2021) (Tables 11.2 and 11.3).

11.5 CONCLUSION

A large number of biologically active compounds are found to be present in dietary polyphenols obtained from plants. This is an interesting and upcoming field of scientific research. The studies discussed in this review focus mostly on the potential of polyphenols to delay skin aging under stress conditions. They also explained the protective effect of polyphenols against cutaneous injuries and oxidative stress and protect against various skin complications. The special emphasis on nano-formulation for dermatological purposes serves the motive to improve the water solubility and instability of the drug. They seem to be very effective against skin disorders such as aging, wounds, chloasma, acne vulgaris, facial redness, and so forth, and polyphenols need to be further evaluated for their bioavailability, which is a major area of concern to achieving the efficacy of the drug. Various clinical studies are available showing the potential of dietary polyphenols. Still, future research must be done to evaluate the potential effect of nanocarrier-assisted topical delivery of dietary polyphenols for the management of various skin conditions. A better route to deliver dietary (oral or topical) polyphenols needs to be thoroughly studied. The long-term safety profile needs to be considered in various age groups such as children, adults, and elderly people.

CONFLICTS OF INTEREST

None of the co-authors have declared any conflicts of interest.

REFERENCES

Abbas, M., Hussain, T., Arshad, M., Ansari, A.R., Irshad, A., Nisar, J., Hussain, F., Masood, N., Nazir, A. and Iqbal, M. 2019. Wound healing potential of curcumin cross-linked chitosan/polyvinyl alcohol. *Int J BiolMacromol* 140: 871–876.

Abdellatif, A.A., Alawadh, S.H., Bouazzaoui, A., Alhowail, A.H. and Mohammed, H.A. 2021. Anthocyanins rich pomegranate cream as a topical formulation with anti-aging activity. *J Dermatol Treat* 32: 983–990.

Aleynova, O.A., Suprun, A.R., Ananev, A.A., Nityagovsky, N.N., Ogneva, Z.V., Dubrovina, A.S. and Kiselev, K.V. 2022. Effect of calmodulin-like gene (cml) overexpression on stilbene biosynthesis in cell cultures of *Vitis amurensis Rupr. Plants* 11: 171.

Amic, D., Davidovic-Amic, D., Beslo, D., Rastija, V., Lucic, B. and Trinajstic, N. 2007. SAR and QSAR of the antioxidant activity of flavonoids. *Curr med chem* 14: 827–845.

Arts, I.C. and Hollman, P.C. 2005. Polyphenols and disease risk in epidemiologic studies. *Am J Clin Nutr* 81: 317s–325s.

Avadhani, K.S., Manikkath, J., Tiwari, M., Chandrasekhar, M., Godavarthi, A., Vidya, S.M., Hariharapura, R.C., Kalthur, G., Udupa, N. and Mutalik, S. 2017. Skin delivery of epigallocatechin-3-gallate (EGCG) and hyaluronic acid loaded nano-transferosomes for antioxidant and anti-aging effects in UV radiation induced skin damage. *Drug deliv* 24: 61–74.

Barry, B.W. 2001. Novel mechanisms and devices to enable successful transdermal drug delivery. *Eur J Pharm Sci* 14: 101–114.

Beckman, C.H. 2000. Phenolic-storing cells: keys to programmed cell death and periderm formation in wilt disease resistance and in general defence responses in plants? *Physiol Mol Plant Pathol* 57: 101–110.

Benson, H. A. and Watkinson, A. C. 2012. Topical and transdermal drug delivery: *Principles and practice*, John Wileyand Sons.

Bhubhanil, S., Talodthaisong, C., Khongkow, M., Namdee, K., Wongchitrat, P., Yingmema, W., Hutchison, J.A., Lapmanee, S. and Kulchat, S. 2021. Enhanced wound healing properties of guar gum/curcumin-stabilized silver nanoparticle hydrogels. *Sci Rep* 11: 21836.

Brown, M.B., Martin, G.P., Jones, S.A. and Akomeah, F.K. 2006. Dermal and transdermal drug delivery systems: current and future prospects. *Drug deliv* 13: 175–187.

Caddeo, C., Nacher, A., Vassallo, A., Armentano, M.F., Pons, R., Fernàndez-Busquets, X., Carbone, C., Valenti, D., Fadda, A.M. and Manconi, M. 2016. Effect of quercetin and resveratrol co-incorporated in liposomes against inflammatory/oxidative response associated with skin cancer. *Int J Pharm* 513: 153–163.

Chen, G., He, L., Zhang, P., Zhang, J., Mei, X., Wang, D., Zhang, Y., Ren, X. and Chen, Z. 2020. Encapsulation of green tea polyphenol nanospheres in PVA/alginate hydrogel for promoting wound healing of diabetic rats by regulating PI3K/AKT pathway. *Mater Sci Eng C* 110: 110686.

Choi, B.H., Kim, J.M. and Kwak, M.-K. 2021. The multifaceted role of NRF2 in cancer progression and cancer stem cells maintenance. *Arch Pharm Res,* 44: 263–280.

Chong, J., Poutaraud, A. and Hugueney, P. 2009. Metabolism and roles of stilbenes in plants. *Plant sci* 177: 143–155.

Corona, G., Tzounis, X., Assunta Dessì, M., Deiana, M., Debnam, E. S., Visioli, F. and Spencer, J.P. 2006. The fate of olive oil polyphenols in the gastrointestinal tract: implications of gastric and colonic microflora-dependent biotransformation. *Free Radic Res* 40: 647–658.

D'archivio, M., Filesi, C., Di Benedetto, R., Gargiulo, R., Giovannini, C. and Masella, R. 2007. Polyphenols, dietary sources and bioavailability. *Ann Ist Super Sanita,* 43: 348–361.

Dang, C., Prasad, Y., Boulton, A. and Jude, E. 2003. Methicillin-resistant Staphylococcus aureus in the diabetic foot clinic: a worsening problem. *Diabet Med* 20: 159–161.

Daré, R.G., Nakamura, C.V., Ximenes, V.F. and Lautenschlager, S.O.S. 2020. Tannic acid, a promising anti-photoaging agent: Evidences of its antioxidant and anti-wrinkle potentials, and its ability to prevent photodamage and MMP-1 expression in L929 fibroblasts exposed to UVB. *Free RadicBiol Med* 160: 342–355.

Das, A.K., Islam, M.N., Faruk, M.O., Ashaduzzaman, M. and Dungani, R. 2020. Review on tannins: Extraction processes, applications and possibilities. *S Afr J Bot* 135: 58–70.

De Groot, H.D. and Rauen, U. 1998. Tissue injury by reactive oxygen species and the protective effects of flavonoids. *Fundam Clin Pharmacol* 12: 249–255.

Deli, M. A. 2009. Potential use of tight junction modulators to reversibly open membranous barriers and improve drug delivery. *BiochimBiophys Acta* 1788: 892–910.

Dhalaria, R., Verma, R., Kumar, D., Puri, S., Tapwal, A., Kumar, V., Nepovimova, E. and Kuca, K. 2020. Bioactive compounds of edible fruits with their anti-aging properties: A comprehensive review to prolong human life. *Antioxidants,* 9: 1123.

Działo, M., Mierziak, J., Korzun, U., Preisner, M., Szopa, J. and Kulma, A. 2016. The potential of plant phenolics in prevention and therapy of skin disorders. *Int J Mol Sci* 17: 160–160.

Dzulhi, S., Anwar, E. and Nurhayati, T. 2018. Formulation, characterization and in vitro skin penetration of green tea (Camellia sinensis L.) leaves extract-loaded solid lipid nanoparticles. *J Appl Pharm Sci* 8: 057–062.

El Maghraby, G.M., Barry, B.W. and Williams, A.C. 2008. Liposomes and skin: from drug delivery to model membranes. *Eur J Pharm Sci* 34: 203–222.

Ghuman, S., Ncube, B., Finnie, J., Mcgaw, L., Njoya, E.M., Coopoosamy, R. and Van Staden, J. 2019. Antioxidant, anti-inflammatory and wound healing properties of medicinal plant extracts used to treat wounds and dermatological disorders. *S Afr J Bot* 126: 232–240.

Graf, B.A., Milbury, P.E. and Blumberg, J.B. 2005. Flavonols, flavones, flavanones, and human health: epidemiological evidence. *J Med Food* 8: 281–290.

Guimarães, I., Baptista-Silva, S., Pintado, M. and L Oliveira, A. 2021. Polyphenols: a promising avenue in therapeutic solutions for wound care. *Appl Sci* 11: 1230.

Guo, J.W., Pu, C.M., Liu, C.Y., Lo, S.L. and Yen, Y.H. 2020. Curcumin-loaded self-microemulsifying gel for enhancing wound closure. *Skin PharmacolPhysiol* 33: 300–308.

Gupta, A., Briffa, S.M., Swingler, S., Gibson, H., Kannappan, V., Adamus, G., Kowalczuk, M., Martin, C. and Radecka, I. 2020. Synthesis of silver nanoparticles using curcumin-cyclodextrins loaded into bacterial cellulose-based hydrogels for wound dressing applications. *Biomacromolecules* 21: 1802–1811.

Han, X., Shen, T. and Lou, H. 2007. Dietary polyphenols and their biological significance. *Int J Mol Sci* 8: 950–988.

Han, Y., Huang, M., Li, L., Cai, X., Gao, Z., Li, F., Rakariyatham, K., Song, M., Fernández Tomé, S. and Xiao, H. 2019. Non-extractable polyphenols from cranberries: potential anti-inflammation and anti-colon-cancer agents. *Food Funct* 10: 7714–7723.

Harwansh, R.K., Mukherjee, P.K., Kar, A., Bahadur, S., Al-Dhabi, N.A. and Duraipandiyan, V. 2016. Enhancement of photoprotection potential of catechin loaded nanoemulsion gel against UVA induced oxidative stress. *J PhotochemPhotobiol B: Bio* 160: 318–329.

Hassanpour, S., Maherisis, N. and Eshratkhah, B. 2011. Plants and secondary metabolites (Tannins): A review. *Int. J. For. Soil Eros* 1(1): 47–53.

He, S., Wu, B., Pan, Y. and Jiang, L. 2008. Stilbene oligomers from Parthenocissus laetevirens: isolation, biomimetic synthesis, absolute configuration, and implication of antioxidative defence system in the plant. *J Org Chem* 73: 5233–5241.

Heenatigala Palliyage, G., Singh, S., Ashby, C.R., Jr., Tiwari, A.K. and Chauhan, H. 2019. Pharmaceutical topical delivery of poorly soluble polyphenols: Potential role in prevention and treatment of melanoma. *AAPS PharmSciTech* 20: 250.

Herrmann, K. and Nagel, C.W. 1989. Occurrence and content of hydroxycinnamic and hydroxybenzoic acid compounds in foods. *Crit Rev Food Sci andNutr* 28: 315–347.

Hu, G. and Zhou, X. 2021. Gallic acid ameliorates atopic dermatitis-like skin inflammation through immune regulation in a mouse model. *Clin CosmetInvestig Dermatol* 14: 1675.

Ichikawa, D., Matsui, A., Imai, M., Sonoda, Y. and Kasahara, T. 2004. Effect of various catechins on the IL-12p40 production by murine peritoneal macrophages and a macrophage cell line, J774.1. *Biol Pharm Bull,* 27: 1353–1358.

Incocciati, A., Di Fabio, E., Boffi, A., Bonamore, A. and Macone, A. 2022. Rapid and simultaneous determination of free aromatic carboxylic acids and phenols in commercial juices by GC-MS after ethyl chloroformate derivatization. *Separations* 9: 9.

Jain, A., Prajapati, S.K., Kumari, A., Mody, N. and Bajpai, M. 2020. Engineered nanosponges as versatile biodegradable carriers: An insight. *J Drug Deliv Sci Technol* 57: 101643.

Jain, S., Prajapati, S.K., Jain, S., Jain, S. and Jain, A. 2018. Propylene glycol-liposome for anticoagulant drug delivery through skin. *J Bionanosci* 12: 721–727.

Jeandet, P., Delaunois, B., Conreux, A., Donnez, D., Nuzzo, V., Cordelier, S., Clément, C. and Courot, E. 2010. Biosynthesis, metabolism, molecular engineering, and biological functions of stilbene phytoalexins in plants. *Biofactors* 36: 331–341.

Kaparekar, P.S., Pathmanapan, S. and Anandasadagopan, S.K. 2020. Polymeric scaffold of gallic acid loaded chitosan nanoparticles infused with collagen-fibrin for wound dressing application. *Intl J BiolMacromol* 165: 930–947.

Katiyar, S.K., Elmets, C.A., Agarwal, R. and Mukhtar, H. 1995. Protection against ultraviolet-B radiation-induced local and systemic suppression of contact hypersensitivity and edema responses in C3H/HeN mice by green tea polyphenols. *PhotochemPhotobiol* 62: 855–861.

Katsoulis, E., Bissell, I. and Hargreaves, D.G. 2006. MRSA pyogenic flexor tenosynovitis leading to digital ischaemic necrosis and amputation. *J Hand Surg* 31: 350–352.

Khoddami, A., Wilkes, M.A. and Roberts, T.H. 2013. Techniques for analysis of plant phenolic compounds. *Molecules* 18: 2328–2375.

Kobayashi, M. and Yamamoto, M. 2005. Molecular mechanisms activating the Nrf2-Keap1 pathway of antioxidant gene regulation. *Antioxid Redox Signal* 7: 385–394.

Kojima, T., Akiyama, H., Sasai, M., Taniuchi, S., Goda, Y., Toyoda, M. and Kobayashi, Y. 2000. Anti-allergic effect of apple polyphenol on patients with atopic dermatitis: a pilot study. *Allergol Int* 49: 69–73.

Lachman, J., Proněk, D., Hejtmánková, A., Pivec, V. and Faitová, K. 2003. Total polyphenol and main flavonoid antioxidants in different onion (Allium cepa L.) varieties. *Sci Hortic* 30: 142–147.

Li, M., Li, P., Tang, R. and Lu, H. 2022. Resveratrol and its derivates improve inflammatory bowel disease by targeting gut microbiota and inflammatory signalling pathways. *Food Sci Hum Wellness* 11: 22–31.

Li, Y. and Zhang, H. 2017. Soybean isoflavones ameliorate ischemic cardiomyopathy by activating Nrf2-mediated antioxidant responses. *Food Funct* 8: 2935–2944.

Liakopoulou, A., Mourelatou, E. and Hatziantoniou, S. 2021. Exploitation of traditional healing properties, using the nanotechnology's advantages: The case of curcumin. *Toxicol Rep* 8: 1143–1155.

Liakos, I., Rizzello, L., Hajiali, H., Brunetti, V., Carzino, R., Pompa, P., Athanassiou, A. and Mele, E. 2015. Fibrous wound dressings encapsulating essential oils as natural antimicrobial agents. *J Mater Chem B* 3: 1583–1589.

Liu, X., Kruger, P., Maibach, H., Colditz, P.B. and Roberts, M.S. 2014. Using skin for drug delivery and diagnosis in the critically ill. *Adv Drug Deliv Rev* 77: 40–49.

Loke, W.M., Proudfoot, J.M., Stewart, S., Mckinley, A.J., Needs, P.W., Kroon, P.A., Hodgson, J.M. and Croft, K.D. 2008. Metabolic transformation has a profound effect on anti-inflammatory activity of flavonoids such as quercetin: Lack of association between antioxidant and lipoxygenase inhibitory activity. *BiochemPharmacol* 75: 1045–1053.

Lu, M., Cai, Y., Fang, J., Zhou, Y., Liu, Z. and Wu, L. 2002. Efficiency and structure-activity relationship of the antioxidant action of resveratrol and its analogs. *Die Pharmazie* 57: 474–478.

Luo, J., Si, H., Jia, Z. and Liu, D. 2021. Dietary anti-aging polyphenols and potential mechanisms. *Antioxidants* 10: 283.

Madison, K.C. 2003. Barrier function of the skin: "la raison d'être" of the epidermis. *J Invest Dermato* 121: 231–241.

Maghimaa, M. and Alharbi, S.A. 2020. Green synthesis of silver nanoparticles from Curcuma longa L. and coating on the cotton fabrics for antimicrobial applications and wound healing activity. *J PhotochemPhotobiol B: Biol* 204: 111806.

Manach, C., Scalbert, A., Morand, C., Rémésy, C. and Jiménez, L. 2004. Polyphenols: food sources and bio-availability. *Am J Clin Nutr* 79: 727–747.

Manach, C., Williamson, G., Morand, C., Scalbert, A. and Rémésy, C. 2005. Bioavailability and bio efficacy of polyphenols in humans. I. Review of 97 bioavailability studies. *Am J Clin Nutr* 81: 230s–242s.

Meydani, M. 2009. Potential health benefits of avenanthramides of oats. *Nutr Rev* 67: 731–735.

Monteiro E Silva, S.A., Calixto, G.M.F., Cajado, J., De Carvalho, P.C.A., Rodero, C F., Chorilli, M. and Leonardi, G.R. 2017. Gallic acid-loaded gel formulation combats skin oxidative stress: development, characterization and ex vivo biological assays. *Polymers* 9: 391.

Morais, R.P., Hochheim, S., De Oliveira, C.C., Riegel-Vidotti, I.C. and Marino, C.E. 2022. Skin interaction, permeation, and toxicity of silica nanoparticles: challenges and recent therapeutic and cosmetic advances. *Int J Pharms* 614: 121439.

Morganti, P., Bruno, C., Guarneri, F., Cardillo, A., Del Ciotto, P. and Valenzano, F. 2002. Role of topical and nutritional supplement to modify the oxidative stress. *Int J Cosmet Sci* 24: 331–339.

Nain, A., Tseng, Y.T., Gupta, A., Lin, Y.F., Sangili, A., Huang, Y.F., Huang, C.C. and Chang, H.T. 2021. Antimicrobial/oxidative/inflammatory nanogels accelerate chronic wound healing. *Smart MaterMed* 3: 148–158.

Nan, W., Ding, L., Chen, H., Khan, F. U., Yu, L., Sui, X. and Shi, X. 2018. Topical use of quercetin-loaded chitosan nanoparticles against ultraviolet B radiation. *Front Pharmacol* 9: 826.

Nasir, A., Bullo, M.M.H., Ahmed, Z., Imtiaz, A., Yaqoob, E., Jadoon, M., Ahmed, H., Afreen, A. and Yaqoob, S. 2020. Nutrigenomics: Epigenetics and cancer prevention: A comprehensive review. *Crit Rev Food Sci Nutr* 60: 1375–1387.

Nenadis, N., Wang, L.-F., Tsimidou, M. and Zhang, H.-Y. 2004. Estimation of scavenging activity of phenolic compounds using the ABTS^{+} assay. *J Agric Food Chem* 52: 4669–4674.

Orduña, L., Li, M., Navarro-Payá, D., Zhang, C., Santiago, A., Romero, P., Ramšak, Ž., Magon, G., Höll, J. and Merz, P. 2022. Direct regulation of shikimate, early phenylpropanoid and stilbenoid pathways by subgroup 2 R2R3-MYBs in grapevine. *Plant J* https://doi.org/10.1111/tpj.15686.

Pandey, K.B. and Rizvi, S.I. 2009. Plant polyphenols as dietary antioxidants in human health and disease. *Oxid Med Cell longev* 2: 270–278.

Park, S.G., Li, M.-X., Cho, W.K., Joung, Y.K. and Huh, K.M. 2021. Thermosensitive gallic acid-conjugated hexanoyl glycol chitosan as a novel wound healing biomaterial. *CarbohydrPolym* 260: 117808.

Pathan, I.B., Munde, S.J., Shelke, S., Ambekar, W. and MallikarjunaSetty, C. 2019. Curcumin loaded fish scale collagen-HPMC nanogel for wound healing application: Ex-vivo and In-vivo evaluation. *Int J Polym Mater PolymBiomater* 68: 165–174.

Pinnell, S.R. 2003. Cutaneous photodamage, oxidative stress, and topical antioxidant protection. *J Am Acad Dermatol* 48(1):1–19.

Prajapati, S.K. and Jain, A. 2020. Dendrimers for advanced drug delivery. In: Nayak, A.K. and Hasnain, M.S. (eds) *Advanced Biopolymeric Systems for Drug Delivery.* Cham: Springer International Publishing, 339–360.

Prajapati, S.K., Jain, A., Jain, A. and Jain, S. 2019. Biodegradable polymers and constructs: A novel approach in drug delivery. *EurPolym J* 120: 109191.

Prajapati, S.K., Malaiya, A., Kesharwani, P., Soni, D. and Jain, A. 2020. Biomedical applications and toxicities of carbon nanotubes. *Drug Chem Toxicol* 45: 435–450.

Prajapati, S.K., Mishra, G., Malaiya, A., Jain, A., Mody, N. and Raichur, A.M. 2021. Antimicrobial application potential of phytoconstituents from turmeric and garlic. In: Pal D., Nayak A.K. (eds) *Bioactive Natural Products for Pharmaceutical Applications. Springer* 140: 409–435.

Prasanthi, D. and Lakshmi, P. 2012. Vesicles-mechanism of transdermal permeation: a review. *Asian J Pharm Clin Res* 5: 18–25.

Rahman, M.M., Rahaman, M.S., Islam, M.R., Rahman, F., Mithi, F.M., Alqahtani, T., Almikhlafi, M.A., Alghamdi, S.Q., Alruwaili, A.S., Hossain, M.S., Ahmed, M., Das, R., Emran, T.B. and Uddin, M.S. 2022. Role of phenolic compounds in human disease: current knowledge and future prospects. *Molecules* 27: 233.

Raza, K., Singh, B., Lohan, S., Sharma, G., Negi, P., Yachha, Y. and Katare, O.P. 2013. Nano-lipoidal carriers of tretinoin with enhanced percutaneous absorption, photostability, biocompatibility and anti-psoriatic activity. *Int J Pharm* 456: 65–72.

Roy, S., Sannigrahi, S., Vaddepalli, R.P., Ghosh, B. and Pusp, P. 2012. A novel combination of methotrexate and epigallocatechin attenuates the overexpression of pro-inflammatory cartilage cytokines and modulates antioxidant status in adjuvant arthritic rats. *Inflammation* 35: 1435–1447.

Rushworth, S.A., Ogborne, R.M., Charalambos, C.A. and O'Connell, M.A. 2006. Role of protein kinase C δ in curcumin-induced antioxidant response element-mediated gene expression in human monocytes. *BiochemBiophys Res Commun* 341: 1007–1016.

Saleem, M., Maddodi, N., Zaid, M.A., Khan, N., Bin Hafeez, B., Asim, M., Suh, Y., Yun, J.-M., Setaluri, V. and Mukhtar, H. 2008. Lupeol inhibits growth of highly aggressive human metastatic melanoma cells in vitro and in vivo by inducing apoptosis. *Clin Cancer Res* 14: 2119–2127.

Scalbert, A., Manach, C., Morand, C., Remesy, C. and Jiménez, L. 2005. Dietary polyphenols and the prevention of diseases. *Crit Rev Food Sci Nutr* 45: 287–306.

Scalbert, A., Morand, C., Manach, C. and Remesy, C. 2002. Absorption and metabolism of polyphenols in the gut and impact on health. *Biomed Pharmacother* 56: 276–282.

Schoellhammer, C.M., Blankschtein, D. and Langer, R. 2014. Skin permeabilization for transdermal drug delivery: recent advances and future prospects. *Expert Opin Drug Deliv* 11: 393–407.

Seigler, D. 2006. Basic pathways for the origin of allelopathic compounds. In: Reigosa M., Pedrol N., González L. (eds) *Allelopathy: A Physiological Process with Ecological Implications*. Springer: Dordrecht: 11–21.

Sellappan, S. and Akoh, C.C. 2002. Flavonoids and antioxidant capacity of Georgia-grown vidalia onions. *J Agric Food Chem* 50: 5338–5342.

Shah, Z.A., Li, R.-C., Ahmad, A.S., Kensler, T.W., Yamamoto, M., Biswal, S. and Doré, S. 2010. The flavanol (−)-epicatechin prevents stroke damage through the Nrf2/HO1 pathway. *J Cereb Blood Flow Metab* 30: 1951–1961.

Shahidi, F. and Ambigaipalan, P. 2015. Phenolics and polyphenolics in foods, beverages and spices: Antioxidant activity and health effects – A review. *J Funct Foods* 18: 820–897.

Shamsi, A., Anwar, S., Mohammad, T., Shahwan, M., Hassan, M. and Islam, A. 2021. Therapeutic potential of polyphenols in Alzheimer's therapy: Broad-spectrum and minimal side effects as key aspects. In: Md Ashraf G., Alexiou A. (eds) *Autism Spectrum Disorder and Alzheimer's Disease.* Springer 111–133.

Sharma, A., Kaur, M., Katnoria, J.K. and Nagpal, A.K. 2018. Polyphenols in food: cancer prevention and apoptosis induction. *Curr Med Chem* 25: 4740–4757.

Sharma, K., Kesharwani, P., Prajapati, S.K., Jain, A., Jain, D., Mody, N. and Sharma, S. 2021. An insight into anticancer bioactives from punica granatum (pomegranate). *Anti-cancer Agents Med Chem* 22(4): 694–702.

Shetty, P.K., Manikkath, J., Tupally, K., Kokil, G., Hegde, A.R., Raut, S.Y., Parekh, H.S. and Mutalik, S. 2017. Skin delivery of EGCG and silibinin: potential of peptide dendrimers for enhanced skin permeation and deposition. *AAPS Pharm Sci Tech* 18: 2346–2357.

Shirmohammadli, Y., Efhamisisi, D. and Pizzi, A. 2018. Tannins as a sustainable raw material for green chemistry: A review. *Ind Crops Prod* 126: 316–332.

Singh, S., Roy, L.D. and Giri, S. 2015. Curcumin protects metronidazole and X-ray induced cytotoxicity and oxidative stress in male germ cells in mice. *Prague Med Rep* 114: 92–102.

Singla, R.K., Dubey, A.K., Garg, A., Sharma, R.K., Fiorino, M., Ameen, S.M., Haddad, M.A. and Al-Hiary, M. 2019. Natural polyphenols: chemical classification, definition of classes, subcategories, and structures. *J AOAC Int* 102: 1397–1400.

Spencer, J.P., Abd El Mohsen, M.M., Minihane, A.-M. and Mathers, J.C. 2008. Biomarkers of the intake of dietary polyphenols: strengths, limitations and application in nutrition research. *Br J Nutr* 99: 12–22.

Spergel, J.M. 2010. Epidemiology of atopic dermatitis and atopic march in children. *Immunol Allergy Clin North Am* 30: 269–280.

Stalikas, C.D. 2007. Extraction, separation, and detection methods for phenolic acids and flavonoids. *J Sep Sci* 30: 3268–3295.

Tao, G., Wang, Y., Cai, R., Chang, H., Song, K., Zuo, H., Zhao, P., Xia, Q. and He, H. 2019. Design and performance of sericin/poly (vinyl alcohol) hydrogel as a drug delivery carrier for potential wound dressing application. *Mater Sci Eng C* 101: 341–351.

Torres-Contreras, A.M., Garcia-Baeza, A., Vidal-Limon, H.R., Balderas-Renteria, I., Ramírez-Cabrera, M.A. and Ramirez-Estrada, K. 2022. Plant secondary metabolites against skin photo damage: Mexican plants, a potential source of UV-radiation protectant molecules. *Plants* 11: 220.

Uygur, R., Yagmurca, M., Alkoc, O., Genc, A., Songur, A., Ucok, K. and Ozen, O. 2014. Effects of quercetin and fish n-3 fatty acids on testicular injury induced by ethanol in rats. *Andrologia* 46: 356–369.

Venza, I., Venza, M., Visalli, M., Lentini, G., Teti, D. and D'alcontres, F.S. 2021. ROS as regulators of cellular processes in melanoma. *Oxid Med Cell Longev*: 1208690.

Vestita, M., Tedeschi, P. and Bonamonte, D. 2022. Anatomy and physiology of the skin. In: Maruccia M., Giudice G. (eds) *Textbook of Plastic and Reconstructive Surgery*. Springer Cham 3–13.

Williams, A. C. and Barry, B. W. 2012. Penetration enhancers. *Adv Drug Deliv Rev* 64: 128–137.

Wood, J.G., Rogina, B., Lavu, S., Howitz, K., Helfand, S. L., Tatar, M. and Sinclair, D. 2004. Sirtuin activators mimic caloric restriction and delay ageing in metazoans. *Nature* 430: 686–689.

Wruck, C., Claussen, M., Fuhrmann, G., Römer, L., Schulz, A., Pufe, T., Waetzig, V., Peipp, M., Herdegen, T. and Götz, M. 2007. Luteolin protects rat PC 12 and C6 cells against MPP+ induced toxicity via an ERK dependent Keapl-Nrf2-ARE pathway. In: Gerlach M., Deckert J., Double K., Koutsilieri E. (eds) *Neuropsychiatric disorders: An integrative approach.* Springer: 57–67.

Wu, C., Hsu, M., Hsieh, C., Lin, J., Lai, P. and Wung, B. 2006. Upregulation of heme oxygenase-1 by Epigallocatechin-3-gallate via the phosphatidylinositol 3-kinase/Akt and ERK pathways. *Life Sci* 78: 2889–2897.

Xu, J., Xu, J. J., Lin, Q., Jiang, L., Zhang, D., Li, Z., Ma, B., Zhang, C., Li, L. and Kai, D. 2020. Lignin-incorporated nanogel serving as an antioxidant biomaterial for wound healing. *ACS Appl Bio Mater* 4: 3–13.

Xu, S., Chang, L., Hu, Y., Zhao, X., Huang, S., Chen, Z., Ren, X. and Mei, X. 2021. Tea polyphenol modified, photothermal responsive and ROS generative black phosphorus quantum dots as nanoplatforms for promoting MRSA infected wounds healing in diabetic rats. *J Nanobiotechnology* 19: 362.

Yan, L., Guo, M.-S., Zhang, Y., Yu, L., Wu, J.-M., Tang, Y., Ai, W., Zhu, F.-D., Law, B.Y.-K. and Chen, Q. 2022. Dietary plant polyphenols as the potential drugs in neurodegenerative diseases: Current evidence, advances, and opportunities. *Oxid Med Cell Longev* 2022: 5288698.

Yang, D.J., Moh, S.H., Son, D.H., You, S., Kinyua, A.W., Ko, C.M., Song, M., Yeo, J., Choi, Y.-H. and Kim, K.W. 2016. Gallic acid promotes wound healing in normal and hyperglucidic conditions. *Molecules* 21: 899.

Zahiri, M., Khanmohammadi, M., Goodarzi, A., Ababzadeh, S., Sagharjoghi Farahani, M., Mohandesnezhad, S., Bahrami, N., Nabipour, I. and Ai, J. 2020. Encapsulation of curcumin loaded chitosan nanoparticle within poly (ε-caprolactone) and gelatinfiber mat for wound healing and layered dermal reconstitution. *Int J BiolMacromol* 153: 1241–1250.

Zhao, H., Xia, C., He, P. and Tu, Y. 2021. Effects of tea polyphenols on anti-inflammation and promotion of wound healing and its mechanisms. *J Zhejiang Univ (Agric Life Sci)* 47(1): 118–126.

12 Dietary Polyphenols from Spices and their Impact on Human Health

*Neelesh Kumar Nema, Nayana Rajan, Merin Babu,
Sachithra Sabu, Swapnil Devidas Khamborkar,
Smitha Sarojam, Linson Cheruveettil Sajan, Aneena Peter,
Baby Kumaranthara Chacko, and Viju Jacob
*E-mail: neeleshk@synthite.com; merinbabu@synthite.com;
sachithra@synthite.com; swapnil@synthite.com; smithas@synthite.com;
linson@synthite.com; aneena@synthite.com; babykc@synthite.com;
viju@synthite.com

CONTENTS

DOI: 10.1201/9781003251538-12

12.1　INTRODUCTION

Spices and condiments have played a significant role in the preparation of food and culinary preparations from antiquity to the present (Del Baño et al., 2003; Sharangi and Acharya, 2018). Folk and traditional medicine all over the world use many of these as a remedy for digestive health. Some are used in medical and religious rituals, and some are for cosmetic purposes. Their active phytochemicals provide potential therapeutic benefits at the molecular level (Gidwani et al., 2022).This chapter covers spices, their classifications, and kinds, as well as polyphenols, signaling pathways, possible therapies, and futuristic ideas that may be used in the food processing, nutraceutical, and pharmaceutical sectors.

12.2　SPICES

A worldwide federation of national standards bodies "ISO" (the International Organization for Standardization) recognizes spices and condiments as:

"Vegetable products or mixtures thereof free from extraneous matter, used for flavouring, seasoning and imparting aroma in foods," by the International Organization for Standardization, a global federation of national standards bodies. In addition to this, culinary herbs and spices are defined by the European Spice Associations (ESA) as "edible parts of plants that are traditionally added to foodstuffs for their natural flavouring, aromatic, and visual properties," whereas the Codex Committee on Spices and Culinary Herbs (CCSCH), which was established in 2013 with the participation of over a hundred countries – India as the host country and the Spices Board (Ministry of Commerce and Industry) as the Secretariat – defines herbs and spices as they are "usually derived from botanical sources, and maybe dried or dehydrated, and either ground or whole and cracked or crushed forms." CCSCH also described its standards, quality control, certification, and policy matters. Spices come in a variety of forms, including underground parts, that is, roots, rhizomes, and aerial parts, that is, bark, leaves, buds, flowers, fruits, seeds, and complete plants (European Spice Association (ESA), 2018).

12.2.1　CLASSIFICATION AND TYPES

Although the list of spices, condiments, and culinary herbs is very enormous, ISO identified 109 spices, which differ from country to country; the Spices Board of India mentioned 52 spices (Table 12.1); the American Spices Trade Association (ASTA) enlisted 41 items; and ESA declared around 86 species. India is recognized as a "Land of Spices" due to the obtainability of wide varieties of quality spices in several states of the country, grown in varied climates, soil conditions, and geographical locations.

12.2.2　SPICES IN FUNCTIONAL FOODS/CULINARY PREPARATION

Spices are utilized in nearly all cuisines across the world. They are cultivated in varied climatic conditions, specific temperatures, altitudes, and atmospheres. The qualities of the spices are influenced by many handling factors, that is, time of collection, picking, and processing time. The therapeutic impact is also influenced by the active ingredients, and organoleptic and chemical qualities. In our body, spices work as appetizers through characteristics expression. Psycho-physiological sensory attributes influence the body to take quality food. In the gastrointestinal tract, they affect the digestion process by triggering Human Electroencephalographic Responses (HER) (Premavalli and Wadikar, 2011; Sowndhararajan and Kim, 2016; Wadikar DD, 2018).

TABLE 12.1
List of Spices Used Worldwide in Food and Culinary Preparations

Sl. No.	Family	No. of Species	Scientific Name	English/Common Name	Plant Part	ISO	India
1	Anacardiaceae	3	*Mangifera indica*	Mango	Immature Fruit (Rind)	√	-
2			*Schinusmolle*	American pepper, Californian pepper tree	Fruit-Wall (Rind)	√	-
3			*Schinusterebenthifolius*	Brazilian pepper	Fruit	√	-
4	Annonaceae	1	*Xylopiaaethiopica*	Negro pepper, Guinean pepper	Fruit	√	-
5	Apiaceae	18	*Anethumgraveolens*	Dill	Fruit	√	√
6			*Anethumsowa*	Indian dill	Leaf, Seed	√	-
7			*Angelica archangelica*	Garden angelica	Root, Seed, Fruit	√	-
8			*Anthriscuscereifolium*	Chervil	Leaf	√	-
9			*Apiumgraveolens*	Celery	Fruit, Root, Aerial Part	√	√
10			*Apiumgraveolens*	Celeriac	Fruit, Root, Leaf	√	-
11			*Buniumpersicum*	Black caraway	Seed, Tuber	√	-
12			*Carumbulbocastanum*	Black caraway	Fruit, Bulb	√	-
13			*Carumcarvi*	Caraway	Fruit	√	√
14			*Coriandrumsativum*	Coriander	Leaf, Seed	√	√
15			*Cuminumcyminum*	Cumin	Fruit	√	√
16			*Ferula assafoetida*	Asafoetida	Rhizome Exudate	√	√
17			*Foeniculumvulgare*	Fennel (Bitter)	Fruit, Leaf	√	-
18			*Foeniculumvulgare*	Fennel (Sweet)	Fruit, Leaf,	√	√
19			*Levisticum officinale*	Lovage	Leaf	√	√
20			*Petroselinumcrispum* Syn. *Petroselinumsativum*	Parsley	Seed, Leaf, Root	√	√
21			*Pimpinellaanisum*	Aniseed	Fruit	√	√
22			*Trachyspermumammi*	Bishop's Weed/ ajwain	Fruit	√	√
23	Araceae	1	*Acoruscalamus*	Sweet flag, Myrtie flag, Calamus	Rhizome	√	√
24	Arteraceae	1	*Artemisia dracunculus*	Tarragon, Estragon	Leaf	√	√
25	Averrhoaceae	2	*Averrhoabilimbi*	Belimbing, Bilimbi cucumber tree	Fruit	√	-
26			*Averrhoacarambola*	Carambola, Caramba	Fruit	√	-
27	Brassicaceae	4	*Armoraciarusticana*	Horse-radish	Rhizome	√	√
28			*Brassica juncea*	Indian mustard	Seed	√	√
29			*Brassica nigra*	Black mustard	Seed	√	-
30			*Sinapis alba*	White mustard, Yellow mustard	Seed	√	-
31	Capparidaceae	1	*Capparis spinosa*	Caper, Caper-bush	Fruit, Root, Floral Bud	√	√
32	Cesalpiniaceae	1	*Tamarindus indica*	Tamarind	Fruit Pod	√	√
33	Clusiaceae	2	*Garcinia cambogia*	Cambodge, Garcinia	Fruit Pericarp	√	√
34			*Garcinia indica*	Kokkam	Fruit Pericarp	√	√
35	Cupressaceae	1	*Juniperuscommunis*	Juniper berry	Berry	√	√
36	Fabaceae	1	*Trigonellafoenum-graecum*	Fenugreek	Seed, Leaf	√	√

(*continued*)

TABLE 12.1 (Continued)
List of Spices Used Worldwide in Food and Culinary Preparations

Sl. No.	Family	No. of Species	Scientific Name	English/Common Name	Plant Part	ISO	India
37	Illiciaceae	1	*Illiciumverum*	Star anise, Chinese anise	Fruit	√	√
38	Iridaceae	1	*Crocus sativus*	Saffron	Stigma	√	√
39	Lamiaceae	29	*Hyssopus officinalis*	Hyssop	Leaf	√	√
40			*Melissa officinalis*	Lemon balm, Melissa	Leaf, Shoot	√	-
41			*Menthaarvensis*	Japanese/Corn mint	Leaf, Shoot	√	-
42			*Mentha citrate*	Bergamol	Leaf, Shoot	√	-
43			*Mentha x piperita*	Peppermint	Fruit, Leaf	√	√
44			*Menthaspicata*	Spearmint, garden mint	Leaf	√	-
45			*Rosmarinus officinalis*	Rosemary	Shoot, Leaf	√	√
46			*Salvia officinalis*	Garden sage	Shoot, Leaf	√	√
47			*Saturejahortensis*	Summer savory	Shoot, Leaf	√	-
48			*Saturejamontana* Syn. *Saturejathymbra*	Winter savory	Leaf, Twig	√	√
49			*Thymus serpyllum*	Wild thyme	Leaf, Twig	√	-
50			*Thymus vulgaris*	Thyme	Leaf	√	√
51			*Ocimum basilicum*	Sweet basil	Leaf	√	√
52			*Origanummajorana* Syn. *Marjoranahortensis*	Sweet marjoram	Leaf	√	√
53			*Origanumvulgare*	Oregano, Origan	Leaf	√	√
54			*Cinnamomumaromaticum* Syn. *Cinnamomum cassia*	Cassia	Bark, Leaf	√	√
55			*Cinnamomumburmanii*	Indonesian cassia	Bark	√	-
56			*Cinnamomumloureirii*	Vietnamese cassia	Bark	√	-
57			*Cinnamomumtamala*	Tejpat, Indian cassia	Bark, Leaf	√	√
58			*Cinnamomumzeylanicum* Blume, Syn. *Cinnamomumverum*	Sri Lankan/ Indian cinnamon	Bark, Leaf	√	√
59			*Laurusnobilis*	Bayleaf	Leaf	√	√
60			*Allium ascalonicum*	Shallot	Bulb	√	-
61			*Allium cepa*	Onion	Bulb	√	-
62			*Allium cepa var. aggregatum*	Potato onions	Bulb	√	-
63			*Allium tuberosum*	Chinese chive	Bulb	√	-
64			*Allium fistulosum*	Japanese bunching onion	Bulb	√	-
65			*Allium porrum*	Leek, winter leek	Bulb, Leaf	√	-
66			*Allium sativum*	Garlic	Bulb	√	√
67			*Allium schoenoprasum*	Chive	Leaf	√	-
68	Myristicaceae	2	*Myristicaargentea*	Papuan nutmeg/ mace	Seed, Kernel	√	-
69			*Myristicafragrans*	Indonesian/ Siauwnutmeg Indonesian/Siauwmace	Seed, Kernel	√	√
70	Myrtaceae	3	*Pimentadioica*	Pimento, Allspice, Jamaica Pepper	Fruit, Seed, Leaf	√	√
71			*Pimentaracemosa*	West Indian bay	Fruit, Leaf	√	-
72			*Syzygiumaromaticum*	Clove	Unopened Flower Bud	√	√

TABLE 12.1 (Continued)
List of Spices Used Worldwide in Food and Culinary Preparations

Sl. No.	Family	No. of Species	Scientific Name	English/Common Name	Plant Part	ISO	India
73	Orchidaceae	3	*Vanilla planifolia* Syn. *Vanilla fragrans*	Vanilla pod	Beans, Fruit (Pod)	√	√
74			*Vanilla tahitensis*	Vanilla	Beans, Fruit (Pod)	√	-
75			*Vanilla pompona*	Pompona vanilla	Beans, Fruit (Pod)	√	-
76	Pandanaceae	1	*Pandanusamaryllifolius* Syn. *Pandanuslatifolius*	Pandanwangi	Leaf	√	-
77	Papaveraceae	1	*Papaversomniferum*	Poppy, Blue maw	Seed	√	√
78	Pedaliaceae	1	*Sesamumindicum*	Sesame, Gingelly	Seed	√	-
79	Piperaceae	3	*Piper guineense*	Benin pepper	Fruit	√	-
80			*Piper longum*	Indian longpepper	Fruit	√	√
81			*Piper nigrum*	Black/White/Green pepper	Fruit	√	√
82	Poaceae	2	*Cymbopogon citratus*	West Indian lemongrass	Leaf	√	-
83			*Cymbopogon nardus*	Sri Lankan citronella	Leaf	√	-
84	Punicaseae	1	*Punica granatum*	Pomegranate seed	Seed (Dried with Flesh)	√	√
85	Ranunculaceae	2	*Nigella damascene*	Damas black cumin	Seed	√	-
86			*Nigella sativa* L.	Black cumin	Seed	√	-
87	Rutaceae	4	*Murrayakoenigii*(L.) Sprengel	Curry leaf	Leaf	√	√
88			*Zanthoxylum bungei* Syn. *Zanthoxylum bungeanum*	Sichuan pepper	Fruit	√	-
89			*Zanthoxylum acanthopodium*	Chinese pepper	Fruit	√	-
90			*Zanthoxylum piperitum*	Japanese pepper	Fruit	√	-
91	Solanaceae	2	*Capsicum annum*	Chilli/Capsicum/ Paprika	Fruit	√	√
92			*Capsicum frutescens*	Bird eye chili	Fruit	√	√
93	Verbanaceae	1	*Lippiagraveolens* Syn. *Lippiaberlandieri*	Mexican oregano	Leaf	√	-
94	Zingiberaceae	16	*Aframomumangustifolium*	Madagascar cardamom	Fruit, Seed	√	-
95			*Aframomumhanburyi*	Cameroon cardamom	Fruit, Seed	√	-
96			*Aframomumkorarima*	Korarima cardamom	Fruit, Seed	√	-
97			*Aframomummelegueta*	Guineagrains	Fruit, Seed	√	-
98			*Alpiniagalanga*	Galangal(Greater)	Rhizome	√	√
99			*Alpiniaofficinarum*	Galangal(Lesser)	Rhizome	√	-
100			*Amomumaromaticum*	Bengal cardamom	Fruit, Seed	√	-
101			*Amomumkepulaga* Syn. *Amomumcompactum*	Indonesian/Java cardamom	Fruit, Seed	√	-
102			*Amomumkrevanh*	Cambodian cardamom	Fruit, Seed	√	-
103			*Amomumsubulatum*	Indian cardamom	Fruit, Seed	√	√
104			*Amomumtsao-ko*	Tsao-ko cardamom	Fruit, Seed	√	-
105			*Curcuma longa*	Turmeric	Rhizome, Leaf	√	√
106			*Elettariacardamomum*	Small cardamom	Fruit, Seed	√	√
107			*Elettariacardamomum*	Sri Lankan cardamom	Fruit, Seed	√	-
108			*Kaempferia galangal*	Galangal	Rhizome	√	-
109			*Zingiber officinale*	Ginger	Rhizome	√	√

12.3 POLYPHENOLS: AN IMPORTANT SECONDARY METABOLITE

Secondary metabolites that derived from spices, for example, phenolics (flavonoids and non-flavonoids), alkaloids, terpenoids, steroids, stilbenoids, and polysaccharides among others, play a significant role in order to promote human well-being (Viuda-Martos et al., 2010; Leja and Czaczyk, 2016) against various diseases, that is, diabetes, cardiovascular, arthritis, Alzheimer's and cancer. Phyto-constituents from spices exhibit a wide range of unique properties, that is, (1) organoleptic – colour, aroma, flavour, and taste; (2) chemical- preserving the food items, antimicrobial (Rudrapal and Chetia, 2016); therapeutic – antioxidant, anti-inflammatory, antidiabetic, cardioprotective, and anti-cancer as well as other attributes. Natural phenol/polyphenol from spices work as strong antioxidants and slow down enzymatic oxidation reaction by acting as reducing substances. In addition to preserving organoleptic qualities, phenolic compounds extend the shelf life of foods by avoiding rancidity, improving the stability of polyunsaturated fatty acids (PUFAs) in fats and oils, and limiting oxidation-related changes in food. (Kähkönen et al., 1999).

12.3.2 STRUCTURE AND DERIVATIVES OF POLYPHENOLS

Phenolic compounds and polyphenols are structurally hydroxyl group's riches, naturally occurring secondary metabolites, generated through shikimic acid/phenylpropanoid and/or the polyketide (PK) pathways that are principally found in the fruits, vegetables, cereals, beverages, and spices. Phenols are aromatic compounds that contain hydroxyl group substituted benzene ring; chemically phenolic moieties are divided into the four separate categories based on hydroxyl group: (I) Pyrocatechol, (II) Resorcinol, (III) Pyrogallol, and (IV) Phloroglucinol (Abdel-Shafy and Mansour, 2017). Polyphenols – "poly" is derived from the Greek word (polus), which means "many, much" – are generated and exist in plant tissues as glycosides or aglycones. They are aromatic compounds with more than one phenolic group close to a neighbouring pyran ring or C-Skelton that are widely dispersed. Phenolic and polyphenol compounds (phenolic aglycones) were initially classified by two scientists J. B. Harborne and N. W. Simmonds in 1964 (Ollis, 1965) based on several carbons in the molecule and number of phenol rings, that is, C6 are phenolic acids and similar chemicals; C6-C1 are phenolic acids and associated derivatives; C15 are flavonoids and anthocyanins structure (Vermerris and Nicholson, 2008).

Flavonoids are one of the most prevalent and extensively distributed plant phenolics, utilized primarily in the human diet for health benefits with desired therapeutic effect (Kumar and Pandey, 2013). Structurally, flavonoids have the general arrangement of a 15- number carbon skeleton ([C6-C3-C6 framework)] in which, A and B are two heterocyclic phenyl rings (C6), while C is a heterocyclic closed pyran ring with one oxygen embedded in it. Flavonoids are divided into the following subgroups: that is,

(a) Flavones: Double bond at C- 2 and 3 positions
(b) Flavanones: No double bond at C- 2 and 3 positions
(c) Flavonols: Double bond at C- 2 and 3 positions; OH group at C- 3 position
(d) Flavanonols: No double bond at C- 2 and 3 positions; OH group at C- 3 position
(e) Flavanols (Flavans): No double bond in C- 2 and 3 positions; OH group at C- 3 positions + Absence of keto group at C- 4 positions.
(f) Anthocyanins: Chromenylium cation with substitution at C- 2 position

The flavonoids are often hydroxylated at specific positions in Ring A- 5, 7; Ring B- 3', 4', and 5'; and in Ring C- 3.

According to IUPAC, flavonoids are classified into three broad categories based on the linkage of B ring adjacent to the chromane moiety as flavonoids or bioflavonoids, isoflavonoids, and neoflavonoids (Rauter et al., 2018).

Figure 12.1 summarizes the general categorization and parent structures of polyphenols.

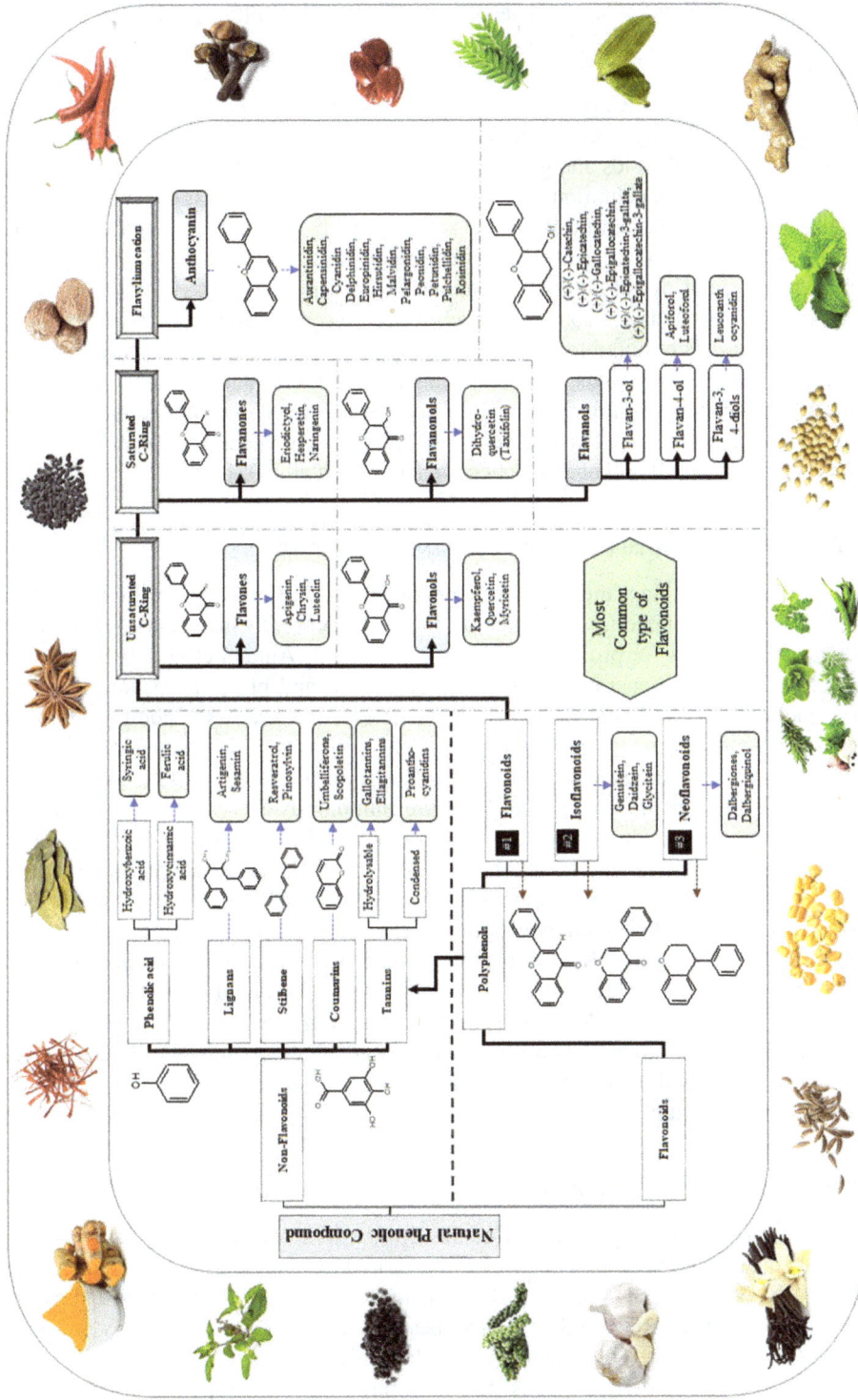

FIGURE 12.1   Categorization and parent structures of polyphenols.

Historically, polyphenols were known globally as "vegetable tannins" globally for the plant materials that were used in leather preparation from animal skins by precipitating gelatin. In 1957 Theodore White defined "tannins" as polyphenol materials that have sufficient phenolic groups with molecular weight between 500 and 3000 Da and were able to form cross-linked complexes with collagen molecules through hydrogen-bonding mechanisms. Plant polyphenols were defined in 1962 by two British phytochemists, E.C. Bate-Smith and Tony Swain, in 1962 as water-soluble chemicals with molecular weights ranging from 500 to 3000 Da that engaged in common phenolic reactions. They can also precipitate alkaloids and gelatin protein, which are the major metabolites. (Quideau et al., 2011). Subsequently, the definition of Bate-Smith, Swain, and White was expanded by Haslam. White–Bate-Smith–Swain–Haslam (WBSSH) described widely accepted common characteristics of polyphenol. According to them, polyphenols are moderately water-soluble compounds with a range of molecular weight of 500 to 4000 containing >12 phenolic hydroxyl groups on 5–7 aromatic rings per 1000 Da. Today, a comprehensive, widely accepted, and scientifically sound definition of "polyphenols" is required to define it properly as so many activities, research work, and experiments are going on in various scientific fields (Abdel-Shafy and Mansour, 2017). So far, around 8,000 polyphenol compounds from different species have been identified (Pandey and Rizvi, 2009). According to a recent study that looked at 79 phenolic components, the spices clove and allspice have the greatest overall polyphenol concentration. (Ali et al., 2021a).

12.3.3 PATHWAYS AND DISTRIBUTION OF POLYPHENOLS IN THE SPICES

The distribution and circulation of the phenolic compound are not uniform at the cellular, subcellular, and tissue levels. Soluble phenolics are generally found in vacuoles whereas insoluble are found in the cell wall of the plant for providing mechanical strength (Ahmad et al., 2019). Phenolics, including polyphenols, are generated through shikimic acid and phenylpropanoid metabolism biosynthetic pathways. A shikimic acid pathway generates the precursor compounds phenylalanine and tyrosine of the phenylpropanoid metabolic pathway. P-coumaroyl-CoA is formed when phenylalanine is combined with malonyl-CoA to form the true foundation of flavonoids. (Petrussa et al., 2013; Liu et al., 2021). Phenylalanine ammonia-lyase (PAL), anthocyanidin synthase (ANS), cinnamate 4-hydroxylase (C4H), 4-coumaroyl CoA ligase (4CL), STS (stilbene synthase), chalcone synthase (CHS), coumarate 3-hydroxylase (C3H), chalcone isomerase (CHI), flavanone 3'5'-hydroxylase (F3'5'H), flavonoid 3'-hydroxylase (F3'H), flavonoid 3-hydroxylase (F3H), isoflavone synthase (FNS), flavonol synthase (FLS), dihydroflavonol-4 reductase (DFR), leucoanthocyanidin reductase (LAR), leucoanthocyanidin oxidase (LDOX), anthocyanidin synthase (ANS), anthocyanidin reductase (ANR), UDP-glucose flavonoid 3-Oglucosyltransferase (UFGT), O-methyl transferases (OMT) are some important enzymes that contribute to generate polyphenols in the phenylpropanoid pathway. A wide range of compounds is generated when the location and measure of methoxyl and hydroxyl groups on the benzene ring change, even though the fundamental structure stays the same.

12.3.4 LEADING POLYPHENOLS FROM SPICES: ANALYSIS AND DRUG DISCOVERY

The Nutrient Data Laboratory is a division of the United States Department of Agriculture. In 2015, the Department of Agriculture compiled and prepared a database of "Flavonoid Content of Selected Foods" based on detection of HPLC-UV data together with HPLC-MS data, and published a report with the identifier "Release 3.2." Cacao beans had the greatest flavonoids content (+)-Gallocatechin (8262.0 mg per 100 g), according to the analysis. Anthocyanin Cyanidin, found in higher concentrations in raspberries, plums, and elderberries, is also responsible for antioxidant defense. The highest quantities of flavonoids content were identified in dried parsley, with apigenin and isorhamnetin content of 4503.50 and 331.24 mg per 100 g, respectively, followed by Mexican dried oregano, with luteolin content of 1028.75 mg per 100 g and naringenin content of 372.00 mg per 100 g. Celery, caper, saffron, lovage, dill, and other spices were found in descending sequence. Flavonols, that is, quercetin and kaempferol, were found in a high frequency as compared to other flavonoids. Individually, the highest amount of apigenin was found in dried parsley, followed by celery and oregano (Mexican). Isorhamnetin content was also high in the same spice. Luteolin was observed to be high in oregano (Mexican) followed by celery and thyme. The main constituents of the caper plant are kaempferol and quercetin. Quercetin (expressed as aglycones) was the most prevalent phyto constituent in another research on dietary flavonoids, followed by kaempferol, luteolin, and apigenin.(Cook and Samman, 1996; Carlsen et al., 2010; Yashin et al., 2017). Other spices also have a good amount of flavonoids responsible for health benefits (Table 12.2).

12.4 POTENTIAL MECHANISMS

Various factors accompanying our daily lifestyle generate free radicals that are scavenged by polyphenols, especially flavonoids. Flavonoids work through numerous mechanisms of action inside the body. Flavonoids exhibit a wide range of therapeutic properties, including anti-inflammatory, anti-allergic, aAnti-CVD, and anti-cancerous efficacy along with their antioxidant characteristics (Rudrapal et al., 2022).

12.4.1 Absorption and interaction inside the body

Polyphenols commonly exist with their multiple O- and C-glycosidic derivative forms. Glycosidic polyphenols (GPs) have polar and hydrophilic properties, making intestinal absorption difficult with diffusion through intestinal glucose transporters, specifically GLUTs and SGLT receptors. Once GPs is hydrolyzed by the catalytic enzyme "lactase phlorizin hydrolase (LPH)" released by an apical membrane, unconjugated polyphenols are then diffused from the small intestine. Polyphenols with only one or two hydroxyl groups are easily absorbed; polyphenols with more than two hydroxyl groups become hydrophobic due to hydrogen bonds between the 4-keto oxygen atom and a nearby hydroxyl group. Flavonoids, that contain more than 4-6 hydroxyl groups are more hydrophobic and are sufficiently transported from the enterocyte. Although aglycones have very high hydrophobicity their overall uptake is limited and varies due to their structural configuration (Ahmad et al., 2019). After entering, a multi-drug resistance transporter (MDR-2) is activated and some polyphenols are transported back to the intestinal lumen. Since flavonoids metabolize quickly and eliminate rapidly in the body, their intrinsic biological activities and potentialities are affected to a maximum extent (Manach et al., 2004). Methylation, sulfation, and glucuronidation are the three important conjugation processes by which polyphenol acts inside a body and reacts with cytochrome P450 (CYP) enzyme, that is, CYP1, 2E1, 3A4, and 19 (Meskin et al., 2003).

12.4.2 Significant signaling pathways

Dietary polyphenols cross the epithelial cell membranes of GIT and trigger various signaling pathways, that is, single-stranded non-coding RNA molecule (NC-RNAs) known as "microRNAs (miRNAs)," the dependent pathway, epidermal growth factor (EGF), protease enzymes-caspases pathway, β cell lymphoma-2 pathways, phosphatidylinositol-3-kinase (PI3K)/Akt/mTOR pathways, mitogen-activated protein kinase (MAPK), Matrix Metalloproteinase enzymatic pathways, transcription factor -nuclear factor (NF)-κB pathways, Deacetylate histones Sirtuin 1 (SIRT1) activating pathway, and so forth (Chung et al., 2010; Sun et al., 2019a).

TABLE 12.2

ISO Spices and "Total Flavonoid Content" Database from the U.S. Department of Agriculture (Bhagwat et al., 2015) with "Total ORAC Values" (Haytowitz and Bhagwat, 2010)

Sl. No	English/ Common Name	Plant Part	Flavones		Flavonols				Flavanones			Total Flavonoid Content (mg per 100 g)	Total-ORAC Value µmol TE/100 g
			Api.	Lut.	Isor.	Kaem.	Myr.	Quer.	Eriod.	Hesp.	Narin.		
1	Parsley	Seed, Leaf, Root	4503.5	19.75	331.24	-	-	-	-	-	-	4523.25	73670
2	Mexican oregano	Terminal Shoot	17.71	1028.75	-	-	-	42	85.33	-	372	1545.79	175295
3	Celery	Fruit, Root, Leaf	78.65	762.4	-	259.19	-	-	-	-	-	841.05	552
4	Caper, Caper-bush	Fruit, Root, Floral Bud	-	-	-	-	-	233.84	-	-	-	493.03	-
5	Saffron	Stigma	-	-	-	205.48	-	-	-	-	-	205.48	-
6	Lovage	Leaf	-	-	-	7	-	170	-	-	-	177.00	-
7	Dill	Fruit	-	-	43.5	13.33	0.7	55.15	-	-	-	112.68	4392
8	Indian mustard	Seed	-	-	16.2	38.2	-	8.8	-	-	-	63.20	-
9	Peppermint	Fruit, Leaf	5.39	12.66	-	-	-	-	30.92	10.16	-	59.13	13978
10	Juniper berry	Berry	7.26	51.4	-	-	-	-	-	-	-	58.66	-
11	Chilli, Capsicum, Paprika	Fruit	-	6.93	-	-	-	50.63	-	-	-	57.56	-
12	Coriander	Leaf, Seed	-	-	-	-	-	52.9	-	-	-	52.90	5141
13	Thyme	Leaf	2.5	45.25	-	-	-	-	-	-	-	47.75	157380
14	Ginger	Rhizome	-	-	-	33.6	-	-	-	-	-	33.60	14840
15	Rosemary	Terminal Shoot, Leaf	0.55	2	-	-	-	-	-	-	24.86	27.41	165280
16	Onion	Bulb	0.01	0.02	5.01	0.65	0.03	20.3	-	-	-	26.02	913
17	Stony leek, Japanese bunching onion	Bulb, Leaf	-	-	-	24.95	-	-	-	-	-	24.95	-

No.	Common name	Part used	Api.	Lut.	Isor	Kaem.	Myr.	Quer.	Eriod.	Hesp.	Narin.	Total	TFC
18	Chive	Leaf	-	0.15	6.75	10	-	4.77	-	-		21.67	2094
19	Chilli, Bird eye chilli	Fruit	1.4	3.87	-	-	1.2	14.7	-	-		21.17	-
20	Garden sage	Terminal Shoot, Leaf	1.2	16.7	-	-	-	-	-	-		17.90	32004
21	Oregano, Origan	Leaf, Flowering top	2.57	1	-	-	2.1	7.3	-	-		12.97	13970
22	Pimento, Allspice, Jamaica Pepper	Fruit, Seed, Leaf	-	10.36	-	-	-	-	-	-		10.36	-
23	Bay leaf	Leaf Fresh	-	-	-	4.82	-	3.19	-	-		8.01	-
24	Turmeric	Rhizome, Leaf	-	-	-	-	2.04	4.92	-	-		6.96	127068
25	Garlic	Bulb	-	-	-	0.26	1.61	1.74	-	-		3.61	5708
26	Sweet marjoram	Leaf, Flower Top	3.5	-	-	-	-	-	-	-		3.5	27297
27	Celeriac	Fruit, Root, Leaf	2.41	-	-	-	-	0.18	-	-		2.59	-
28	Mango	Immature Fruit (Rind)	0.01	0.02	-	0.05	0.06	-	-	-		1.86	1300
			(+)-Catechin 1.72 & Cyanidin 0.10, Delphinidin 0.02, Pelargonidin 0.02										
29	Sweet fennel	Fruit, Leaf, Twig	-	-	-	-	-	-	-	-		1.31	307
30	Pomegranate seed	Seed Dried with Flesh	(+)-Catechin 0.40, (-)-Epicatechin 0.08, (-)-Epigallocatechin 0.16, (+)-Gallocatechin 0.17	0.23	1.08				-			0.81	4479

Api.-Apigenin, Lut.-Luteolin, Isor- Isorhamnetin, Kaem.-Kaempferol, Myr.-Myricetin, Quer.-Quercetin, Eriod.-Eriodictyol, Hesp.-Hesperetin, Narin.-Naringenin;

Amount in mg/100g; Gray Shaded Area- Non-Indian Spices

Total Flavonoids Content; database from U.S. Department of Agriculture, Release 3.2 (2015) by Nutrient Data Laboratory, Beltsville, Maryland

12.5 POLYPHENOLS AND THE POTENTIAL HEALTH EFFECTS

Spices and spice oleoresins are used as folk medicine in traditional cultures worldwide. Apart from flavour and aroma, spice's polyphenols give nutritional requirements and beneficial therapeutic effects to the body by participating in multifunctional enzymatic reactions and targets (Sachan et al., 2018). They are well- known and recognized for antimicrobial efficacy against viruses, bacteria, fungi, yeasts, and microbial toxins synthesis. Spice polyphenols are well-known for their antioxidant properties, similar to the antiviral, antibacterial, and antifungal properties of essential oils and terpenoids components. Flavonoids involve in oxidation and reduction reactions. They enhance the absorption of vitamin C and influence to digestion process (Leja and Czaczyk, 2016). They also act as a natural preservative and help to promote good health promotion (Shahidi and Hossain, 2018).

12.5.1 HEALTH ASSISTANCES: THERAPEUTIC AND PHARMACOLOGICAL PROPERTIES

Spices have their aromatic values in terms of aroma, flavour, taste, and potential health benefits. Apart from these individual properties, they can work synergistically and impart organoleptic properties to culinary preparation and functional foods. Several phytoconstituents, including polyphenols, work inside the body individually and/or in a combination for getting desired multiple therapeutic effects to human beings (Viuda-Martos et al., 2010; Leja and Czaczyk, 2016; Torre Torres et al., 2017). Spice polyphenols and their therapeutic activities are presented in Table 12.3.

12.5.1.1 Antioxidant activity and their impression on human health

Antioxidants are substances that neutralize free radicals, and eliminate other reactive oxygen together with reactive nitrogen species, which are responsible for most chronic diseases due to erroneous routine, food practices, stressful lifestyle, tobacco smoke, pollution/contaminants, radiations like UV and X-ray, xeno-biotics, and so forth. They slow down the autoxidation process of essential compounds and alter the biochemical characteristic to support the longevity of cells, repair DNA, and live a healthy life (Carlsen et al., 2010; Mehta and Gowder, 2015; Mamdapur et al., 2021). They are widely used in food processing industries and extensively are extensively added to several foodstuffs for stabilizing food components to prevent spoilage (Mamdapur et al., 2021).

Safety, acceptability, and market opinion, spices are one of the most important categories of interest for the exploration of natural antioxidants. Spices include a wide range of phenolic and polyphenol chemicals that are now widely employed to counteract oxidation processes caused by free radicals. Presently, spices play a dual role in the food- processing industry. On one side, they enhance the organoleptic properties of foodstuff and on another side, show strong antioxidant potential concerning their corresponding total polyphenol concentrations. Keeping the balance between these two required dimensions, compounds derived from spices are now exploring up to the maximum extent (Yanishlieva et al., 2006; Wojdyło et al., 2007; Srinivasan, 2014). Polyphenols have been shown to inhibit lipid peroxidation, platelet aggregation, capillary permeability, fragility, and the enzyme activities of cyclooxygenase and lipoxygenase, among other things. They also reported an inhibiting number of enzymes like hydrolases, hyaluronidase, alkaline phosphatase, arylsulphatase, cAMP, phosphodiesterase, lipase, α-glucosidase, kinase, and so on (Sandhar et al., 2011; Deepa et al., 2013). Active flavones isoscutellarein 7-O-glucoside and genkwanin are the other important polyphenols from the rosemary herb (*Rosmarinus Officinalis*) found maximally in young stages of the leaf (Del Baño et al., 2003) and shows strong antioxidant activity.

12.5.1.2 Anti-inflammatory activity

Inflammation is a sequential body response against various pathophysiological conditions and diseases that are classified as either acute or chronic type and affect the quality of life subjectively. In response to tissue injury, in any pathogen invasion, or infectious attack, the body produces

TABLE 12.3
Polyphenols and Therapeutic Activity from the Selected Spices

Sl. No.	Spices	Plant Part	Phyto-Component	Ref.	Polyphenol Activities
1.	Aniseed	Seed	Apigenin-7-glucoside	(Shojaii and Abdollahi Fard, 2012; Rebey et al., 2018; Sun et al., 2019b)	Antioxidant, Antibacterial (Amer and Aly, 2019)
2.			Epicatechin-3-gallate		
3.			Luteolin		
4.			Naringenin		
5.			Quercetin-3-glucuronide		
6.			Rutin		
7.			Isoorientin	(Shahidi and Hossain, 2018)	
8.			Isovitexin		
9.	Basil	Ariel Parts	Apigenin-7-O-glucoside	(Shahidi and Hossain, 2018)	Antimicrobial
10.			Luteolin-5-O-glucoside	(Grayer et al., 2002; Picos-Salas et al., 2021)	
11.			Luteolin 7-O-glucoside		
12.	Black pepper	Fruit	Apigenin	(Shahidi and Hossain, 2018; Ali et al., 2021a; Ashokkumar et al., 2021)	Antioxidant, ROS-Free Radical Quenching Efficacy
13.			Hydroxytyrosol 4-O-glucoside		
14.			Kaempferol		
15.			Rhoifolin		
16.			Scopoletin		
17.			Sesamin		
18.			Luteolin 8-C-glucoside		
19.	Cardamom	Fruit	Catechin	(Cárdenas Garza et al., 2021)	Digestive disorder, Hypertension
20.			Kaempferol		
21.			Luteolin		
22.			Myricetin		
23.			Quercetin		
24.			(+)-Gallocatechin	(Ali et al., 2021a)	
25.			Carnosic acid		
26.			Dihydroxyphenylacetic acid		
27.			Hesperetin 3 0 -O-glucuronide		
28.			Pelargonidin (An anthocyanin)		
29.	Celery	Seed	Apiin (Diglycoside of apigenin)	(Muzolf-Panek and Stuper-Szablewska, 2021)	Antioxidant CVD treatment
30.			Chrysoeriol glycoside (3'-o-methylated flavonoids)		

(*continued*)

TABLE 12.3 (Continued)
Polyphenols and Therapeutic Activity from the Selected Spices

Sl. No.	Spices	Plant Part	Phyto-Component	Ref.	Polyphenol Activities
31.			Luteolin glycosides		
32.	Cinnamon	Bark	Catechin	(Rao and Gan, 2014)	Natural insulin producer, Antidiabetic, Antihyperlipidemic, Atherosclerosis treatment,
33.			Isorhamnetin		
34.			Kaempferol		
35.			Quercetin		
36.			Rutin		
37.	Clove	Bud	Ombutin 3-O-β-D-glucopyranoside	(Johannah et al., 2015; El-Maati et al., 2016)	Antioxidant, Arthritis, CVD diseases
38.			Quercetin		
39.			Tamarixetin 3-O-β-D-glucopyranoside		
40.	Coriander	Ariel Parts	Apigenin	(Mechchate et al., 2021)	Diabetes(Mechchate et al., 2021)
41.			Kaempferol		
42.			Quercetin		
43.			Rhamnetin (7-methoxy quercetin)		
44.		Seed	Apigenin		
45.			Kaempferol		
46.			Luteolin		
47.			Naringin		
48.			Quercetin dihydrate		
49.			Quercetin-3-rhamnoside		
50.			Rutin trihydrate		
51.	Cumin	Seed	Amentoflavone (Biflavonoid)	(Johri, 2011; Gangadharappa et al., 2017; Gotmare and Tambe, 2018)	Antioxidant, Digestive wellness
52.			Apigenin		
53.			Catechin		
54.			Luteolin		
55.			Quercetin		
56.	Curry leaf	Leaf	Catechin	(Shahidi and Hossain, 2018)	Diabetics, Cancers, Cardiovascular diseases, Obesity(Abeysinghe et al., 2021)
57.			Epicatechin		
58.			Myricetin		
59.			Naringin		
60.			Quercetin		
61.			Quercetin-3-glycoside		
62.			Rutin		

TABLE 12.3 (Continued)
Polyphenols and Therapeutic Activity from the Selected Spices

Sl. No.	Spices	Plant Part	Phyto-Component	Ref.	Polyphenol Activities
63.	Dill	Seed	Isorhamnetin	(Ksouri et al., 2015)	Inhibition of AGEs formation; Atherosclerosis treatment, Diabetes (Goodarzi et al., 2016)
64.			Kaempferol		
65.			Myricetin		
66.			Quercetin		
67.	Fennel	Seed	Apigenin	(Rather et al., 2016)	Digestive wellness
68.			Quercetin		
69.			Rutin		
70.	Fenugreek	Seed	Isovitexin	(Kannappan and Anuradha, 2009; Wani and Kumar, 2018; Akbari et al., 2020; Syed et al., 2020)	Diabetes
71.			Naringin		
72.			Quercetin		
73.			Rutin		
74.			Tricin (An O-methylated flavone)		
75.			Vitexin (apigenin flavone glucoside)		
76.	Garlic	Bulb	hydroxycinnamic acid	(Steiner, 1997)	Protection against oxidative damage, CVD Treatment
77.			Kaempferol		
78.			Myricetin		
79.			Quercetin		
80.	Ginger	Rhizome	6-Gingerol, 8-Gingerol and 10-Gingerol	(Asamenew et al., 2019; Idris et al., 2019)	CVD Treatment
81.			6-Shogaol, 8-Shogaol and 10-Shogaol		
82.			Kaempferol		
83.	Kokam	Fruit rind	Cyanidin-3-glucoside	(Singh et al., 2022)	Antioxidant
84.			Cyanidin-3-sambubioside		
85.			Garcinol Cyanidin-3-glucoside		
86.	Mustard	Seed	Isorhamnetin	(Martinović et al., 2020)	Oxidative stress (Tian and Deng, 2020)
87.			Kaempferol		
88.			p-Hydroxybenzoic		
89.			Quercetin		
90.			Sinapic acids		
91.	Oregano	Leaves	Apigenin	(Gutiérrez-Grijalva et al., 2018)	Antioxidant
92.			Eriodictyol		
93.			Luteolin		
94.			Naringenin		

(continued)

TABLE 12.3 (Continued)
Polyphenols and Therapeutic Activity from the Selected Spices

Sl. No.	Spices	Plant Part	Phyto-Component	Ref.	Polyphenol Activities
95.			Quercetin		
96.	Parsley	Seed	Apigenin	(Stan et al., 2012)	Anticancer (Ross and Kasum, 2002)
97.			Kaempferol		
98.			Luteolin		
99.			Quercetin		
100.	Rosemary	Ariel Parts	Apigenin	(Petiwala et al., 2013)	Anticancer(Allegra et al., 2020)
101.			Luteolin		
102.			Naringenin		
103.	Saffron	Petal	Kaempferol	(Troiano and Reuter, 2014; Jadouali et al., 2017, 2019)	Cognitive Treatment, Antidepressant
104.			Kaempferol soforoside		
105.			Quercetin rutinoside		
106.	Sage	Ariel Parts	Apigenin	(Roby et al., 2013)	Antioxidant
107.			Luteolin		
108.	Thyme	Ariel Parts	Apigenin	(Roby et al., 2013; Szilvaśsy et al., 2013)	Antioxidant
109.			Chrysin (5,7-dihydroxyflavone)		
110.			Diosmetin (An O-methylated flavone)		
111.			Luteolin		
112.	Turmeric	Rhizome	Curcumin (A flavonoid polyphenol)	(Malik and Mukherjee, 2014)	Oxygen scavenger

biochemical markers, hormonal metabolites, and immunological components (Brøchner and Toft, 2009) like reactive oxygen species (ROS), that is, hydroxyl radical ($\cdot$OH), peroxyl radicals (ROO$\cdot$), perhydroxyl radicals (HO$_2\cdot$), alkoxy radical (RO$\cdot$), superoxide ion (O$_2\cdot^-$); hydrolytic enzymes, growth factors (i.e., angiotensin II), and inflammatory cytokines, that is, interferon-gamma (IFN-g), IL-2, IL-4, IL-6, IL-8, and IL-17) and others (Hayyan et al., 2016). Polyphenols regulate the expression of several pro-inflammatory cytokines, lipoxygenase, inducible nitric oxide synthases (iNOS), cyclooxygenase, neutrophil degranulation, protein kinase C, immune cells by interfering interleukins productions and gene expression, and so forth (Sandhar et al., 2011).

Curcuminoids that is, curcumin, desmethoxycurcumin, and bisdesmethoxycurcumin constitute the type of natural polyphenol compounds, derived from the exotic spice Indian "Solid Gold - Turmeric" also known as "Indian saffron." Curcuminoids are yellow-pigmented curcuminoids derived from the rhizome of Curcuma longa, which has stronger antioxidant and anti-inflammatory activity and is currently widely utilized as a viable candidate for the prevention and/or treatment of chronic illnesses such as cancer (Nema et al., 2021). Curcumin is one of the key curcuminoids that modulate cytokines mRNA expressions via the PTGS2 gene-encoded enzyme "Prostaglandin-endoperoxide synthase 2" (commonly known as cyclooxygenase-2 or COX-2) and regulate inflammatory reactions.

It also affects pro-inflammatory cytokines, as well as the length of muscle contractions, when combined with myokine, which is secreted by myocytes. Curcumin inhibits numerous inflammatory signaling pathways at both RNA and protein levels, that is, pro-inflammatory transcription factor NF-κB (Siwak et al., 2005; B. Aggarwal et al., 2011), the release of NLRP3 (Gong et al., 2015) that is a key factor in any inflammatory pathways (Yin et al., 2018), metalloproteinase-9 (MMP-9), MAPK/ERK, STAT3-transcription 3 and interleukinsi.e.IL-6 and Tumour Necrosis Factor-alpha (TNF alpha), and so forth (Siwak et al., 2005; Aggarwal and Harikumar, 2009; Cavaleri, 2018; Hahn et al., 2018; Paulraj et al., 2019). Curcumin polyphenol inhibits the pro-inflammatory factor HMGB-1 nuclear protein (Sambasivarao, 2013), which binds to extracellular mediators, (Damage-associated Molecular Pattern) DAMP, and (Pathogen-associated Molecular Pattern) PAMPs to assist the endocytosis process in conjunction with transmembrane receptor RAGE (Receptor for Advanced Glycation Endproducts) also known as AGER. As a result, numerous triggered inflammatory cytokines, such as IL-1, IL-6, IL-18, and TNF-α, are produced. (Willingham et al., 2009; Kim et al., 2011).

Quercetin is another intriguing chemical with a wide range of medicinal potential. Quercetin's anti-inflammatory benefits are attributable to its inherent anti-oxidant capabilities. and its capabilities to interfere with peculiar biological pathways. Its analgesic activity is due to the inhibitory effect of neurochemical and mechanical properties, that is, allodynia (Carullo et al., 2017). Spice cinnamon has Quercetin-3-glucoside at very high concentrations.

12.5.1.3 Anticancer effect

Flavonoids have been intensively researched for their chemopreventive properties in a variety of malignancies. In the targeted cells, maximum polyphenols stop cell cycle progression in the late G1 and G2/M phases and activate apoptotic signal transduction pathways (caspase 8). Due to its unusual conjugated structure with two methoxy phenols and an enol form of diketone, curcumin has anticancer and anti-oxidant capabilities (Giordano and Tommonaro, 2019). Among all flavonoids, quercetin is one of the most studied due to its strong antioxidant properties, which work through a variety of mechanisms, anti-inflammatory properties, cell cycle regulation, inhibition of cancer cell proliferation, regulation of detoxification of carcinogens, apoptosis induction, suppressing oncogenes, and hormone/growth factor activity modulation, among others. Many preclinical and clinical studies show that progression of cancer is caused by dysregulation of cancer-sensitive microRNA (miRNA) function. Quercetin inflect miRNAs and defend cancer growth. Several *in vitro* and *in vivo* investigations have demonstrated quercetin's anti-cancer, anti-inflammatory, and anti-proliferative activities. Furthermore, some research has shown that it can be used as a pain reliever for cancer-related discomfort. According to reports, nano-micellar formulations of quercetin have a higher water solubility than free quercetin, resulting in improved bioavailability and consequently stronger tumor-suppressing effects. (Zheng et al., 2016; Kim et al., 2019). Caper (*Capparis spinosa*) anaromatics spice in the Mediterranean kitchen is belongs to the family Capparidaceae. Due to the presence of polyphenols, vitamin C, and vitamin E, aerial portions and unopened flowers are often utilized for nutritional value. (Tlili et al., 2010). According to the USDA database, kaemferol, and quercetin are the two major constituents present in the plant as 259.19 and 233.84 mg per 100 g, respectively. Hydroalcoholic extract of caper enriched with quercetin content (10.06%) showed significant anti-tumor efficacy (Moghadamnia et al., 2019). Another putative anticancer possibility, kaempferol (Imran et al., 2019), operates through the phosphoinositide 3-kinase (PI3K)/protein kinase B (AKT) pathways, has to be proven with more research. In addition to other, well-recognized polyphenols, resveratrol has been explored in the last few decades due to its pharmacological and therapeutic effects. Resveratrol has antitumor, anti-proliferative, and apoptosis properties against different types of cancerous cells mediated by various signaling pathways and modes of action for cancer management. Other flavonoids, such as genistein and silymarin, interfere with the G1 phase by reducing the activities of cyclin-dependent kinases. LPS-induced degradation of endogenous

IkBa is blocked by rosemary carnosol. Many spices, including celery, parsley, bay leaves, and oregano, are high in polyphenols and glycosides, which can aid in cancer prevention (Sun et al., 2019a).

12.5.1.4   Cardio-protective effects

Multiple variables, such as excessive cholesterol, hypertension, impaired fibrinolysis, elevated platelet aggregation, blood-clotting delay, and other associated risk factors all contribute to cardiovascular disease (CVD) (Khurana et al., 2013; Goszcz et al., 2017) that originatesd from the complications of atherosclerosis in the majority of cases (Habauzit and Morand, 2012). Both oxidative stress and reactive oxygen species (ROS) are linked to endothelial damage, which causes endothelium membrane proteins such as vascular endothelial adhesion molecule-1 (VCAM-1), intercellular cell adhesion molecule-1 (ICAM-1), and E-selectin to interact with monocytes and T lymphocytes, allowing them to adhere to the endothelium surface; this leads to the formation of early atherosclerotic plaque (Libby, 2006). ROS inhibits nitric oxide (NO)-dependent vasorelaxation in a variety of ways. First, reactive oxygen species diminish the bioavailability of NO by reducing the expression of an enzyme responsible for NO manufacture, endothelial nitric oxide synthase (eNOS), and second, ROS directly decompose NO. (Lubos, 2008; Khurana et al., 2013).

Polyphenols consumption can reduces the risk of cardiovascular disease (CVD) up to 46 percent (Tresserra-Rimbau et al., 2014) by improving plasma lipid profiles and endothelial function, modulating abnormal platelet aggregation, reducing pro-inflammatory cytokines, and so forth (Lubos, 2008), and limiting the effects of cellular aging (Khurana et al., 2013). Flavonoids in the diet, particularly flavan-3-ols and proanthocyanidins, lowered the risk of primary and secondary cardiovascular disease through modifying several preventive pathways (Schroeter et al., 2010).

12.5.1.5   Effects against cognitive and neurodegenerative disease

Spices and phytoconstituents generated from them might be promising possibilities for treating Alzheimer's disease (AD) (Satheeshkumar et al., 2016). Curcuminoids reduced the cholinergic enzyme acetylcholinesterase (AChE), which might be useful in the treatment of Alzheimer's disease. Curcuminoids may be utilized for cognitive management since they have memory-enhancing properties (Ahmed and Gilani, 2009; Dhakal et al., 2019).

12.5.1.6   Anti-diabetic property

Spices are well-known candidates for controlling the blood sugar level. One of the well-recognized spices, cinnamon is used as an aromatic condiment and as a flavouring agent from antiquity. Cinnamon has been reported for its improved insulin sensitivity and antidiabetic activity (Anderson et al., 2004; Cao et al., 2007; Anderson, 2008) due to the presence of rutin (>90%), catechin, quercetin, kaempferol, and isorhamnetin (Rao and Gan, 2014). Insulin-potentiating properties due to polyphenols were attracted to researchers and scientists for further deep research. In 2004, Anderson et al. (2004) isolated and characterized insulin-like molecules called "polyphenol type-A polymers" (procyanidin oligomers of the flavonoid catechins/epicatechins) from cinnamon and concluded that an aqueous extract of cinnamon rich in polyphenols activates insulin receptors (IR) by increasing tyrosine phosphorylation activity and decreasing phosphatase activity (Anderson et al., 2004). Cinnamon polyphenols have also been shown to increase the number of insulin-receptor and glucose transporter type 4 (GLUT4) proteins, as well as the number of other proteins involved in insulin signaling. (Cao et al., 2007; Anderson, 2008) because of chalcone polyphenols. In a meta-analysis, Davis and Yokoyama found that cinnamon consumption reduced fasting blood glucose (FBG) levels in persons with type 2 diabetes and prediabetes (Davis and Yokoyama, 2011). When 30 common herbs and spices with a library of 2,300 compounds were screened virtually with the DIA-DB web server, cinnamon was found the most efficacious spice against diabetes, even in a very high diluted concentration (Broadhurst et al., 2000). Cinnamon was revealed to be the most effective spice against diabetes (Pereira et al., 2019) even at a very high diluted concentration when 30 popular herbs and spices with a library of 2,300 components were tested digitally with the DIA-DB webserver (Pereira et al., 2019). Cinnamon, allspice, bay leaves, nutmeg, and cloves were revealed to have substantial antidiabetic potential if the study simply looked at spices (Pereira et al., 2019).

12.5.2 As dietary supplements: Overlook for health benefits

The dietary supplement and nutraceutical businesses are becoming interested in spices and their phytoconstituents. Many firms are creating value-added nutraceuticals from spices and their oleoresins in a range of dosage forms, including pills, capsules, syrup, and sachets, among others. Ginger, curcumin, coriander, capsicum, and other spices are among the best-suited kitchen treasures available on the market as a healthy regime. Curcuminoids rich turmeric extract suppresses inflammatory responsible factors, reduces cardiovascular complications, and promotes optimal heart health. It is powerful combination with omega-3 fatty acids. It is a potent antioxidant that protects cells from oxidative damage, as well as an anti-inflammatory that fights infections, promotes heart health, and functions as a natural immunity booster. CurQmeg-3 is a good example of a commercial product that may be used in this way. On the other hand, ginger is beneficial for digestion and constipation. It also aids in the alleviation of cramps and motion sickness. Green tea polyphenols are recognized for their slimming and detoxifying properties. In the same way, many more polyphenols may be acquired from herbs and spices for health benefits and can be incorporated into our daily diet.

12.5.3 Futuristic approaches: opportunities for food and pharmaceuticals

Synthetic antioxidants used in the food service/processing industry include butylated hydroxyanisole (BHA) in edible fats and fat-containing foods, butylated hydroxytoluene (BHT) in convenience/canned foods, prepared cereal products, and snacks, propyl gallate (PG) in meats, chicken soup base, and tert-butyl hydroquinone (TBHQ) in pre-made frozen foods and packaged dinners (Lourenço et al., 2019). Natural antioxidants are becoming increasingly popular across the world. Foods, drinks, and spices containing phenol and polyphenol act as dietary antioxidants for human health management (Yanishlieva et al., 2006; Pandey and Rizvi, 2009) by scavenging ROS and reactive nitrogen species (Tsao, 2010). Preservatives containing polyphenol that produces using nanotechnology can enhance food packaging and preserve food quality, including appearance, distinctive flavors, novel textures, fresh sensations, reduced fat, improved nutrient absorption, and microbiological safety (Chaudhry et al., 2008; Yu et al., 2018). Flavonoid nanotechnology may be of relevance in the realm of food technology application (Rupasinghe, 2020).

Carvacrol, coumaric acid, trans-cinnamaldehyde, eugenol, gallic acid, and rosmarinic acid, as well as its phenolic derivatives such as carvacrol, coumaric acid, trans-cinnamaldehyde, eugenol, gallic acid, and rosmarinic acid, exhibit antibacterial action against foodborne pathogenic microorganisms. Natural phenol-rich extracts can not only extend the shelf life of foods by reducing bacterial contamination, but they can also coexist in food systems with probiotic Lactic acid bacteria (LAB) to provide humans with additional health advantages (Nazzaro et al., 2020). Polyphenol-containing prebiotics enhances the impact of gut microbes by altering the microbial metabolites that make them available to the host (Chan et al., 2018; Nazzaro et al., 2020). Unknown is how unconsumed flavonoids interact with the colon microbiota and the metabolites that result, as well as their role in disease prevention and therapy. Microbiome and flavonoid pharmacology, as well as flavonoid-inspired therapies in association with microbiota, might be a future priority for the industry.

Polyphenols derived from spices can be used as natural antioxidants in dairy industries at low concentrations (Tzima et al., 2020). Aqueous rosemary extract may be utilized in yogurt without altering any of the desired sensory attributes (Ali et al., 2021b). In another study, natural antioxidants such as chamomile and fennel decoctions were shown to increase the oxidative stability of yogurts when compared to the synthetic preserved ingredient potassium sorbate (Caleja et al., 2016; Tzima et al., 2020).

Based on statistics from 163 countries, including total cases, total fatalities, and total recovered cases, it was predicted that spice consumption had a function and potential to battle COVID-19 (Elsayed and Khan, 2020). Phytoconstituents from spices can be utilized to build novel medications to decrease inflammation and boost immunity, according to recent Molecular Docking research of important bioactive chemicals. Apigenin, curcumin, and quercetin have a stronger therapeutic potential than the usual medication methotrexate in the treatment of arthritis. Quercetin is more effective at inhibiting the inflammatory marker cyclooxygenase (COX), which is important in cancer biology than aspirin or celecoxib. As a result, bioactive therapies for arthritis can be considered safe and effective (Ghate and Kulkarni, 2021). Quercetin and curcumin are examples of polyphenols found in spices have been shown to have beneficial effects in obese people, possibly by reducing intracellular oxidative stress, chronic inflammation, interfering adipogenesis, and lipogenesis, in addition to suppressing preadipocyte differentiation, and so forth. Spice polyphenols, which have yet to be examined, may shortly prove to be useful and effective in the treatment of obesity.

12.6   CONCLUSION AND FUTURE PROSPECTIVE

Spices are an important part of our daily life and are used in our daily cuisine. Spices are something that everyone is familiar with, and it would be hard to cook a meal without them. Spices are high in secondary metabolites, such as essential polyphenols, which have a wide range of applications in the food and pharmaceutical industries, including food processing, packaging, and storage, dietary supplements/nutraceutical for health benefits against cancer type 2 diabetes, obesity, neurodegenerative diseases, and so on. This section included spice classifications and kinds, as well as polyphenols, cell signaling, prospective therapeutics, and potential ways for use in the food and pharmaceutical sectors. Polyphenols and their glycosides are most typically found in plant-based foods like spices, and their powerful antioxidant properties allow them to be used in a range of applications. However, except for a few well-known spices and their associated polyphenols, many of them have yet to be studied. They should be incorporated into existing nutrition-education efforts and standards to encourage healthy living.

ACKNOWLEDGMENTS

The authors acknowledge the management of Synthite Industries, India, for their support and inspiration.

AUTHOR DISCLOSURE STATEMENT

The authors declare that there is no conflict of interest.

REFERENCES

Abdel-Shafy, H.I., and Mansour, M.S.M. 2017. Polyphenols: Properties, occurrence, content in food and potential effects. In *Environ Sci & Engg: Toxicology*, ed. B.R. Gurjar. Studium Press: 232–261.

Abeysinghe, D.T., Kumara, K.A.H., Kaushalya, K.A.D., Chandrika, U.G., and Alwis, D.D.D.H. 2021. Phytochemical screening, total polyphenol, flavonoid content, in vitro antioxidant and antibacterial activities of Sri Lankan varieties of Murraya koenigii and Micromelum minutum leaves. *Heliyon* 7: e07449.

Aggarwal, B.B., and Harikumar, K.B. 2009. Potential therapeutic effects of curcumin, the anti-inflammatory agent, against neurodegenerative, cardiovascular, pulmonary, metabolic, autoimmune and neoplastic diseases. *Int J Biochem Cell Biol* 41: 40–59.

Aggarwal, B., Prasad, S., Reuter, S., Kannappan, R., R. Yadav, V., Park, B., et al. 2011. Identification of novel anti-inflammatory agents from Ayurvedic medicine for prevention of chronic diseases. *Curr Drug Targets* 12: 1595–1653.

Ahmad, R., Hameed, A., Sameen, A., Imran, A., Ahmad, M.H., Yasmin, A., et al. 2019. Polyphenols; biosynthesis, classification and pharmacological utilization. In *Food Nutr* (India: Avid Science), 1–31.

Ahmed, T., and Gilani, A.H. 2009. Inhibitory effect of curcuminoids on acetylcholinesterase activity and attenuation of scopolamine-induced amnesia may explain medicinal use of turmeric in Alzheimer's disease. *Pharmacol Biochem Behav* 91: 554–559.

Akbari, S., Nour, A.H., and Yunus, R.M. 2020. Determination of phenolics and saponins in fenugreek seed extracted via microwave-assisted extraction method at the optimal condition. *IOP Conf Ser Mater Sci Eng* 736: 022024.

Ali, A., Wu, H., Ponnampalam, E.N., Cottrell, J.J., Dunshea, F.R., and Suleria, H. A.R. 2021a. Comprehensive profiling of most widely used spices for their phenolic compounds through lc-esi-qtof-ms2 and their antioxidant potential. *Antioxidants* 10: 721.

Ali, H.I., Dey, M., Alzubaidi, A.K., Alneamah, S.J.A., Altemimi, A.B., and Pratap-Singh, A. 2021b. Effect of rosemary (*Rosmarinus officinalis* l.) supplementation on probiotic yogurt: Physicochemical properties, microbial content, and sensory attributes. *Foods* 10: 2393.

Allegra, A., Tonacci, A., Pioggia, G., Musolino, C., and Gangemi, S. 2020. Anticancer activity of *Rosmarinus officinalis* L.: Mechanisms of action and therapeutic potentials. *Nutrients* 12: 1739.

Amer, A., and Aly, U. 2019. Antioxidant and antibacterial properties of anise (*Pimpinella anisum* L.). *Egypt Pharm J* 18: 68–73.

Anderson, R.A. 2008. Chromium and polyphenols from cinnamon improve insulin sensitivity. *Proc Nutr Soc* 67: 48–53.

Anderson, R.A., Broadhurst, C.L., Polansky, M.M., Schmidt, W.F., Khan, A., Flanagan, V.P., et al. 2004. Isolation and characterization of polyphenol type-A polymers from cinnamon with insulin-like biological activity. *J Agric Food Chem* 52: 65–70.

Asamenew, G., Kim, H.-W., Lee, M.-K., Lee, S.-H., Kim, Y.J., Cha, Y.-S., et al. 2019. Characterization of phenolic compounds from normal ginger (*Zingiber officinale* Rosc.) and black ginger (*Kaempferia parviflora* Wall.) using UPLC–DAD–QToF–MS. *Eur Food Res Technol* 245: 653–665.

Ashokkumar, K., Murugan, M., Dhanya, M.K., Pandian, A., and Warkentin, T.D. 2021. Phytochemistry and therapeutic potential of black pepper – *Piper nigrum* (L.)] essential oil and piperine: A review. *Clin Phytoscience* 7: 2–11.

Bhagwat, S., Haytowitz, D.B., and Holden, J.M. 2015. USDA Database for the Flavonoid Content of Selected Foods, Release 3.2. Beltsville, MD. Available at: https://data.nal.usda.gov/dataset/usda-database-flavon oid-content-selected-foods-release-32-november-.

Broadhurst, C.L., Polansky, M.M., and Anderson, R.A. 2000. Insulin-like biological activity of culinary and medicinal plant aqueous extracts in vitro. *J Agric Food Chem* 48: 849–852.

Brøchner, A.C., and Toft, P. 2009. Pathophysiology of the systemic inflammatory response after major accidental trauma. *Scand J Trauma Resusc Emerg Med* 17: 1–10.

Caleja, C., Barros, L., Antonio, A.L., Carocho, M., Oliveira, M.B.P.P., and Ferreira, I.C.F.R. 2016. Fortification of yogurts with different antioxidant preservatives: A comparative study between natural and synthetic additives. *Food Chem* 210: 262–268.

Cao, H., Polansky, M.M., and Anderson, R.A. 2007. Cinnamon extract and polyphenols affect the expression of tristetraprolin, insulin receptor, and glucose transporter 4 in mouse 3T3-L1 adipocytes. *Arch Biochem Biophys* 459: 214–222.

Cárdenas Garza, G.R., Elizondo Luévano, J.H., Bazaldúa Rodríguez, A.F., Chávez Montes, A., Pérez Hernández, R.A., Martínez Delgado, A.J., et al. 2021: Benefits of cardamom (Elettaria cardamomum (L.) maton) and turmeric (*Curcuma longa* L.) extracts for their applications as natural anti-inflammatory adjuvants. *Plants* 10: 1908.

Carlsen, M.H., Halvorsen, B.L., Holte, K., Bøhn, S.K., Dragland, S., Sampson, L., et al. 2010. The total antioxidant content of more than 3100 foods, beverages, spices, herbs and supplements used worldwide. *Nutr J* 9: 1–11.

Carullo, G., Cappello, A.R., Frattaruolo, L., Badolato, M., Armentano, B., and Aiello, F. 2017. Quercetin and derivatives: Useful tools in inflammation and pain management. *Future Med Chem* 9: 79–93.

Cavaleri, F. 2018. Presenting a new standard drug model for turmeric and its prized extract, curcumin. *Int J Inflam* 2018: 1–18.

Chan, C.L., Gan, R.Y., Shah, N.P., and Corke, H. 2018. Polyphenols from selected dietary spices and medicinal herbs differentially affect common food-borne pathogenic bacteria and lactic acid bacteria. *Food Control* 92: 437–443.

Chaudhry, Q., Scotter, M., Blackburn, J., Ross, B., Boxall, A., Castle, L., et al. 2008. Applications and implications of nanotechnologies for the food sector. *Food Addit Contam – Part A Chem Anal Control Expo Risk Assess* 25: 241–258.

Chung, S., Yao, H., Caito, S., Hwang, J., Arunachalam, G., and Rahman, I. 2010. Regulation of SIRT1 in cellular functions: Role of polyphenols. *Arch Biochem Biophys* 501: 79–90.

Cook, N.C., and Samman, S. 1996. Flavonoids-chemistry, metabolism, cardioprotective effects, and dietary sources. *J Nutr Biochem* 7: 66–76.

Davis, P.A., and Yokoyama, W. 2011. Cinnamon intake lowers fasting blood glucose: Meta-analysis. *J Med Food* 14: 884–889.

Deepa, G., Ayesha, S., Nishtha, K., and Thankamani, M. 2013. Comparative evaluation of various total antioxidant capacity assays applied to phytochemical compounds of Indian culinary spices. *Int Food Res J* 20: 1711–1716.

Del Baño, M.J., Lorente, J., Castillo, J., Benavente-García, O., Del Río, J.A., Ortuño, A., et al. 2003. Phenolic diterpenes, flavones, and rosmarinic acid distribution during the development of leaves, flowers, stems, and roots of *Rosmarinus officinalis*. Antioxidant activity. *J Agric Food Chem* 51: 4247–4253.

Dhakal, S., Kushairi, N., Phan, C.W., Adhikari, B., Sabaratnam, V., and Macreadie, I. 2019. Dietary polyphenols: A multifactorial strategy to target Alzheimer's disease. *Int J Mol Sci* 20: 1–40.

El-Maati, M.F.A., Mahgoub, S.A., Labib, S.M., Al-Gaby, A.M.A., and Ramadan, M.F. 2016. Phenolic extracts of clove (*Syzygium aromaticum*) with novel antioxidant and antibacterial activities. *Eur J Integr Med* 8: 494–504.

Elsayed, Y., and Khan, N.A. 2020. Immunity-boosting spices and the Novel Coronavirus. *ACS Chem Neurosci* 11: 1696–1698.

European Spice Association (ESA) 2018. List of culinary herbs and spices. *ESA* 2018: 1–5.

Gangadharappa, H.V., Mruthunjaya, K., and Singh, R.P. 2017. *Cuminum cyminum* – A popular spice: An updated review. *Pharmacogn J* 9: 292–301.

Ghate, U., and Kulkarni, H. 2021. Polyphenols, spices and vegetarian diet for immunity and anti-inflammatory drug design. In *Bioactive Compounds*, ed. Zepka L., Queiroz, Nascimento, T.C.D. and Jacob-Lopes, E. (IntechOpen, London), 1–14.

Gidwani, B., Bhattacharya, R., Shukla, S.S., and Pandey, R.K. 2022. Indian spices: past, present and future challenges as the engine for bio-enhancement of drugs: impact of COVID-19. *J Sci Food Agric*, 102: 1–13.

Giordano, A., and Tommonaro, G. 2019. Curcumin and cancer. *Nutrients* 11: 1–20.

Gong, Z., Zhou, J., Li, H., Gao, Y., Xu, C., Zhao, S., et al. 2015. Curcumin suppresses NLRP3 inflammasome activation and protects against LPS-induced septic shock. *Mol Nutr Food Res* 59: 2132–2142.

Goodarzi, M.T., Khodadadi, I., Tavilani, H., and Abbasi Oshaghi, E. 2016. The role of *Anethum graveolens* L. (Dill) in the management of diabetes. *J Trop Med* 201: 1–11.

Goszcz, K., Duthie, G.G., Stewart, D., Leslie, S.J., and Megson, I.L. 2017. Bioactive polyphenols and cardiovascular disease: Chemical antagonists, pharmacological agents or xenobiotics that drive an adaptive response? *Br J Pharmacol* 174: 1209–1225.

Gotmare, S.R., and Tambe, E.A. 2018. Chemical characterization of cumin seed oil (*Cuminum Cyminum*) by gcms and its comparative study. *Int J Sci Res Biol Sci* 5, 36–45.

Grayer, R.J., Kite, G.C., Veitch, N.C., Eckert, M.R., Marin, P.D., Senanayake, P., et al. 2002. Leaf flavonoid glycosides as chemosystematic characters in Ocimum. *Biochem Syst Ecol* 30: 327–342.

Gutiérrez-Grijalva, E.P., Picos-Salas, M.A., Leyva-López, N., Criollo-Mendoza, M. S., Vazquez-Olivo, G., and Heredia, J.B. 2018. Flavonoids and phenolic acids from oregano: Occurrence, biological activity and health benefits. *Plants* 7: 1–23.

Habauzit, V., and Morand, C. 2012. Evidence for a protective effect of polyphenols-containing foods on cardiovascular health: An update for clinicians. *Ther Adv Chronic Dis* 3: 87–106.

Hahn, Y.Il, Kim, S.J., Choi, B.Y., Cho, K.C., Bandu, R., Kim, K.P., et al. 2018. Curcumin interacts directly with the Cysteine 259 residue of STAT3 and induces apoptosis in H-Ras transformed human mammary epithelial cells. *Sci Rep* 8: 1–14

Haytowitz, D., and Bhagwat, S. 2010. USDA database for the oxygen radical absorbance capacity (ORAC) of selected foods, Release 2. Beltsville, MD.

Hayyan, M., Hashim, M.A., and Alnashef, I.M. 2016. Superoxide ion: Generation and chemical implications. *Chem Rev* 116: 3029–3085.

Idris, N.A., Yasin, H.M., and Usman, A. 2019. Voltammetric and spectroscopic determination of polyphenols and antioxidants in ginger (*Zingiber officinale* Roscoe). *Heliyon* 5: e01717.

Imran, M., Salehi, B., Sharifi-rad, J., Gondal, T.A., Arshad, M.U., Khan, H., et al. 2019. Kaempferol: A key emphasis to its anticancer potential. *Molecule* 24: 1–16.

Jadouali, S.M., Atifi, H., Mamouni, R., Majourhat, K., Bouzoubaâ, Z., Laknifli, A., et al. 2019. Chemical characterization and antioxidant compounds of flower parts of *Moroccan crocus sativus* L. *J Saudi Soc Agric Sci* 18, 476–480.

Jadouali, S.M., Bouzoubaâ, Z., Majourhat, K., Mamouni, R., Gharby, S., and Atifi, H. 2017. Polyphenols content, flavonoids and antioxidant activity of petals, stamens, styles and whole flower of *Crocus sativus* of Taliouine. *Acta Hortic* 1184: 301–308.

Johannah, N.M., Renny, R.M., Gopakumar, G., Maliakel, B., Sureshkumar, D., and Krishnakumar, I.M. 2015. Beyond the flavour: A de-flavoured polyphenol rich extract of clove buds (*Syzygium aromaticum* L) as a novel dietary antioxidant ingredient. *Food Funct* 6: 3373–3382.

Johri, R.K. 2011. *Cuminum cyminum* and *Carum carvi*: An update. *Pharmacogn Rev* 5: 63–72.

Kähkönen, M.P., Hopia, A.I., Vuorela, H.J., Rauha, J.-P., Pihlaja, K., Kujala, T.S., et al. 1999. Antioxidant activity of plant extracts containing phenolic compounds. *J Agric Food Chem* 47: 3954–3962.

Kannappan, S., and Anuradha, C.V. 2009. Insulin sensitizing actions of fenugreek seed polyphenols, quercetin and metformin in a rat model. *Indian J Med Res* 129: 401–408.

Khurana, S., Venkataraman, K., Hollingsworth, A., Piche, M., and Tai, T. 2013. Polyphenols: Benefits to the cardiovascular system in health and in aging. *Nutrients* 5, 3779–3827.

Kim, D.C., Lee, W., and Bae, J.S. 2011. Vascular anti-inflammatory effects of curcumin on HMGB1-mediated responses in vitro. *Inflamm Res* 60: 1161–1168.

Kim, D.H., Khan, H., Ullah, H., Hassan, S.T.S., Šmejkal, K., Efferth, T., et al. 2019. MicroRNA targeting by quercetin in cancer treatment and chemoprotection. *Pharmacol Res* 147: 104346.

Ksouri, A., Dob, T., Belkebir, A., Lamari, L., Krimat, S., and Metidji, H. 2015. Total phenolic, antioxidant, antimicrobial activities and cytotoxicity study of wild *Anethum graveolens* L. *Int J Pharmacogn Phytochem Res* 7: 1025–1032.

Kumar, S., and Pandey, A.K. 2013. Chemistry and biological activities of flavonoids: An overview. *Sci World J* 2013e: 1–16.

Leja, K.B., and Czaczyk, K. 2016. The industrial potential of herbs and spices – A mini-review. *Acta Sci Pol Technol Aliment* 15: 353–365.

Libby, P. 2006. Inflammation and cardiovascular disease mechanisms. *Am J Clin Nutr* 83: 456S-460S.

Liu, W., Feng, Y., Yu, S., Fan, Z., Li, X., Li, J., et al. 2021. The flavonoid biosynthesis network in plants. *Int J Mol Sci* 22: 1–18.

Lourenço, S.C., Moldão-Martins, M., and Alves, V.D. 2019. Antioxidants of natural plant origins: From sources to food industry applications. *Molecules* 24: 14–16.

Lubos, E. 2008. Role of oxidative stress and nitric oxide in atherothrombosis. *Front Biosci* 13: 5323–5344.

Malik, P., and Mukherjee, T.K. 2014. Structure-function elucidation of antioxidative and prooxidative activities of the polyphenolic compound curcumin. *Chinese J Biol* 2014: 1–8.

Mamdapur, G.M.N., Nema, N.K., Chacko, B.K., and Paul, J. 2021. Bibliometric analysis of the 200 most cited articles on antioxidant from 1976–2020. DigitalCommons@University of Nebraska – Lincoln. *Library Philosophy and Practice (e-journal)* 2021: 5358.

Manach, C., Scalbert, A., Morand, C., Rémésy, C., and Jiménez, L. 2004. Polyphenols: Food sources and bioavailability. *Am J Clin Nutr* 79: 727–747.

Martinović, N., Polak, T., Ulrih, N.P., and Abramovič, H. 2020. Mustard seed: phenolic composition and effects on lipid oxidation in oil, oil-in-water Emulsion and oleogel. *Ind Crops Prod* 156: 1–8.

Mechchate, H., Es-safi, I., Amaghnouje, A., Boukhira, S., A. Alotaibi, A., Al-zharani, M., et al. 2021. Antioxidant, anti-inflammatory and antidiabetic proprieties of LC-MS/MS identified polyphenols from coriander seeds. *Molecules* 26: 487.

Mehta, S.K., and Gowder, S.J.T. 2015. Members of antioxidant machinery and their functions. *Basic Princ Clin Significance Oxidative Stress* (Intech Open Science) 59–85.

Meskin, M.S., Bidlack, W.R., Davies, A.J., Lewis, D.S., and Randolph, R.K. 2003. *Phytochemicals: Mechanisms of action.* (CRC Press). 1–203.

Moghadamnia, Y., Kani, S.N.M., Ghasemi-Kasman, M., Kani, M.T.K., and Kazemi, S. 2019. The anti-cancer effects of *Capparis spinosa* hydroalcoholic extract. *Avicenna J Med Biotechnol* 11: 43–47.

Muzolf-Panek, M., and Stuper-Szablewska, K. 2021. Comprehensive study on the antioxidant capacity and phenolic profiles of black seed and other spices and herbs: Effect of solvent and time of extraction. *J Food Meas Charact* 15: 4561–4574.

Nazzaro, F., Fratianni, F., De Feo, V., Battistelli, A., Da Cruz, A.G., and Coppola, R. 2020. Polyphenols, the new frontiers of prebiotics, *Adv Food Nutr Res*. 94: 35–89.

Nema, N.K., Mamdapur, G.M.N., Sarojam, S., Khamborkar, S.D., Sajan, L.C., Sabu, S., et al. 2021. Preventive medicinal plants and their phytoconstituents against SARS-CoV-2/COVID-19. *Pharmacognosy Res* 13: 173–191.

Ollis, W.D. 1965. Biochemistry of phenolic compounds. *J Am Chem Soc* 87: 1157–1158.

Pandey, K.B., and Rizvi, S.I. 2009. Plant polyphenols as dietary antioxidants in human health and disease. *Oxid Med Cell Longev* 2: 270–278.

Paulraj, F., Abas, F., Lajis, N.H., Othman, I., and Naidu, R. 2019. Molecular pathways modulated by curcumin analogue, diarylpentanoids in cancer. *Biomolecules* 9: 1–14.

Pereira, A.S.P., Banegas-Luna, A.J., Peña-García, J., Pérez-Sánchez, H., and Apostolides, Z. 2019. Evaluation of the anti-diabetic activity of some common herbs and spices: Providing new insights with inverse virtual screening. *Molecules* 24: 4030.

Petiwala, S.M., Puthenveetil, A.G., and Johnson, J.J. 2013. Polyphenols from the Mediterranean herb rosemary (*Rosmarinus officinalis*) for prostate cancer. *Front Pharmacol* 4: 1–4.

Petrussa, E., Braidot, E., Zancani, M., Peresson, C., Bertolini, A., Patui, S., et al. 2013. Plant flavonoids-biosynthesis, transport and involvement in stress responses. *Int J Mol Sci* 14: 14950–14973.

Picos-Salas, M.A., Heredia, J.B., Leyva-López, N., Ambriz-Pérez, D.L., and Gutiérrez-Grijalva, E.P. 2021. Extraction processes affect the composition and bioavailability of flavones from lamiaceae plants: A comprehensive review. *Processes* 9: 1–38.

Premavalli, K.S., and Wadikar, D.D. 2011. *Appetite regulation and role of appetizers*. Ed. S.R. Mitchell. Nova Science.

Quideau, S., Deffieux, D., Douat-Casassus, C., and Pouységu, L. 2011. Plant polyphenols: Chemical properties, biological activities, and synthesis. *Angew Chemie – Int Ed* 50: 586–621.

Rao, P.V., and Gan, S.H. 2014. Cinnamon: A multifaceted medicinal plant. *Evidence-Based Complement Altern Med* 2014: 1–12.

Rather, M.A., Dar, B.A., Sofi, S.N., Bhat, B.A., and Qurishi, M.A. 2016. Foeniculum vulgare: A comprehensive review of its traditional use, phytochemistry, pharmacology, and safety. *Arab J Chem* 9: S1574–S1583.

Rauter, A.P., Ennis, M., Hellwich, K.H., Herold, B.J., Horton, D., Moss, G.P., et al. 2018. *Nomenclature of flavonoids (IUPAC Recommendations 2017)*.

Rebey, B.I., Bourgou, S., Wannes, W.A., Selami, I.H., Tounsi, M.S., Marzouk, B., et al. 2018. Comparative assessment of phytochemical profiles and antioxidant properties of Tunisian and Egyptian anise (*Pimpinella anisum* L.) seeds. *Plant Biosyst – An Int J Deal with all Asp Plant Biol* 152: 971–978.

Roby, M.H.H., Sarhan, M.A., Selim, K.A.H., and Khalel, K.I. 2013. Evaluation of antioxidant activity, total phenols and phenolic compounds in thyme (*Thymus vulgaris* L.), sage (*Salvia officinalis* L.), and marjoram (*Origanum majorana* L.) extracts. *Ind Crops Prod* 43: 827–831.

Ross, J.A., and Kasum, C.M. 2002. Dietary flavonoids: Bioavailability, metabolic effects, and safety. *Annu Rev Nutr* 22: 19–34.

Rudrapal, M., and Chetia, D. 2016. Plant flavonoids as potential source of future antimalarial leads. *Syst Rev Pharm* 8: 13–18.

Rudrapal, M., Khairnar, S.J., Khan, J., Dukhyil, A.B., Ansari, M.A., Alomary, M.N., et al. 2022. Dietary polyphenols and their role in oxidative stress-induced human diseases: Insights into protective effects, antioxidant potentials and mechanism(s) of action. *Front Pharmacol* 13: 1–15.

Rupasinghe, H.P.V. 2020. Special Issue: Flavonoids and their disease prevention and treatment potential: Recent advances and future perspectives. *Molecules* 25: 4746.

Sachan, A.K., Kumar, S., Kumari, K., and Singh, D. 2018. Medicinal uses of spices used in our traditional culture: World wide. *J Med Plants Stud* 6: 116–122.

Sambasivarao, S.V. 2013. Cellular electroporation induces dedifferentiation in intact newt limbs. *J Immunol* 18: 1199–1216.

Sandhar, H.K., Kumar, B., Prasher, S., Tiwari, P., Salhan, M., and Sharma, P. 2011. A review of phytochemistry and pharmacology of flavonoids. *Int Pharm Sci* 1: 25–41.

Satheeshkumar, N., Vijayan, R.S.K., Lingesh, A., Santhikumar, S., and Vishnuvardhan, C. 2016. Spices: Potential therapeutics for Alzheimer's disease. *Adv Neurobiol* 57–78.

Schroeter, H., Heiss, C., Spencer, J.P.E., Keen, C.L., Lupton, J.R., and Schmitz, H.H. 2010. Recommending flavanols and procyanidins for cardiovascular health: Current knowledge and future needs. *Mol Aspects Med* 31: 546–557.

Shahidi, F., and Hossain, A. 2018. Bioactives in spices, and spice oleoresins: Phytochemicals and their beneficial effects in food preservation and health promotion. *J Food Bioact* 3: 8–75.

Sharangi, A.B. and Acharya, S.K. 2018. Spices in India and beyond: The origin, history, tradition and culture. In *Ind Spices: The Legacy, Production and Processing of India's Treasured Export*, ed. A.B. Sharangi (Springer: Cham), 1–12.

Shojaii, A., and Abdollahi Fard, M. 2012. Review of pharmacological properties and chemical constituents of pimpinella anisum. *ISRN Pharm* 2012: 1–8.

Singh, P., Roy, T.K., Kanupriya, C., Tripathi, P.C., Kumar, P., and Shivashankara, K.S. 2022. Evaluation of bioactive constituents of *Garcinia indica* (kokum) as a potential source of hydroxycitric acid, anthocyanin, and phenolic compounds. *LWT* 156: 112999.

Siwak, D.R., Shishodia, S., Aggarwal, B.B., and Kurzrock, R. 2005. Curcumin-induced antiproliferative and proapoptotic effects in melanoma cells are associated with suppression of IκB kinase and nuclear factor κB activity and are independent of the B-Raf/mitogen-activated/extracellular signal-regulated protein kinase pat. *Cancer* 104: 879–890.

Sowndhararajan, K., and Kim, S. 2016. Influence of fragrances on human psychophysiological activity: With special reference to human electroencephalographic response. *Sci Pharm* 84: 724–751.

Srinivasan, K. 2014. Antioxidant potential of spices and their active constituents. *Crit Rev Food Sci Nutr* 54: 352–372.

Stan, M., Soran, M.L., Varodi, C., Lung, I., and Lazar, M.D. 2012. Extraction and identification of flavonoids from parsley extracts by HPLC analysis. In *AIP Conference Proceedings*: 50–52.

Steiner, M. 1997. The role of flavonoids and garlic in cancer prevention. *Food Factors Cancer Prev* 222–225.

Sun, L.R., Zhou, W., Zhang, H.M., Guo, Q. S., Yang, W., Li, B.J., et al. 2019a. Modulation of multiple signaling pathways of the plant-derived natural products in cancer. *Front Oncol* 9: 1–15.

Sun, W., Shahrajabian, M.H., and Cheng, Q. 2019b. Anise (*Pimpinella anisum* L.), a dominant spice and traditional medicinal herb for both food and medicinal purposes. *Cogent Biol* 5: 1673688.

Syed, Q.A., Rashid, Z., Ahmad, M.H., Shukat, R., Ishaq, A., Muhammad, N., et al. 2020. Nutritional and therapeutic properties of fenugreek (*Trigonella foenum-graecum*): A review. *Int J Food Prop* 23: 1777–1791.

Szilvaśsy, B., Rak, G., Saŕosi, S., Novák, I., Pluhaŕ, Z., and Abrańko, L. 2013. Polyphenols in the aqueous extracts of garden thyme (*Thymus vulgaris*) chemotypes cultivated in Hungary. *Nat Prod Commun* 8: 605–608.

Tian, Y., and Deng, F. 2020. Phytochemistry and biological activity of mustard (*Brassica juncea*): A review. *CYTA – J Food* 18: 704–718.

Tlili, N., Khaldi, A., Triki, S., and Munné-Bosch, S. 2010. Phenolic compounds and vitamin antioxidants of caper (*Capparis spinosa*). *Plant Foods Hum Nutr* 65: 260–265.

Torre Torres, J.E.D.L., Gassara, F., Kouassi, A.P., Brar, S.K., and Belkacemi, K. 2017. Spice use in food: Properties and benefits. *Crit Rev Food Sci Nutr* 57: 1078–1088.

Tresserra-Rimbau, A., Rimm, E.B., Medina-Remón, A., Martínez-González, M.A., de la Torre, R., Corella, D., et al. 2014. Inverse association between habitual polyphenol intake and incidence of cardiovascular events in the PREDIMED study. *Nutr Metab Cardiovasc Dis* 24: 639–647.

Troiano, R., and Reuter, W.M. 2014. Liquid chromatography/Analysis of polyphenols in saffron petal extracts with UHPLC / UV / MS, Perkin Elmer, Waltham, MA: 1–5

Tsao, R. 2010. Chemistry and biochemistry of dietary polyphenols. *Nutrients* 2: 1231–1246.

Tzima, K., Brunton, N.P., Choudhary, A., and Rai, D.K. 2020. Potential applications of polyphenols from herbs and spices in dairy products as natural antioxidants. *Herbs, Spices Med Plants*, 283–299.

Vermerris, W., and Nicholson, R. 2008. Families of phenolic compounds and means of classification, in *Phenolic Compound Biochemistry*, W. Vermerris and R. Nicholson, eds. (Dordrecht: Springer), 1–34.

Viuda-Martos, M., Ruiz-Navajas, Y., Fernández-López, J., and Pérez-Álvarez, J.A. 2010. Spices as functional foods. *Crit Rev Food Sci Nutr* 51: 13–28.

Wadikar, D.D., Patki, P.P., and Sharma, R.K. 2018. Appetizer: A food category or food adjective? *Indian J Nutr* 5: 1–4.

Wani, S.A., and Kumar, P. 2018. Fenugreek: A review on its nutraceutical properties and utilization in various food products. *J Saudi Soc Agric Sci* 17: 97–106.

Willingham, S.B., Allen, I.C., Bergstralh, D.T., Brickey, W.J., Huang, M.T.-H., Taxman, D. J., et al. 2009. NLRP3 (NALP3, Cryopyrin) facilitates in vivo caspase-1 activation, necrosis, and HMGB1 release via inflammasome-dependent and independent pathways. *J Immunol* 183: 2008–2015.

Wojdyło, A., Oszmiański, J., and Czemerys, R. 2007. Antioxidant activity and phenolic compounds in 32 selected herbs. *Food Chem* 105: 940–949.

Yanishlieva, N.V., Marinova, E., and Pokorný, J. 2006. Natural antioxidants from herbs and spices. *Eur J Lipid Sci Technol* 108: 776–793.

Yashin, A., Yashin, Y., Xia, X., and Nemzer, B. 2017. Antioxidant activity of spices and their impact on human health: A review. *Antioxidants* 6: 2–18.

Yin, H., Guo, Q., Li, X., Tang, T., Li, C., Wang, H., et al. 2018. Curcumin suppresses IL-1β secretion and prevents inflammation through inhibition of the NLRP3 Inflammasome. *J Immunol* 200: 2835–2846.

Yu, H., Park, J.Y., Kwon, C.W., Hong, S.C., Park, K.M., and Chang, P.S. 2018. An overview of nanotechnology in food science: Preparative methods, practical applications, and safety. *J Chem* 2018: 1–11.

Zheng, J., Zhou, Y., Li, Y., Xu, D.P., Li, S., and Li, H.B. 2016. Spices for prevention and treatment of cancers. *Nutrients* 8: 1–35.

13 Dietary Polyphenols in Drug Discovery by Drug Repurposing and Computational Screening

Ipsa Padhy, Aastha Mahapatra, Fahima Dilnawaz, and Tripti Sharma**
*E-mail: ip.sper@buodisha.edu.in; aashume1008@gmail.com; fahimadilnawaz@gmail.com; triptisharma@soa.ac.in

CONTENTS

13.1 INTRODUCTION

Dietary polyphenols are the eclectic range of naturally occurring bioactive principles, chemically recognized as derivatives of flavonoids and other polyphenols. The regular dietary intake of polyphenols extend up to 1 gram per day basically by consumption of fruits, tea, green tea, coffee,

DOI: 10.1201/9781003251538-13

cereals and legumes (Han et al., 2007; Scalbert et al., 2005). Figure 13.1 below showcases major implications of polyphenols as they reduce the risk of various age-associated diseases, such as osteoporosis, neurodegeneration, cancer, obesity and cardiovascular disease. They also play a major role in age-related issues while regulating the gut microbiota (Gowd et al., 2019; Wu et al., 2021). Phenolic compounds are known to be strong antioxidants, and various studies have proved positive correlation between content of polyphenol in fruit and their antioxidant activity. It has been successfully proven that polyphenolic compounds protect the cellular structures against oxidative damage by modulating the antioxidant response by increasing the antioxidant enzyme activity by enabling smooth functioning of the mitochondria (Tober et al., 2019). For example sweet potato leaf polyphenols showed ~ 0.84 and ~ 6. 20 times more free radical scavenging activity than that of grape seed polyphenols and tea polyphenols, respectively, at the concentration of ~ 20 μg/mL in a study carried out in a certain set ranges of pH temperature and light intensities (Sun et al., 2017). Bio-availability values obtained in a study of polyphenols concluded that the colonic metabolic products contribute more than expected in their absorption. A number of bioactive effects are induced by the catabolic products of colonic microflora in the circulatory system of the human body (Teng et al., 2019). Last two decades have witnessed lots of meticulous efforts to obtain novel drug candidates from plants and deep-sea flora to combat MDR (multi-drug resistance) (David et al., 2015; Tortorella et al., 2018). Polyphenols are used to design new strategies to prevent the harmful effects of synthetic drugs and have shown clinical potentiality (Rimbau et al., 2018). Secondary metabolites of plants are very much useful for various therapeutic activities. In the last decade a plethora of ethanopharmacological resource has been accessible from various therapeutic guidebooks, for example, Ayurvedic Pharmacopoeia of India is one such manual (Franco et al., 2015). These manuals have been very helpful over the decade, but they do not provide any apt approaches for *in silico* screening of the plant-based metabolites. Therefore, over time scientists have established several databases that should serve as informative depots. These technologies have been very helpful in bringing in new manifold avenues

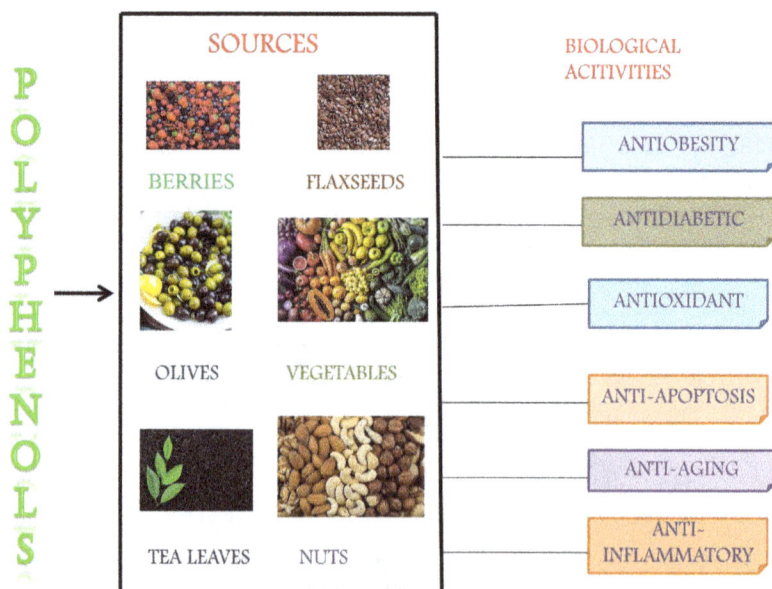

FIGURE 13.1   Different sources of polyphenols and their biological activities.

facilitating higher levels of data intricacy. Various libraries are being created by computational chemists based on natural bioactive products for providing admirable means for filtering and selecting compounds for research interest (Baxevanis et al., 2015; Prachayasittikul et al., 2019). The pipeline of drug development is being exploited utilizing various arrays of phytocompounds employing chemoinformatic tools, such as molecular docking-based high-throughput screening (Lagunin et al., 2014). Repurposing of drugs is a tactic for ascertaining newer usages for permitted or trial drugs. The approach provides several pluses in comparison to conventional drug development, such as lowering research cost, shorter research time and most importantly less failure rates (Pushpakom et al., 2019).

13.2 STRUCTURAL CLASSIFICATION AND CHEMISTRY OF POLYPHENOLS

The abundant amount of micronutrients present in our diet, such as polyphenols, has a preventative role towards diseases like cancer and cardiovascular diseases (Scalbert et al., 2005). Polyphenols are the phenolic compounds that are available in the plant kingdom. So far 8,000 phenolic structures are known. Polyphenols are the products of the secondary metabolism in plants that are mainly produced by the shikimate and acetate pathway. The polyphenols mainly occur in conjugated form having one or more than one residue linked to the hydroxyl group, also a direct linkage of sugar units with an aromatic carbon atom. The linked sugar units can be monosaccharides, disaccharides, or oligosaccharides. Glucose, rhamnose, galactose, arabinose, xylose, glucuronic and galacturonic acids are common sugars found in the polyphenol structures. The compounds are also associated with carboxylic acid, organic acids, amines, lipids, and linkages with other phenols. The broad distribution range of polyphenols starts from phenolic acid like simple molecules to highly polymerized compounds such as tannins (Bravo et al., 1998). In terms of classification, polyphenols can be classified either as flavonoids and non-flavonoids or sub-classed into various parts based on the substituent group, phenol unit on the molecular structure and linkages between the phenol units (Cardona et al., 2013). Figure 13.2 displays classification of polyphenols in accordance with their structural frameworks.

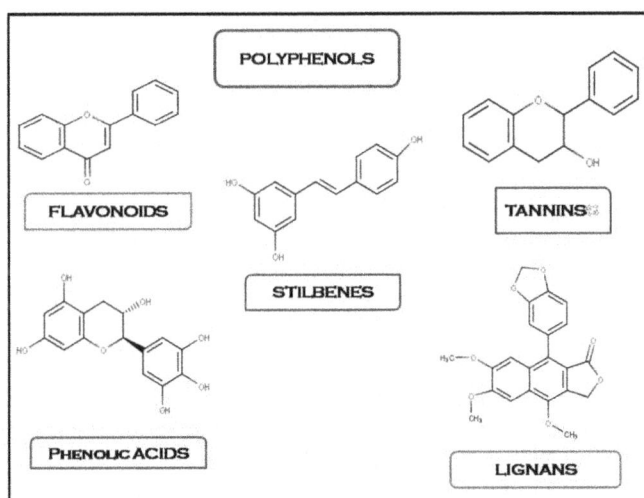

FIGURE 13.2   Classification of polyphenols.

13.2.1 Flavonoids

Flavonoids are an ubiquitous group of plant phenolics as they are an important part of human health and food organoleptics (Tapas et al., 2008). Flavonoids are widely distributed in vegetables, berries, nuts, tea, wine and so forth (Middleton et al., 1998). Flavonoids have been categorized based on their molecular structure such as flavanones, flavanols, isoflavones, flavonols, flavones, and anthocyanins (Brodwaska et al., 2017; Rice et al., 1996). They quench the reactive oxygen species and lipid radicals. Since long flavonoids have been used for maintaining good health of blood vessels and marketed as anti-spasmodic and anti-inflammatory medications. More importantly flavonoid derivatives and their metal complexes play a significant role in industrial, agriculture, and pharmaceutical chemistry (Kumar et al., 2009; Strissel et al., 2005). The main examples of useful flavonoids are luteolin, quercetin, chrysin, myricetin and genistein.

13.2.2 Tannins

Tannins are water-soluble polyphenols with a wide natural occurrence and have been classified into hydrolysable (HT) and condensed tannins (CT) with reference to their chemical structure and properties (Anthansaidu et al., 2001). The tannins are biosynthesized basically from quinic acid and p-coumaroyl CoA (Sieniawaska et al., 2017). The prominent biological activities of tannins include hydroxyl radical scavenging activity, pro-oxidant, oxidative and an ability to form chelation of ferrous ions (Moilanen et al., 2016).

13.2.3 Phenolic acids

Phenolic acids are a group of natural constituents mainly found in cereal grains. The bran, which is the outer part of the kernel, has the highest concentration of phenolic acids (Kim et al., 2006). The structure of phenolic acid varies due to the difference in number and position of the hydroxyl group on the aromatic ring. These naturally occurring groups are found to be strong antioxidants against the Reactive Oxygen Species (ROS) and various free radicals (Andreasen et al., 2001). The antioxidant potential of phenolic compounds are mainly linked to reducing properties as hydrogen or electron donating groups which leads to free radical scavenging activity. Furthermore, they possess the ability of metal chelation mainly in the case of metals like iron and copper, which suppress the metal catalyzed free radical formation. The metal chelation and radical scavenging activity of the phenolic acid is highly influenced by the type of structure. The number of hydroxyl group and their position with respect to the carboxyl group has a prominent influence on the antioxidant activity (Vuolo et al., 2019). Phenolic acids are mainly classified into benzoic acid and cinnamic acid with carbon atoms of C1–C6 and C3–C6, respectively. These compounds usually exist as hydroxybenzoates of gallic acid, salicylic acid, protocatechuic acid, ellagic acid, and gentisic acid and hydroxycinnamates of p-coumaric acid, caffeic acid, ferulic acid, chlorogenic acid, and sinapic acids or in conjugated forms (Saibabu et al., 2015).

13.2.4 Lignans

A set of long-known natural products widely distributed among the plant kingdoms are called Lignans. A set of more than 200 compounds has been identified in this general class. A variety has been observed in the chemical assembly of the two phenylpropanold units, which are a major characteristic of liganans. An apparent diversity is also observed based on the degree of oxidation and types of substituent (MacRae et al., 1984). Lignans are formed from the shikimic acid biosynthetic pathway and the main scaffold is a combination of two or more phenyl propanoid units (Ayres et al., 1990; Topeneo et al., 2016). Based on the molecular linkage between them they are designated as classical lignans and neolignans. When the molecular linkage of monomers occurs between positions β-β′ (also referred to as an 8-8′), these compounds are designated as "classical lignans." In contrast, the compounds are grouped into "neolignans" if the main structural units are coupled in any other way (non β-β′ linkage) (Cui et al., 2020). Various in silico predictions along with in vitro assay on lignans suggest that they are biologically active and possess prominent antibacterial, anticholinesterasic, antioxidant, and cytotoxic activities. Furthermore, ethanopharmacological uses of lignans were also investigated for potential clinical use. The Absorption, Distribution, Metabolism, Excretion and Toxicity (ADMET) profiles of various sub-classes of lignans were assessed, and it was established that it can be used for future therapeutics. Traditionally, health benefits attributed to lignans have included a lowered risk of heart disease, menopausal symptoms, osteoporosis, and breast cancer (Barker et al., 2019).

13.2.5 STILBENES

A class of phenolic metabolites found in various edible plants like peanuts, berries and grapevine. These compounds have been studied extensively for their bioactivity and benefits to human health (Khawand et al., 2018). There has not been much research in the natural sources and safe amounts of total intake of stilbenes, as the procedure for the qualitative and quantitative analysis is very complex. It is difficult to estimate the safe quantity of stilbenes that can be consumed by humans and, in fact, the dietary intake of stilibenes differs all around the world according to the type of diet. These are synthesized by shikimate and phenylalanine ammonia lyase pathways and can be consumed as food or medicinal plant entities (Dubrovina et al., 2017; Jandet et al., 2010). The derivatives of stilbenes are known as stilbenoids, and they are usually thermally and chemically stable. According to the chemical structure, stilbenoids belong to a group of polyphenolic compounds known as phenylpropanoids. The most commonly known and tested stilbenes are resveratrol, derivatives of naturally occurring combretastatin A1, OXi4503 and pterostilbene (E-3,5-dimethoxy-4'-hydroxystilbene) (Krawczyk et al., 2019). Antiobesity activity of stilibenes has been widely researched and it has been discovered that in animal studies it shows beneficial effects in reductions of lipid accumulation, regulation of glucose homeostasis, inflammation alleviation, and modulation of gut microbiota (Chou et al., 2018). Currently, there are many studies going on to repurpose stilbene analog which show other biological activities for CoV-2 spike protein and human ACE2 receptor complex (Wahedi et al., 2021).

13.3   COMPUTATIONAL-BASED APPROACHES FOR REPURPOSING DIETARY POLYPHENOL

Computational approaches for repurposing have an advantage over conventional statistical strategies and are diversified to include data mining, semantic inferencing, collaborative filtering of datasets, application of machine-learning techniques, use of deep neural networking and further network analysis strategies (Jarada et al., 2020). The application of artificial intelligence has also heightened the drug-repurposing strategies (Selvaraj et al., 2021). The different computational methods of drug designing, such as virtual screening, pharmacophore modelling Quantitative structure activity relationship (QSAR) modelling and other ligand-based approaches are elaborated under specific sections of this chapter.

13.3.1   VIRTUAL SCREENING

The virtual screening methods are either structure-based or ligand-based. Protein-ligand docking is one of the basically used methods for structure-based virtual screening in combination of molecular dynamic simulations. The ligand-based approaches rely upon QSAR and pharmacophore modelling, shape-based similarity and molecular fingerprint similarity methods (Sharma et al., 2019).

There is a great paradigm shift seen in the virtual screening approaches; basically integration of molecular fingerprint similarity searching with structure-based virtual screening (Cereto-Massague et al., 2015). The molecular fingerprints can be two- or three-dimensional and are grouped into topological, circular, substructural, hybrid, and pharmacophoric fingerprints. Similarity searching of compound decks by evaluating intrinsic similarity between desired molecules is a major visual screening approach using molecular fingerprints (Muegge and Mukherjee 2016). The amalgamation of diversified virtual screening methods through data fusion and applying combinatorial structure-based strategies of consensus scoring is trending (Willet et al., 2013). Virtual screening of compound libraries for drug/lead likeliness rely upon Lipinski's "rule of five" and more recently being modified into "rule of three" that specify required physicochemical properties quantitatively (Lavecchia et al., 2013). Table 13.1 displays different databases available as repositories of phytochemicals.

A number of web-based tools and programs like ChemMapper (Gong et al., 2013), iDrug-target (Chen et al., 2020), Swiss Target Prediction (Daina et al., 2019), CMAP (Lamb et al., 2006), MeSHDD (Brown and Patel, 2017), DASPfind (Alawi et al., 2016) and so forth are available for similarity searching of drug databases predicting drug target interactions, linking drugs and drug-induced gene expressions to diseases.

13.3.2 PHARMACOPHORE MODELING

A pharmacophore is an ensemble of molecular and atomic features for a ligand necessitating its interaction with a biomolecular target and thus eliciting a biological response. These features are depicted as hydrogen bond donors and acceptors, positive and negative groups, hydrophobic features, aromatic rings and exclusion volumes, the latter is essential in avoiding a clash of the ligand molecule onto the target surface while mapping. Pharmacophore model generation can be ligand-based, structure-based or ligand receptor-complex-based (Choudhury and Sastry, 2019). Ligand-based pharmacophore modelling methods predict the most suitable three-dimensional molecular feature necessary for interacting in the binding pocket of target protein. In this context conformational search for the most suitable ligand molecule or calculating all possible conformations of ligand with minimum energy is a prerequisite. In the structure-based approach the complementary topology of three dimensional targets is elucidated. Calculation of molecular force fields, defining chemical features with respect to active ligand/active aligned molecules and retrieving substructural patterns are employed strategies for receptor-based or receptor-ligand based pharmacophore modelling (Sharma et al., 2021). Figure 13.3 depicts a general workflow of pharmacophore modelling.

Major applications of pharmacophore modelling include lead optimization, molecular profiling, target fishing, *Denovo* drug design, and prediction of pharmacokinetic features (Muhammed and Yalicin, 2021). The pharmacophores generated are virtually screened against molecule libraries using either fingerprint-based or three-dimensional alignment approaches (Schaller et al., 2020). In Table 13.2 often used software tools for pharmacophore modelling are listed.

13.3.3 MOLECULAR DOCKING

Molecular docking is an imperative structure-based drug design tool displaying and defining the ligand-receptor intermolecular interactions through suitable computer-generated algorithms by different available software programs for docking studies. (Ferreira et al., 2015). The scoring functions generated can be force-field based, empirical, semiemperical, knowledge-based and consensus-based (Tripathi and Bankaitis, 2017).The intermolecular interactions dealt in the docking studies include electrostatic and electrodynamic atomic features, steric forces, and solvation features (Chaudhary and Mishra, 2016). Docking can be rigid body docking, flexible docking or flexible ligand docking majorly. Different computational algorithms employed in molecular docking programs are fragment-based, Monte Carlo, genetic, fast-shape matching, incremental

TABLE 13.1

Commonly referenced databases for phytochemicals

Databases	Information retrieved	Description	Availability	Reference
IMPPAT	Phytochemicals(9500) from Indian medicinal plants (1700)	2D, 3D chemical structures, ADMET prediction, protein target prediction, assessment of drug likeliness.	Open access for academic purposes	Mohanraj et al., 2019
Phytochemica	963 phytochemicals	Plant sources, chemical class, 3D chemical structure, IUPAC names, SMILES notations, physicochemical features.	Open access for academic purposes and pharma industries	Pathania et al., 2015
BindingDB	1,023,385 molecules (synthetic and natural) with	Protein-ligand interaction features.	Free web access	Gilson et al., 2015
ChEMBL	2,105,465 bioactive molecules	Chemical structures, protein structures, target sequences, bioactivity data.	Manually curated open access	Gaulton et al., 2017; Mendez. et al., 2019
MAPS	2200 phytocompounds	Chemical structures, bioactivities	Free access	Ashfaq et al., 2013
Pubchem	111million compounds including synthetic and natural	2D and 3D chemical structures, physicochemical data, bioassay generated data, ADMET profiles	Free access	Kim et al., 2021
Dr Duke's Phytochemical and ethnobotanical database	Phytocompounds	Phytochemicals from specific plant sources, biological activities, toxicity data, ethnobotanical uses.	Free access to public, license maintained by Creative Commons CCZero, USDA	USDA, 2016
Phenol explorer (2.0, 3.0, 3.6)	More than 500 dietary polyphenols	Polyphenol contents in dietary sources as well in processed foods; pharmacokinetic profile of polyphenols.	Free access	Neveu et al., 2010
KNApSAcK family databases	50,048 plant based metabolites	Molecular formula and masses of plant metabolites, mass spectra details, ethnobotanical uses.	commercial	Afendi et al., 2012
MPD3	7062 phytochemicals.	Chemical structures, test targets, bioactivities.	Free access	Mumtaz et al., 2017
NeMedPlant	Phytochemicals from aromatic and medicinal plants of north east India.	Chemical structures, physiochemical features, plant sources of phytochemicals and bioactivities.	Free access	Meetei et al., 2012
HMDB	220,945 small molecule metabolites	Chemical structures, NMR, GCMS, LCMS molecular spectras, metabolic pathways, target sequences	Free access	Wishart et al., 2018
COCONUT	406,747 phytochemicals	Chemical structures, molecular formulas	Free access	Soronika et al., 2021
CMAUP	47,635 phytochemicals	Chemical structures, physiochemical properties, target associations, gene ontologies	Free access	Zeng et al., 2019

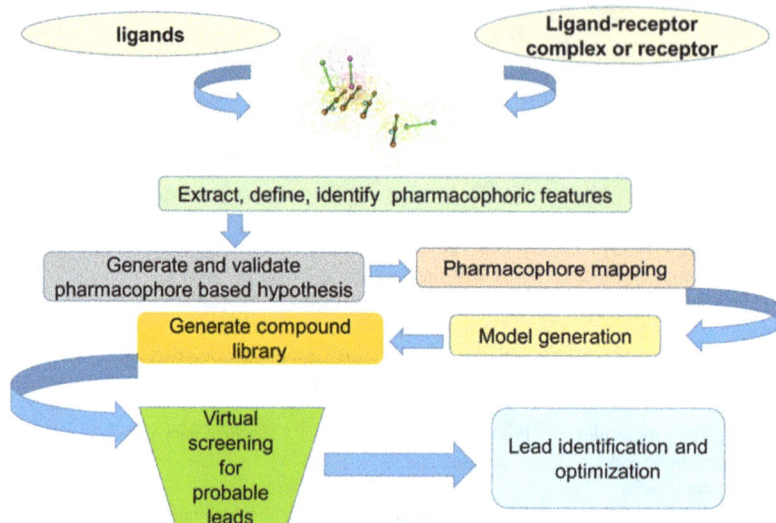

FIGURE 13.3   Pharmacophore modelling workflow.

construction, distance geometry, simulated annealing and molecular-dynamic based (Azevedo et al., 2012; Agarwal and Mehrotra, 2016). Converging to phytochemicals, molecular docking approaches are very dynamic and robust encompassing advanced feature-based pharmacophoric and molecular dynamic studies owing to their nonspecific and multitargeted physiological effects (Temml and Schuster, 2021). Table 13.3 lists extensively used molecular docking programs.

13.3.4   QSAR APPROACHES

Ligand-based drug-designing strategies basically comprise QSAR modelling, pharmacophore modelling, and similarity searching, and currently equipped with advanced molecular and quantum mechanical techniques (Shim et al., 2011). QSAR/QSPR (quantitative structure property relationship) mathematically relates structural features/molecular descriptors of a compound to its biological activity (Liu and Long, 2009). The molecular descriptors are classified as topological (Weiner, Balaban, Randic index), constitutional (molecular weight, atom, and bond counts), geometrical (WHIM, GETAWAY); weighted holistic invariant molecular (WHIM) and geometry topology atomic weights assembly (GETAWAY), thermodynamic and electronic (Hammett constant and dissociation constant) (Khan et al., 2016). The molecular descriptors are grouped into different dimensionalities depending upon the level of complexities, such as 1D, 2D, 3D, 4D and 5D (Wang et al., 2015). A QSAR model is built by computing a set of descriptors for a test set of compounds through suitable algorithms and validating by statistically fitting methods – though some good-quality QSAR models have been built without computing descriptors by using long-short term memory neural networks (Chakravarti and Alla 2019). PaDEL (Yap, 2011); ChemDes (Dong et al., 2015); Modred (Moriwaki et al., 2018); QuBiLSMIDAS (Jacas et al., 2014); Pydescriptor (Masand et al., 2017); alvaDesk (Mauri, 2020) and so forth are some open source software for computing different types of molecular descriptors.

   With the progress, in time Cloud-based web servers have also been developed to amalgamate force-field calculations, molecular structure generation, molecular alignments and so forth. One such example is Cloud-3D QSAR (http://chemyang.ccnu.edu.cn/ccb/server/cloud3dQSAR/) available freely (Wang et al., 2021). Some standalone software has also been proposed for

TABLE 13.2
Pharmacophore modelling software tools

Software tools	Utility	Approach	Availability	URL
Ligandscout	Creation of 3D pharmacophore model	Ligand based, ligand-receptor complex based, structure based	commercial	www.inteligand.com/ligandscout/
Align it/Silicos-it/ Pharao	Small molecule alignment based on 3D Gaussian spheres	Ligand based	Free	http://silicos.it.be
PHASE (SCHRODINGER)	Pharmacophore based virtual screening, lead optimization	Ligand based, receptor based	Commercial	www.schrodinger.com/Phase/
Zinc Pharmer	Virtual screening	Structure based	Free	http://zincpharmer.csb.pitt.edu
MOE	Pharmacophore modelling and virtual screening	Ligand based	Commercial	www.chemcomp.com/MOE-Pharmacophore_Discovery.html
Catalyst (BIOVIA DISCOVERY STUDIO)	Pharmacophore mapping, virtual screening	Ligand based	Commercial	http://accelerys.com/produ cts/discovery studio/ pharmacophore.html

TABLE 13.3
Molecular docking software

Docking software	Utility	Availability	Reference
AutodockVina	Protein ligand docking based on genetic algorithm	Open source	Trott and Olson 2010
Glide	High-throughput virtual screening and docking, molecular dynamic simulations	Commercial	Friesner et al., 2006
Ligandfit	Monte Carlo and shape based algorithms for ligand-protein docking	Commercial	Venkatachalam et al., 2003
Molgero virtual docker	Integrated docking program based on heuristic search algorithm	Open source	Thomsen et al., 2006

multitasking QSAR modelling, such as QSAR-Co (Ambure et al., 2019). Chemometric tools statistically validate generated QSAR models such as regression-based (multiple regression, partial least squares), classification-based (linear discriminant analysis, cluster analysis) and machine-learning based (support vector machine, artificial neural network, random forest, k-nearest neighbor, Naïve bayes, principal component analysis, and so forth) (Lo et al., 2018; Roy et al., 2015). 3D QSAR modelling methods are widely used in the drug-discovery pipeline such as CoMFA, CoMSIA, CSP-SAR,AFMoc, CoMMA, GERM, SoMFA, Compass, HQSAR, MSA, RSA, MQSM, and so forth (Acharya et al.,2011; Roy et al., 2015).

13.3.4.1 Case study 1
Jena et al. envisaged an in silico study for the therapeutic targeting of two polyphenols, catechin and curcumin against SARS-CoV-2 infections. The structures of targets were the spike protein of the SARS-CoV-2 virus, ACE2 receptor of host cell, and the complex of receptor-binding domain

of spike protein and ACE2 receptor were retrieved from PDB database. ClusPro 2.0, an automated rigid docking tool was used for docking of the two proteins. The 2D canonical of the curcumin (1) and epigallocatechin (2) were retrieved from PubChem database and converted to 3D forms using the CHIMERA program. Ligand-protein docking (AutoDockVina), visualization and analysis (Discovery studio Client R2) were performed. Three-dimensional low energy conformations of ligand-protein complexes were generated using LigPrep and GLIDE tools. Molecular dynamic simulations, RMSD, and RMSF were done using Desmond software program. The binding energy values for curcumin and epigallocatechin were found to be -7.9kcal/mol and -10.5kcal/mol, respectively for spike protein; -7.8kcal/mol and -8.9kcal/mol for ACE2 receptor and -7.6kcal/mol and -9.1kcal/mol for receptor-binding domain/ACE2 complex. The molecular docking and dynamic studies reveal binding of both curcumin and epigallocatechin at the interface of RBD/ACE2 complex therefore destabilizing it and advocating for their repurposing for Covid-19 infections (Jena et al., 2021).

13.3.4.2   Case study 2

Verma et al. advocated for the repurposing potential of some phytochemicals against SARS-CoV-2 infections. For the study, a systematic review of literature regarding antiviral medicinal plants was carried out using following PRISMA principles. PubChem-NCBI, DrugBank and ZINC databases were used for retrieving structures of natural compounds following their molecular and geometrical optimization using quantum chemistry method (PM3). The crystal structures of target proteins (spike protein receptor binding domain, RdRp, ACE2, human furin, main protease) were downloaded from PDB source. Pharmacophore modeling was carried out by Ligandscout program. Machine-learning algorithms were applied for scoring of ligand-based pharmacophores. A molecular similarity search was done using PharmGist webserver. Molecular docking studies and further visualization and analysis were done using Molegro virtual docker, CHIMERA and Discovery studio suites, respectively. Molecular dynamic simulations were carried out employing GROMACS software along with binding affinity estimations based on MM-PBSA method. The molecular dynamic simulations corroborated machine learning algorithm analysis (assay of central Bayesian model) deduced curcumin, quercetin (3) and gingerol (4) (comparable to remdesivir) binding strongly with host cell ACE2, human furin; spike protein, RdRp and main protease, respectively (Verma et al., 2021).

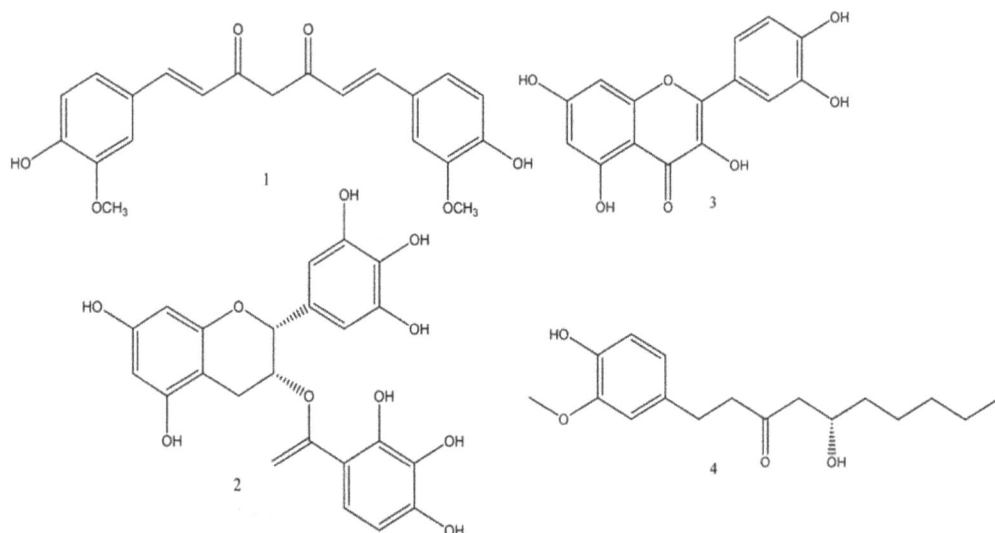

13.4   DIETARY POLYPHENOLS AS LEADS IN DRUG DESIGN AND DISCOVERY

There are a plethora of plant polyphenols with huge diversified therapeutic potential. Presently continuous efforts are made to propel their journey from screening hits to marketed drugs in adjunction with computational, biotechnological and pharmacological interventions (Atansov et al., 2015). Many computational approaches have been cited in literature reviewing repurposing these wonder phytochemicals for many diseases (Rallabandi et al., 2020). Herein, we provide a detailed discussion regarding the in silico strategies undertaken to repurpose selected dietary polyphenols as potential leads for drug design and discovery.

13.4.1   CURCUMIN

Diferuloyl methane or curcumin is naturally obtained from rhizomes of *Curcuma longa* (Zingiberaceae) via various extraction techniques (Doldolva et al., 2021). Much scientific literature enumerates multivariate pharmacological activity of curcumin exemplifying such as antimalarial (Correa et al., 2017); antidengue (Balasubramanian et al., 2019); antibacterial (Xue et al., 2020); anticancer (Ahsan et al., 2013); antidiabetic (Kalaycıoğlu et al., 2017) and so forth. Because of current advances in bioinformatics and computational tools, the biological activity horizon of curcumin seems to have widened via repurposing towards new disease indications. Wang et al., using a structure-based design strategy to synthesize curcumin analogs with a piperidine ring as potent leads against histone lysine specific demethylase-1(LSD-1). Molecular docking studies, in vitro enzyme inhibition assay and study of the anticancer effect on A549 human lung cancer cell lines and U87-MG human glioblastoma cell lines concluded compounds (5) and (6) as most potent inhibitors of LSD-1. Compound 5 exhibited the highest binding affinity (binding) -7.33kcal/mol towards the target (Wang et al., 2019).

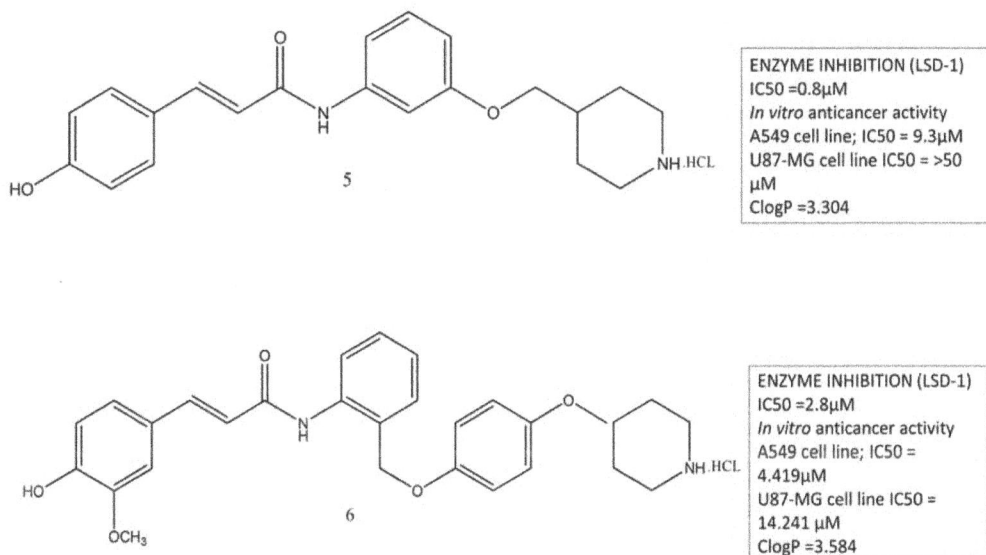

ENZYME INHIBITION (LSD-1)
IC50 =0.8μM
In vitro anticancer activity
A549 cell line; IC50 = 9.3μM
U87-MG cell line IC50 = >50 μM
ClogP =3.304

ENZYME INHIBITION (LSD-1)
IC50 =2.8μM
In vitro anticancer activity
A549 cell line; IC50 = 4.419μM
U87-MG cell line IC50 = 14.241 μM
ClogP =3.584

Current investigations revealed curcumin to allosterically inhibit dengue virus NS2B-NS3 protease by distorting its active conformation. An enzyme assay (in vitro) showed curcumin (IC_{50}= 7.18±0.62μM; inhibition constant, K_i=4.35±0.02μM) to be a non-competitive inhibitor of NS2B-NS3 protease by changing V_{max} without altering the values of Michaelis-Mentens K_m value. Molecular dynamic simulations advocated for the disruption of closed (active) conformation of the protease by unfolding the β-hairpin of C-terminal of NS2B chain in complexation with curcumin in a cavity (conserved in many flaviviral proteases) on allosteric site (Lim et al., 2020). Using a structure-based drug approach, Kesharwani et al. designed (Ligbuilder tools) curcumin analogs as Human Topoisomerase ll inhibitors. Docking simulations and ADME filtering (Schrodinger suite) revealed about 293 out of 1000 curcumin analogs showing good binding affinity towards the selected target in comparison to standard antineopastic compound Salvicine (dock binding affinity, -8.24 kcal/mol) especially top five ligands with binding affinities -16.70 kcal/mol, -16.03 kcal/mol, -15.49 kcal/mol, -15.43 kcal/mol and -15.09 kcal/mol, respectively (Kesharwani et al., 2018). In another study Chaudhary et al. designed and synthesized eleven pyrazoline-curcumin analogs as possible inhibitors of human β kinase (IKKβ). Molecular docking and molecular dynamic simulations revealed the chloro-bromo analog (7) with higher binding affinity (-11.534kcal/mol) than curcumin (-7.12kcal/mol) and forming a stable complex with target protein, respectively. Except for molecular weight, all other physiochemical parameters were found under acceptable limits advocating for their druggability. In vitro cytotoxic assay on HeLa human cervical cancer cell lines disclosed the chloro-bromo analog (IC_{50}; 8.7μg/ml) being more cytotoxic than curcumin (IC_{50}; 42.4μg/ml) but less cytotoxic than standard, paclitaxel (IC_{50}; 0.0087μg/ml). Caspase-3 enzyme cleavage revealed chloro-bromo analog significantly ($p < 0.05$) apoptotic (70.5%) than curcumin (19.9%) (Chaudhary et al., 2020).

13.4.2 RESVERATROL

Resveratrol, a naturally occurring stilbene from dietary polyphenol family possesses numerous health benefits in being an antidiabetic (Yang et al., 2018), anticancer (Jadhav et al., 2016), antieplileptic (Nieocyzm et al., 2019), anti-inflammatory (Walker et al., 2014) and as an antioxidant (Oh et al., 2018). Owing to the immense pharmacological aspects of resveratrol, many scientific expeditions are being done by computational approaches to unravel its new-fangled therapeutic aura. A novel series of flavone-resveratrol hybrids were designed and synthesized as potent inhibitors of p38-α MAPK as a therapeutic target as revealed in reverse docking studies. In vitro anti-inflammatory studies and immunoblotting assay discovered compound (8) as a potent interceptors of pro-inflammatory cytokines presenting it as promising scaffold for future development of anti-inflammatory agents. Percentage inhibitions of pro-inflammatory markers in LPS induced RAW 264.7 macrophages at low 1μM range were determined as nitric oxide (44.76); prostaglandin E2 (35.71); interleukine-6 (53.48); tumor necrosis factor-α (23.39) and interleukine-1β (41.02) (Hassan et al., 2019).

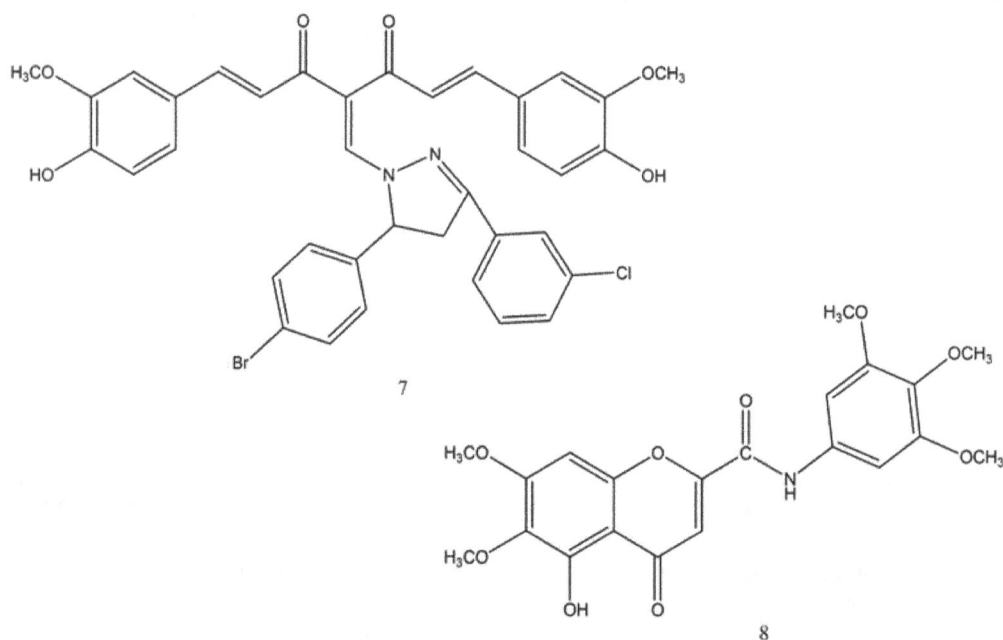

7

8

13.4.3 Catechins

Catechins are a pharmacologically versatile group of polyphenols abundantly found in nature, majorly in tea, apples, persimmons, berries, cacaos, (Isemura et al., 2019) and other plant species (Albuquerque et al., 2017; Lopez-Miranda et al., 2016; Zhan et al., 2014). Catechins have been explored with many biological activities, most interestingly as an antioxidant (Mahindrakar et al., 2020), antiobesity (Jiang et al., 2019), antidiabetic (Nazir et al., 2021), antiepileptic (Ahmed et al., 2021), anticancer (Sun et al., 2020), antiviral (Yang et al., 2014), and antifungal agent (Ning et al., 2015). This diversified pharmacological nature of catechins has been exploited to find their new therapeutic potentials. In a study, Azam et al. investigated the curative role of green tea polyphenols in Parkinson's disease. Molecular docking simulations (Autodock 4.2) revealed monoamine oxidase-B (PDB ID:2V5Z) as the most promising target for the catechins out of 18 target receptors selected. Table 13.4 depicting the analysis of docking results (Discovery studio visualizer 2.5 and PyMOL tools) and evaluation of ADMET properties (Molinspiration and OSIRIS web tools) suggested epigallocatechin gallate as the most promising lead for design and development of multitarget inhibitors for Parkinsonism (Azam et al., 2015).

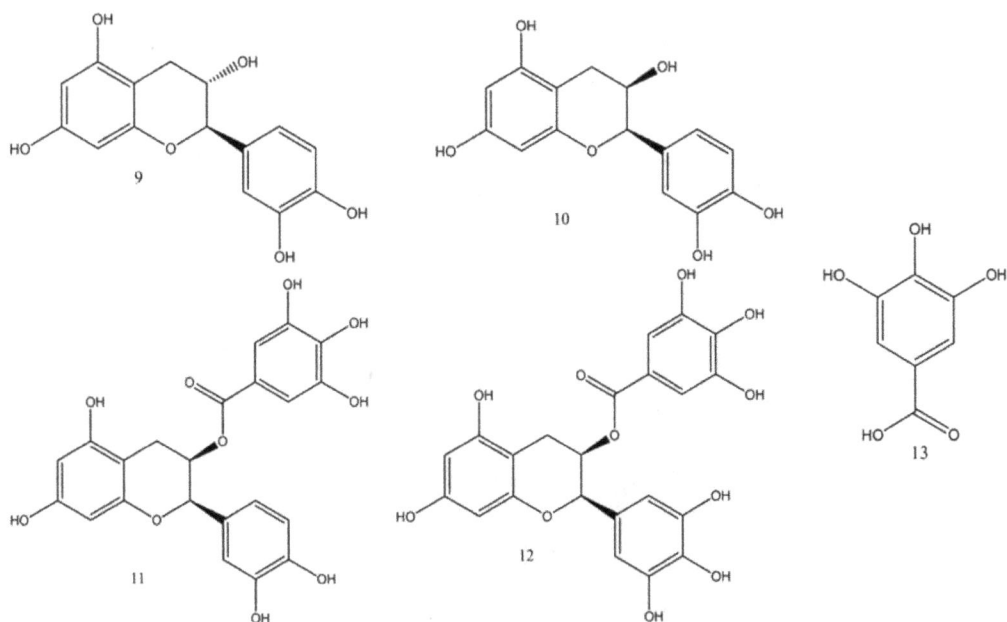

13.4.4 Rutin

Rutin (14) is one of the widely distributed dietary flavonoids extracted from diverse plant species (Habtamariam et al., 2015; Horosanskaia et al., 2017; Kraujalis et al., 2015). It is being reported as an antimicrobial (Garro et al., 2015), antioxidant (Jeszka et al., 2015), anticancer (Caparica et al., 2020), anti-inflammatory (Yoo et al., 2014) and antidiabetic (Vilhena et al., 2020). Many approaches have been taken to delve deep into the therapeutic potential of rutin and establish it as a potent lead against many diseases by exploring novel therapeutic targets. Azam et al. investigated the interaction of rutin with monoamine oxidase-B, a therapeutic target for Parkinson's disease, by implementing molecular docking, molecular dynamics and quantum chemistry calculations. Binding energies, docking inhibition constant and root mean square distance of rutin and standard drug safinamide were determined as -11.79kcal/mol, 2.28nM, 1.81Å and -18.59kcal/mol, 23.55nM and 1.11Å, respectively. The results obtained were significant in contrast to reference MAO-B inhibitor,

TABLE 13.4
Molecular docking (for monoamine oxidase B) and ADMET parameters computed for the catechins

Compound	Binding free energy (Kcal/mol)	Docking inhibition constant (nM)	Root mean square deviation (Å)	Lipinski's rule of five violation	mutagenic	irritant	tumorigenic	Reproductive effect
Catechin (9)	-8.84	0.33	2.43	No	No	No	No	No
Epicatechin (10)	-8.80	0.36	2.63	No	No	Yes	No	No
Epicatechingallate (11)	-10.67	0.02	7.42	No	No	Yes	No	No
Epigallocatechin (2)	-8.72	0.41	2.01	No	No	No	No	No
Epigallocatechin gallate (12)	-10.55	0.02	7.61	No	No	No	No	No
Gallic acid (13)	-5.05	197.4*	8.29	No	Yes	No	No	Yes

* data given in µM

safinamide (15), thereby apprehending rutin as a potent lead in the discovery of novel inhibitors of MAO-B ameliorating Parkinson's disease (Azam et al., 2019).

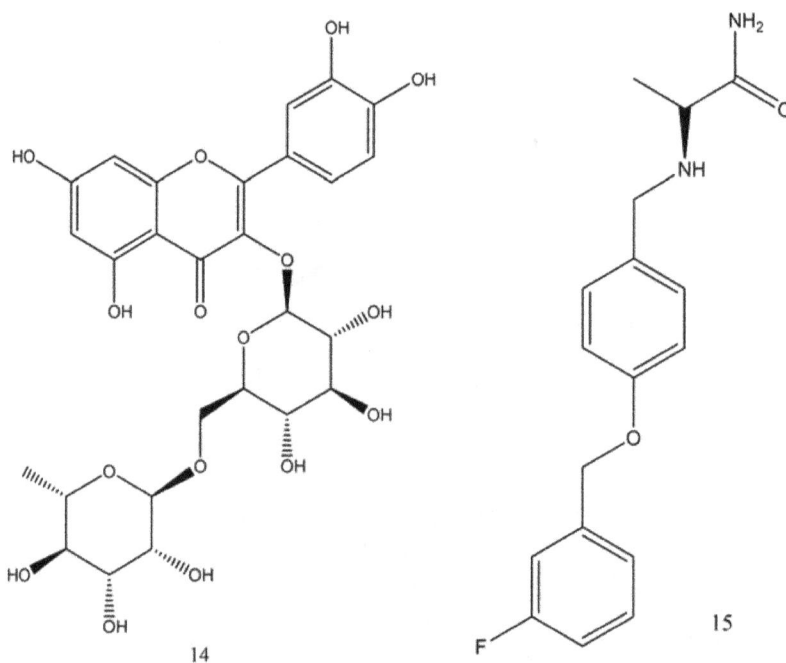

In another study presented an in silico approach acknowledging repurposing of rutin as a potent dual inhibitor for main protease and spike protein of the novel SARS-CoV-2 virus by bringing about conformational changes in the targets as revealed by cluster and secondary structure analysis. High dock scores (binding energies of -8.367kcal/mol and -11.553kcal/mol for spike protein and main protease, respectively) decreased values of radius of gyration (4nM and 2.23nM for spike-rutin and main protease-rutin complexes) and constant solvent accessible surface area (485nM/S²/N and 152nM/S^2/N for spike-rutin and main protease-rutin complexes) of proteins explained for high binding affinity for two important targets (Kumari et al., 2022).

13.4.5 Quercetin

Quercetin (3) is a widely distributed polyphenol, its major dietary sources being apples, grapes, blueberries, nuts, onion, tea, and so forth (Shah et al., 2016). It has also been extracted from plant sources by a number of methods (Dai et al., 2019; Ghoreishi et al., 2016; Hong et al., 2013; Lekar et al., 2014; Ranjbari et al., 2012; Zhu et al., 2019). The pharmacological activities of this polyphenol are diverse in being an antihypertensive (Grande et al., 2016), antidiabetic (Srinivasan et al., 2018), antiobesity (Pangeni et al., 2017), antibacterial (Wang et al., 2018), antiviral (Sun et al., 2021) and antifungal (Gao.et al., 2016). In a study, Liu et al. carried out virtual screening of traditional Chinese medicine-based small molecules: Quercetin and chlorogenic acid (16) for exploring their anti-influenza A H1N1 infections. The H1N1 A/PR/8/34 subtype (neuramidase protein) was selected as target, and its structure was derived via homology modeling and structural alignment (Clustal Omega tool) based on the three-dimensional structure of template subtype HINI (A/Brewig Mission/1/1918). Molecular docking (Autodock tools) of quercetin, chlorogenic acid (16), and zanamivir (17) (as reference) with H1N1 A/PR/8/34 subtype revealed the almost equivalent dock scores by computing all internal molecular interactions as -10.23 kcal/mol, -11.05 kcal/mol and -11.24 kcal/mol, respectively. Chemioinformatic analysis by FAF-Drugs2 server (Lagorce et al., 2011) revealed quercetin and chlorogrnic acid as having high drug likeliness, low toxicity and good bioavailability obeying Lipinski's rule of five. The in silico results were corroborated by in vitro and in vivo results. In vitro tests comprised of cytotoxic assay (methyl thiazolyl tetrazolium assay), cytopathic effect assay, effect rate estimation and enzymatic assay on Madin-Darbey canine kidney cells infected with influenza A H1N1 (A/PR/8/34). The cytotoxic assay revealed the maximum nontoxic dose of zanamivir, quercetin and chlorogenic acid to be 71.31 μg/ml, 61.54 μg/ml and 73.72 μg/ml respectively. Zanamivir and quercetin significantly inhibited the virus-infected cells at 1.56 μg/ml while chlorogenic acid had similar inhibition at 3.13 μg/ml. The in vivo studies analyzed mean lung parameters (lung index and lung index inhibition) and survival rate in BALB/c mice model at three doses (240, 480 and 960mg/kg/d respectively) for the test and reference candidates revealing quercetin equipotent to zanamivir in inhibiting neuramidase protein and can be a promising lead for discovery of anti-influenza drugs (Liu et al., 2016).

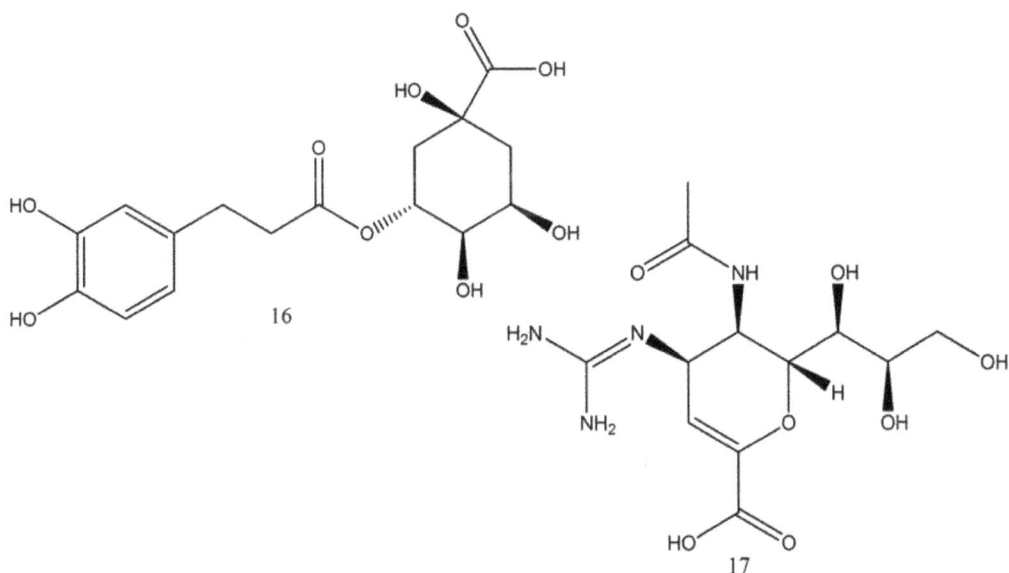

13.4.6 KAEMPFEROL

Kaempherol is ubiquitous naturally occurring dietary polyphenol with immense health benefits (Yang et al., 2021). Its numerous bioactivities are experimentally proven to be like anti-diabetics (Sharma et al., 2020); anti-cancers (Gao et al., 2018); antioxidants (Sharma et al., 2021); antivirals (Zhu et al., 2018) and so forth. Kaempferol has been extracted from many plant sources by different methods (Altemimi et al., 2020; Huang et al., 2014; Montano et al., 2011; Zeka et al., 2015). Many in silico and in vitro approaches have been employed to design and synthesize kaempferol analogues as leads. Molecular modelling studies explored mTOR1 as a target for kaempferol against hepatocellular carcinoma. Molecular docking-based virtual screening (Biovia Discovery studio tools) and ADMET prediction (ADMET SAR Tools) of drug likeness properties revealed kaempferol as a potent inhibitor of two selected targets; AKt1 (dock score of 33.4) and FRB binding domain of mTOR (dock score of 27.3), which was found superior to standard drug Everolimus, having dock scores of 24.4 and 20.1 for AKt1 and FRB binding domain of mTOR, respectively (Siniprasad et al., 2020). In another study, Babu et al. discovered kaempferol-3-O-β-glucopyranoside (18), kaempferol-3-O-α-L-rhamnopyranoside (19) and 5-hydroxy-7,4'-dimethoxy-6,8-di-C-methylflavone (20) that are isolated as (methanolic extracts) from fruits of *Syzgium alternifolium* (Myrtaceae) as novel inhibitors of HER2 receptors against gastric cancers implementing molecular docking, molecular dynamic simulations and pharmacophore-based virtual screening, the results being complemented by in vitro evaluation.

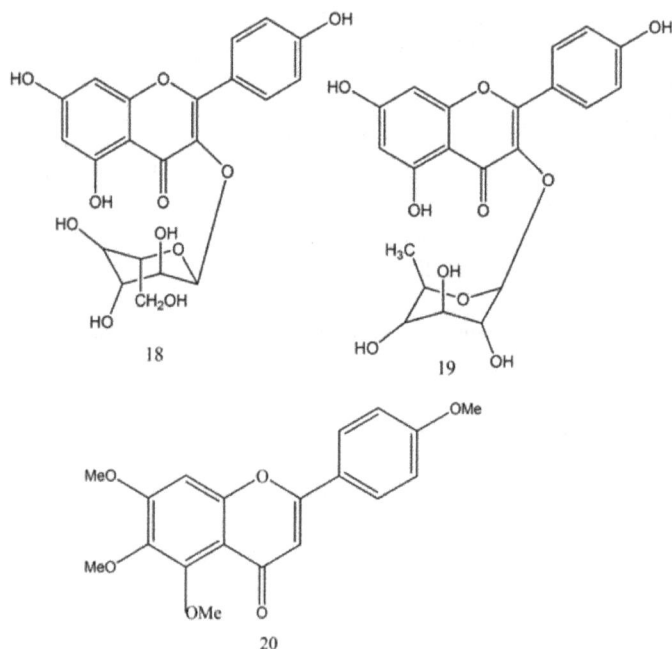

The data presented in Table 13.5 clearly indicates compound 19 as a potent hit as compared to the screened similar hits ZINC69703192 (21), ZINC59763389 (22) and ZINC85816423 (23).

TABLE 13.5
Molecular docking and molecular dynamics simulation of bioactive compounds with HER2 receptor

Compound	Binding energy (kcal/mol)	Docking sore	Binding affinity (pKi)
18	-33.28	-28.2	10.29
19	-30.10	-22.8	6.93
20	-19.2	-15.2	5.493
21*	-40.13	-32.2	11.21
22*	-34.24	-28.4	10.33
23*	-26.23	-27.2	9.62

* Similar screened hits from ZINC database based on compound 2-HER2complex selected as query dataset.

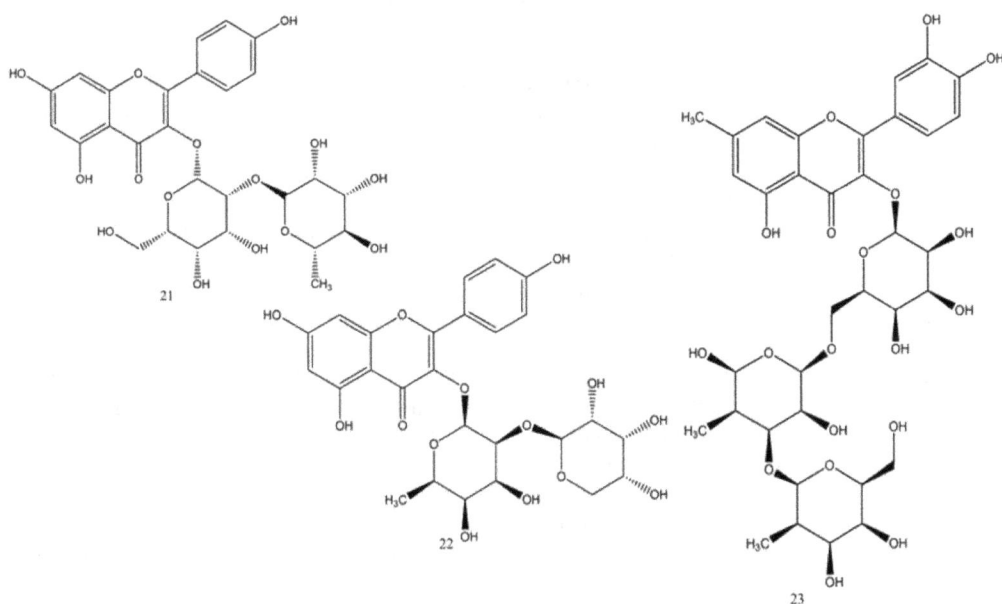

In vitro anticancer activities of the three natural compounds were evaluated by MTT cytotoxic assay and cell cycle progression assay. The cytotoxicity assay revealed compounds 19 and 20 (IC$_{50}$ values 30.37±4.10μg/ml and 35.53±5.03μg/ml, respectively) were more toxic to treated AGS gastric cancer cell lines than compound 18 (IC$_{50}$ 57.09±2.96μg/ml). Pharmacophore modelling revealed compound 19 as strong inhibitor of HER2 receptor the compounds 18, 19 and 20 along with best two screened hits were subjected to drug likeness and ADMET prediction using 3D QSAR descriptors (OSRIS and MOLINSPIRATION servers). The studies revealed ZINC67903192 (21) as potent inhibitor of HER2 receptor, thereby paving way for novel natural compound-derived scaffolds for gastric cancers (Babu et al., 2016).

13.5 CONCLUSION AND FUTURE PERSPECTIVES

Polyphenols are widespread natural compounds with a multitude of chemical structures, biosynthetic origin and enormous versatile pharmacological attributes. Huge progress has been observed in the field of nutritional and medicinal research in a quest for screening candidates for ameliorating diabetes mellitus, neurodegenerative malfunctions, microbial infections, obesity and cancers. There are substantial evidences in favor of different polyphenolic compounds, such as curcumin, resveratrol, rutin, quercetin, catechins, kaempferol being better prospects in comparison to synthetic molecules for prevention and management of chronic diseases. Computational approaches may be of help in understanding the mechanism, prediction of pharmacokinetic properties, and possible undesirable consequences of these dietary polyphenols. However, further studies on human subjects are required to authorize the pharmacological mechanisms and health effects of polyphenols. Moreover, the in vitro and in vivo studies have utilized higher doses than commonly present in human diets and so the level at which polyphenols can be safely and beneficially consumed remains unclear. Further research is needed to understand whether and how the same benefits from polyphenols consumed in whole foods can be derived from isolated forms.

REFERENCES

Acharya, C., Coop, A., Polli, J.E., and MacKerell, A.D. 2011. Recent advances in ligand-based drug design: Relevance and utility of the conformationally sampled pharmacophore approach. *Curr Comput Aided Drug Des* 7: 10.

Afendi, F.M., Okada, T., Yamazaki, M., Hirai-Morita, A., Nakamura, Y., Nakamura, K., Ikeda, S., Takahashi, H., Altaf-Ul-Amin, M., Darusman, L.K., and Saito, K. 2012. KNApSAcK family databases: Integrated metabolite–plant species databases for multifaceted plant research. *Plant Cell Physiol* 53: e1.

Agarwal, S., and Mehrotra, R. 2016. An overview of molecular docking. *JSM Chem* 4: 1024.

Ahmed, H., Khan, M.A., Amtul, Z., Zaidi, S.A.A., and Muhammad, S. 2021. In-silico and In-vivo: Evaluating the therapeutic potential of Kaempferol, Quercetin and Catechin to treat chronic epilepsy in a rat model. *Front Bioeng Biotechnol* 9: 938.

Ahsan, M.J., Khalilullah, H., Yasmin, S., Jadav, S.S., and Govindasamy, J. 2013. Synthesis, characterisation, and in vitro anticancer activity of curcumin analogues bearing pyrazole/pyrimidine ring targeting EGFR tyrosine kinase. *BioMed Res Int* 2013: 23954.

Albuquerque, B.R., Prieto, M.A., Barreiro, M.F., Rodrigues, A., Curran, T.P., Barros, L., and Ferreira, I.C. 2017. Catechin-based extract optimization obtained from *Arbutus unedo* L. fruits using maceration/microwave/ultrasound extraction techniques. *Ind Crop Prod* 95: 404.

Altemimi, A.B., Mohammed, M.J., Yi-Chen, L., Watson, D.G., Lakhssassi, N., Cacciola, F., and Ibrahim, S.A. 2020. Optimization of ultrasonicated kaempferol extraction from ocimum basilicum using a box – Behnken Design and its densitometric validation. *Foods* 9: 1379.

Ambure, P., Halder, A.K., Diaz, H.G., and Cordeiro, M.N.D. 2019. QSAR-Co: An open source software for developing robust multitasking or multitarget classification-based QSAR models. *J Chem Inf Model* 59: 2538.

Andreasen, M.F., Kroon, P.A., Williamson, G., and Garcia-Conesa, M.T. 2001. Esterase activity able to hydrolyze dietary antioxidant hydroxycinnamates is distributed along the intestine of mammals. *J Agric Food Chem* 49: 5679.

Ashfaq, U.A., Mumtaz, A., ul Qamar, T., and Fatima, T. 2013. MAPS Database: Medicinal plant activities, phytochemical and structural database. *Bioinformation* 9: 993.

Atanasov, A.G., Waltenberger, B., Pferschy-Wenzig, E.M., Linder, T., Wawrosch, C., Uhrin, P., Temml, V., Wang, L., Schwaiger, S., Heiss, E.H., and Rollinger, J.M. 2015. Discovery and resupply of pharmacologically active plant-derived natural products: A review. *Biotechnol Adv* 33: 1582.

Athanasiadou, S., Kyriazakis, I., Jackson, F., and Coop, R.L. 2001. Direct anthelmintic effects of condensed tannins towards different gastrointestinal nematodes of sheep: In vitro and in vivo studies. *Vet Parasitol* 99: 205.

Ayres, D.C., Ayres, D.C., and Loike, J.D. 1990. *Lignans: Chemical, Biological and Clinical Properties.* Cambridge University Press.

Azam, F., Abodabos, H.S., Taban, I.M., Rfieda, A.R., Mahmood, D., Anwar, M.J., Khan, S., Sizochenko, N., Poli, G., Tuccinardi, T., and Ali, H.I. 2019. Rutin as promising drug for the treatment of Parkinson's disease: An assessment of MAO-B inhibitory potential by docking, molecular dynamics and DFT studies. *Mol Simul* 45: 1563.

Azam, F., Mohamed, N., and Alhussen, F. 2015. Molecular interaction studies of green tea catechins as multitarget drug candidates for the treatment of Parkinson's disease: Computational and structural insights. *Netw Comput Neural Syst* 26: 97.

Ba-Alawi, W., Soufan, O., Essack, M., Kalnis, P., and Bajic, V.B. 2016. DASPfind: New efficient method to predict drug–target interactions. *J Cheminformatics* 8: 1.

Babu, T.M.C., Rammohan, A., Baki, V.B., Devi, S., Gunasekar, D., and Rajendra, W. 2016. Development of novel HER2 inhibitors against gastric cancer derived from flavonoid source of *Syzygium alternifolium* through molecular dynamics and pharmacophore-based screening. *Drug Des Devel Ther* 10: 3611.

Barker, D. 2019. Lignans. *Molecules*. 24:1424.

Balasubramanian, A., Pilankatta, R., Teramoto, T., Sajith, A.M., Nwulia, E., Kulkarni, A., and Padmanabhan, R. 2019. Inhibition of dengue virus by curcuminoids. *Antivir. Res.* 162: 171.

Baxevanis, A.D., and Bateman, A. 2015. The importance of biological databases in biological discovery. *Curr Protoc Bioinform* 50: 1.

Bravo, L. 1998. Polyphenols: Chemistry, dietary sources, metabolism, and nutritional significance. *Nutr Rev* 56: 317.

Brodowska, K.M. 2017. Natural flavonoids: Classification, potential role, and application of flavonoid analogues. *Eur J Biol Res* 7: 108.

Brown, A.S., and Patel, C.J. 2017. MeSHDD: Literature-based drug-drug similarity for drug repositioning. *J Am Med Inform Assoc* 24: 614.

Calderon-Montano, J., Burgos-Morón, E., Pérez-Guerrero, C., and López-Lázaro, M. 2011. A review on the dietary flavonoid kaempferol. *Mini Rev Med Chem* 11: 298.

Caparica, R., Júlio, A., Araújo, M.E.M., Baby, A.R., Fonte, P., Costa, J.G., and Santos de Almeida, T. 2020. Anticancer activity of rutin and its combination with ionic liquids on renal cells. *Biomolecules* 10: 233.

Cardona, F., Andrés-Lacueva, C., Tulipani, S., Tinahones, F.J., and Queipo-Ortuño, M.I. 2013. Benefits of polyphenols on gut microbiota and implications in human health. *J Nutr Biochem* 24: 1415.

Cereto-Massagué, A., Ojeda, M.J., Valls, C., Mulero, M., Garcia-Vallvé, S., and Pujadas, G. 2015. Molecular fingerprint similarity search in virtual screening. *Methods* 71: 58.

Chakravarti, S.K., and Alla, S.R.M. 2019. Descriptor free QSAR modeling using deep learning with long short-term memory neural networks. *Front Artif Intell Appl* 2: 17.

Chaudhary, K.K., and Mishra, N. 2016. A review on molecular docking: Novel tool for drug discovery. *Databases* 3: 1029.

Chaudhary, M., Kumar, N., Baldi, A., Chandra, R., Babu, M.A., and Madan, J. 2020. 4-Bromo-4'-chloro pyrazoline analog of curcumin augmented anticancer activity against human cervical cancer, HeLa cells: In silico-guided analysis, synthesis, and in vitro cytotoxicity. *J Biomol Struct Dyn* 38: 1335.

Chen, H., Cheng, F., and Li, J. 2020. iDrug: Integration of drug repositioning and drug-target prediction via cross-network embedding. *PLoS Comput Biol* 16: e1008040.

Choudhury, C., and Sastry G.N. 2019. Pharmacophore modelling and screening: concepts, recent developments and applications in rational drug design. In *Structural Bioinformatics: Applications in Preclinical Drug Discovery Process. Challenges and Advances in Computational Chemistry and Physics*. Vol. 27. ed. Mohan. C. Springer, Cham: 25–53.

Chou, Y.C., Ho, C.T., and Pan, M.H. 2018. Stilbenes: Chemistry and molecular mechanisms of anti-obesity. *Curr Pharmacol Rep* 4: 202.

Cui, Q., Du, R., Liu, M., and Rong, L. 2020. Lignans and their derivatives from plants as antivirals. *Molecules* 25: 183.

Dai, Y., and Row, K.H. 2019. Application of natural deep eutectic solvents in the extraction of quercetin from vegetables. *Molecules* 24: 2300.

Daina, A., Michielin, O., and Zoete, V. 2019. Swiss Target Prediction: Updated data and new features for efficient prediction of protein targets of small molecules. *Nucleic Acids Res* 47: W357.

David, B., Wolfender, J.L., and Dias, D.A. 2015. The pharmaceutical industry and natural products: Historical status and new trends. *Phytochem Rev* 14: 299.

Doldolova, K., Bener, M., Lalikoğlu, M., Aşçı, Y.S., Arat, R., and Apak, R. 2021. Optimization and modeling of microwave-assisted extraction of curcumin and antioxidant compounds from turmeric by using natural deep eutectic solvents. *Food Chem* 353: 129337.

Dong, J., Cao, D.S., Miao, H.Y., Liu, S., Deng, B.C., Yun, Y.H., Wang, N.N., Lu, A.P., Zeng, W.B., and Chen, A.F. 2015. ChemDes: an integrated web-based platform for molecular descriptor and fingerprint computation. *J Cheminformatics* 7: 1.

Dubrovina, A.S., and Kiselev, K.V. 2017. Regulation of stilbene biosynthesis in plants. *Planta* 246: 597.

El Khawand, T., Courtois, A., Valls, J., Richard, T., and Krisa, S. 2018. A review of dietary stilbenes: Sources and bioavailability. *Phytochem Rev* 17: 1007.

Ferreira, L.G., Dos Santos, R.N., Oliva, G., and Andricopulo, A.D. 2015. Molecular docking and structure-based drug design strategies. *Molecules* 20: 13384.

Friesner, R.A., Murphy, R.B., Repasky, M.P., Frye, L.L., Greenwood, J.R., Halgren, T.A., Sanschagrin, P.C., and Mainz, D.T. 2006. Extra precision glide: Docking and scoring incorporating a model of hydrophobic enclosure for protein–ligand complexes. *J Med Chem* 49: 6177.

Gao, M., Wang, H., and Zhu, L. 2016. Quercetin assists fluconazole to inhibit biofilm formations of fluconazole-resistant Candida albicans in in vitro and in vivo antifungal managements of vulvovaginal candidiasis. *Cell Physiol Biochem* 40: 727.

Gao, Y., Yin, J., Rankin, G.O., and Chen, Y.C. 2018. Kaempferol induces G2/M cell cycle arrest via checkpoint kinase 2 and promotes apoptosis via death receptors in human ovarian carcinoma A2780/CP70 cells. *Molecules* 23: 1095.

García-Jacas, C.R., Marrero-Ponce, Y., Acevedo-Martínez, L., Barigye, S.J., Valdés-Martiní, J.R., and Contreras-Torres, E. 2014. QuBiLS-MIDAS: A parallel free-software for molecular descriptors computation based on multilinear algebraic maps. *J Comput Chem* 35:1395.

Garro, M.F., Ibáñez, A.G.S., Vega, A.E., Sosa, A.C.A., Pelzer, L., Saad, J.R., and Maria, A.O. 2015. Gastroprotective effects and antimicrobial activity of Lithraea molleoides and isolated compounds against Helicobacter pylori. *J Ethnopharmacol* 176: 469.

Gaulton, A., Hersey, A., Nowotka, M., Bento, A.P., Chambers, J., Mendez, D., Mutowo, P., Atkinson, F., Bellis, L.J., Cibrián-Uhalte, E. and Davies, M., 2017. The ChEMBL database in 2017. *Nucleic Acids Res* 45: D945.

Gilson, M.K., Liu, T., Baitaluk, M., Nicola, G., Hwang, L., and Chong, J. 2016. BindingDB in 2015: A public database for medicinal chemistry, computational chemistry and systems pharmacology. *Nucleic Acids Res* 44: D1045.

Ghoreishi, S.M., Hedayati, A., and Mousavi, S.O. 2016. Quercetin extraction from Rosa damascena Mill via supercritical CO2: Neural network and adaptive neuro fuzzy interface system modeling and response surface optimization. *J Supercrit Fluids* 112: 57–66.

Gong, J., Cai, C., Liu, X., Ku, X., Jiang, H., Gao, D., Li, H. 2013. ChemMapper: a versatile web server for exploring pharmacology and chemical structure association based on molecular 3D similarity method. *Bioinformatics*. 29: 1827–1829.

Gowd, V., Karim, N., Shishir, M.R.I., Xie, L., and Chen, W. 2019. Dietary polyphenols to combat the metabolic diseases via altering gut microbiota. *Trends Food Sci Technol* 93: 81.

Grande, F., Parisi, O.I., Mordocco, R.A., Rocca, C., Puoci, F., Scrivano, L., Quintieri, A.M., Cantafio, P., Ferla, S., Brancale, A., and Saturnino, C. 2016. Quercetin derivatives as novel antihypertensive agents: Synthesis and physiological characterization. *Eur J Pharm Sci* 82: 161.

Habtemariam, S., and Varghese, G.K. 2015. Extractability of rutin in herbal tea preparations of *Moringa stenopetala* leaves. *Beverages* 1: 169.

Han, X., Shen, T., and Lou, H. 2007. Dietary polyphenols and their biological significance. *Int J Mol Sci* 8: 950.

Hassan, A.H., Yoo, S.Y., Lee, K.W., Yoon, Y.M., Ryu, H.W., Jeong, Y., Shin, J.S., Kang, S.Y., Kim, S.Y., Lee, H.H., and Park, B.Y. 2019. Repurposing mosloflavone/5, 6, 7-trimethoxyflavone-resveratrol hybrids: Discovery of novel p38-α MAPK inhibitors as potent interceptors of macrophage-dependent production of proinflammatory mediators. *Eur J Med Chem* 180: 253.

Hong, Y., and Chen, L. 2013. Extraction of quercetin from Herba Lysimachiae by molecularly imprinted-matrix solid phase dispersion. *J Chromatogr B* 941: 44.

Horosanskaia, E., Minh Nguyen, T., Dinh Vu, T., Seidel-Morgenstern, A., and Lorenz, H. 2017. Crystallization-based isolation of pure rutin from herbal extract of *Sophora japonica* L. *Org Process Res Dev* 21: 1769.

Huang, J.X., Zhang, J., Zhang, X.R., Zhang, K., Zhang, X., and He, X.R. 2014. *Mucor fragilis* as a novel source of the key pharmaceutical agents podophyllotoxin and kaempferol. *Pharm Biol* 52: 1237.

Isemura, M. 2019. Catechin in human health and disease. *Molecules* 24: 528.

Jadhav, P., Bothiraja, C., and Pawar, A. 2016. Resveratrol-piperine loaded mixed micelles: Formulation, characterization, bioavailability, safety and in vitro anticancer activity. *RSC Adv* 6: 112795.

Jarada, T.N., Rokne, J.G., and Alhajj, R. 2020. A review of computational drug repositioning: Strategies, approaches, opportunities, challenges, and directions. *J Cheminformatics* 12: 1.

Jeandet, P., Delaunois, B., Conreux, A., Donnez, D., Nuzzo, V., Cordelier, S., Clément, C., and Courot, E. 2010. Biosynthesis, metabolism, molecular engineering, and biological functions of stilbene phytoalexins in plants. *Biofactors* 36: 331.

Jena, A.B., Kanungo, N., Nayak, V., Chainy, G.B.N., and Dandapat, J. 2021. Catechin and curcumin interact with S protein of SARS-CoV2 and ACE2 of human cell membrane: Insights from computational studies. *Sci Rep* 11: 1.

Jeszka-Skowron, M., Krawczyk, M., and Zgoła-Grześkowiak, A. 2015. Determination of antioxidant activity, rutin, quercetin, phenolic acids and trace elements in tea infusions: Influence of citric acid addition on extraction of metals. *J Food Compos Anal* 40: 70.

Jiang, Y., Ding, S., Li, F., Zhang, C., Sun-Waterhouse, D., Chen, Y., and Li, D. 2019. Effects of (+)-catechin on the differentiation and lipid metabolism of 3T3-L1 adipocytes. *J Funct Foods* 62: 103558.

Kalaycıoğlu, Z., Gazioğlu, I., and Erim, F.B. 2017. Comparison of antioxidant, anticholinesterase, and antidiabetic activities of three curcuminoids isolated from *Curcuma longa* L. *Nat Prod Res* 31: 2914.

Kesharwani, R.K., Singh, D.B., Singh, D.V., and Misra, K. 2018. Computational study of curcumin analogues by targeting DNA topoisomerase II: A structure-based drug designing approach. *Netw Model Anal Health Inform Bioinform* 7: 1.

Khan, A.U. 2016. Descriptors and their selection methods in QSAR analysis: Paradigm for drug design. *Drug Discov Today* 21: 1291.

Kim, K.H., Tsao, R., Yang, R., and Cui, S.W. 2006. Phenolic acid profiles and antioxidant activities of wheat bran extracts and the effect of hydrolysis conditions. *Food Chem* 95: 466.

Kim, S., Chen, J., Cheng, T., Gindulyte, A., He, J., He, S., Li, Q., Shoemaker, B.A., Thiessen, P.A., Yu, B., and Zaslavsky, L. 2021. PubChem in 2021: New data content and improved web interfaces. *Nucleic Acids Res* 49: D1388.

Kraujalis, P., Venskutonis, P.R., Ibanez, E., and Herrero, M. 2015. Optimization of rutin isolation from Amaranthus paniculatus leaves by high pressure extraction and fractionation techniques. *J Supercrit Fluids* 104: 234.

Krawczyk, H. 2019. The stilbene derivatives, nucleosides, and nucleosides modified by stilbene derivatives. *Bioorg Chem* 90: 103073.

Kumar, S., Dhar, D.N., and Saxena, P.N. 2009. Applications of metal complexes of Schiff bases – A review. *J Sci Ind Res* 68: 181.

Kumari, A., Rajput, V.S., Nagpal, P., Kukrety, H., Grover, S., Grover, A. 2022. Dual inhibition of SARS-CoV-2 spike and main protease through a repurposed drug, rutin. *J Biomol Struct Dyn.* 40: 4987–4999.

Lagorce, D., Maupetit, J., Baell, J., Sperandio, O., Tufféry, P., Miteva, M.A., Galons, H., and Villoutreix, B.O. 2011. The FAF-Drugs2 server: A multistep engine to prepare electronic chemical compound collections. *Bioinformatics* 27: 2018.

Lagunin, A.A., Goel, R.K., Gawande, D.Y., Pahwa, P., Gloriozova, T.A., Dmitriev, A.V., Ivanov, S.M., Rudik, A.V., Konova, V.I., Pogodin, P.V., and Druzhilovsky, D.S. 2014. Chemo-and bioinformatics resources for in silico drug discovery from medicinal plants beyond their traditional use: A critical review. *Nat Prod Rep* 31: 1585.

Lamb, J., Crawford, E.D., Peck, D., Modell, J.W., Blat, I.C., Wrobel, M.J., Lerner, J., Brunet, J.P., Subramanian, A., Ross, K.N., and Reich, M. 2006. The Connectivity Map: Using gene-expression signatures to connect small molecules, genes, and disease. *Science* 313: 1929.

Lavecchia, A., and Di Giovanni, C. 2013. Virtual screening strategies in drug discovery: A critical review. *Curr Med Chem* 20: 2839.

Lekar, A.V., Borisenko, S.N., Vetrova, E.V., Sushkova, S.N., and Borisenko, N.I. 2014. Extraction of quercetin from Polygonum hydropiper L. by subcritical water. *Am J Agric Biol Sci* 9:1.

Lim, L., Dang, M., Roy, A., Kang, J., and Song, J. 2020. Curcumin allosterically inhibits the dengue NS2B-NS3 protease by disrupting its active conformation. *ACS Omega* 5: 25677.

Liu, P., and Long, W. 2009. Current mathematical methods used in QSAR/QSPR studies. *Int J Mol Sci.* 10:1978-1998.

Liu, Z., Zhao, J., Li, W., Shen, L., Huang, S., Tang, J., Duan, J., Fang, F., Huang, Y., Chang, H., and Chen, Z. 2016. Computational screen and experimental validation of anti-influenza effects of quercetin and chlorogenic acid from traditional Chinese medicine. *Sci Rep* 6: 1.

López-Miranda, S., Serrano-Martínez, A., Hernández-Sánchez, P., Guardiola, L., Pérez-Sánchez, H., Fortea, I., Gabaldón, J.A., and Núñez-Delicado, E. 2016. Use of cyclodextrins to recover catechin and epicatechin from red grape pomace. *Food Chem* 203: 379.

Lo, Y.C., Rensi, S.E., Torng, W., and Altman, R.B. 2018. Machine learning in chemoinformatics and drug discovery. *Drug Discov Today* 23: 1538.

MacRae, W.D., and Towers, G.N. 1984. Biological activities of lignans. *Phytochemistry* 23: 1207.

Mahindrakar, K.V., and Rathod, V.K. 2020. Ultrasonic assisted aqueous extraction of catechin and gallic acid from Syzygiumcumini seed kernel and evaluation of total phenolic, flavonoid contents and antioxidant activity. *Chem Eng Process* 149: 107841.

Martinez-Correa, H.A., Paula, J.T., Kayano, A.C.A., Queiroga, C.L., Magalhães, P.M., Costa, F.T., and Cabral, F.A. 2017. Composition and antimalarial activity of extracts of Curcuma longa L. obtained by a combination of extraction processes using supercritical $CO2$, ethanol and water as solvents. *J Supercrit Fluids* 119: 122.

Masand, V.H., and Rastija, V. 2017. PyDescriptor: A new PyMOL plugin for calculating thousands of easily understandable molecular descriptors. *Chemom Intell Lab Syst* 169: 12.

Mauri, A. 2020. alvaDesc: A tool to calculate and analyze molecular descriptors and fingerprints. In: *Ecotoxicological QSARs. Methods in Pharmacology and Toxicology*, ed. Roy, K. Humana, New York: 801–820.

Medina-Franco, J.L. 2015. Discovery and development of lead compounds from natural sources using computational approaches. In *Evidence-based Validation of Herbal Medicine*, ed. P.K. Mukherjee, 455–475. Elsevier.

Meetei, P.A., Singh, P., Nongdam, P., Prabhu, N.P., Rathore, R.S., and Vindal, V. 2012. NeMedPlant: A database of therapeutic applications and chemical constituents of medicinal plants from north-east region of India. *Bioinformation* 8: 209.

Middleton, E. 1998. Effect of plant flavonoids on immune and inflammatory cell function. In *Flavonoids in the Living System. Advances in Experimental Medicine and Biology*, ed. Manthey, J.A., and Buslig, B.S. Springer, Boston: 175–182.

Mohanraj, K., Karthikeyan, B.S., Vivek-Ananth, R.P., Chand, R.P., Aparna, S.R., Mangalapandi, P., and Samal, A. 2018. IMPPAT: A curated database of Indian medicinal plants, phytochemistry and therapeutics. *Sci Rep* 8: 1.

Moilanen, J., Karonen, M., Tähtinen, P., Jacquet, R., Quideau, S., and Salminen, J.P. 2016. Biological activity of ellagitannins: Effects as anti-oxidants, pro-oxidants and metal chelators. *Phytochemistry* 125: 165.

Moriwaki, H., Tian, Y.S., Kawashita, N., and Takagi, T. 2018. Mordred: A molecular descriptor calculator. *J. Cheminformatics* 10: 1.

Muegge, I., and Mukherjee, P. 2016. An overview of molecular fingerprint similarity search in virtual screening. *Expert Opin Drug Discov* 11: 137.

Muhammed, M.T., and Esin, A.Y. 2021. Pharmacophore modeling in drug discovery: Methodology and current status. *J Turkish Chem Soc* 8: 749.

Mumtaz, A., Ashfaq, U.A., ul Qamar, M.T., Anwar, F., Gulzar, F., Ali, M.A., Saari, N., and Pervez, M.T. 2017. MPD3: A useful medicinal plants database for drug designing. *Nat Prod Res* 31: 1228.

Nazir, N., Zahoor, M., Ullah, R., Ezzeldin, E., and Mostafa, G.A. 2021. Curative effect of catechin isolated from *Elaeagnus umbellata* Thunb. berries for diabetes and related complications in streptozotocin-induced diabetic rats model. *Molecules* 26: 137.

Neveu, V., Perez-Jiménez, J., Vos, F., Crespy, V., du Chaffaut, L., Mennen, L., Knox, C., Eisner, R., Cruz, J., Wishart, D., and Scalbert, A. 2010. Phenol-Explorer: An online comprehensive database on polyphenol contents in foods. *Database* 2010: bap024

Nieoczym, D., Socała, K., Jedziniak, P., Wyska, E., and Wlaź, P. 2019. Effect of pterostilbene, a natural analog of resveratrol, on the activity of some antiepileptic drugs in the acute seizure tests in mice. *Neurotox Res* 36: 859.

Ning, Y., Ling, J., and Wu, C.D. 2015. Synergistic effects of tea catechin epigallocatechin gallate and antimycotics against oral Candida species. *Arch Oral Biol* 60: 1565.

Oh, W.Y., and Shahidi, F. 2018. Antioxidant activity of resveratrol ester derivatives in food and biological model systems. *Food Chem* 261: 267.

Pangeni, R., Kang, S.W., Oak, M., Park, E.Y., and Park, J.W. 2017. Oral delivery of quercetin in oil-in-water nanoemulsion: In vitro characterization and in vivo anti-obesity efficacy in mice. *J Funct Foods* 38: 571.

Pathania, S., Ramakrishnan, S.M., and Bagler, G. 2015. Phytochemica: A platform to explore phytochemicals of medicinal plants. *Database* 2015: bav075

Prachayasittikul, V., Worachartcheewan, A., Shoombuatong, W., Songtawee, N., Simeon, S., Prachayasittikul, V., and Nantasenamat, C. 2015. Computer-aided drug design of bioactive natural products. *Curr Top Med Chem* 15: 1780.

Pushpakom, S., Iorio, F., Eyers, P.A., Escott, K.J., Hopper, S., Wells, A., Doig, A., Guilliams, T., Latimer, J., McNamee, C., and Norris, A. 2019. Drug repurposing: Progress, challenges and recommendations. *Nat Rev Drug Discov* 18: 41.

Rallabandi, H.R., Mekapogu, M., Natesan, K., Saindane, M., Dhupal, M., Swamy, M.K., and Vasamsetti, B.M.K. 2020. Computational Methods Used in Phytocompound-Based Drug Discovery. In *Plant-derived Bioactives*, ed. Swamy, M. Springer, Singapore: 549–573.

Ranjbari, E., Biparva, P., and Hadjmohammadi, M.R. 2012. Utilization of inverted dispersive liquid–liquid microextraction followed by HPLC-UV as a sensitive and efficient method for the extraction and determination of quercetin in honey and biological samples. *Talanta* 89: 117.

Rice-Evans, C.A., Miller, N.J., and Paganga, G. 1996. Structure-antioxidant activity relationships of flavonoids and phenolic acids. *Free Radic Biol Med* 20: 933.

Roy, K., Kar, S., and Das, R.N. 2015. Statistical methods in QSAR/QSPR. In *A Primer on QSAR/QSPR Modeling*, ed. Roy, K. Springer, Cham: 37–59.

Roy, K., Kar, S., and Das, R.N. 2015. *Understanding the Basics of QSAR for Applications in Pharmaceutical Sciences and Risk Assessment*. Academic Press.

Saibabu, V., Fatima, Z., Khan, L.A., and Hameed, S. 2015. Therapeutic potential of dietary phenolic acids. *Adv Pharmacol Sci 2015*.

Santana Azevedo, L., Pretto Moraes, F., Morrone Xavier, M., Ozorio Pantoja, E., Villavicencio, B., Aline Finck, J., Menegaz Proenca, A., Beiestorf Rocha, K., and Filgueira de Azevedo, W. 2012. Recent progress of molecular docking simulations applied to development of drugs. *Curr Bioinform* 7: 352.

Scalbert, A., Manach, C., Morand, C., Rémésy, C., and Jiménez, L. 2005. Dietary polyphenols and the prevention of diseases. *Crit Rev Food Sci Nutr* 45: 287.

Schaller, D., Šribar, D., Noonan, T., Deng, L., Nguyen, T.N., Pach, S., Machalz, D., Bermudez, M., and Wolber, G. 2020. Next generation 3D pharmacophore modeling. *Wiley Interdiscip Rev Comput Mol Sci* 10: e1468.

Selvaraj, G., Kaliamurthi, S., Peslherbe, G.H., and Wei, D. 2021. Application of artificial intelligence in drug repurposing: A mini-review. *Curr Chin Sci* 1:1.

Shah, P.M., Priya, V.V., and Gayathri, R. 2016. Quercetin-a flavonoid: A systematic review. *J Pharm Sci Res* 8: 878.

Sharma, D., Tekade, R.K., and Kalia, K. 2020. Kaempferol in ameliorating diabetes-induced fibrosis and renal damage: An in vitro and in vivo study in diabetic nephropathy mice model. *Phytomedicine* 76: 153235.

Sharma, N., Biswas, S., Al-Dayan, N., Alhegaili, A.S., and Sarwat, M. 2021. Antioxidant role of kaempferol in prevention of hepatocellular carcinoma. *Antioxidants* 10: 1419.

Sharma, S., Gupta, J., Prabhakar, P.K., Gupta, P., Solanki, P., and Rajput, A. 2019 Phytochemical repurposing of natural molecule: Sabinene for identification of novel therapeutic benefits using in silico and in vitro approaches. *Assay Drug Dev Technol* 17: 339.

Sharma, V., Wakode, S., and Kumar, H. 2021. Structure-and ligand-based drug design: Concepts, approaches, and challenges. In *Chemoinformatics and Bioinformatics in the Pharmaceutical Sciences*, ed. Sharma, N., Ojha, H., Raghav, P.K., and Goyal, R.K. Academic Press: 27–53.

Shim, J., and MacKerell Jr, A.D. 2011. Computational ligand-based rational design: role of conformational sampling and force fields in model development. *Med Chem Comm* 2: 356.

Sieniawska E., and Baj, T. 2017. Tannins. In *Pharmacognosy*, Badal, S., and Delgoda, R. eds. Academic Press: 199–232.

Siniprasad, P., Nair, B., Balasubramaniam, V., Sadanandan, P., Namboori, P.K., and ReghuNath, L. 2020. Evaluation of kaempferol as AKT dependent mTOR regulator via targeting FKBP-12 in hepatocellular carcinoma: An in silico approach. *Lett Drug Des Discov* 17: 1401.

Sorokina, M., Merseburger, P., Rajan, K., Yirik, M.A., and Steinbeck, C. 2021. COCONUT online: collection of open natural products database. *J Cheminformatics* 13: 1.

Srednicka-Tober, D., Ponder, A., Hallmann, E., Głowacka, A., and Rozpara, E. 2019. The profile and content of polyphenols and carotenoids in local and commercial sweet cherry fruits (Prunusavium L.) and their antioxidant activity in vitro. *Antioxidants* 8: 534.

Srinivasan, P., Vijayakumar, S., Kothandaraman, S., and Palani, M. 2018. Anti-diabetic activity of quercetin extracted from *Phyllanthus emblica* L. fruit: In silico and in vivo approaches. *J Pharm Anal* 8: 109.

Strissel, Pamela L., and Strick, R. 2005. Multiple effects of bioflavonoids on gene regulation, cell proliferation and apoptosis: natural compounds move into the lime light of cancer research. *Leuk Res* 29: 859.

Sun, H., Yin, M., Hao, D., and Shen, Y. 2020. Anti-cancer activity of catechin against A549 lung carcinoma Cells by induction of cyclin kinase inhibitor p21 and suppression of cyclin E1 and P–AKT. *Appl Sci* 10: 2065.

Sun, H.N., Mu, T.H., and Xi, L.S. 2017. Effect of pH, heat, and light treatments on the antioxidant activity of sweet potato leaf polyphenols. *Int J Food Prop* 20: 318.

Sun, Y., Li, C., Li, Z., Shangguan, A., Jiang, J., Zeng, W., Zhang, S., and He, Q. 2021. Quercetin as an antiviral agent inhibits the Pseudorabies virus *in vitro* and *in vivo*. *Virus Res* 305: 198556.

Su, X.Z., and Miller, L.H. 2015. The discovery of artemisinin and the Nobel Prize in physiology or medicine. *Sci China Life Sci* 58: 1175.

Tanase, C., Coşarcă, S., and Muntean, D.L. 2019. A critical review of phenolic compounds extracted from the bark of woody vascular plants and their potential biological activity. *Molecules* 24: 1182.

Tapas, A. R., Sakarkar, D. M., and Kakde R. B. 2008. Flavonoids as nutraceuticals: a review. *Trop J Pharm Res* 7: 1089.

Temml, V., and Schuster, D. 2021. Molecular Docking for Natural Product Investigations: Pitfalls and Ways to Overcome Them. In *Molecular Docking for Computer-Aided Drug Design*, ed. Coumar, M.S. Academic Press, 391–405.

Teng, H., and Chen, L. 2019. Polyphenols and bioavailability: An update. *Crit Rev Food Sci* Nutr 59: 2040.

Teponno, R.B., Kusari, S., and Spiteller, M. 2016. Recent advances in research on lignans and neolignans. *Nat Prod Rep* 33: 1044.

Thomsen, R., and Christensen, M.H. 2006. MolDock: A new technique for high-accuracy molecular docking. *J Med Chem* 49: 3315.

Tortorella, E., Tedesco, P., Palma Esposito, F., January, G.G., Fani, R., Jaspars, M., and De Pascale, D. 2018. Antibiotics from deep-sea microorganisms: Current discoveries and perspectives. *Mar Drugs 16*: 355.

Tresserra-Rimbau, A., Lamuela-Raventos, R.M., and Moreno, J.J. 2018. Polyphenols, food and pharma. Current knowledge and directions for future research. *Biochem Pharmacol* 156: 186.

Tripathi, A., and Bankaitis, V.A. 2017 Molecular docking: From lock and key to combination lock. *J Mol Med Clin Appl*. 2

Trott, O., and Olson, A.J. 2010. AutoDockVina: improving the speed and accuracy of docking with a new scoring function, efficient optimization, and multithreading. *J Comput Chem* 31: 455.

U.S. Department of Agriculture, Agricultural Research Service. 1992–2016. *Dr. Duke's Phytochemical and Ethnobotanical Databases*. Home Page, http://phytochem.nal.usda.gov/

Venkatachalam, C.M., Jiang. X., Oldfield, T., and Waldman, M. 2003. LigandFit: a novel method for the shape-directed rapid docking of ligands to protein active sites. *J Mol Graph Model* 21: 289.

Verma, A.K., Kumar, V., Singh, S., Goswami, B.C., Camps, I., Sekar, A., Yoon, S., and Lee, K.W. 2021. Repurposing potential of Ayurvedic medicinal plants derived active principles against SARS-CoV-2 associated target proteins revealed by molecular docking, molecular dynamics and MM-PBSA studies. *Biomed Pharmacother* 137: 111356.

Vilhena, R.O., Figueiredo, I.D., Baviera, A.M., Silva, D.B., Marson, B.M., Oliveira, J.A., Peccinini, R.G., Borges, I.K., and Pontarolo, R. 2020. Antidiabetic activity of Musa x paradisiaca extracts in streptozotocin-induced diabetic rats and chemical characterization by HPLC-DAD-MS. *J Ethnopharmacol* 254: 2666.

Vuolo, M.M., Lima, V.S., and Maróstica Junior, M.R. 2019. Phenolic compounds: Structure, classification, and antioxidant power. In *Bioactive Compounds*, ed. Campos, M.R.S. Woodhead Publishing: 33–50.

Wahedi, H.M., Ahmad, S., and Abbasi, S.W. 2021. Stilbene-based natural compounds as promising drug candidates against COVID-19. *J Biomol Struct Dyn* 39: 3225.

Walker, J., Schueller, K., Schaefer, L.M., Pignitter, M., Esefelder, L., and Somoza, V. 2014. Resveratrol and its metabolites inhibit pro-inflammatory effects of lipopolysaccharides in U-937 macrophages in plasma-representative concentrations. *Food Funct* 5: 74.

Wang, J., Zhang, X., Yan, J., Li, W., Jiang, Q., Wang, X., Zhao, D., and Cheng, M. 2019. Design, synthesis and biological evaluation of curcumin analogues as novel LSD1 inhibitors. *Bioorg Med Chem Lett* 29: 126683.

Wang, S., Yao, J., Zhou, B., Yang, J., Chaudry, M.T., Wang, M., Xiao, F., Li, Y., and Yin, W. 2018. Bacteriostatic effect of quercetin as an antibiotic alternative in vivo and its antibacterial mechanism in vitro. *J Food Prot* 81: 68.

Wang, T., Wu, M.B., Lin, J.P., and Yang, L.R. 2015. Quantitative structure–activity relationship: Promising advances in drug discovery platforms. *Expert Opin Drug Discov* 10: 1283.

Wang, Y.L., Wang, F., Shi, X.X., Jia, C.Y., Wu, F.X., Hao, G.F., and Yang, G.F. 2021. Cloud 3D-QSAR: A web tool for the development of quantitative structure–activity relationship models in drug discovery. *Brief Bioinform* 22: bbaa276.

Willett, P. 2013. Fusing similarity rankings in ligand-based virtual screening. *Comput Struct Biotechnol J* 5: e201302002.

Wishart, D.S., Feunang, Y.D., Marcu, A., Guo, A.C., Liang, K., Vázquez-Fresno, R., Sajed, T., Johnson, D., Li, C., Karu, N., and Sayeeda, Z. 2018. HMDB 4.0: The human metabolome database for 2018. *Nucleic Acids Res* 46: D608.

Wu, M., Luo, Q., Nie, R., Yang, X., Tang, Z., and Chen, H. 2021. Potential implications of polyphenols on aging considering oxidative stress, inflammation, autophagy, and gut microbiota. *Crit Rev Food Sci Nutr* 61: 2175.

Xue, B., Huang, J., Zhang, H., Li, B., Xu, M., Zhang, Y., Xie, M., and Li, X. 2020. Micronized curcumin fabricated by supercritical CO_2 to improve antibacterial activity against Pseudomonas aeruginosa. *Artif Cells Nanomed Biotechnol* 48: 1135.

Yang, D.K., and Kang, H.S. 2018. Anti-diabetic effect of cotreatment with quercetin and resveratrol in streptozotocin-induced diabetic rats. *Biomol Ther* 26: 130.

Yang, L., Gao, Y., Bajpai, V.K., El-Kammar, H.A., Simal-Gandara, J., Cao, H., Cheng, K.W., Wang, M., Arroo, R.R., Zou, L., and Farag, M.A. 2021. Advance toward isolation, extraction, metabolism and health benefits of kaempferol, a major dietary flavonoid with future perspectives. *Critical Reviews in Food Sci Nutr* 1.

Yang, Z.F., Bai, L.P., Huang, W.B., Li, X.Z., Zhao, S.S., Zhong, N.S., and Jiang, Z.H. 2014. Comparison of in vitro antiviral activity of tea polyphenols against influenza A and B viruses and structure–activity relationship analysis. *Fitoterapia 93*: 47.

Yap, C.W. 2011. PaDEL-descriptor: An open source software to calculate molecular descriptors and fingerprints. *J. Comput Chem* 32: 1466.

Yee, L.C., and Wei, Y.C. 2012. Current modeling methods used in QSAR/QSPR. *Assessment* 10:11.

Yoo, H., Ku, S.K., Baek, Y.D., and Bae, J.S. 2014. Anti-inflammatory effects of rutin on HMGB1-induced inflammatory responses *in vitro* and *in vivo*. *Inflamm Res* 63: 197.

Zhang, H., Tang, B., and Row, K. 2014. Extraction of catechin compounds from green tea with a new green solvent. *Chem Res Chin* 30: 37.

Zeng, X., Zhang, P., Wang, Y., Qin, C., Chen, S., He, W., Tao, L., Tan, Y., Gao, D., Wang, B., and Chen, Z. 2019. CMAUP: A database of collective molecular activities of useful plants. *Nucleic Acids Res* 47: D1118.

Zeka, K., Ruparelia, K.C., Continenza, M.A., Stagos, D., Vegliò, F., and Arroo, R.R. 2015. Petals of *Crocus sativus* L. as a potential source of the antioxidants crocin and kaempferol. *Fitoterapia* 107:128.

Zhu, L., Wang, P., Yuan, W., and Zhu, G. 2018. Kaempferol inhibited bovine herpesvirus 1 replication and LPS-induced inflammatory response. *Acta Viro* 62: 220.

Zhu, Y., Yu, J., Jiao, C., Tong, J., Zhang, L., Chang, Y., Sun, W., Jin, Q., and Cai, Y. 2019. Optimization of quercetin extraction method in *Dendrobium officinale* by response surface methodology. *Heliyon* 5: e02374.

Index

Note: Page numbers in *italics* indicate figures and in **bold** indicate tables on the corresponding pages.

For Product Safety Concerns and Information please contact our EU
representative GPSR@taylorandfrancis.com
Taylor & Francis Verlag GmbH, Kaufingerstraße 24, 80331 München, Germany

www.ingramcontent.com/pod-product-compliance
Lightning Source LLC
Chambersburg PA
CBHW082109220326
41598CB00066BA/5853